MOLECULAR BIOLOGY
INTELLIGENCE
UNIT

Intracellular Pathogens in Membrane Interactions and Vacuole Biogenesis

Jean-Pierre Gorvel, Ph.D.
Centre d'Immunologie
INSERM-CNRS-Universite de la Méditereanée
Marseille, France

LANDES BIOSCIENCE / EUREKAH.COM
GEORGETOWN, TEXAS
U.S.A.

KLUWER ACADEMIC / PLENUM PUBLISHERS
NEW YORK, NEW YORK
U.S.A.

Intracellular Pathogens in Membrane Interactions and Vacuole Biogenesis

Molecular Biology Intelligence Unit

Eurekah.com / Landes Bioscience
Kluwer Academic / Plenum Publishers

Printed in the U.S.A.

Kluwer Academic / Plenum Publishers, 233 Spring Street, New York, New York, U.S.A. 10013
http://www.wkap.nl/

Please address all inquiries to the Publishers:
Eurekah.com / Landes Bioscience, 810 South Church Street
Georgetown, Texas, U.S.A. 78626
Phone: 512/ 863 7762; FAX: 512/ 863 0081
www.Eurekah.com
www.landesbioscience.com

Intracellular Pathogens in Membrane Interactions and Vacuole Biogenesis, edited by Jean-Pierre Gorvel, Landes / Kluwer dual imprint / Landes series: Molecular Biology Intelligence Unit

ISBN: 0-306-47833-1

While the authors, editors and publisher believe that drug selection and dosage and the specifications and usage of equipment and devices, as set forth in this book, are in accord with current recommendations and practice at the time of publication, they make no warranty, expressed or implied, with respect to material described in this book. In view of the ongoing research, equipment development, changes in governmental regulations and the rapid accumulation of information relating to the biomedical sciences, the reader is urged to carefully review and evaluate the information provided herein.

Library of Congress Cataloging-in-Publication Data

Intracellular pathogens in membrane interactions and vacuole biogenesis
/ [edited by] Jean-Pierre Gorvel.
p. ; cm. -- (Molecular biology intelligence unit)
Includes bibliographical references.
ISBN 0-306-47833-1
1. Endocytosis. 2. Host-bacteria relationships.
[DNLM: 1. Cytoplasmic Vesicles--immunology. 2. Immunity, Natural--physiology. 3. Bacterial Infections--physiopathology. 4. Bacterial Proteins--immunology. 5. Cell Membrane Structures--immunology. 6. Membrane Transport Proteins--immunology. QW 541 I66 2003] I. Gorvel, Jean-Pierre. II. Molecular biology intelligence unit (Unnumbered)
QH634.I55 2003
571.9'36--dc21

2003012216

To my parents Jean-Marie and Gilberte,
Vilma, my wife, and my sons
Alexandre-Manuel, Cyril and Laurent.

CONTENTS

EDITOR

Jean-Pierre Gorvel, Ph.D.
Centre d'Immunologie
INSERM-CNRS-Universite de la Méditerranée
Marseille, France

CONTRIBUTORS

Carmen Alvarez-Dominguez
Servicio de Inmunología
Hospital Universitario "Marqués de Valdecilla"
Santander, Cantabria, Spain
Chapter 4

Fredrik Bäckhed
Microbiology and Tumor Biology Center
Karolinska Institutet
Stockholm, Sweden
Chapter 17

Oddmund Bakke
Department of Biology
Division of Molecular Cell Biology
University of Oslo
Oslo, Norway
Chapter 16

Gurdyal S. Besra
School of Microbiological, Immunological and Virological Sciences
The Medical School
University of Newcastle-upon-Tyne
Newcastle-upon-Tyne, U.K.
Chapter 15

Oliver Billker
Max-Planck-Institut für Infektionsbiologie
Abteilung Molekulare Biologie
Berlin, Germany
Chapter 13

Juan S. Bonifacino
Cell Biology and Metabolism Branch
National Institutes of Health
Bethesda, Maryland, U.S.A.
Chapter 2

Steve Caplan
Department of Biochemistry and Molecular Biology
University of Nebraska Medical Center
Omaha, Nebraska, U.S.A.
Chapter 2

Rosângela P. da Silva
Sir William Dunn School of Pathology
University of Oxford
Oxford, U.K.
Chapter 5

Alice Dautry-Varsat
Unité de Biologie des Intéractions Cellulaires
Institut Pasteur
Paris, France
Chapter 12

Chantal de Chastellier
Department of Medical Biochemistry
Faculty of Health Sciences
University of Cape Town
Observatory, South Africa
Chapter 10

Jean-Michel Escola
Department of Biochemistry, Sciences II
University of Geneva
Geneva, Switzerland
Chapter 1

Siamon Gordon
Sir William Dunn School of Pathology
University of Oxford
Oxford, U.K.
Chapter 5

Tone F. Gregers
Department of Biology
Division of Molecular Cell Biology
University of Oslo
Oslo, Norway
Chapter 16

Jean Gruenberg
Department of Biochemistry, Sciences II
University of Geneva
Geneva, Switzerland
Chapter 1

Sigrid Heinsbroek
Sir William Dunn School of Pathology
University of Oxford
Oxford, U.K.
Chapter 5

Jonathan C. Howard
Institute for Genetics
University of Cologne
Cologne, Germany
Chapter 18

Isabelle Jutras
Unité de Biologie des Intéractions Cellulaires
Institut Pasteur
Paris, France
Chapter 12

Jonathan C. Kagan
Yale University School of Medicine
Section of Microbial Pathogenesis
Boyer Center for Molecular Medicine
New Haven, Connecticut, USA
Chapter 9

Stéphane Méresse
Centre d'Immunologie de Marseille-Luminy
Marseille, France
Chapter 8

Thomas F. Meyer
Max-Planck-Institut für Infektionsbiologie
Abteilung Molekulare Biologie
Berlin, Germany
Chapter 13

Edgardo Moreno
Programa de Investigación en Enfermedades Tropicales
Escuela de Medicina Veterinaria
Universidad Nacional
Heredia, Costa Rica
Chapter 7

Ignacio Moriyón
Departamento de Microbiologia
Universidad de Navarra
Pamplona, Spain
Chapter 14

Tommy W. Nordeng
Department of Biology
Division of Molecular Cell Biology
University of Oslo
Oslo, Norway
Chapter 16

Bongi Ntolosi
Sir William Dunn School of Pathology
University of Oxford
Oxford, U.K.
Chapter 5

Carla Peña-Macarro
Servicio de Inmunología
Hospital Universitario "Marqués de Valdecilla"
Santander, Cantabria, Spain
Chapter 4

Lorena Perrone
Dipartimento di Biologia e Patologia Cellulare e Molecolare
University of Naples "Federico II"
Medical School
Naples, Italy
Chapter 3

Jean Pieters
Biozentrum
Department of Biochemistry
Basel, Switzerland
Chapter 11

Javier Pizarro-Cerdá
Unité des Interactions Bactéries-cellulaires
Department de Bactériologie
Institut Pasteur
Paris, France
Chapter 7

Andreas Popp
Max-Planck-Institut für Infektionsbiologie
Abteilung Molekulare Biologie
Berlin, Germany
Chapter 13

Steven A. Porcelli
Department of Microbiology
and Immunology
Albert Einstein College of Medicine
Bronx, New York, USA
Chapter 15

Amaya Prada-Delgado
Servicio de Inmunología
Hospital Universitario "Marqués
de Valdecilla"
Santander, Cantabria, Spain
Chapter 4

Agneta Richter Dahlfors
Microbiology and Tumor Biology Center
Karolinska Institutet
Stockholm, Sweden
Chapter 17

Craig R. Roy
Section of Microbial Pathogenesis
Yale University School of Medicine
Boyer Center for Molecular Medicine
New Haven, Connecticut, USA
Chapter 9

Philippe J. Sansonetti
Unité de Pathogénie Microbienne
Moléculaire
INSERM U389
Institut Pasteur
Paris, France
Chapter 6

Olivia Steele-Mortimer
Laboratory of Intracellular Parasites
National Institute of Allergy
and Infectious Disease
Rocky Mountain Laboratories
Hamilton, Montana, U.S.A.
Chapter 8

Agathe Subtil
Unité de Biologie des Intéractions
Cellulaires
Institut Pasteur
Paris, France
Chapter 12

Lutz Thilo
Department of Medical Biochemistry
Faculty of Health Sciences
University of Cape Town
Observatory, South Africa
Chapter 10

Guy Tran Van Nhieu
Unité de Pathogénie Microbienne
Moléculaire
INSERM U389
Institut Pasteur
Paris, France
Chapter 6

Benjamin Wyplosz
Unité de Biologie des Interactions
Cellulaires
Institut Pasteur
Paris, France
Chapter 12

Chiara Zurzolo
Dipartimento di Biologia e Patologia
Cellulare e Molecolare
University of Naples "Federico II"
Medical School
Naples, Italy
Chapter 3

PREFACE

Infectious diseases have been considered part of human history, and in industrialised countries we believed that tuberculosis, typhoid fever, typhus, smallpox, and many other epidemic diseases were under control. After a flurry of discoveries that included antibiotics and antivirals, very few new bug killers have been discovered during the last 25 years. Perhaps as a consequence, the last quarter century has been characterised by the emergence of new diseases and fear of ancient infectious diseases has returned. Now, it is time to better understand the molecular interactions between pathogenic bacteria and host cells. The last five years have witnessed the publication of remarkable original scientific and review articles devoted to cellular microbiology, a discipline first termed in 1996 by P. Cossart and colleagues that cross-links cell biology and microbiology. The fundamental cellular mechanisms that can fall prey to bacteria or bacterial products are: signal transduction, membrane trafficking and organelle biogenesis, cytoskeletal dynamics, cell adhesion, and the regulation of cell survival versus cell death. The goal of this book is to associate cellular microbiology and immunology to tackle the complex molecular mechanisms that underly host-bacteria relationship. Understanding these fundamental mechanisms is a clear challenge, in order to gain knowledge that will certainly be the basis for new vaccine design. Should we never forget Louis Pasteur's famous maxim "Messieurs, c'est les microbes qui auront le dernier mot"?

This book offers advanced information on the molecular interactions between host cell organelles and pathogens, which have found a way to stay inside the cell. The adoption of an intracellular life-style confers several advantages to pathogenic bacteria. Indeed, they become insensitive to humoral responses and complement-mediated killing and do not need to maintain adhesion properties with the external plasma membrane. However, they have to invent strategies to get inside the cell and, furthermore, to reach the so-called replication niche.

Phagocytosis is essential for the uptake and the eventual degradation of dying cells, inert particles and live infectious agents. It plays a critical role in essential biological functions such as inflammation, immunity and development. Phagocytosis occurs in professional phagocytes (macrophages and neutrophils), but also to a lesser extent in non-professional phagocytes such as fibroblasts, endothelial and epithelial cells. After internalization in professional phagocytes, inert particles are found in a membrane-bound compartment, the nascent phagosome, which undergoes a maturation process into an hydrolases-rich phagolysosome. In contrast, once within the host cells, intracellular pathogens control the fate of their nascent membrane-bound compartments, circumventing host defences, further degradation and providing nutritional environment. Pathogens have evolved several different ways of finding a successful intracellular replication niche. Some pathogens such as *Shigella, Listeria* and *Rickettsiae* escape from its nascent membrane-bound compartment and replicate in the cytoplasm, while others, such as *Mycobacteria, Salmonella, Brucella, Chlamydia* and *Legionella*, replicate within an idiosyncratic niche, the parasitic vacuole. The nature of these vacuoles is very diverse indicating that pathogens control the biogenesis of their vacuole. Most the vacuoles containing pathogens first interact with the endocytic pathway. During endocytosis, cell surface receptors and their ligands, particles and solutes are taken up by vesicles, which form at the plasma membrane and deliver their content upon fusion with early endosomes. Macromolecules destined for degradation are then targeted to late endosomes and finally to lysosomes. Membrane homeostasis in endocytosis is maintained both by the organization of membranes in subcompartments such as lipid rafts and a constant contact with the exocytic pathway. Therefore it is not surprising to find that vacuoles containing pathogens may display Golgi apparatus molecules (*Chlamydia*) or endoplasmic reticulum proteins (*Brucella, Legionella*).

Another general feature of host-pathogen interaction is the host response to intracellular microorganisms. Microbial infections can be controlled by several effector systems that are brought into play during infection. Two broad responses take place between the host and the pathogen. Responses independent of lymphocytes have been termed "innate responses" such as the macrophagic phagocyte system, which are mobilized and activated quickly without the direct participation of T lymphocytes. The cells of the innate response cooperate with those of the adaptive response, in part by presenting peptide antigens to specific T lymphocytes and by releasing cytokines and chemokines. Upon T cell activation, the cellular response is activated. T cells are involved not only in the activation of macrophages but also in the killing of infected cells and in regulating B cells for antibody expression.

The chapters in this book are grouped in five sections. I *Endocytosis and phagocytosis*, collectively, the chapters in this section constitute the basis of the major intracellular organelles that are implicated in the membrane interactions with vacuoles containing pathogens. II *Professional and non professional phagocytes*, here, we describe the major differences between the two host cell types which can be infected by microorganisms. III *Maturation pathways of bacteria-containing vacuoles*, in this part are described the molecular interactions between vacuoles and intracellular organelles leading to the search of the holy Grail, the replication niche. IV *Host response*, host cells are able to react against intruders and eventually mount host responses. In this chapter the various types of host response mechanisms against intracellular intruders are reviewed. V *Co-evolution*, in this final chapter, will advances in knowledge on bacteria-host cell interactions be fast enough to find tools, important for controlling microorganism development? This is certainly a matter of co-evolution.

We hope that this book will appeal to scientists interested in cell biology, microbiology and immunology, as well as to clinically-oriented investigators concerned with infectious diseases. I am deeply grateful to the contributors to this book, as well as to all the scientists who contribute new ideas and concepts to the immunology and cell biology of host-pathogen interactions.

Jean-Pierre Gorvel

CHAPTER 1

Membrane Traffic in the Endocytic Pathway

Jean-Michel Escola and Jean Gruenberg

Membrane Trafficking in Eukaryotic Cells

Eukaryotic cells need to be in constant communication with their environment in order to perform most of their functions, including transmission or reception of metabolic and proliferative signals, uptake of nutrients, and adhesion. Segregation of proteins and lipids into discrete organelles is fundamental for cell homeostasis: this heterogeneity is established and maintained despite continuous membrane flow through transport pathways. To preserve their biochemical identity, compartments must efficiently sort and remove cargo proteins and lipids while ensuring retention of resident molecules.

In the biosynthetic pathway, both soluble proteins destined for secretion and membrane proteins of the vacuolar apparatus are synthesized in the endoplasmic reticulum (ER) membrane, and, once correctly folded, transported through the Golgi apparatus to the trans-Golgi network (TGN). From the TGN, they are sorted, packaged into specific vesicles and forwarded to their cellular destination (endosomes or the plasma membrane). Fusion of TGN-derived vesicles with the plasma membrane (exocytosis) can be either constitutive or regulated. All eukaryotic cells carry out constitutive exocytosis but only a small subset of cells, specialized in the secretion of hormones, neurotransmitters or digestive enzymes, display a regulated secretory pathway. The endocytic pathway functions as a mirror image of the biosynthetic pathway. Cell surface proteins, including receptors, lipids and solutes are endocytosed and delivered to early endosomes. From there, proteins are sorted either to be rapidly recycled back to the cell surface, at least in part via recycling endosomes, or are selectively transported towards late endosomes and lysosomes for degradation.

Endosomes provide a central station in the vacuolar apparatus, at the cross-road between reutilization of membrane components via recycling pathways and degradation of down-regulated molecules along the route leading to lysosomes. As a consequence, the endocytic pathway plays a crucial role in physiological and pathological processes, including nutrient uptake, signaling, adhesion, cellular membrane turnover and defense against pathogenic agents. In the immune response, antigen presentation by specialized antigen presenting cells involves endocytosis of the foreign antigens and their processing in degradative compartments. In addition to the normal physiological function of the endocytic pathway, endocytosis is used by several bacterial pathogens and viruses to gain entry into the host cell. Some bacteria, such as *Listeria* or *Shigella*, induce their own internalization by the host cell, and then escape from the endocytic system into the cytoplasm where they multiply. Others, such as *Mycobacterium* or *Leishmania*, after being internalized, modify the environment of the vacuole in order to make themselves a decent home.

Ways of Entry into the Cell

Entry into mammalian cell can occur via different pathways. While phagocytosis is responsible for the uptake of solid particles, one often refers to pinocytosis for the uptake of extracellular

Intracellular Pathogens in Membrane Interactions and Vacuole Biogenesis, edited by Jean-Pierre Gorvel.

fluid, macromolecules and other solutes. Pinocytosis can be further subdivided into macropinocytosis, which is associated with the formation of relatively large vacuoles that are heterogeneous in size and shape (0.5-200 nm), and micropinocytosis which occur through the generation of small vesicles (50-150 nm). Evidence is accumulating that micropinocytosis also occurs via more than one pathways. In addition to clathrin-dependent endocytosis, which is responsible for the internalization of most, but not all, known receptors and their ligands (receptor-mediated endocytosis), other less-well characterized pathways also exist, at least in some cell-types, and are collectively referred to as clathrin-independent endocytosis.

Clathrin-dependent endocytosis is responsible for the internalization of most, but not all (see below), cell surface receptors that have been characterized, including receptors with housekeeping functions that cycle constitutively between endosomes and the plasma membrane, and receptors for growth factors and hormones whose entry process is regulated by ligand-binding. Once formed, clathrin-coated vesicles specifically dock onto and fuse with early endosomes, thereby delivering their membrane and content. Many components regulating protein traffic and sorting in this pathway have been identified and characterized, in some cases at the atomic level, including the adaptor complex 2 (AP2, see below).[1] Much less is known about clathrin-independent endocytosis, which in itself may comprise different mechanisms. This route is involved in retrieval of membrane after induced exocytosis in adrenal chromaffin cells[2] and is responsible for the uptake of desmosome.[3] Interestingly, recent studies revealed that the interleukin 2 receptor (IL2-R) is internalized by a pathway, which does not depend on Eps15, a key-component of the clathrin pathway but might involve cell surface microdomains rich in cholesterol and glycosphingolipids ("rafts").[4] Several lines of evidence, indeed, suggest that such microdomains can provide an additional entry route, in particular caveolae, which containing the protein caveolin, at least in some cell-types.[5] In addition, internalization might also occur via a separate pathways(s) that does not depend on clathrin/AP2 or rafts/caveolae.[6] It is generally believed that that clathrin-dependent and -independent routes then meet in early endosome, the first station of the endocytic pathway, consistently with the fact that IL2-R follows the classical degradation pathway to lysosomes. However, it is also possible that other entry routes by-pass early endosomes. SV40 was recently shown to enter cells via caveolae and then to reside within caveosomes that do not contain endocytic tracers.[7] Whether this pathway is a specialized, virus-induced route of entry remains to be investigated.

Both phagocytosis and macropinocytosis are triggered in specialized cells in response to specific stimuli. Macropinocytose is a non-clathrin route of entry related to phagocytosis and associated with areas where membrane spreading and ruffling takes place. This route is modulated and used by some pathogenic bacteria.[8] It is stimulated by the epidermal growth factor (EGF),[9] and by Ras.[10] Macropinocytosis involves remodeling of the cortical actin cytoskeleton, involving Ras, Rac, and cdc42[11] and Rho proteins, implicated in macropinocytosis downregulation in dendritic cells during maturation.[12] Phagocytosis is a process which allows the internalization of large particulate material. It is initiated by binding of the particle to cell surface receptors, triggers a reorganization of the plasma membrane and its cortical cytoskeletal elements, leading to particle engulfment and formation of the phagosome. Over the last few years, it is now believed that phagosomes undergo progressive changes through multiple exchanges with endocytic organelles to form phagolysosomes. Indeed, during phagolysosome biogenesis, phagosomes intersect the biosynthetic pathway and fuse sequentially with early endosomes, late endosomes and lysosomes.[13] These interactions, facilitated by phagosome binding and movement along cytoskeletal elements,[14] and controlled in part by small GTPases of the rab family,[15] allow the acquisition of molecules conferring new functions to maturing phagosomes. One of the recruited proteins on phagosomes is TACO.[16] Active retention of TACO on phagosomes by living mycobacteria allows them to survive within macrophages.[16] One of the challenges in the field will be to understand the precise molecular mechanisms which regulate interactions between endosomes, lysosomes and thus control the biogenesis of phagolysosomes. The recent characterization of phagosome proteins (>140) using a proteomic approach[17] may provide new insights into phagosome functions.

The Endocytic Pathway

After internalization into animal cells, cell surface proteins and lipids, as well as solutes, first appear in peripheral early endosomes, sometimes referred to as sorting endosome.[18] Depending on their fate, these proteins can either be recycled back to the cell surface for reutilization, at least in part via recycling endosomes, or transported to late endosomes and then lysosomes to be degraded. The recycling and lysosomal pathways exhibit major differences with respect to membrane organization and dynamics.

Early Endosomes

Early endosomes consist of tubular and cisternal elements associated to vesicular regions of 0.3-0.5 µm diameter and are located in the cell periphery. The tubular part resembles recycling endosomes and is involved in recycling back to the cell surface. In contrast, the vesicular part accumulates membrane invaginations, hence a characteristic multivesicular appearance, but it is not clear to what extent these then become free vesicles in the endosomal lumen. These vesicular regions correspond to forming multivesicular bodies (MVBs), which will mediate transport to late endosomes, and are thus also referred to as endosomal carrier vesicles (ECVs). Here, we will refer to these vesicles as ECV/MVB to avoid confusion with late endosomes that also contain internal membranes. The mildly acidic lumenal pH (6.0-6.2) of the early endosome is due to the presence of the vacuolar proton pump[19] and facilitates uncoupling of ligands from their receptors. Then, recycling receptors are very rapidly (half-life < 3 min) segregated into the tubular portion and are recycled back to the plasma membrane for another round of internalization.[18] In contrast, most ligands accumulate in the vesicular part, presumably due to a high volume/surface ratio in forming ECV/MVBs, and together with solutes are delivered to late endosome and lysosomes.

While there is no doubt that sorting occurs in early endosomes, surprisingly little is known about sorting signals and mechanisms involved. Sequence motives in the cytoplasmic domains of recycling receptors that may regulate early endosomal sorting, have not been identified, leading to the notion that transport along this pathway may occur by default. However, proteins destined to be degraded, including down-regulated cell surface receptors, must be sorted away from recycling receptors, since the former proteins are very efficiently incorporated into forming ECV/MVBs and transported to lysosomes. Sequence motives have been found in the cytoplasmic domains of P-selectin,[20,21] IL2-R,[22] AIDS-encoded Nef during down-regulation of the Nef-CD4 complex,[23] as well as perhaps in the EGF receptor.[24,25] However, these signals bear little resemblance to each other, and a consensus has not emerged until now. Recent studies also suggest that protein ubiquitination may play a critical role in protein sorting at this, as well as other, intracellular transport steps.[26] In addition, it is also becoming apparent that endocytosed lipids are not stochastically redistributed to recycling and degradation pathways.[27] In fact, biophysical constraints may facilitate incorporation of some lipids at the neck of forming recycling tubules, while other lipids may be preferentially incorporated into invaginations of nascent ECV/MVBs.[28] As a consequence, proteins that tend to partition into a given lipid environment may be thus follow the corresponding lipids, suggesting that early endosomal sorting may be in part lipid-based, as was proposed for the plasma membrane and TGN.

The Recycling Pathway

Recycling endosomes consist of networks of very thin tubules (50-60 nm in diameter and up to several µm in length), organized around the microtubule-organizing center in some, but not all, cell types. These structures are similar to the tubular regions of early endosomes which are often interconnected into separate networks.[29] The pH of this compartment is slighlty higher (6.4) than that of the sorting endosome and this organelle does not contain material destined to degradation (e.g., LDL). Although several lines of evidence show that recycling can occur vis a rapid and a slow route, it is not clear whether passage through the recycling endosomes is an obligatory route for all recycling molecules, or whether rapid recycling back to the cell surface also occurs directly from early endosome. In addition, there is no doubt protein sorting must occur in recycling endosomes of polarized epithelial cells, since both transcytosed and recycling proteins follow the same route, before being transported to opposite plasma membrane domain.[30] In non polarized cells, the situation is less clear. However, electron microscopy studies revealed the existence of budding profiles on endosomes, many containing clathrin and other coat proteins,[31] and endosome-derived small vesicles have been identified that contain recycling membrane proteins and might be involved in direct or indirect transport to the plasma membrane.[32] Finally, recents studies show that RME-1, a protein of of the family of Eps15-homology (EH)- proteins with characteristics of endocytic accessory proteins, is associated to recycling endosomes and may be involved in exit from the compartment.[33] Components that regulate traffic along the recycling pathway have been identified, including small GTPases and SNAREs,[34,35] as well as components and regulators of the actin cytoskeleton,[36] but their precise functions are not always clear.

The Degradation Pathway

Components that regulate the biogenesis of ECV/MVBs have been identified, including an endosomal COP complex (see below). In addition, studies in yeast suggest that class E vps genes are involved in the biogenesis of yeast multivesicular bodies.[37] Some of these genes have mammalian counterparts,[38] but their precise functions in ECV/MVB biogenesis are unclear. A word of caution may be needed when extrapolating from mammalian to yeast endocytic pathways and vice-versa. A striking features of mammalian endocytosis is the capacity to reutilize cell surface components efficiently and rapidly via recycling routes, while equivalent recycling routes, if they exist, do not appear to be as efficient in yeast cells. In addition, yeast cells do not appear to contain the complex network of membrane invaginations of mammalian endosomes in the degradation pathway. Small and regularly shaped vesicles were observed in the vacuole of yeast strains with impaired vacuolar hydrolase activity.[39] In addition to the basic machinery of yeast cells, mammalian cells are likely to have evolved a more elaborate endosomal membrane system to ensure optimal regulation and reutilization of proteins and lipids. Once formed on early endosomes, ECV/MVBs are transported on microtubules towards late endosomes with which they fuse.[40] This transport step, as well as perhaps transport to lysosomes, may depend on the small GTPase Rab7.[41] In addition, a number of SNAREs involved in docking/fusion have also been implicated at different steps along the degradation pathway, but precise localization and functions are not always clear.[34]

Late endosomes are generally believed to provide the last sorting station before lysosomes, from where proteins can be recycled, for example to the TGN. They also exhibit a complex morphological organization with cisternal, tubular and multivesicular regions, and exhibit a protein and lipid composition different from that of early endosomes.[29] In addition, major lipid remodelling occurs in late endosomes. Their internal membrane invaginations accumulate large amounts of the unique phospholipid lysobisphosphatidic acid (LBPA), and these LBPA-membranes regulate both protein and lipid traffic through late endosomes (see below).[42,43] In addition to LBPA, some proteins are known to distribute preferentially within internal membranes. These include down-regulated EGF receptor,[44] but also recycling MHC class II molecules in transit,[45] mannose 6-phosphate receptor[46] and tetraspan proteins.[47] In contrast, lysosomal glycoproteins as Lamp-1 or Lamp-2 are to be restricted to the limiting membrane of late endosomes.[42,47] While it is attractive to believe that some proteins may be preferentially incorported within LBPA-membranes or excluded from the limiting membrane, the mechanisms that regulate protein transport into and from late endosome internal membranes remain to be elucidated.

Lysosomes are usually regarded as the terminal degradation compartment of the endocytic pathway,[48] and play important roles in the degradation of phagocytosed material,[49] autophagy,[50] crinophagy[51] and proteolysis of cytosolic proteins transported across the lysosomal membrane by a carrier-mediated mechanism.[52] However, the relationships between late endosomes and lysosomes remain mysterious. Morphologically, lysosomes and late endosomes can often be distinguished, because the latter compartment exhibits multilamellar/multivesicular structures as well as tubular and cisternal regions, while lysosomes appear simpler, spherical and more electron-dense. Both compartments share the same abundant proteins (so-called lysosomal glycoproteins) as major membrane constituents, and contain lysosomal enzymes.[48] In fact, no protein has been shown until now to distribute exclusively to lysosomes. From a compositional viewpoint, lysosomes can only be distinguished from late endosomes by the fact that they do not contain proteins present in late endosomes, including mannose-6-phosphate receptors (MPRs) in transit and the small GTPases rab7 and rab9. While late endosomes and lysosomes clearly exhibit different functions (sorting and degradation, respectively), recent studies, in fact, show that they can interact transiently to form a hybrid intermediate. This hybrid organelle exhibits intermediate properties of both compartments, implying that continued efflux of material from this hybrid is required for the ultimate reformation of dense lysosomes.[53-55] Finally, several studies suggest the existence of a retrograde transport out of lysosomes. When the intracellular Ca^{2+} concentration is elevated to 1-5 μM,[56] lysosomes have been shown to undergo exocytosis. It has been shown protein traffic out of the lysosome-like MHC class II compartment is possible for MHC class II molecules when the degradation of the invariant chain is blocked.[57]

Coat Components

Clathrin, Adaptors and Associated Proteins

Clathrin-coated pits form as shallow membrane invaginations at the plasma membrane. These pits then invaginate, presumably because of clathrin polymerization, eventually leading to fission and formation of a clathrin-coated vesicle. Clathrin-coated pits function as efficient sorting devices at the plasma membrane. Most membrane proteins that are destined to be endocytosed, including receptors, are recruited into pits by interactions of well characterized diLeu or Tyr-based motives in the protein cytoplasmic domains with the μ2 chain of the heterotetrameric adaptor complex AP2.[58] This complex, in turn, recruits clathrin, through interactions with the μ2 AP2 subunit. Internalization of the β-adrenergic receptor, however, bypasses the need for AP2, via direct interactions between a non-visual arrestin and the clathrin head domain.[59] Adaptor complexes are also involved in protein sorting at other transport steps. Major progress has been made in our understanding of these interactions at the atomic level by solving the structure of some of the components involved.[1] AP-2 also recruits various accessory molecules and enzymes, including amphiphysin, Eps15 and auxilin, which are all necessary for clathin-vesicle formation, although their precise functions remain to be eluciated. Eps15 belongs to a growing family of adaptor proteins containing EH (Eps15 homology) domains that play a role in clathrin-mediated endocytosis.[60] It is ubiquitously and constitutively associated with AP-2 clathrin adaptor protein complex within coated pits. Inhibition of the AP-2/Eps15 interaction inhibits endocytosis both in vivo and in vitro, showing that Eps15 is required for the early steps of clathrin-mediated endocytosis.[61] Epsin was recently identified in screens for proteins that interact with the eps15 homology (EH) domain of eps15[62] and with clathrin.[63] Intersectin has been shown to contain EH domains and localizes to the plasma membrane at clathrin-coated pits[64,65] and binds components of the endocytic complex, including dynamin, Eps15, and Epsin.[64-67] In *Drosophila*, it has been established that Numb which determines cell fate during nervous and muscle system development, physically interacts with the receptor Notch,[68,69] in vivo and in vitro with Eps15,[70] the appendage domain of α-adaptin of AP-2[71] and localize to the endocytic pathway.[71]

At present, four distinct heterotetrameric adaptor complexes have been identified, designated AP-1 to AP-4. Each is composed of a related 100-kDa β subunit, a unique subunit of 100-160 kDa (designated γ for AP-1, α for AP-2, δ for AP-3, and ε for AP-4), a related μ subunit of 50 kDa, and a related σ subunit of 20 kDa. AP-1 mediates transport of lysosomal enzymes bound to Man6P-R from the TGN to endosomes.[72] Knock-out of the μ1 chain of AP-1 in mouse might suggests that AP-1 may also function in retrograde transport from endosomes back to the TGN.[73] Recent studies also show that basolateral targeting in epitheliaal cells is mediated by a novel form of the AP-1 complex.[74] AP-3 coat protein mediates the transport of the major late endosomal and lysosomal glycoproteins (Lamp-1 and Lamp-2) from TGN to endosomes/lysosomes, but the precise pathway is still unclear in mammalian cells. Mutations in AP-3 subunits in mice result in coat color defects and bleeding disorders,[75] and in *Drosophila*, result in defects in eye pigmentation.[76] In humans, mutations in the AP-3 β3A subunit cause an inherited disorder, Hermandsky-Pudlak syndrome (HPS) in which patients show deficiencies in skin and eye pigmentation and a complete lack of dense granules in platelets, resulting in impaired blood clotting.[77] These sorting events are likely to involve interaction between the AP-3 complex and both dileucine-based[78] and tyrosine-based[79] motifs in the cytoplasmic domains of the membrane proteins present in these organelles. Finally, a fourth adaptor complex AP-4 has been discovered. It is expressed ubiquitously and at low abundance, but its function is unknown.[79]

Clathrin forms the outer surface of the vesicle coat and is composed of a heterohexameric complex composed of three clathrin heavy chains, joined at their C-termini, and three clathrin light chains, which a three-legged appearance (triskelion).[80] Clathrin triskelia assemble into flat or spherical polyhedral lattice on membranes, and pure clathrin can polymerize into empty cages that resemble the clathin lattice on membranes. Hence, it is generally believed that clathrin acts a molecular scaffold and that its polymerization drives the budding process. Vesicle fission requires a member of the dynamin family, which contains functionally diverse, high molecular mass GTPases (100 kDa) that have atypically low affinities for GTP and high intrinsic rates of GTP hydrolysis. Three isoforms have been identified. Dynamin-1 is neuron-specific, dynamin-2 is ubiquitously expressed, and dynamin-3 is expressed in brain, testes, and lungs. Originally, dynamin was found to function in endocytosis and was assumed to be involved in the formation of coated vesicles from plasma membrane,[81] but its precise function in vesicle formation remains controversial.[82] Amphiphysin plays a key role in endocytosis by recruiting dynamin to clathrin coated pits.[83] Amphiphysin I is found exclusively in brain. Amphiphysin II can be found associated with dynamin.[84] Amphyphisin IIm, a novel Amphiphysin II isoform has been shown to recruit dynamin to the phagosome membrane in a PI3-kinase dependent way.[85] Once formed the free vesicle rapidly uncoats, perhaps through the action of uncoated ATPase (hsc70),[86] and then docks onto and fuses with early endosomes, in a process that depends on the small GTPase Rab5 and its effectors.[35]

COP-I

COPI coats are well documented to be responsible for retrograde transport at early steps of the biosynthetic pathway. COPI exists as an equimolar heteroheptamer consisting of α (160 kDa), β (107 kDa), β' (102 kDa), γ (100 kDa), δ (60 kDa), ε (36 kDa) and ζ (20 kDa).[87] These subunits are believed to be associated in a complex (the coatomer), whether cytosolic or membrane-associated. COPI-coated vesicles can be readily distinguished from clathrin-coated vesicles, as their coat appears thinner and less regular by classical electron microscopy. Recruitment of the coat onto Golgi membranes and coat depolymerization depends on the small GTP-binding protein ARF1. An endosomal COPI complex, lacking the γ and δ subunits, has been described on early endosomes.[88,89] In vitro and in vivo experiments indicate that transport from early to late endosomes and formation of ECV/MVBs from early endosomal membranes depend both on this COPI subcomplex and on endosome acidification. Both mechanisms are likely to be related functionally, since COP recruitment onto endosomal membranes is itself pH-dependent. COP-I was also shown to be required for vesicle retrieval from phagosomes, but not for phagosome formation nor maturation.[90] Finally, sorting of downregulated CD4 bound to AIDS-encoded Nef in early endosomes involves βCOP.[23] The precise mechanism of COP action in the endocytic pathway is still unclear, but it is tempting to speculate that COPI proteins participate in the selective incorporation of proteins into forming ECV/MVBs.

SNAREs and Rab Proteins

SNARE Proteins

Major progress has been made in understanding the mechanisms that regulate specific docking and subsequent fusion of membranes, although the precise role of many components are still unclear. Members of the SNARE family clearly play a critical role in docking and presumably in fusion.[34] SNAREs (soluble N-ethylene maleimide-sensitive factor attachment protein receptors)[91] represent families of small and mostly membrane-bound proteins that are located on the cytoplasmic surfaces of membranes of the secretory and endocytic pathway. Based on their preferred localization, SNARE proteins were classified originally as either v-SNAREs (present on vesicle membranes), or t-SNAREs (present on target membranes). Specific interactions between t- and v-SNAREs are thought to play a critical role in the fidelity of vesicle docking and fusion events.[92] Upon bilayer fusion, all SNAREs are embedded in the same membrane,[93,94] and the ATPase NSF/sec18, together with its soluble cofactor SNAP/sec17, dissociates the SNARE complex, allowing SNARE recycling and reactivation.[95] Several v- and t-SNAREs have been found in the endocytic pathway, along both recycling and degratation routes.[34] However, the precise distribution of some endocytic SNAREs, as well as their site of action, is often unclear or controversial. Although a given SNARE will be most abundant in a certain compartment, during vesicular transport, SNAREs inevitably spread through several compartments. Thus, docking/fusion specificity cannot be solely determined by the distribution of endosomal SNAREs.

Rab Proteins

Rab proteins are small GTPases anchored to the cytoplasmic surface of specific intracellular membrane compartments via geranyl-geranylation group of C-terminal Cys residues, and are important regulators of membrane traffic on the biosynthetic and endocytic pathways.[96] Much like Ras, a Rab GTPase interacts with effectors in the active, GTP-bound state, and GTP hydrolysis faciliated by a GAP protein inactivates the signaling process. Reactivation requires GDP-GTP exchange through a specific exchange factor. Accumulated evidence suggests that Rab GTPases recruit tethering and docking factors to establish firm contact between the membranes to fuse, after which SNAREs become involved and complete the fusion process.[97]

Several Rab proteins have been localized to endosomes.[96] Rab5 is found on the plasma membrane and early endosomes, and regulates both clathrin-coated vesicle internalization and endosome fusion. This GTPase can interact with numerous effectors, including rabaptin5 bound to the Rab5 exchange factor itself and PI-3 kinases. Moreover, two Rab5 effectors EEA1 and Rabenosyn-5, which are both required for early endosome fusion, contain the conserved phosphoinositol-3-phosphate (PI3P)-binding FYVE motif. These observations have lead to the notion that active, GTP-bound Rab5 builds a specific effector platform on the membrane, which could integrate different mechanisms that regulate transport, including membrane fusion, membrane budding and interaction with cytoskeletal components.[96] Rab5 might also contribute to the spatial organization of docking/fusion sites for vesicles arriving from different pathways, as its effector EEA1 can interact with the SNAREs syntaxin-6,[98] which has been implicated in TGN to early endosome traffic, and syntaxin-13, which is required for endosome fusion.[99] The findings that the Rab5 effector rabaptin5 also interacts with Rab4,[100] a small GTPase involved in recycling, and that the Rab4 effector Rabip4 contains a FYVE motive and localizes to early endosomes,[101] raise the possibility that these Rab proteins are functionally coupled, perhaps reflecting the existence of a physical link between different platforms. It will be interesting to determine whether cross talk between Rab effectors also exists along other pathways.

Both Rab4 and Rab11 have been sequentially implicated in the recycling pathway from early endosomes back to the plasma, but their precise functions are still unclear.[96,102] Recently, Rab15 was shown to colocalize with Rab4 and Rab5 and Rab11, and to be involved in endocytosis, perhaps by differentially regulating early endocytic trafficking as compared to Rab5.[103,104] Rab7 is involved in the degradation pathway leading to lysosomes endocytic traffic, including transport from early to late endosomes, late endosome fusion and late endosome-lysosome interactions.[105-108] Moreover, Rab7 is essential for cellular vacuolation induced by the *Helicobacter pylori* cytotoxin VacA[108] and for the maturation of *Salmonella typhimurium*-containing vacuoles.[109] However, the exact role of Rab7 and its mechanism of action in late endocytic traffic are still not known. Rab9 regulates recycling of Man6P-receptor back to the TGN, and recent studies suggest that this GTPase may be selectively involved in sorting of Man6P-receptor molecules via interactions with the Rab8 effector TIP47 (tail-interacting protein of 47 kDa).[110] Its amino acid sequence is highly similar to the lipid droplet protein ADRP (adipose-differentiation-related protein) and has been shown to be associated with lipid droplets.[111] Studies in yeast indicate that recycling from endosomes/vacuole to the TGN depends on the retromer complex, which contains the PX-protein Vps5p (see below).[112] Mammalian homologues of the retromer complex have been identified,[113] and it has been proposed that TIP47/Rab9 and the retromer complex may function in protein sorting and cargo selection on different late endosomal populations.[114]

Lipids and Protein-Lipid Microdomains

Originally, it has been proposed that the lipid bilayer functioned as a neutral two-dimensional solvent, having little influence on membrane protein function. A turning point came when the lipid raft hypothesis was formulated as consisting of dynamic assemblies of cholesterol and sphingolipids in the exoplasmic leaflet of the bilayer.[115,116] Rafts are believed to play a critical role in a number of essential cellular functions, including signaling, protein sorting and pathogen infection.[116,117] Raft components are found in the early endocytic circuit, and may play a role in endosomal protein sorting.[28,118] Since rafts have been extensively reviewed recently[116] they will not be further discussed here. In fact, it appears that different membrane domains, including some with well-defined properties, coexist in endocytic compartments. The dynamic interplay between these domains may provide a driving force responsible both for the specific organization of each compartment and for the movement of cargo molecules.

Phosphoinositides and Partners

Evidence in yeast and mammalian cells is accumulating that phosphoinositides play a direct role in membrane transport, in addition to their classical signaling functions.[119,120] Biochemical efforts to identify proteins interacting with PI-3 kinases led to the discovery of a family of protein containing a zinc finger domain now referred to as FYVE domain (22 in mammals and 5 in yeast; ref. 121), including the Rab5 effectors EEA1 and Rabenosyn-5, which are both required for early endosome fusion.[122,123] EEA1 is in fact restricted to early endosomes,[118,124,125] as well perhaps other FYVE proteins.[101,126] Among the latter proteins, Hrs (Hepatocyte growth factor-regulated tyrosine kinase substrate) was shown to be required for ventral morphogenesis after targeted gene disruption in mice.[127] Enlarged endosomes were also observed, suggesting that ventral folding defects may result from impaired endosomal function.

Recent studies have also uncovered the existence of another PI3P binding domain (PX or PHOX homology) in PHOX 40 and 47,[128,129] which are involved in reactive oxygen production by neutrophils, in the t-SNARE Vam7p involved in transport to the yeast vacuole,[130] and in the sorting nexin SNX3.[131] The PX domain is shared by a relatively large number of proteins, including PLD1 and PI-3 kinase, but also other SNX proteins. Interestingly, SNX1, which also interacts with the FYVE protein Hrs,[132] was identified as enhancing EGF-receptor degradation.[133] Its yeast homologue Vps5p is necessary for endosome-TGN recycling, as a component of the retromer complex. These observations might suggest that Snx1 functions are broader in mammalian than in yeast cells. While, in addition to SNX1, other SNX proteins interact with receptor tyrosine kinases.[134,135] SNX3 and SNX15 are involved in endosomal transport and their overexpression disrupts endosome morphology.[131,136] The precise functions of SNX proteins are still unclear, as are the relationships between endosomes, PX-proteins, FYVE-proteins and the PI3P pathway, but these findings clearly broaden even further the interplay between PI3P and endosomal functions.

In addition, PI3P was also found on internal membranes of ECV/MVBs after labeling cryosections with a probe containg a double FYVE motive.[137] Although it is not clear whether the presence of PI3P on these internal membranes reflects a role for this lipid in the biogenesis of invaginations, or simply the metabolism of the lipid, several lines of evidence suggest that PI3P is also involved in the degradation pathway leading to lysosomes. The yeast FYVE protein Fab1p, which is a PI3P-5-OH kinase that generates PI(3,5)P2, is essential for maintenance of normal vacuolar morphology and was proposed to regulate cargo-selective sorting into the vacuole lumen.[37] Wortmannin, a drug that inhibits PI-3-kinase was reported to inhibit multivesicular body formation in mammalian cells.[138] At the same time, in yeast, evidence is accumulating that signaling through the PI-3-kinase Vps34p and PI3P has a crucial role in the TGN–vacuole transport.[139] It thus appears that, in addition to its function in early endosomal dynamics, PI3P might also function in the biogenesis of multivesicular endosomes[139] underscoring the close connection between early endosome function and morphogenesis.

Annexin II-Cholesterol

Annexins seem to be endowed with the intrinsic ability to self-organize at the membrane surface into bidimensional ordered arrays.[140] It is attractive to speculate that annexins, using a variety of membrane association mechanisms, might form platforms on the membrane of different sub-cellular compartments, which would specifically interact with cytosolic components, including the cytoskeleton. Annexin II localizes to the plasma membrane and on early endosomes, and it was shown to be involved in the dynamics of early and/or recycling endosomes.[141,142] Annexin II membrane binding uses an unconventional, cholesterol-dependent mechanism,[143] but perhaps not at the plasma membrane.[144] Annexin II also interacts with proteins of the cortical actin cytoskeleton, and it seems to be non-randomly distributed on early endosomal membranes, and concentrated in areas, from which actin-like filaments seem to emanate.[143] It is thus possible that cholesterol-rich region of early endosomal membranes that interact with annexin II play a role in the general organization and dynamics of this compartment.

LBPA

Within late endosomes, but not ECV/MVBs, internal membrane invaginations accumulate large amounts of the unique phospholipid lysobisphosphatidicacid (LBPA), corresponding to ≥15% of the total late endosomal phospholipids.[42] This lipid is believed to be exhibit an unusal stereochemical configuration, and thus to be a poor substrate for phospholipases and resistant to degradation.[145] LBPA is presumably synthesized in situ,[146] and has an inverted cone shape. This structure may facilitate the formation of the invaginations that form the multivesicular elements of late endosomes.[41]

Sandhoff and collaborators have demonstrated using an in vitro liposome assay that negatively-charged phospholipids, in particular LBPA, greatly facilitate the degradation of several glycolipids.[147] As LBPA itself is poorly degradable, one function of LBPA-membranes could be to present lipids and proteins that need to be degraded to the hydrolytic machinery. However, internal membranes of late endosomes do not only contain molecules destined to be degraded, but also tetraspanins, MHC class II and Man6P-R in transit (see above). Moreover, endocytosed antibodies against LBPA accumulate in late endosomes upon binding to their antigen, and specifically inhibit Man6P-receptor transport.[42] Conversely, loss of Man6P-receptor expression promotes LBPA accumulation in multilamellar bodies.[148] Endocytosed anti-LBPA antibodies also cause cholesterol accumulation in late endosomes, mimicking the cholesterol-storage disorder Niemann-Pick type C (NPC), leading to the idea that LBPA-membranes serve as collecting and distribution device for LDL-derived cholesterol.[43] In turn, cholesterol accumulation in late endosomes interferes with the Man6P-receptor cycle in several cell types,[43,149] and with CD63 and P-selectin cycling to Weibel-Palade bodies in endothelial cells.[150] Sphingolipids accumulate in multivesicular compartments in tissue from NPC patients, and, conversely, cholesterol accumulates in late endocytic compartments in several sphingolipid storage disorders.[151,152] Altogether, these observations indicate that LBPA-rich membranes may have turnpike functions in the late endosomal network, as they seem to be involved both in protein and lipid transport and lipid degradation.

Conclusions

Major progress has been made over the past decade in our understanding of the mechanisms that control endocytic membrane traffic, and several components that play a role in endocytic traffic are linked to human diseases and involved in development.[153] A number of key components have been identified and characterized, as well membrane domains or protein complexes that interact with membrane lipids. Important new tools are now available for studying endocytic traffic, including critical mutations in key factors regulating traffic and the combined use of tagged-proteins and video-microscopy. In parallel, the study of infections has developed tremendously in recent years, in particular our understanding of the molecular basis of interactions between cells and pathogenic agents, which very rapidly evolve novel strategies to alter host cells. These studies have already revealed pathways and molecular mechanisms that were unsuspected or that could not be easily studied by classical methods. Although some hints are being provided by studies that reveal the existence of a close interplay between signaling and transport,[38] an important challenge will be to understand how these different protein complexes and dynamic domains are integrated to support both transport and organelle functions.

References

1. Marsh M, McMahon HT. The structural era of endocytosis. Science 1999; 285:215-220.
2. Artalejo CR, Henley JR, McNiven MA et al. Rapid endocytosis coupled to exocytosis in adrenal chromaffin cells involves Ca2+, GTP, and dynamin but not clathrin. Proc Natl Acad Sci USA 1995; 92:8328-8332.
3. Demlehner MP, Schafer S, Grund C et al. Continual assembly of half-desmosomal structures in the absence of cell contacts and their frustrated endocytosis: a coordinated Sisyphus cycle. J Cell Biol 1995; 131:745-760.
4. Lamaze C, Dujeancourt A, Baba T et al. Interleukin 2 receptors and detergent-resistant membrane domains define a clathrin-independent endocytic pathway. Mol Cell 2001; 7:661-671.
5. van der Goot FG, Harder T. Raft membrane domains: from a liquid-ordered membrane phase to a site of pathogen attack. Semin Immunol 2001; 13:89-97.
6. Rodal SK, Skretting G, Garred O et al. Extraction of cholesterol with methyl-beta-cyclodextrin perturbs formation of clathrin-coated endocytic vesicles. Mol Biol Cell 1999; 10:961-974.
7. Pelkmans L, Kartenbeck J, Helenius A. Caveolar endocytosis of simian virus 40 reveals a new two-step vesicular-transport pathway to the ER. Nat Cell Biol 2001; 3:473-483.
8. Sansonetti PJ. Microbes and microbial toxins: paradigms for microbial-mucosal interactions III. Shigellosis: From symptoms to molecular pathogenesis. Am J Physiol Gastrointest Liver Physiol 2001; 280:G319-323.
9. Haigler HT, McKanna JA, Cohen S. Rapid stimulation of pinocytosis in human carcinoma cells A-431 by epidermal growth factor. J Cell Biol 1979; 83:82-90.
10. Bar-Sagi D, Feramisco JR. Induction of membrane ruffling and fluid-phase pinocytosis in quiescent fibroblasts by ras proteins. Science 1986; 233:1061-1068.
11. Bishop AL, Hall A. Rho GTPases and their effector proteins. Biochem J 2000; 348:241-255.
12. Nobes C, Marsh M. Dendritic cells: New roles for Cdc42 and Rac in antigen uptake? Curr Biol 2000; 10:R739-741.
13. Jahraus A, Tjelle TE, Berg T et al. In vitro fusion of phagosomes with different endocytic organelles from J774 macrophages. J Biol Chem 1998; 273:30379-30390.

14. Defacque H, Egeberg M, Habermann A et al. Involvement of ezrin/moesin in de novo actin assembly on phagosomal membranes. EMBO J 2000; 9:199-212.
15. Duclos S, Diez R, Garin J et al. Rab5 regulates the kiss and run fusion between phagosomes and endosomes and the acquisition of phagosome leishmanicidal properties in RAW 264.7 macrophages. J Cell Sci 2000; 113:3531-3541.
16. Ferrari G, Langen H, Naito M et al. A coat protein on phagosomes involved in the intracellular survival of mycobacteria. Cell 1999; 97:435-47.
17. Garin G, Diez R, Kieffer S et al. The phagosome proteome: Insight into phagosome functions. J Cell Biol 2001; 152:165-180.
18. Yamashiro DJ, Maxfield FR. Acidification of morphologically distinct endosomes in mutant and wild-type Chinese hamster ovary cells. J Cell Biol 1987; 105:2723-2733.
19. Gruenberg J, Maxfield FR. Membrane transport in the endocytic pathway. Curr Opin Cell Biol 1995; 7:552-563.
20. Green SA, Setiadi H, McEver RP et al. The cytoplasmic domain of P-selectin contains a sorting determinant that mediates rapid degradation in lysosomes. J Cell Biol 1994; 124:435-448.
21. Straley KS, Daugherty BL, Aeder SE et al. An atypical sorting determinant in the cytoplasmic domain of P-selectin mediates endosomal sorting. Mol Biol Cell 1998; 9:1683-1694.
22. Subtil A, Rocca A, Dautry-Varsat A. Molecular characterization of the signal responsible for the targeting of the interleukin 2 receptor beta chain toward intracellular degradation. J Biol Chem 1998; 273:29424-29429.
23. Piguet V, Gu F, Foti M et al. Nef-induced CD4 degradation: a diacidic-based motif in Nef functions as a lysosomal targeting signal through the binding of beta-COP in endosomes. Cell 1999; 97:63-73.
24. Kornilova E, Sorkina T, Beguinot L et al. Lysosomal targeting of epidermal growth factor receptors via a kinase-dependent pathway is mediated by the receptor carboxyl-terminal residues 1022-1123. J Biol Chem 1996; 271:30340-30346.
25. Kil SJ, Hobert M, Carlin C. A leucine-based determinant in the epidermal growth factor receptor juxtamembrane domain is required for the efficient transport of ligand-receptor complexes to lysosomes. J Biol Chem 1999; 274:3141-3150.
26. Hicke L. Ubiquitin-dependent internalization and down-regulation of plasma membrane proteins. FASEB J 1997; 11:1215-1226.
27. Kobayashi T, Hirabayashi Y. Lipid membrane domains in cell surface and vacuolar systems. Glycoconj J 2000; 17:163-171.
28. Mukherjee S, Maxfield FR. Role of membrane organization and membrane domains in endocytic lipid trafficking. Traffic 2000; 1:203-211.
29. Gruenberg J, Maxfield FR. Membrane transport in the endocytic pathway. Curr Opin Cell Biol 1995; 7:552-563.
30. Apodaca G, Katz LA, Mostov KE. Receptor-mediated transcytosis of IgA in MDCK cells is via apical recycling endosomes. J Cell Biol 1994; 125:67-86.
31. Stoorvogel W, Oorschot V, Geuze HJ. A novel class of clathrin-coated vesicles budding from endosomes. J Cell Biol. 1996; 132:21-33.
32. Lim SN, Bonzelius F, Low SH et al. Identification of discrete classes of endosome-derived small vesicles as a major cellular pool for recycling membrane proteins. Mol Biol Cell 2001; 12:981-995.
33. Lin SX, Grant B, Hirsh D et al. Rme-1 regulates the distribution and function of the endocytic recycling compartment in mammalian cells. Nat Cell Biol 2001; 3:567-572.
34. Chen YA, Scheller RH. SNARE-mediated membrane fusion. Nat Rev Mol Cell Biol 2001; 2:98-106.
35. Zerial M, McBride H. Rab proteins as membrane organizers. Nat Rev Mol Cell Biol 2001; 2:107-117.
36. Taunton J. Actin filament nucleation by endosomes, lysosomes and secretory vesicles. Curr Opin Cell Biol 2001; 13:85-91.
37. Odorizzi G, Babst M, Emr SD. Fab1p PtdIns(3)P 5-kinase function essential for protein sorting in the multivesicular body. Cell 1998; 95:847-858.
38. Cavalli V, Corti M, Gruenberg J. Endocytosis and signaling cascades: A close encounter. FEBS Lett 2001; 498:190-196.
39. Wurmser AE, Emr SD. Phosphoinositide signaling and turnover: PtdIns(3)P, a regulator of membrane traffic, is transported to the vacuole and degraded by a process that requires lumenal vacuolar hydrolase activities. EMBO J 1998; 17:4930-4942.
40. Aniento F, Emans N, Griffiths G et al. Cytoplasmic dynein-dependent vesicular transport from early to late endosomes. J Cell Biol 1993; 123:1373-1387.
41. Meresse S, Gorvel JP, Chavrier P. The rab7 GTPase resides on a vesicular compartment connected to lysosomes. J Cell Sci 1995; 108:3349-3358.

42. Kobayashi T, Stang E, Fang KS et al. A lipid associated with the antiphospholipid syndrome regulates endosome structure and function. Nature 1998; 392:193-197.
43. Kobayashi T, Beuchat MH, Lindsay M et al. Late endosomal membranes rich in lysobisphosphatidic acid regulate cholesterol transport. Nat Cell Biol 1999; 1:113-118.
44. Felder S, Miller K, Moehren G et al. Kinase activity controls the sorting of the epidermal growth factor receptor within the multivesicular body. Cell 1990; 61:623-634.
45. Peters PJ, Neefjes JJ, Oorschot V et al. Segregation of MHC class II molecules from MHC class I molecules in the Golgi complex for transport to lysosomal compartments. Nature 1991; 349:669-676.
46. Griffiths G, Hoflack B, Simons K et al. The mannose 6-phosphate receptor and the biogenesis of lysosomes. Cell 1988; 52:329-341.
47. Escola JM, Kleijmeer MJ, Stoorvogel W et al. Selective enrichment of tetraspan proteins on the internal vesicles of multivesicular endosomes and on exosomes secreted by human B-lymphocytes. J Biol Chem 1998; 273:20121-20127.
48. Kornfeld S, Mellman I. The biogenesis of lysosomes. Annu Rev Cell Biol 1989; 5:483-525.
49. Funato K, Beron W, Yang CZ et al. Reconstitution of phagosome-lysosome fusion in streptolysin O-permeabilized cells. J Biol Chem 1997; 272:16147-16151.
50. Lawrence BP, Brown WJ. Autophagic vacuoles rapidly fuse with pre-existing lysosomes in cultured hepatocytes. J Cell Sci 1992; 102:515-526.
51. Noda T, Farquhar MG. A non-autophagic pathway for diversion of ER secretory proteins to lysosomes. J Cell Biol 1992; 119:85-97.
52. Cuervo AM, Dice JF. A receptor for the selective uptake and degradation of proteins by lysosomes. Science 1996; 273:501-503.
53. Storrie B, Desjardins M (1996). The biogenesis of lysosomes: Is it a kiss and run, continuous fusion and fission process? Bioessays 18, 895-903.
54. Luzio JP, Rous BA, Bright NA, et al. Lysosome-endosome fusion and lysosome biogenesis. J Cell Sci 2000; 113:1515-1524.
55. Pryor PR, Mullock BM, Bright NA et al. The role of intraorganellar Ca(2+) in late endosome-lysosome heterotypic fusion and in the reformation of lysosomes from hybrid organelles. J Cell Biol 2000; 149:1053-1062.
56. Rodriguez A, Webster, P, Ortego, J et al. Lysosomes behave as Ca2+-regulated exocytic vesicles in fibroblasts and epithelial cells. J Cell Biol 1997; 137:93-104.
57. Brachet V, Raposo G, Amigorena S et al. Ii chain controls the transport of major histocompatibility complex class II molecules to and from lysosomes. J Cell Biol 1997; 137:51-65.
58. Kirchhausen T. Adaptors for clathrin-mediated traffic. Annu Rev Cell Dev Biol 1999; 15:705-732.
59. Lin FT, Krueger KM, Kendall HE et al. Clathrin-mediated endocytosis of the beta-adrenergic receptor is regulated by phosphorylation/dephosphorylation of beta-arrestin1. J Biol Chem 1997; 272:31051-31057.
60. Di Fiore PP, Pelicci PG, Sorkin A. EH: a novel protein-protein interaction domain potentially involved in intracellular sorting. Trends Biochem Sci 1997; 22:411-413.
61. Benmerah A, Lamaze C, Begue B et al. AP-2/Eps15 interaction is required for receptor-mediated endocytosis. J Cell Biol 1998; 140:1055-1062.
62. McPherson PS, de Heuvel E, Phillie J et al. EH domain-dependent interactions between Eps15 and clathrin-coated vesicle protein p95. Biochem Biophys Res Commun 1998; 244:701-705.
63. Drake MT, Downs MA, Traub LM. Epsin binds to clathrin by associating directly with the clathrin-terminal domain. Evidence for cooperative binding through two discrete sites. J Biol Chem 2000; 275:6479-6489.
64. Hussain NK, Yamabhai M, Ramjaun AR et al. Splice variants of intersectin are components of the endocytic machinery in neurons and nonneuronal cells. J Biol Chem 1999; 274:15671-15677.
65. Sengar AS, Wang W, Bishay J et al. The EH and SH3 domain Ese proteins regulate endocytosis by linking to dynamin and Eps15. EMBO J 1999; 18:1159-1171.
66. Roos J, Kelly RB. Dap160, a neural-specific Eps15 homology and multiple SH3 domain-containing protein that interacts with Drosophila dynamin. J Biol Chem 1998; 273:19108-19119.
67. Yamabhai M, Hoffman NG, Hardison NL et al. Intersectin, a novel adaptor protein with two Eps15 homology and five Src homology 3 domains. J Biol Chem 1998; 273:31401-31407.
68. Berezovska O, McLean P, Knowles R et al. Notch1 inhibits neurite outgrowth in postmitotic primary neurons. Neuroscience 1999; 93:433-439.
69. Sestan N, Artavanis-Tsakonas S, Rakic P. Contact-dependent inhibition of cortical neurite growth mediated by notch signaling. Science 1999; 286:741-746.
70. Salcini AE, Confalonieri S, Doria M et al. Binding specificity and in vivo targets of the EH domain, a novel protein-protein interaction module. Genes Dev 1997; 11:2239-2249.

71. Santolini E, Puri C, Salcini AE et al. Numb is an endocytic protein. J Cell Biol 2000; 151:1345-1352.
72. Le Borgne R, Hoflack B. Protein transport from the secretory to the endocytic pathway in mammalian cells. Biochim Biophys Acta 1998; 1404:195-209.
73. Meyer C, Zizioli D, Lausmann S et al. mu1A-adaptin-deficient mice: lethality, loss of AP-1 binding and rerouting of mannose 6-phosphate receptors. EMBO J 2000; 19:2193-2203.
74. Fölsch H, Pypaert M, Schu P et al. Distribution and function of AP-1 clathrin adaptor complexes in polarized epithelial cells. J Cell Biol 2001; 152:595-606.
75. Kantheti P, Qiao X, Diaz ME et al. Mutation in AP-3 delta in the mocha mouse links endosomal transport to storage deficiency in platelets, melanosomes, and synaptic vesicles. Neuron 1998; 21:111-122.
76. Mullins C, Hartnell LM, Wassarman DA et al. Defective expression of the mu3 subunit of the AP-3 adaptor complex in the Drosophila pigmentation mutant carmine. Mol Gen Genet 1999; 262:401-412.
77. Dell'Angelica EC, Shotelersuk V, Aguilar RC et al. Altered trafficking of lysosomal proteins in Hermansky-Pudlak syndrome due to mutations in the beta 3A subunit of the AP-3 adaptor. Mol Cell 1999; 3:11-21.
78. Darsow T, Burd CG, Emr SD. Acidic di-leucine motif essential for AP-3-dependent sorting and restriction of the functional specificity of the Vam3p vacuolar t-SNARE. J Cell Biol 1998; 142:913-922.
79. Dell'Angelica EC, Mullins C, Bonifacino JS. AP-4, a novel protein complex related to clathrin adaptors. J Biol Chem 1999; 274:7278-7285.
80. Kirchhausen T. Clathrin. Annu Rev Biochem 2000; 69:699-727.
81. Schmid SL, McNiven MA, De Camilli P. Dynamin and its partners: a progress report. Curr Opin Cell Biol 1998; 10:504-512.
82. Sever S, Damke H, Schmid SL. Garrotes, springs, ratchets, and whips: putting dynamin models to the test. Traffic 2000; 1:385-392.
83. Wigge P, McMahon HT. The amphiphysin family of proteins and their role in endocytosis at the synapse. Trends Neurosci 1998; 21:339-344.
84. Leprince C, Romero F, Cussac D et al. A new member of the amphiphysin family connecting endocytosis and signal transduction pathways. J Biol Chem 1997; 272:15101-15105.
85. Gold ES, Morrissette NS, Underhill DM et al. Amphiphysin IIm, a novel amphiphysin II isoform, is required for macrophage phagocytosis. Immunity 2000; 12:285-292.
86. Newmyer SL, Schmid SL. Dominant-interfering Hsc70 mutants disrupt multiple stages of the clathrin-coated vesicle cycle in vivo. J Cell Biol 2001; 152:607-620.
87. Kreis TE, Lowe M, Pepperkok R. COPs regulating membrane traffic. Annu Rev Cell Dev Biol 1995; 11:677-706.
88. Whitney JA, Gomez M, Sheff D et al. Cytoplasmic coat proteins involved in endosome function. Cell 1995; 83:703-713.
89. Aniento F, Gu F, Parton RG et al. An endosomal beta COP is involved in the pH-dependent formation of transport vesicles destined for late endosomes. J Cell Biol 1996; 133:29-41.
90. Botelho RJ, Hackam DJ, Schreiber AD et al. Role of COPI in phagosome maturation. J Biol Chem 2000; 275:15717-15727.
91. McNew JA, Weber T, Parlati F et al. Close is not enough: SNARE-dependent membrane fusion requires an active mechanism that transduces force to membrane anchors. J Cell Biol 2000; 150:105-117.
92. Söllner T, Whiteheart SW, Brunner M et al. SNAP receptors implicated in vesicle targeting and fusion. Nature 1993; 362:318-324.
93. Hanson PI, Roth R, Morisaki H et al. Structure and conformational changes in NSF and its membrane receptor complexes visualized by quick-freeze/deep-etch electron microscopy. Cell 1997; 90:523-535.
94. Lin RC, Scheller RH. Structural organization of the synaptic exocytosis core complex. Neuron 1997; 19:1087-1094.
95. Söllner T, Bennett MK, Whiteheart SW et al. A protein assembly-disassembly pathway in vitro that may correspond to sequential steps of synaptic vesicle docking, activation, and fusion. Cell 1993; 75:409-418.
96. Zerial M, McBride H. Rab proteins as membrane organizers. Nat Rev Mol Cell Biol 2001; 2:107-117.
97. Pfeffer SR. Transport-vesicle targeting: Tethers before SNAREs. Nat Cell Biol 1999; 1:E17-E22.
98. Simonsen A, Gaullier JM, D'Arrigo A et al. The Rab5 effector EEA1 interacts directly with syntaxin-6. J Biol Chem 1999; 274:28857-28860.

99. McBride HM, Rybin V, Murphy C et al. Oligomeric complexes link Rab5 effectors with NSF and drive membrane fusion via interactions between EEA1 and syntaxin 13. Cell 1999; 98:377-386.
100. Stenmark H, Vitale G, Ullrich O et al. Rabaptin-5 is a direct effector of the small GTPase Rab5 in endocytic membrane fusion. Cell 1995; 83:423-432.
101. Cormont M, Mari M, Galmiche A, et al. A FYVE-finger-containing protein, Rabip4, is a Rab4 effector involved in early endosomal traffic. Proc Natl Acad Sci USA 2001; 98:1637-1642.
102. van der Sluijs P, Hull M, Webster P et al. The small GTP-binding protein rab4 controls an early sorting event on the endocytic pathway. Cell 1992; 70:729-740.
103. Zuk PA, Elferink LA. Rab15 mediates an early endocytic event in Chinese hamster ovary cells. J Biol Chem 1999; 274:22303-22312.
104. Zuk PA, Elferink LA. Rab15 differentially regulates early endocytic trafficking. J Biol Chem 2000; 275:26754-26764.
105. Feng Y, Press B, Wandinger-Ness A. Rab 7: an important regulator of late endocytic membrane traffic. J Cell Biol 1995; 131:1435-1452.
106. Chavrier P, Parton RG, Hauri HP et al. Localization of low molecular weight GTP binding proteins to exocytic and endocytic compartments. Cell 1990; 62:317-329.
107. Meresse S, Gorvel JP, Chavrier P. The rab7 GTPase resides on a vesicular compartment connected to lysosomes. J Cell Sci 1995; 108:3349-3358.
108. Papini E, Satin B, Bucci C et al. The small GTP binding protein rab7 is essential for cellular vacuolation induced by Helicobacter pylori cytotoxin. EMBO J 1997; 16:15-24.
109. Meresse S, Steele-Mortimer O, Finlay BB et al. The rab7 GTPase controls the maturation of Salmonella typhimurium-containing vacuoles in HeLa cells. EMBO J 1999; 18:4394-4403.
110. Diaz E, Pfeffer SR. TIP47: A cargo selection device for mannose 6-phosphate receptor trafficking. Cell 1998; 93:433-443.
111. Wolins NE, Rubin B, Brasaemle DL. TIP47 associates with lipid droplets. J Biol Chem 2001; 276:5101-5108.
112. Seaman MN, McCaffery JM, Emr SD. A membrane coat complex essential for endosome-to-Golgi retrograde transport in yeast. J Cell Biol 1998; 142:665-681.
113. Haft CR, de la Luz Sierra M, Bafford R et al. Human orthologs of yeast vacuolar protein sorting proteins Vps26, 29, and 35: assembly into multimeric complexes. Mol Biol Cell 2000; 11:4105-4116.
114. Pfeffer SR. Membrane transport: Retromer to the rescue. Curr Biol 2001; 11:R109-R111.
115. Simons K, van Meer G. Lipid sorting in epithelial cells. Biochemistry 1988; 27:6197-6202.
116. Simons K, Ikonen E. Functional rafts in cell membranes. Nature 1997; 387:569-572.
117. Schiavo G, van der Goot FG. The bacterial toxin toolkit. Nat Rev Mol Cell Biol 2001; 2:530-537.
118. Gagescu R, Demaurex N, Parton RG et al. The recycling endosome of Madin-Darby canine kidney cells is a mildly acidic compartment rich in raft components. Mol Biol Cell 2000; 11:2775-2791.
119. Toker A, Cantley LC. Signalling through the lipid products of phosphoinositide-3-OH kinase. Nature 1997; 387:673-676.
120. De Camilli P, Emr SD, McPherson PS et al. Phosphoinositides as regulators in membrane traffic. Science 1996; 271:1533-1539.
121. Stenmark H, Aasland R. FYVE-finger proteins—Effectors of an inositol lipid. J Cell Sci 1999; 112:4175-4183.
122. Simonsen A, Lippe R, Christoforidis S et al. EEA1 links PI(3)K function to Rab5 regulation of endosome fusion. Nature 1998; 394:494-498.
123. Nielsen E, Christoforidis S, Uttenweiler-Joseph S et al. Rabenosyn-5, a novel Rab5 effector, is complexed with hVPS45 and recruited to endosomes through a FYVE finger domain. J Cell Biol 2000; 151:601-612.
124. Mu FT, Callaghan JM, Steele-Mortimer O et al. EEA1, an early endosome-associated protein. EEA1 is a conserved alpha-helical peripheral membrane protein flanked by cysteine "fingers" and contains a calmodulin-binding IQ motif. J Biol Chem 1995; 270:13503-13511.
125. Verges M, Havel RJ, Mostov KE. A tubular endosomal fraction from rat liver: biochemical evidence of receptor sorting by default. Proc Natl Acad Sci USA. 1999; 96:10146-10151.
126. Komada M, Masaki R, Yamamoto A et al. Hrs, a tyrosine kinase substrate with a conserved double zinc finger domain, is localized to the cytoplasmic surface of early endosomes. J Biol Chem 1997; 272:20538-20544.
127. Hayakawa A, Kitamura N. Early endosomal localization of hrs requires a sequence within the proline- and glutamine-rich region but not the FYVE finger. J Biol Chem 2000; 275:29636-29642.
128. Ellson CD, Gobert-Gosse S, Anderson KE et al. PtdIns(3)P regulates the neutrophil oxidase complex by binding to the PX domain of p40phox. Nat Cell Biol 2001; 3:679-682.

129. Kanai F, Liu H, Field SJ et al. The PX domains of p47phox and p40phox bind to lipid products of PI(3)K. Nat Cell Biol 2001; 3:675-678.
130. Cheever ML, Sato TK, de Beer T et al. Phox domain interaction with PtdIns(3)P targets the Vam7 t-SNARE to vacuole membranes. Nat Cell Biol 2001; 3:613-618.
131. Xu Y, Hortsman H, Seet L et al. SNX3 regulates endosomal function through its PX-domain-mediated interaction with PtdIns(3)P. Nat Cell Biol 2001; 3:658-666.
132. Chin LS, Raynor MC, Wei X et al. Hrs interacts with sorting nexin 1 and regulates degradation of epidermal growth factor receptor. J Biol Chem 2001; 276:7069-7078.
133. Kurten RC, Cadena DL, Gill GN. Enhanced degradation of EGF receptors by a sorting nexin, SNX1. Science 1996; 272:1008-1010.
134. Haft CR, de la Luz Sierra M, Barr VA et al. Identification of a family of sorting nexin molecules and characterization of their association with receptors. Mol Cell Biol 1998; 18:7278-7287.
135. Parks WT, Frank DB, Huff C et al. Sorting nexin 6, a novel SNX, interacts with the transforming growth factor-beta family of receptor serine-threonine kinases. J Biol Chem 2001; 276:19332-19339.
136. Barr VA, Phillips SA, Taylor SI et al. Overexpression of a novel sorting nexin, SNX15, affects endosome morphology and protein trafficking. Traffic 2000; 1:1904-916.
137. Gillooly DJ, Morrow IC, Lindsay M et al. Localization of phosphatidylinositol 3-phosphate in yeast and mammalian cells. EMBO J 2000; 19:4577-4588.
138. Fernandez-Borja M, Wubbolts R, Calafat J et al. Multivesicular body morphogenesis requires phosphatidyl-inositol 3-kinase activity. Curr Biol 1999; 9:55-58.
139. Simonsen A, Wurmser AE, Emr SD et al. The role of phosphoinositides in membrane transport. Curr Opin Cell Biol 2001; 13:485-492.
140. Oling F, Bergsma-Schutter W, Brisson A. Trimers, dimers of trimers, and trimers of trimers are common building blocks of annexin a5 two-dimensional crystals. J Struct Biol 2001; 133:55-63.
141. Harder T, Gerke V. The subcellular distribution of early endosomes is affected by the annexin II2p11(2) complex. J Cell Biol 1993; 123:1119-1132.
142. Emans N, Gorvel JP, Walter C et al. Annexin II is a major component of fusogenic endosomal vesicles. J Cell Biol 1993; 120:1357-1369.
143. Harder T, Kellner R, Parton RG et al. Specific release of membrane-bound annexin II and cortical cytoskeletal elements by sequestration of membrane cholesterol. Mol Biol Cell 1997; 8:533-545.
144. Babiychuk EB, Draeger A. Annexins in cell membrane dynamics. Ca(2+)-regulated association of lipid microdomains. J Cell Biol 2000; 150:1113-1124.
145. Brotherus J, Renkonen O, Fischer W et al. Novel stereoconfiguration in lyso-bis-phosphatidic acid of cultured BHK-cells. Chem Phys Lipids 1974; 13:178-182.
146. Amidon B, Brown A, Waite M. Transacylase and phospholipases in the synthesis of bis(monoacylglycero)phosphate. Biochemistry 1996; 35:13995-14002.
147. Wilkening G, Linke T, Sandhoff K. Lysosomal degradation on vesicular membrane surfaces. Enhanced glucosylceramide degradation by lysosomal anionic lipids and activators. J Biol Chem 1998; 273:30271-30278.
148. Reaves BJ, Row PE, Bright NA et al. Loss of cation-independent mannose 6-phosphate receptor expression promotes the accumulation of lysobisphosphatidic acid in multilamellar bodies. J Cell Sci 2000; 113:4099-4108.
149. Galve-de Rochemonteix B, Kobayashi T, Rosnoblet C et al. Interaction of anti-phospholipid antibodies with late endosomes of human endothelial cells. Arterioscler Thromb Vasc Biol 2000; 20:563-574.
150. Kobayashi T, Vischer UM, Rosnoblet C et al. The tetraspanin CD63/lamp3 cycles between endocytic and secretory compartments in human endothelial cells. Mol Biol Cell 2000; 11:1829-1843.
151. Simons K, Gruenberg J. Jamming the endosomal system: Lipid rafts and lysosomal storage diseases. Trends Cell Biol 2000; 10:459-462.
152. Pagano RE, Puri V, Dominguez M et al. Membrane traffic in sphingolipid storage diseases. Traffic 2000; 1:807-815.
153. Aridor M, Hannan LA. Traffic jam: A compendium of human diseases that affect intracellular transport processes. Traffic 2000; 1:836-851.

CHAPTER 2

Lysosomes

Steve Caplan and Juan S. Bonifacino

The term lysosome, which means "lytic body', was first used in the early 1950s by Christian De Duve to describe a newly characterized membrane-bound degradative organelle.[1] It is clear today that the lysosome serves as a major site for degradation of both extracellular and intracellular materials in eukaryotic cells. While mammalian cells typically contain several hundred lysosomes, yeast and plants generally have a single, large structure known as the vacuole.

Function and Composition

Over 45 hydrolases (listed in Table 1) within the lysosomal lumen are involved in the degradation process. All of these enzymes are acid hydrolases that require an acidic environment for optimal activity. The lysosomal hydrolases are isolated from the cytoplasm by the limiting membrane of the lysosome, which is highly enriched in a number of highly glycosylated proteins known as "lamps", (for lysosome-associated membrane proteins) or "limps" (for lysosome integral membrane proteins; see Fig. 1). Although the functions of most of these membrane proteins remain enigmatic, it has been suggested that they play a role in protecting the lysosomal membrane from degradation by the hydrolases.[2,3] The membrane also contains a proton pump that derives energy from ATP to transfer protons into the lumen and maintain acidic conditions (pH 4.6-5.0).[4] Lysosome degradation can be inhibited by various agents, some of which affect the pH of the lysosomal lumen (see Table 2). Lysosomal hydrolases and membrane proteins also localize to a certain extent to a family of organelles termed lysosome-related organelles.[5,6] These organelles include lytic granules of T lymphocytes and natural killer cells, Major Histocompatiblity Complex Class II compartments in B lymphocytes and dendritic cells, platelet dense granules in platelets, and melanosomes in melanocytes. Each of these organelles carry out specialized functions, but all share some characteristics with lysosomes.

Morphology

It is difficult to distinguish mammalian lysosomes from other organelles of the endocytic system such as endosomes on the basis of morphology alone. Unlike most other organelles, lysosomes have a very heterogeneous morphology. They are typically spherical, although tubular shaped lysosomes can be discerned in various cell types. The size of lysosomes normally ranges between 0.2 and 0.4μm in diameter, and the total volume of lysosomes within the cell has been estimated at about 1%. Electron microscopy studies demonstrate that many lysosomes have intralumenal electron-dense content, whereas endosomes generally do not show electron density (Fig. 2a). The presence of lamps on the limiting membrane, together with the absence of markers such as the cation-dependent and -independent mannose-6-phosphate receptors which localize to late endosomes, is considered characteristic of the lysosome. Lysosomes are typically distributed throughout the cytoplasm, often with a concentration in the perinuclear region (Fig. 2b). Also characteristic of the lysosome is its high buoyant density compared to other organelles of the endocytic system. This allows purification of lysosomes on

Intracellular Pathogens in Membrane Interactions and Vacuole Biogenesis,
edited by Jean-Pierre Gorvel.

Table 1. Lysosomal hydrolases

Lysosomal Enzymes	Substrates
NUCLEASES	
Acid ribonuclease	RNA
Acid deoxyribonuclease	DNA
PHOSPHATASES	
Acid phosphatase	Most phosphomonoesters
Acid phosphodiesterase	Oligonucleotides and other Phosphodiesters
Phosphatidic acid phosphatase	Phosphatidic acid
Acid pyrophosphatase	ATP, FAD
Phosphoprotein phosphatase	Phosphoproteins
GLYCOPROTEINGLYCOLIPID CARBOHYDRATE HYDROLASES	
α-D-galactosidase	Galactosides
β-D-galactosidase	Galactosides
α-D-glucosidase	Glycogen
β-D-glucosidase	Glycoproteins
α-D-mannosidase	Mannosides, glycoproteins
β-xylosidase	Glycoproteins and mucopolysaccharides
β-D-galactocerebrosidase	Galactocerebroside
α-L-fucosidase	Fucose-containing macromolecules
β-hexosaminidase α subunit	Gangliosides (G_{M2})
β-hexosaminidase β subunit	Gangliosides (G_{M2})
Sialidase	Sialyl oligosaccharides
Glycosylasparaginase	Aspartyl-linked glycoproteins
β-D-glucocerebrosidase	Glucosylceramide
α-N-acetylgalactosaminidase	Proteins containing glycosidic-linked N-acetylgalactosamine
GLYCOSAMINOGLYCAN HYDROLASES	
Lysozyme	Bacterial cell wall and mucopolysaccharides
Arylsulfatase A	Organic sulfates
Arylsulfatase B	Organic sulfates
Hyaluronidase	Hyaluronic acids, Chondroitin sulfates
α-D-glucuronidase	Polysaccharides and mucopolysaccharides
α-L-iduronidase	Mucopolysaccharides
α-L-iduronate-2-sulfate sulfatase	Heparan and dermatan sulfate intermediates
N-acetylgalactosamine-6-sulfate sulfatase	Galactose sulfate
Heparan N-sulfatase	Heparan sulfate
α-N-acetylglucosaminidase	Heparan sulfate
α-glucosaminide acetyltransferase	Heparan sulfate
N-acetylglucosamine-6-sulfatase	Heparan sulfate
PROTEASES AND PEPTIDASES	
Arylamidase	Amino acid arylamides
Cathepsin A	Proteins and peptides
Cathepsin B	Proteins and peptides
Cathepsin C	Peptides

continued on next page

Table 1. Continued

Lysosomal Enzymes	Substrates
Cathepsin D	Proteins
Cathepsin S	Proteins
Collagenase	Collagen
Pepstatin-insensitive peptidase	Peptides
LIPASES	
Acid sphingomyelinase	Sphingomyelin
Lysosomal acid lipase	Cholesterol esters and triglycerols
Esterase	Fatty acyl esters
Phospholipase	Phospholipids
Triglyceride lipase	Triglycerides
Palmitoyl protein thioesterase	Thioester groups

Percoll density gradients, where they typically migrate towards the bottom of the gradient below the other organelles.

Pathways to the Lysosome

A number of the best characterized routes of trafficking to the lysosome are discussed below, and depicted in Figure 3.

Biosynthetic Transport

Newly synthesized lysosome resident proteins are transported from the TGN to either the lysosomal lumen, or the lysosomal limiting membrane. Under abnormal circumstances, certain proteins destined for other organelles may be sent to the lysosome for degradation directly from the TGN. One example is the T-cell antigen receptor complex. Under normal conditions, the seven or more receptor subunits assemble into a complex that traverses the secretory pathway and is expressed on the plasma membrane. However, in the absence of one of its subunits, ζ, the incomplete receptor complexes are delivered to lysosomes for degradation.[7]

Transport of Lysosomal Hydrolases to the Lysosome

Soluble hydrolases are co-translationally inserted into the lumen of the endoplasmic reticulum (ER), where the signal peptide is cleaved, and N-linked oligosaccharides are added. Upon transport to the Golgi complex, the soluble hydrolases are recognized by a phosphotransferase, which transfers N-acetylglucosamine-1-phosphate to mannose residues on the oligosaccharide chains to produce N-acetylglucosamine-1-phospho-6-mannose residues. The N-acetylglucosamine is then cleaved by the enzyme N-acetylglucosamine-1-phosphodiester-α-N-acetylglucosaminidase, and the hydrolase thus becomes "tagged" with a mannose-6-phosphate moiety. The mannose-6-phosphate moieties on the soluble hydrolases are then recognized at the *trans*-Golgi network (TGN) by mannose-6-phosphate receptors, which deliver the hydrolases to an endosomal compartment. After releasing the hydrolases, the receptors return to the TGN for subsequent rounds of enzyme delivery. Two different mannose-6-phosphate receptors are known, both type I transmembrane proteins: a cation-independent receptor with a molecular mass of 300kD, and a cation-dependent receptor of 46kD.[8] The former is a bifunctional receptor, and also binds insulin-like growth factor II, which is involved in development.[9] The latter mannose-6-phosphate receptor is a homodimer that binds only to lysosomal hydrolases. These two receptors sort distinct but overlapping subsets of lysosomal hydrolases.[10]

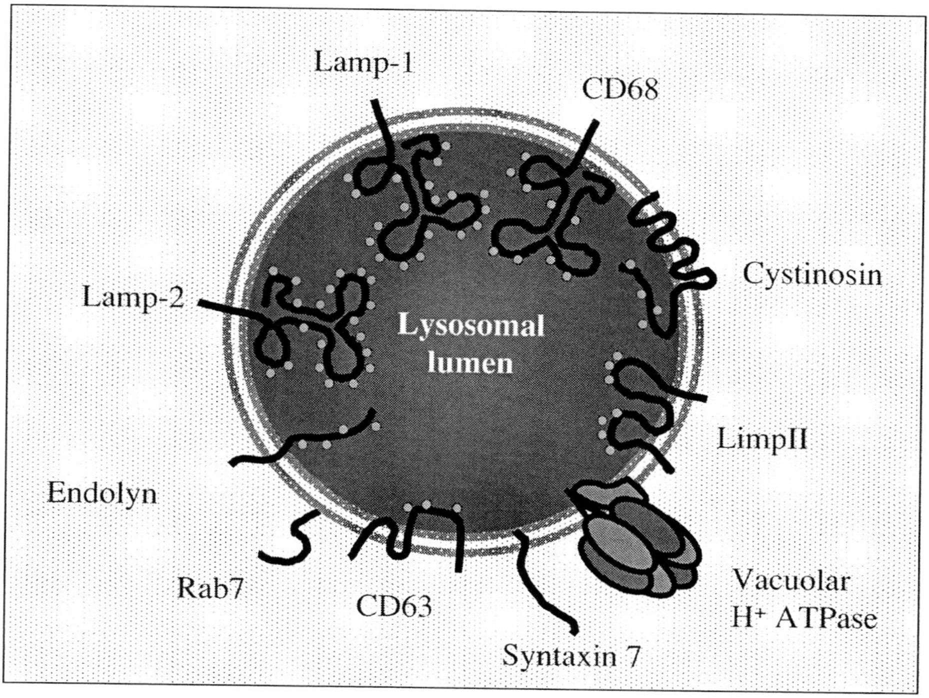

Figure 1. Schematic representation of well characterized lysosomal membrane proteins, depicting their lumenal, transmembrane and cytosolic domains. Putative N-linked glycosylation sites in the lumenal domains are indicated by circles.

A key event in the process of targeting lysosomal hydrolases to the lysosomal lumen is the recognition of these enzymes in the Golgi complex by the phosphotransferase. Since lysosomal enzymes do not appear to share significant homology, it has been proposed that these enzymes display a similar three-dimensional feature or "signal patch" that is recognized by the phosphotransferase.[11]

Transport of Lysosomal Membrane Proteins to the Lysosome

Many integral lysosomal membrane proteins contain tyrosine or dileucine signals in their cytoplasmic tails that mediate sorting to the lysosomal membrane.[12] These tails are recognized by adaptor protein (AP) complexes, which are recruited to the sites where vesicles form and bud from the TGN.[12-14] Four different adaptor complexes have been identified: AP-1, AP-2, AP-3, and AP-4. A role has been demonstrated for AP-2 in the internalization of cell surface receptors via clathrin-coated pits.[15] AP-1 is thought to play a role in the recycling of mannose-6-phosphate receptors from endosomes back to the TGN.[16] The AP-3 complex is involved in the biogenesis of lysosomes and lysosome-like organelles, and proteins destined for the membranes of these organelles are partly misrouted to the plasma membrane in cells deficient in AP-3 subunits.[17-19] Evidence suggests that while lysosomal membrane proteins may be routed directly to lysosomes from the TGN, some proteins arrive at the lysosomes by an indirect pathway. In the latter case, proteins are first routed to the plasma membrane, whereupon they undergo internalization in an AP-2-dependent process, and progress through the endocytic pathway from endosomes to lysosomes.[20] While no clear function has yet been uncovered for the AP-4 complex, its localization to the TGN hints that it could also be involved in sorting to the lysosome.[21,22]

Table 2. Inhibitors of lysosomal degradation

Inhibitor	Mode of Action
PROTEASE INHIBITORS	
Leupeptin	Reversible inhibitor of trypsin and cysteine proteases
Pepstatin A	Aspartyl protease inhibitor
E64	Irreversibly inhibits cysteine proteases
Phenylmethylsulfonyl fluoride	Inhibits serine proteases
ACIDOTROPIC AGENTS	
Ammonium chloride	Water-soluble, permeant weak base that effectively neutralizes the acidic lysosomal lumen
Chloroquine	Tertiary amine that accumulates in endosomes and lysosomes and neutralizes the acidic lysosomal lumen
Primaquine	Tertiary amine that accumulates in endosomes and lysosomes and neutralizes the acidic lysosomal lumen
AMINO ACID METHYL ESTERS	
Methionine methyl ester	Permeates lysosomal membranes and is hydrolyzed by esterases in the lysosomal lumen, where the accumulating methionine induces osmotic swelling of lysosomes
O-methyl-serine dodecylamide hydrochloride (MSDH)	Permeates lysosomal membranes and is hydrolyzed by esterases in the lysosomal lumen, where the accumulating serine induces osmotic swelling of lysosomes
l-leucyl l-leucine methyl ester	Permeates lysosomal membranes and is hydrolyzed by esterases in the lysosomal lumen, where the accumulating leucine induces osmotic swelling of lysosomes
IONOPHORES	
Monensin	A hydrophobic compound that functions as an Na^{+}/H^{+} ionophore by readily incorporating into membranes and binding both ions; neutralizes acidic endocytic compartments
Nigericin	Ionophore that acts as a K^{+}/H^{+} exchange antiporter and neutralizes acidic endocytic compartments
CCCP	Acts as proton ionophore to nullify proton gradients
H^{+} PUMP INHIBITORS	
Bafilomycin A_1	Specifically inhibits vacuolar-type H^{+} ATPases
Concanamycin B	Specifically inhibits vacuolar-type H^{+} ATPases

Endocytosis

Endocytosis is one of the best studied routes by which macromolecules reach the lysosome, and it is vital for cellular function.[23] This process is characterized by the engulfment of fluid (i.e., pinocytosis) or macromolecules (e.g., receptor-mediated endocytosis) by the plasma membrane. While fluid-phase endocytosis occurs constitutively, receptor-mediated endocytosis is generally a regulated process. In both types of endocytosis, the invaginated plasma membrane

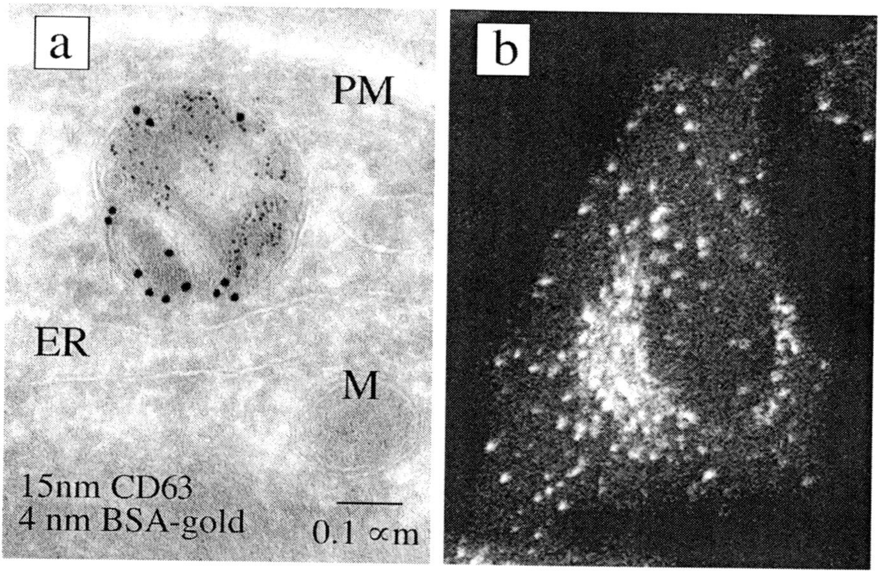

Figure 2. a) Immunoelectron microscopy of a typical lysosome. Human fibroblasts were allowed to internalize BSA-gold, and chased sufficiently to ensure lysosomal localization of this fluid-phase marker. Cells were then fixed and processed for ultra-thin frozen sectioning. Sections were labeled with antibody to the lysosomal membrane protein CD63 and with protein A-gold. Notice the co-localization of internalized BSA-gold and CD63 to the lysosome. The electron micrograph was kindly provided by Lisa M. Hartnell (NIH). b) Lysosomes visualized by immunofluorescence microscopy. Lysosomes in HeLa cells were decorated by indirect immunofluorescence utilizing an antibody to the lysosomal membrane protein, lamp-1 and a secondary antibody conjugated to rhodamine. Lysosomes are evident both in a perinuclear cluster, and in the periphery of the cell.

pinches off from the surface with its macromolecular contents, and forms an endocytic vesicle, known as an early endosome. These organelles may be either vesicular or tubular in structure, and are more commonly localized to the cell periphery. Fluid or macromolecules taken up by endocytosis are usually found in these structures within a few minutes. Five to fifteen minutes later internalized materials can be found within late endosomes, which are often larger and localized to the perinuclear region. Ultimately, the contents of late endosomes reach lysosomes, although the mechanism by which this occurs remains unclear (discussed below, in Models for the Biogenesis of Lysosomes).

Many cell surface receptors are internalized by endocytosis and ultimately undergo degradation in the lysosome. For example, under conditions when the cell needs to down-regulate its input of signals through the TCR the receptor is internalized and targeted to lysosomes. A dileucine signal within the cytosolic tail of the CD3 γ chain is recognized by the AP-2 complex, and the receptor is internalized into clathrin coated pits.[24,25] Ligand binding to the epidermal growth factor receptor induces a similar sequence of events for this receptor, ultimately culminating in its degradation in lysosomes.[26]

Endocytosis is also key for the process of receptor-mediated nutrient uptake. For example, cholesterol uptake by cells is crucial for the biogenesis of new membranes. Most cholesterol molecules in the blood are bound to a large particle known as low-density lipoprotein (LDL).[27] Receptors for LDL localize to the plasma membrane and associate with clathrin-coated pits. When LDL binds to its receptor, it undergoes endocytosis along with its cargo of about 1500

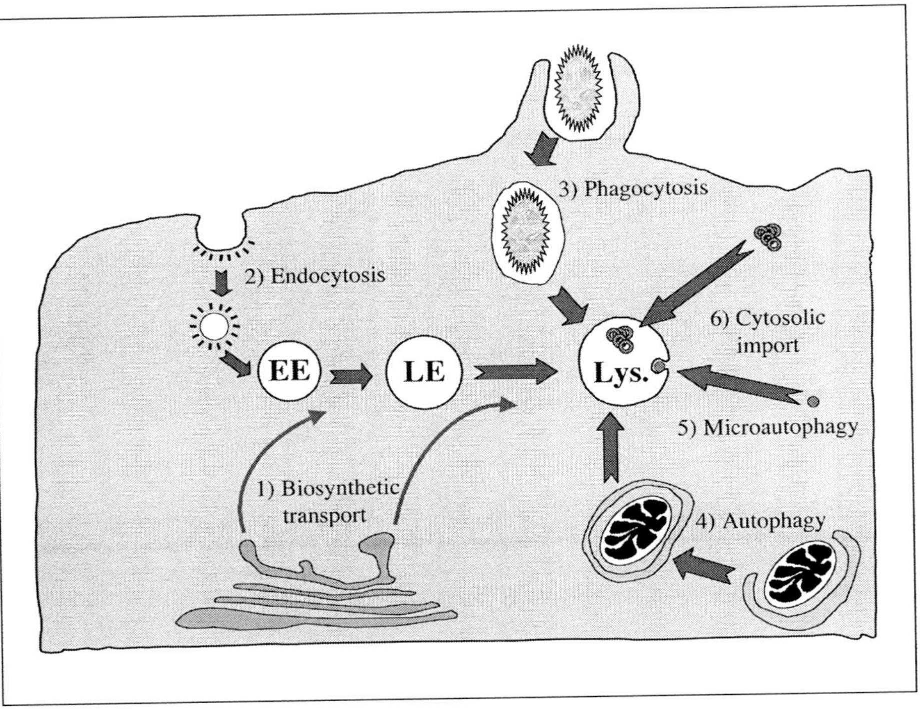

Figure 3. Schematic diagram illustrating six of the best characterized routes by which proteins reach the lysosome.

cholesterol molecules esterified to fatty acids, and 500 non-esterified cholesterol molecules. The LDL and its lipid cargo traverse the endocytic pathway and reach lysosomes, where enzymes hydrolyze the cholesteryl esters and release free cholesterol, while the receptor itself is recycled from an early endosomal sorting compartment to the plasma membrane for additional rounds of endocytosis. Endocytic receptor-mediated nutrient uptake is also exemplified by the transferrin receptor.[28] Ferrotransferrin binds to the transferrin receptor, and the receptor-ligand complex is internalized via clathrin-coated pits. Once the complex reaches early endosomes, the acidic pH causes the release of iron, which is then transported into the cytosol. The apotransferrin-receptor complex is subsequently recycled to cell surface where the neutral pH outside the cell induces the release of the apotransferrin and frees the receptor for subsequent rounds of iron uptake.

Phagocytosis

Phagocytosis is a type of endocytosis by which phagocytic cells (e.g., macrophages, neutrophils, reticulo-endothelial cells) internalize and degrade large particles (e.g., bacteria, senescent erythrocytes).[29] Unlike pinocytosis, which occurs in a constitutive fashion, phagocytosis occurs only once the foreign particle binds to specific receptors on the phagocyte plasma membrane. For example, phagocytes can neutralize opsonized bacteria (i.e., bacteria that have been "coated" by serum immunoglobulins) by binding them via immunoglobulin Fc receptors. Bacteria are then internalized into a phagosome, a structure which is often much larger than typical endosomes derived from pinocytosis. The phagosome eventually fuses with conventional lysosomes to form a structure known as the phagolysosome, and the phagocytosed material is degraded.

Table 3. Lysosomal diseases

Name of Disease	Deficient Protein	Pathogenic Effects at the Cellular Level	Clinical Symptoms
SPHINGOLIPIDOSES			
Tay-Sachs	β-hexosaminidase α subunit	Excessive storage of ganglioside G_{M2}	Progressive neurological degeneration
Sandhoff	β-hexosaminidase β subunit	Excessive storage of ganglioside G_{M2}	Progressive neurological degeneration
G_{M2} activator protein deficiency	G_{M2} activator protein (various small acidic saposins that act in sphingolipid degradation)	Excessive storage of ganglioside G_{M2}, or accumulation of various lipids, depending on the specific saposin deficiency	Progressive neurological degeneration
G_{M1} gangliosidosis and Morquio type B	β-D-galactosidase	Excessive storage of all β-D-galactosidase substrates: G_{M1}, oligosaccharides with terminal galactose, and keratan sulfate (G_{M1} gangliosidosis) or excessive storage of keratan sulfate, but not G_{M1} (Morquio type B)	Neurological symptoms, hepatosplenomegaly, bone deformities (G_{M1} gangliosidosis) or progressive skeletal dysplasias (Morquio type B)
Galactosialidosis	Protective protein: A protein that complexes with: β-D-galactosidase, and N-acetyl-α-neuraminidase	β-D-galactosidase, and N-acetyl-α-neuraminidase are rendered inactive, leading to excessive storage of oligosaccharides with terminal sialic acid and/or galactose	Neurologic disease, skeletal dysplasias
Fabry disease	α-D-galactosidase	Storage of glycosphingolipids with a terminal α-glycosidically linked galactose	Blockage of blood vessels by accumulation of lipids
Niemann Pick disease types A and B	Acid sphingomyelinase	Sphingomyelin accumulates rather than undergoing hydrolysis to ceramide and phosphorylcholine	Psychomotor retardation, visceromegaly
Niemann Pick disease types C and D	NPC-1	Accumulation of cholesterol	Psychomotor retardation visceromegaly
Metachromatic leukodystrophy	Arylsufatase A	Excessive storage of sulfatide, the substrate for arylsulfatase A	Progressive demyelination

continued on next page

Table 3. Continued

Gaucher disease	β-D-glucocerebrosidase	Accumulation of glucosylceramide	Bone marrow, spleen, liver affected
Krabbe disease	β-D-galactocerebrosidase	Accumulation of galactocerebroside	Neurological degeneration
GLYCOPROTEINOSES			
Sialidosis	Sialidase (neuraminidase)	Accumulation of sialyl oligosaccharides	Progressive impaired vision, skeletal dysplasia, mental retardation, hepatosplenomegaly
Fucosidosis	α-L-fucosidase	Accumulation of fucose-containing substrates, such as keratan sulfate and sphingolipids with N-linked	Neurological degeneration oligosaccharide groups
Aspartyl-glucosaminuria	Glycosylasparaginase	Accumulation of aspartyl-linked glycoproteins	Mental retardation and pulmonary complications
Mannosidosis	α-D-mannosidase	Accumulation of N-linked glycoproteins	Progressive mental deterioration
Schindler disease	α-N-acetyl-	Accumulation of substrates containing glycosidic- galactosaminidase	Neurologic degeneration linked N-acetylgalactosamine
MUCOPOLYSACCHARIDOSES			
Hurler-Scheie syndrome (Mucopolysaccharidosis type I)	α-L-iduronidase	Accumulation and urinary secretion of non-degraded mucopolysaccharides	Hepatosplenomegaly, skeletal dysplasias, (severe symptoms-Hurler disease) Joint stiffness, heart valve problems (mild symptoms-Scheie disease)
Hunter syndrome (Mucopolysaccharidosis type II)	α-L-iduronate-2- sulfate sulfatase	Accumulation of heparan and dermatan sulfate degradation intermediates	Similar to Hurler-Scheie syndrome
Sanfilippo syndrome type A (mucopolysaccharidosis type IIIA)	Heparan N-sulfatase	Accumulation of heparan sulfate	Severe neuronal degeneration, hepatosplenomegaly

continued on next page

Table 3. Continued

Sanfilippo syndrome type B (mucopolysaccharidosis type IIIB)	α-N-acetyl-glucosaminidase	Accumulation of heparan sulfate	Severe neuronal degeneration, hepatosplenomegaly
Sanfilippo syndrome type C (mucopolysaccharidosis type IIIC)	Acetyl CoA: α-glucosaminide acetyltransferase	Accumulation of heparan sulfate	Severe neuronal degeneration, hepatosplenomegaly
Sanfilippo syndrome type D (mucopolysaccharidosis IIID)	N-acetyl-glucosamine 6-sulfatase	Accumulation of heparan sulfate	Severe neuronal degeneration, hepatosplenomegaly
Morquio A syndrome (mucopolysaccharidosis type IV)	N-acetylgalactosamine-6-sulfate sulfatase	Accumulation of galactose-sulfate	Severe skeletal deformities, clouding of the cornea, non-neurological symptoms
Maroteaux Lamy syndrome (mucopolysaccharidosis type VI)	Arylsulfatase B	Accumulation of chondroitin-4-sulfate and dermatan sulfate	Similar to Morquio A syndrome
Sly syndrome (mucopolysaccharidosis type VII)	β-D-glucuronidase	Accumulation of chondroitin-sulfates, and glycosaminoglycans such as dermatan and heparan sulfate	Skeletal dysplasia, mental retardation, hepatosplenomegaly
Multiple sulfatase deficiency	All lysosomal sulfatases lack 2-amino-3-oxopropanoic acid	Sulfate groups are poorly cleaved and molecules containing them accumulate	Similar to symptoms observed for other sulfatase deficiencies
		modification	
MUCOLIPIDOSES			
I-cell disease (mucolipidosis type II)	N-acetyl-glucosaminyl-phosphotransferase	Lysosomal hydrolases are not tagged by mannose-6-phosphate, and thus secreted rather than being targeted to the lysosome	Coarse face and skin, skeletal abnormalities, severe psychomotor retardation, death
Pseudo-Hurler polydystrophy	Partial N-acetyl-glucosaminyl-phosphotransferase	Certain lysosomal hydrolases are not tagged by Mannose-6-phosphate, and thus secreted rather than being targeted to the lysosome	Similar to other mucolipidoses, Milder form of I-cell disease
Mucolipidosis type IV	Ganglioside sialidase	Accumulation of gangliosides in lysosomes	Severe neurologic abnormalities

continued on next page

Table 3. Continued

OTHER LYSOSOMAL DISORDERS			
Wolman disease	Lysosomal acid lipase	Accumulation of cholesterol esters and triglycerols	Hepatosplenomegaly, anaemia, adrenal gland calcification
Cholesterol ester storage disease	Lysosomal acid lipase	Accumulation of cholesterol esters and triglycerols	Similar to Wolman disease, but milder with later onset
Pompe disease	α-D-glucosidase	Accumulation of glycogen	Severe skeletal, cardiologic and neurologic problems
Danon disease (x-linked vacuolar cardiomyopathy and myopathy)	Lamp-2	Accumulation of glycogen-containing autophagic vacuoles and impairment of protein degradation	Cardiomyopathy, myopathy and mental retardation
Cystinosis syndrome	Cystinosin	Accumulation of cystine in the lysosome	Kidney failure, form of Fanconi
Non-cytinosis form of Fanconi syndrome	Unknown	Light chains of monoclonal antibodies are incompletely degraded by the renal tubule lysosome, form crystals and interfere with apical membrane transporters	Failure in proximal renal reabsorption, glycosuria, aminoaciduria, hypophosphatemia
Batten disease (neuronal ceroid-lipofuscinosis)	8 loci with 4 cloned genes: *CLN1*- lysosomal palmitoyl-protein thioesterase *CLN2*- lysosomal pepstatin-insensitive peptidase *CLN3*- participates in lysosomal pH homeostasis *CLN5*- unknown lysosomal membrane protein	Accumulation of autofluorescent storage material	Neurodegeneration

continued on next page

Table 3. Continued

Chediak-Higashi syndrome	Mutations in the *CHS1* gene, which codes for the protein	Giant lysosomes, inhibition in degradation of selected proteins	Progressive neurological degeneration, hypopigmentation, severe immune deficiency, bleeding diathesis, accelerated lymphoproliferation
Griscelli syndrome or Griscelli syndrome with Haemophagocytic syndrome	Possible involvement of the *MYO5A* gene (myosin Va) or mutations in Rab27a	Impaired secretion of secretory lysosomes/lytic granules, presumed loss of transfer of melanosomes from melanocytes to keratinocytes, uncontrolled activation of T-lymphocytes and macrophages, abnormal organelle morphology, presumed loss of transfer of melanosomes from melanocytes to keratinocytes	Skin and hair hypopigmentation, immunodeficiency, accelerated lymphoproliferation, reduced immune cell function
Hermansky-Pudlak syndrome type 1	Mutations in the *HPS1* gene	Abnormal lysosome-related organelles, such as melanosomes, and absence of platelet dense granules, accumulation of incompletely degraded lipids in lysosomes (a form of ceroid lipofuscinosis)	Oculocutaneous albinism and prolonged bleeding times, pulmonary fibrosis, ulcerative colitis
Hermansky-Pudlak syndrome type 2	Mutations in the *ADTB3A* gene, which encodes the AP-3 subunit β3A	Abnormal lysosome-related organelles, such as melanosomes, and absence of platelet dense granules, accumulation of incompletely degraded lipids in lysosomes (a form of ceroid lipofuscinosis)	Oculocutaneous albinism and prolonged bleeding times, pulmonary fibrosis, ulcerative colitis

Autophagy

Autophagy is a highly regulated process by which a region of the cytoplasm is sequestered by a membrane of unknown origin termed phagophore. This enclosed structure, known as an autophagosome, is targeted for degradation by the lysosome.[30] The autophagosome intersects with the endocytic system by fusing with late endosomes or lysosomes, and the enclosed cytoplasmic region or organelle is then degraded.

Microautophagy

The process of microautophagy is similar to that of autophagy, although smaller portions of cytosol are sequestered. The major difference between these two degradation routes lies in the mode of regulation. Whereas autophagy appears to be regulated primarily by amino acids and/or insulin,[31] microautophagy does not appear to be acutely regulated.[32]

Cytosolic Import

Another possible route for the transport of proteins to the lysosome for degradation is by direct import across the lysosomal membrane. For example, a number of cytosolic proteins contain "KFERQ-type" sequences that confer transport to the lysosome. Annexin II and annexin VI each contain KFERQ-type signals, and each protein is degraded in lysosomes more rapidly in response to serum starvation.[33] However, other annexins lacking these KFERQ signals remain unaffected by serum starvation. It remains unclear as to whether these proteins adhere to autophagosomes that fuse with lysosomes, or whether they are capable of being imported independently across the lysosomal membrane. In Chinese hamster ovary cells, a 96kD integral lysosomal membrane protein homologous to lamp-2 has been implicated in the selective import of proteins to the lysosome for degradation.[34] Data also suggest that proteins such as dihydrofolate reductase are imported across the lysosomal membrane in an unfolded state, and that this transport is chaperoned by heat shock family proteins.[35]

Models for the Biogenesis of Lysosomes

Originally, it was postulated that lysosomes were formed by budding from the TGN as complete, intact organelles. These newly formed structures were called "primary lysosomes', as opposed to lysosomes that had already fused with endosomes or phagosomes, which were termed "secondary lysosomes'.

Other models for lysosomal biogenesis have been proposed recently, yet there is currently no consensus as to which of these models is the most accurate.[36-38] One model holds that lysosomes are "stable organelles', that utilize both anterograde and retrograde vesicular transport processes to assure that the necessary proteins reach lysosomes and are maintained there. Another model postulates that lysosomes are dynamic organelles that gradually mature from endosomes into lysosomes, and evolve by receiving TGN-derived vesicles carrying specific lysosomal proteins. A third model, known as "fusion-fission", maintains that the formation of mature lysosomes is a result of continuous contact between late endosomes and lysosomes, leading to the formation of a hybrid organelle that later fissions. Some studies suggest that hybrid organelles are not actually formed, but instead there is a more transient "kiss and run" process, where materials may be transferred from organelle to organelle. In either case, it is assumed that this contact may allow lysosomes to acquire both newly synthesized proteins as well as incoming cargo through the endocytic pathways. Neither of these models is mutually exclusive, and it is likely that aspects of all of them contribute to lysosome biogenesis.

Proteins Involved in Targeting and Fusion to the Lysosome

Pre-Fusion Events

A family of small GTP-binding proteins known as Rabs play crucial roles in mediating trafficking throughout the endocytic and secretory pathways.[39,40] These proteins are all comprised of about 200 amino acids, with conserved overall structural similarity to Ras. It is thought that their binding and hydrolysis of GTP regulates the rate of membrane fusion. For example, it has been well documented that Rab5 plays a key role in regulating trafficking to early endosomes.[41] Rab7, on the other hand, has been shown to play a role in regulating trafficking to late endosomes[42] and lysosomes.[43] This is partly based on its homology to the yeast ypt7 protein, which localizes to the vacuole and affects sorting to this organelle.[44] Indeed, studies in mammalian cells utilizing a dominant-negative (inactive) Rab7 indicate that this protein affects lysosomal biogenesis by causing dispersal of lysosomes, inducing the loss of perinuclear clustering,[43] and impairing the delivery of lysosomal enzymes such as Cathepsin D.[45] Recent studies in yeast have shed light on the events that occur prior to homotypic vacuolar fusion, an event that is comparable to fusion with the mammalian lysosome. These studies suggest that a complex of proteins acts at the vacuole to allow docking of apposing membranes, and tethers them together so that fusion may proceed.[46-51] These proteins are all classified as belonging to the vacuolar protein sorting (Vps) family, mutations in each of their genes in yeast leads to a phenotype of missorted vacuolar proteins. Evidence suggests that these proteins, which include Vps39, Vps41, Vps11, Vps16, Vps18 and Vps33, interact with Rab7 and facilitate binding of apposing vacuoles to initiate the process of fusion.†

Fusion

Membrane fusion is thought to be driven primarily by a family of integral membrane proteins known as SNAREs (reviewed in refs. 52, 40). Different SNAREs localize to the membrane of transport vesicles (called v-SNAREs) and the membrane of the target organelles (known as t-SNAREs). The t-SNARE on the target membrane forms a complex with a ubiquitously localized protein, known as SNAP25. When the two membranes are juxtaposed sufficiently near each other, the v-SNARE binds to the t-SNARE/SNAP25 complex, and forms a "prefusion" complex. Homotrimeric N-Ethylmaleimide-Sensitive Fusion protein (NSF) is then thought to bind to the "prefusion complex" to form a 20S complex. NSF is a 76kD protein with two ATP binding and hydrolysis domains. ATP hydrolysis by NSF is linked to complex disassembly, and conformational change in SNARE proteins, preventing reassembly. The conformational change enables membrane fusion, in a mechanism that has not been fully elucidated. However, it has been demonstrated that transport vesicles may contain both t- and v-SNAREs [53-56] and several studies suggest that the action of NSF is more consistent with that of a molecular chaperone, whereby it acts prior to fusion events by changing the conformation of SNAREs. Accordingly, this conformational change is hypothesized to allow dissociation of t-SNAREs and v-SNAREs from each other, allowing for t/v-SNARE interactions between apposing membranes. Once fusion occurs, it is thought that NSF and the $\alpha/\beta/\gamma$ SNAPs allow dissociation of the SNAREs, possibly by catalyzing Rab GTP to GDP exchange.

Thus far, very few SNAREs have been identified that either localize to mammalian lysosomes and/or play a role in fusion with the lysosome. For example, the t-SNARE syntaxin 7 was thought to localize to early endosome membranes,[57] but recent data suggest that it localizes to late endosomes and lysosomes and may mediate heterotypic (late endosome/lysosome) and/or homotypic (lysosome/lysosome) fusion.[58] The v-SNARE vesicle-associated-membrane-protein (Vamp7) has also been localized to late endosomes and lysosomes[57,59,60] and has also been shown to play a role in heterotypic and homotypic lysosome fusion.[59]

Although it has been hypothesized that lysosomes are continuously undergoing cycles of fusion and fission, little is known about the proteins involved in the latter process. One candidate is the human LYST protein.[61] Mutations in LYST (and its murine homolog, BEIGE) produce a disease known as Chediak-Higashi Syndrome, which is characterized by the formation of giant lysosomes. This suggests that LYST plays a physiologic role in lysosome fission.

†A recent study utilizing mammalian cells demonstrates that the human Vps39 homolog, known as hVam6p, serves as a lysosomal tethering/docking factor to promote lysosome clustering and fusion in vivo.[69]

Human Diseases Involving Lysosomes

Another measure of the critical nature of lysosomes is the number of diseases related to lysosome dysfunction. Over 40 different diseases have been described that are directly related to the lysosome (summarized in Table 2, reviewed in refs. 62-65). Most of these diseases involve a deficiency of a catabolic pathway, and are known collectively as "lysosomal storage diseases'. Several lysosomal diseases are not marked by loss of enzymatic activity, but rather by missorting of lysosomal proteins. In almost all instances, the damage induced by lysosomal diseases is primarily due to a massive accumulation of a certain substrate(s) that is normally degraded in the lysosome. One reason why this accumulation is so acute, is that most lysosomal hydrolases carry out their enzymatic activity in an exolytic manner, cleaving sugars, fats, and sulfates one at a time from the ends of larger molecules. Substrates are thus degraded in a sequential fashion, and each enzyme in the chain cannot act without the prior activity of its predecessor. Accordingly, the loss of a single enzymatic activity within this chain of events induces the accumulation of a substrate that cannot be further hydrolyzed by any of the other downstream hydrolases.

There is a high level of variability in the clinical symptoms and severity of these lysosomal storage diseases. Many of the diseases have been classified by the time of onset of symptoms, with a "severe infantile", "intermediate juvenile" and "mild adult" classification. While the type of symptoms of early and late onset are usually similar (notwithstanding the differences in severity), defects in the very same gene have been known to cause very different symptoms. For instance, β-D-galactosidase deficiency induces either G_{M1} gangliosidosis or Morquio type B. In the former, there is excessive storage of all β-D-galactosidase substrates, including GM1, and the primary symptom is neurological degeneration.[66] However, the latter β-D-galactosidase deficiency does not result in excessive G_{M1} storage, and the predominant symptom is the induction of skeletal dysplasias.[67]

While the vast majority of human lysosomal diseases are caused by a deficiency in a lysosomal enzyme, certain types of lysosomal storage disease may be caused by deficiencies in proteins lacking intrinsic enzymatic activity. For example, patients presenting with Hermansky-Pudlak syndrome 1 and Hermansky-Pudlak syndrome 2 have mutations in the genes *HPS1* and *ADTB3A*, respectively.[68,18] Although the function of the protein coding for the former is unknown, the latter gene codes for the β3A subunit of the AP-3 complex. Evidence has accumulated demonstrating that this adaptor complex plays a role in the biogenesis of lysosome-related organelles, thus suggesting that this disease results from aberrant trafficking to lysosomes and related organelles.

Concluding Remarks

Although discovered over forty years ago, great strides are still being made in furthering our understanding of the composition and function of lysosomes. Nevertheless, many important questions remain unanswered. For example, the physiologic roles of many of the lysosome membrane resident glycoproteins have yet to be discovered. Perhaps one of the major outstanding questions relates to the mode of lysosome biogenesis. The discovery of new mammalian lysosomal SNAREs, and proteins that regulate the processes of lysosomal tethering, docking and fusion may prove key to a more complete understanding of lysosome function.

References

1. de Duve C. Lysosomes revisited. Eur J Biochem 1983; 137:391-397.
2. Lewis V, Green SA, Marsh M et al. Glycoproteins of the lysosomal membrane. J Cell Biol 1985; 100:1839-1847.
3. Barriocanal JG, Bonifacino JS, Yuan L et al. Biosynthesis, glycosylation, movement through the Golgi system, and transport to lysosomes by an N-linked carbohydrate-independent mechanism of three lysosomal integral membrane proteins. J Biol Chem 1986; 261:16755-16763.
4. Arai K, Shimaya A, Hiratani N et al. Purification and characterization of lysosomal H(+)-ATPase. An anion-sensitive v-type H(+)-ATPase from rat liver lysosomes. J Biol Chem 1993; 268:5649-5660.

5. Dell'Angelica EC, Mullins C, Caplan S et al. Lysosome-related organelles. Faseb J 2000; 14:1265-1278.
6. Stinchcombe J C, Griffiths GM. Regulated secretion from hemopoietic cells. J Cell Biol 1999; 147:1-5.
7. Bonifacino JS, Lippincott-Schwartz J, Chen C et al. Association and dissociation of the murine T cell receptor associated protein (TRAP). Early events in the biosynthesis of a multisubunit receptor. J Biol Chem 1988; 263:8965-8971.
8. Dahms NM, Lobel P, Kornfeld S. Mannose 6-phosphate receptors and lysosomal enzyme targeting. J Biol Chem 1989; 264:12115-12118.
9. Tong PY, Tollefsen SE., Kornfeld S. The cation-independent mannose 6-phosphate receptor binds insulin-like growth factor II. J Biol Chem 1988; 263:2585-2588.
10. Kornfeld S. Structure and function of the mannose 6-phosphate/insulinlike growth factor II receptors. Annu Rev Biochem 1992; 61:307-330.
11. Baranski TJ, Faust PL, Kornfeld S. Generation of a lysosomal enzyme targeting signal in the secretory protein pepsinogen. Cell 1990; 63:281-291.
12. Kirchhausen T, Bonifacino JS, Riezman H. Linking cargo to vesicle formation: Receptor tail interactions with coat proteins. Curr Op Cell Biol 1997; 9:488-495.
13. Bonifacino JS, Dell'Angelica EC. Molecular bases for the recognition of tyrosine-based sorting signals. J Cell Biol 1999; 145:923-926.
14. Hirst J, Robinson MS. Clathrin and adaptors. Biochim Biophys Acta 1998; 1404:173-193.
15. Nesterov A, Carter RE, Sorkina T et al. Inhibition of the receptor-binding function of clathrin adaptor protein AP-2 by dominant-negative mutant μ2 subunit and its effects on endocytosis, EMBO J 1999;18:2489-2499.
16. Meyer C, Zizioli D, Lausmann S et al. mu1A-adaptin-deficient mice: lethality. loss of AP-1 binding and rerouting of mannose 6-phosphate receptors, Embo J 2000;19:2193-203.
17. Kantheti P, Qiao X, Diaz ME et al. Mutation in AP-3 delta in the mocha mouse links endosomal transport to storage deficiency in platelets, melanosomes, and synaptic vesicles. Neuron 1998; 21:111-122.
18. Dell'Angelica EC, Shotelersuk V, Aguilar RC et al. Altered trafficking of lysosomal proteins in Hermansky-Pudlak syndrome due to mutations in the beta 3A subunit of the AP-3 adaptor. Mol Cell 1999; 3:11-21.
19. Feng L, Seymour AB, Jiang S et al. The beta3A subunit gene (Ap3b1) of the AP-3 adaptor complex is altered in the mouse hypopigmentation mutant pearl, a model for Hermansky-Pudlak syndrome and night blindness. Hum Mol Genet 1999; 8:323-330.
20. Hunziker W, Geuze HJ. Intracellular trafficking of lysosomal membrane proteins. Bioessays 1996; 18:379-389.
21. Dell'Angelica EC, Mullins C, Bonifacino JS. AP-4, a novel protein complex related to clathrin adaptors. J Biol Chem 1999; 274:7278-7285.
22. Hirst J, Bright NA, Rous B et al. Characterization of a fourth adaptor-related protein complex. Mol Biol Cell 1999; 10:2787-2802.
23. Mellman I. Endocytosis and molecular sorting. Annual Rev Cell Dev Biol 1996; 12:575-625.
24. Letourneur F, Klausner RD. A novel di-leucine motif and a tyrosine-based motif independently mediate lysosomal targeting and endocytosis of CD3 chains. Cell 1992; 69:1143-1157.
25. Dietrich J, Hou X, Wegener AM et al. CD3 gamma contains a phosphoserine-dependent di-leucine motif involved in down-regulation of the T cell receptor. Embo J 1994; 13:2156-2166.
26. Carpenter G. The EGF receptor: a nexus for trafficking and signaling. Bioessays 2000; 22:697-707.
27. Brown MS, Goldstein JL. A receptor-mediated pathway for cholesterol homeostasis. Science 1986; 232:34-47.
28. Conrad ME, Umbreit JN, Moore EG. Iron absorption and transport. Am J Med Sci 1999; 318:213-229.
29. Tjelle TE, Lovdal T, Berg T. Phagosome dynamics and function. Bioessays 2000; 22:255-263.
30. Seglen PO, Bohley P. Autophagy and other vacuolar protein degradation mechanisms. Experientia 1992; 48:158-172.
31. Blommaart EF, Luiken JJ, Meijer AJ. Autophagic proteolysis: control and specificity. Histochem J 1997; 29:365-385.
32. Mortimore GE, Poso AR, Lardeux BR. Mechanism and regulation of protein degradation in liver. Diabetes Metab Rev 1989; 5:49-70.
33. Cuervo AM, Gomes AV, Barnes JA et al. Selective degradation of annexins by chaperone-mediated autophagy. J Biol Chem 2000; 275:33329-33335.
34. Cuervo AM, Dice JF. A receptor for the selective uptake and degradation of proteins by lysosomes. Science 1996; 273:501-503.

35. Salvador N, Aguado C, Horst M et al. Import of a cytosolic protein into lysosomes by chaperone-mediated autophagy depends on its folding state. J Biol Chem 2000; 275:27447-456.
36. Gruenberg J, Maxfield FR. Membrane transport in the endocytic pathway. Curr Opin Cell Biol 1995; 7:552-563.
37. Storrie B, Desjardins M. The biogenesis of lysosomes: Is it a kiss and run, continuous fusion and fission process? Bioessays 1996; 18:895-903.
38. Luzio JP, Rous BA, Bright NA et al. Lysosome-endosome fusion and lysosome biogenesis. J Cell Sci 2000; 113:1515-1524.
39. Novick P, Zerial M. The diversity of Rab proteins in vesicle transport. Curr Opin Cell Biol 1997; 9:496-504.
40. Pfeffer SR. Transport-vesicle targeting: Tethers before SNAREs. Nat Cell Biol 1999; 1:E17-22.
41. Bucci C, Parton RG, Mather IH et al. The small GTPase rab5 functions as a regulatory factor in the early endocytic pathway. Cell 1992; 70:715-728.
42. Feng Y, Press B, Wandinger-Ness A. Rab 7: An important regulator of late endocytic membrane traffic. J Cell Biol 1995; 131:1435-1452.
43. Bucci C, Thomsen P, Nicoziani P et al. Rab7: A key to lysosome biogenesis. Mol Biol Cell 2000; 11:467-480.
44. Haas A, Scheglmann D, Lazar T et al. The GTPase Ypt7p of Saccharomyces cerevisiae is required on both partner vacuoles for the homotypic fusion step of vacuole inheritance. Embo J 1995; 14:5258-5270.
45. Press B, Feng Y, Hoflack B et al. Mutant Rab7 causes the accumulation of cathepsin D and cation- independent mannose 6-phosphate receptor in an early endocytic compartment. J Cell Biol 1998; 140:1075-1089.
46. Price A, Seals D, Wickner W et al. The docking stage of yeast vacuole fusion requires the transfer of proteins from a cis-SNARE complex to a Rab/Ypt protein. J Cell Biol 2000; 148:1231-1238.
47. Ungermann C, Price A, Wickner W. A new role for a SNARE protein as a regulator of the Ypt7/Rab-dependent stage of docking. Proc Natl Acad Sci USA 2000; 97:8889-8891.
48. Wurmser AE, Sato TK, Emr SD. New component of the vacuolar class C-Vps complex couples nucleotide exchange on the ypt7 GTPase to SNARE-dependent docking and fusion. J Cell Biol 2000; 151:551-562.
49. Seals DF, Eitzen G, Margolis N et al. A Ypt/Rab effector complex containing the Sec1 homolog Vps33p is required for homotypic vacuole fusion. Proc Natl Acad Sci USA 2000; 97:9402-9407.
50. Eitzen G, Will E, Gallwitz D et al. Sequential action of two GTPases to promote vacuole docking and fusion. Embo J 2000; 19:6713-6720.
51. Sato TK, Rehling P, Peterson MR et al. Class C Vps protein complex regulates vacuolar SNARE pairing and is required for vesicle docking/fusion. Mol Cell 2000; 6:661-671.
52. Sollner TH, Rothman JE. Molecular machinery mediating vesicle budding, docking and fusion. Experientia 1996; 52:1021-1025.
53. Walch-Solimena C, Blasi J, Edelmann L et al. The t-SNAREs syntaxin 1 and SNAP-25 are present on organelles that participate in synaptic vesicle recycling. J Cell Biol 1995; 128:637-645.
54. Nichols BJ, Ungermann C, Pelham HR et al. Homotypic vacuolar fusion mediated by t- and v-SNAREs. Nature 1997; 387:199-202.
55. Tagaya M, Toyonaga S, Takahashi M et al. Syntaxin 1 (HPC-1) is associated with chromaffin granules. J Biol Chem 1995; 270:15930-15933.
56. Steel GJ, Tagaya M, Woodman PG. Association of the fusion protein NSF with clathrin-coated vesicle membranes. Embo J 1996; 15:745-752.
57. Prekeris R, Yang B, Oorschot V et al. Differential roles of syntaxin 7 and syntaxin 8 in endosomal trafficking. Mol Biol Cell 1999; 10:3891-3908.
58. Mullock BM, Smith CW, Ihrke G et al. Syntaxin 7 is localized to late endosome compartments, associates with vamp 8, and Is required for late endosome-lysosome fusion. Mol Biol Cell 2000; 11:3137-3153.
59. Ward DM, Pevsner J, Scullion MA et al. Syntaxin 7 and VAMP-7 are soluble N-ethylmaleimide-sensitive factor attachment protein receptors required for late endosome-lysosome and homotypic lysosome fusion in alveolar macrophages. Mol Biol Cell 2000; 11:2327-2333.
60. Advani RJ, Yang B, Prekeris R et al. VAMP-7 mediates vesicular transport from endosomes to lysosomes. J Cell Biol 1999; 146:765-776.
61. Certain S, Barrat F, Pastural E et al. Protein truncation test of LYST reveals heterogenous mutations in patients with Chediak-Higashi syndrome. Blood 2000; 95:979-983.
62. Neufeld EF. Lysosomal storage diseases. Annu Rev Biochem 1991; 60:257-280.
63. Gieselmann V. Lysosomal storage diseases. Biochim Biophys Acta 1995; 1270:103-136.

64. Karageorgos LE, Isaac EL, Brooks DA et al. Lysosomal biogenesis in lysosomal storage disorders. Exp Cell Res 1997; 234:85-97.
65. Suzuki K, Proia RL. Mouse models of human lysosomal diseases. Brain Pathol 1998; 8:195-215.
66. Suzuki Y, Oshima A. A beta-galactosidase gene mutation identified in both Morquio B disease and infantile GM1 gangliosidosis [letter]. Hum Genet 1993; 91:407.
67. Mancini GM, Hoogeveen AT, Galjaard H et al. Ganglioside GM1 metabolism in living human fibroblasts with beta- galactosidase deficiency. Hum Genet 1986; 73:35-38.
68. Oh J, Bailin T, Fukai K et al. Positional cloning of a gene for Hermansky-Pudlak syndrome, a disorder of cytoplasmic organelles. Nat Genet 1996; 14:300-306.
69. Caplan S, Hartnell LM, Naslavsky N et al. Human Vam6p promotes lysosome clustering and fusion in vivo. J Cell Biol 2001; 154:109-121.

Chapter 3

Lipid Rafts and Host Cell-Pathogen Interactions

Lorena Perrone and Chiara Zurzolo

Introduction

In this Chapter we will describe the structure of membrane microdomains known as rafts that are enriched in glycosphingolipids and cholesterol. We will analyze the role of rafts in different cellular functions, focusing attention on their involvment in the entry and survival of some pathogens in host cells.

Rafts Organization

The Singer-Nicholson model of cell membrane organization describes the lipid bilayer as a neutral two-dimensional solvent. Conversely, in model lipid bilayers, it has been shown that lipids are present in several phases, with increasing fluidity from gel, liquid-ordered (lo) and liquid-disordered (lc) states.[1] The new model of membrane organization proposed in the "raft" hypothesis by Simons and co-workers[2,3] postulates the presence in membranes of dynamic lateral assemblies between specific lipids denoted rafts. According to this hypothesis, long largely saturated sphingolipids can pack together more than phospholipids which are rich in shorter and unsaturated acyl chains. This differential packaging can lead to phase separation and therefore is a key feature of raft lipid organization in the bilayer. Indeed, association of long saturated fatty acyl chains with cholesterol confers a lipid organization similar to the liquid-order state of a model membrane.[1] Thus, the physical interaction between certain lipids is the driving force of membrane domain formation.

The idea that sphingolipids can laterally cluster comes from kinetic studies of the intravesicular transfer rates of different lipids using ultrastructural labeling analysis in which glycosphingolipids were labelled with lectins and toxins (for review see ref. 4). Using liposomes as membrane models it has been demonstrated that only certain sphingolipids can laterally cluster.[5] Moreover, it has been shown in model membranes that cerebrosides and ceramides form microdomains in phosphatydilcholine bilayers.[4]

Cholesterol plays a key role in the physical structure of these microdomains, because in its absence the sphingolipid association leads to a more rigid structure (gel phase) that does not exist in the presence of cholesterol.[4] Indeed, in the presence of cholesterol lipid mixtures do not undergo a gel/liquid disordered phase separation but a liquid ordered/ liquid disordered (lo/lc) phase separation. Sterols facilitate the formation of liquid ordered domains by promoting phase separation between lipid mixtures[1] and cholesterol appears to be required as shown by the fact that cholesterol depletion affects rafts function.[6] Thus lipid rafts are organized in a liquid-order phase in the lipid bilayer, probably surrounded by lipid domains enriched in unsaturated glycerolipids that exist in a liquid disordered phase.

The organization of sphingolipids and cholesterol in liquid ordered phases gives to these microdomains the property of insolubility in non-ionic detergents, which has been used as a

Intracellular Pathogens in Membrane Interactions and Vacuole Biogenesis, edited by Jean-Pierre Gorvel.

tool to isolate raft-containing domains from cells. These domains have been differently called DRM, DIGS, GEMS, TIFF and can associate with specific proteins while excluding others. Proteins with raft affinity are glycosylphosphatidylinositol (GPI)-anchored proteins,[7] doubly acylated proteins such as the α-subunits of heterotrimeric G proteins and Src-family kinases, cholesterol-linked and palmitoylated proteins, and transmembrane proteins, in particular the ones carrying palmitoylation motifs (for review see ref. 6). These proteins can transiently associate with rafts, which modulate their function.

Rafts, Caveolae and Caveolae-Like Structures

Caveolae are small plasma membrane invaginations present at the surface of many different cell types. They are also present intracellularly as clusters of vesicles close to the plasma membrane.[8] A marker protein for caveolae is caveolin 1 (cav1), which is a member of a family that includes another two components, cav2 and cav3.[9,10] Cav1 and cav2 are co-expressed and form hetero-oligomers in many cell types,[11] while cav3 is muscle specific.[12] It has been proposed that cav1 is the driving force for caveolae formation. Through its capability to form homo-oligomers[13] it can form the coat of caveolae invaginations and mediate the organization of rafts in these types of vesicles.[14,15]

Despite initial confusion, the data present in the literature clearly indicate that rafts exist independently from caveolae and caveolae-like structures and that the latter are a specialized form of rafts with a specific coat (in the case of caveolae) and with specific protein and lipid compositions. Therefore, the insertion of caveolins and/or caveolin-like proteins in the membrane probably generates different types of domains.

Cholesterol is another key component of caveolae because it can stabilize the formation of cav1 homo-oligomers. Cholesterol-binding drugs such as nystatin, filipin and β-cyclodestrin inhibit the formation of caveolae.[16] Because of its ability to bind both cholesterol and GSLs, cav1 could stabilize rafts and cause their coalescence into caveolae (see ref. 2 for review).

The N-terminus of cav1 interacts with the protein filamin.[17] This protein colocalizes with cav1 in caveolae and is detected in stress fibers. Potentially, it could therefore associate caveolae with the actin cytoskeleton.[17]

A new protein family proposed to be involved in the formation of caveolae are flotilins, which can associate with cav1, but also appear able to induce intracellular membrane invaginations similar to caveolae in the absence of cav1.[18]

Other proteins capable of forming caveolae-like structures are the MEC-2/stomatin family. These proteins are structurally similar to caveolins and are found in detergent-insoluble complexes, so they might be involved in the formation of non-caveolar rafts.[19]

Also annexins, a family of intracellular Ca-dependent lipid-binding proteins, might have a role in raft domain organization.[20] Annexin II and annexin XIIIb are found in DIG fractions. Annexin XIIIb is myristoylated and it has been found in apical transport vesicles in polarized MDCK cells, suggesting a role for it in protein sorting.[21] Annexin II shows a specific domain responsible for its association with the cytoplasmic leaflet of the membrane. Moreover, upon alteration of cholesterol homeostasis, membrane-bound annexin II has been found associated with actin and cytoskeleton proteins, suggesting that this protein might link the raft domains to the cytoskeleton.[22]

Rafts Detection

Evidence of the existence of rafts has been obtained using different strategies. The first method used to identify raft associated molecules is based on the isolation of detergent-resistant membrane complexes which float to low density fractions in a sucrose gradient at equilibrium centrifugation.[2,3,23] These fractions, that are insoluble in non-ionic detergents, are called detergent-insoluble glycolipid-rich domains (DIG) or detergent-resistant membranes (DRM) and are enriched in glycosphingolipids, sphingomyelin and cholesterol. Nevertheless, this method can lead to some artifacts. For example cytoskeleton associated proteins do not float on sucrose gradients, conversely association of a protein with rafts could be so weak that it is destroyed by the detergent.[7] Moreover, DIGs could contain proteins not raft associated.

A key question in the rafts hypothesis is whether these microdomains exist in living cells, and they exist in membrane prior to detergent extraction. Recently several different methods based on different microscopic techniques have been used to visualize rafts in living cells. By using the single-particle tracking method, gold particles were attached to raft components and confinement zones were determined.[24] Using this technique it has been demonstrated that glycosphingolipids, the ganglioside GM1 and Thy-1 (a GPI-anchored protein) were confined to zones of 200-300 nm in diameter.[25]

Another demonstration of the in vivo existence of rafts derives from the analysis of GPI-anchored proteins clustering on the surface of living cells using a modification of the fluorescent resonance energy transfer technique (FRET).[26] By this method it has been demonstrated that GPI-anchored folate receptors are clustered in domains of smaller than 70 nm in diameter. Similarly, using chemical cross-linking with a bifunctional reagent it has been demonstrated that GPI-anchored proteins are clustered and that they form oligomers of up to 15 molecules which should occupy a small size domain in the membrane.[27] Furthermore, a conventional FRET approach has been used to compare the organization of different GPI-anchored proteins (folate receptor, CD59, and 5'NT) and of GM1 detected by cholera toxin B. This analysis demonstrated that these molecules are not concentrated in large and stable microdomains, but that a small fraction is present in small and unstable rafts.[28]

A different approach to study rafts is to perturb their lipid composition, which might lead to protein dissociation from rafts. Perturbation of lipid rafts occurs following cholesterol sequestration using filipin, nystatin or amphotericin, upon treatment with pore-forming agents such as saponin and digitonin, or following cholesterol depletion using methyl-β-cyclodestrin, or by inhibition of cholesterol biosynthesis with lovastatin (for review see ref. 6). In these various conditions proteins previously found in DRM are excluded from these domains and in some cases they behave differently.

A key question is whether raft-associated proteins are distributed between different rafts or whether different rafts exist in which specific proteins are selectively clustered. Recently it has been demonstrated that different microdomains exist at the plasma membrane.[29] For example the raft associated protein prominin is soluble following TX-100 extraction but insoluble in Lubrol WX, another non-ionic detergent, suggesting that different rafts exist and that proteins are compartmentalized specifically only in some rafts.[30]

Rafts Functions

Because of their property to segregate and concentrate different lipids and proteins, rafts work as a platform to concentrate specific molecules. They are therefore tought to play a major role in the sorting of proteins and lipids in the exocytic and endocytic pathways and in cell signaling.[3,6]

Signal Transduction

A key mechanism regulating signal transduction is the compartmentalization of particular substrates which are rapidly and specifically phosphorylated by select protein kinases. Compartmentalization of molecules involved in a specific signaling cascade is necessary for the proper activation by an external stimulus, because the same signaling molecules can activate diverse groups of effectors in different signaling pathways. Thus, restricted localization in subcellular compartments of a specific kinase and its substrates is important for the specificity and the efficiency of the effects induced by this kinase. It has been demonstrated that rafts play a key role in recruitment of signaling molecules, which show high affinity for raft lipids.[31] Moreover, raft-associated activated receptors can recruit specific downstream signaling molecules in rafts and this compartmentalization plays a key role in regulating signal transduction (for review see ref. 6). In fact, rafts recruit signaling proteins in a specific subcellular micro-

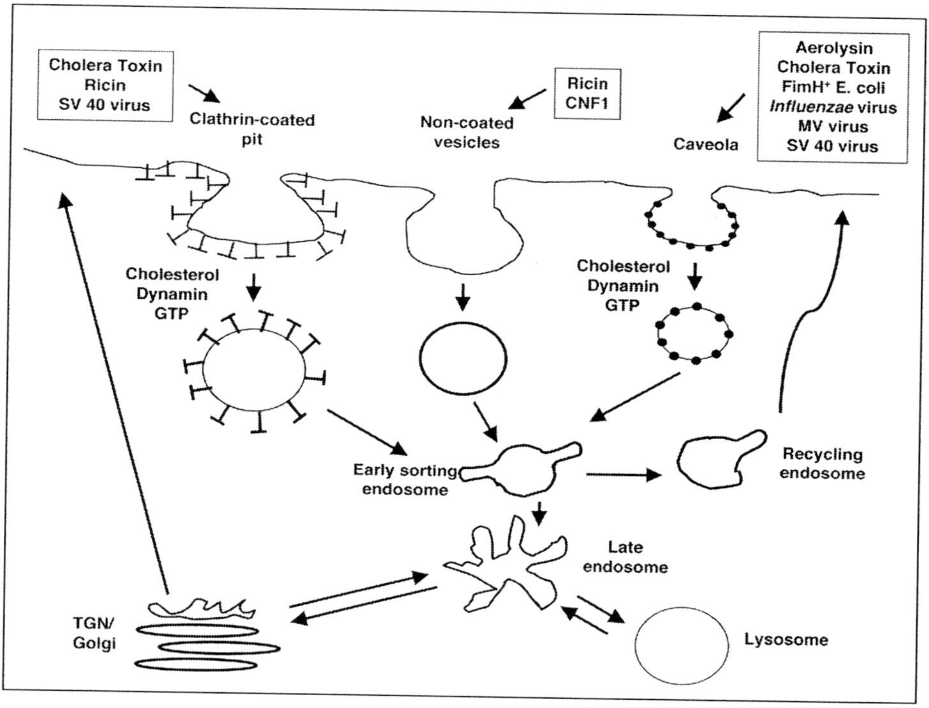

Figure 1. Different endocytic pathways are described: clathrin-dependent, caveolae-mediated, and clathrin/caveolae-independent. Moreover, the sorting pathways of endocytosed molecules in different endocytic compartment are indicated. After internalization, the molecules are found in the early sorting endosomes. From this compartment the molecules can recycle to the plasma membrane (through recycling endosomes), or they can be sorted in late endosomes. Molecules following the degradative pathway are found in the lysosomal compartment, while other molecules recycle continuously from the TGN and late endosomes, as well from lysosomes and late endosomes.

environment in wich their phosphorylation state can be modified by local protein kinases and phosphatases.

The T cell receptor (TCR)-mediated signaling cascade is the best characterized signaling pathway occurring in rafts. TCR is composed of ab- heterodimers associated with the CD3 complex and the z-homodimer. Several GPI-anchored proteins amplify TCR-activated signaling.[32] In T cells activation of a GPI-anchored receptor induces a signal through the membrane which activates the Src-family kinases.[33] Compartmentalization in rafts of molecules involved in TCR-induced signaling has been demonstrated by visualizing raft domains in intact cells using labeled cholera toxin B (CT-B) followed by cross-linking into patches with anti-CT-B antibodies. It has been shown that some molecules (LCK, LAT and TCR) are associated with lipid rafts, while others (CD45) are excluded from these domains, conferring a specific regulation of this pathway.[34]

The GPI-linked GFRa receptor is another example of the relevance of rafts in modulation of signal transduction. GFRa act as a co-receptor for the Ret transmembrane receptor. The latter protein is unable to bind its ligand in the absence of GFRa. Moreover, Ret localizes in lipid rafts only upon interaction with GFRa in a ligand-mediated mechanism. Activation of the signaling cascade is blocked when Ret is not localized in rafts upon stimulation with the ligand, showing that rafts localization is necessary for this signaling pathway.[35,36] Thus, rafts seem to be important to regulate the activity of some transmembrane receptors as well as to recruit downstream signaling molecules such as the Src-family kinases Lyn and Fyn.

Conversely, the activity of a small GTPase is regulated by rafts compartmentalization. H-ras is recruited in rafts because of its palmitoylation. It has been shown that alteration of the rafts microenvironment by depletion of cholesterol inhibits H-ras-mediated Raf activation.[37] Recently the role of rafts in the activity of H-ras has been analyzed in details. GTP binding to H-ras leads to redistribution of this GTPase from rafts into bulk plasma membrane. Dissociation of active H-ras from rafts is necessary for raf activation, thus, active and inactive H-ras are differently distributed at the plasma membrane and this differential distribution plays a key role in the subsequent signaling pathway.[38]

Intracellular Sorting

Rafts would also be expected to exist in the intracellular membranes, especially in the Golgi apparatus and in endosomal compartments. Although no direct evidence have yet been provided, indirect experiments suggest that this is the case. By pulse chase experiments it has been demonstrated that after their synthesis in the endoplasmic reticulum (ER) proteins become associated with detergent-resistent membranes only upon transfer to the Golgi apparatus.[39,40] This could occur because glycosphingolipids are absent in the ER and are synthesized in the Golgi.[39,41] It has been proposed that the specific association with rafts of GPI-anchored and some other apically sorted proteins in the Trans Golgi Network (TGN) might be the key event for delivery of these proteins to the apical domain of the plasma membrane in polarized epithelial cells.[3,40,42] Rafts could play a role also in the sorting of secreted molecules, as recently shown for thyroglobulin in thyroid cells.[43]

In the endosomal compartment rafts might have a sorting capacity in recycling of lipids and GPI-anchored proteins.[44] By electron microscopy Mayor et al demonstrated that the GPI-anchored folate receptor is internalized and during the first step of endocytosis it colocalizes with the transferrin receptor. They have shown that the transferrin receptor, internalized by coated pits, is quickly recycled at the plasma membrane, while the folate receptor is retained in a different endocytic compartment. Furthermore, this retention is regulated by the cholesterol content in the membranes, suggesting that lipid composition is important in the endocytic sorting of raft associated proteins.[44]

Endocytosis

Rafts seem to be involved in modulating the endocytosis of several molecules. Molecules can be internalized via fluid phase internalization, which indicates the cellular uptake of molecules and solutes in small vesicles that bud from the plasma membrane without binding to a receptor. Conversely, in the receptor-mediated internalization process ligands are bound to receptors and concentrated prior to internalization. This mechanism of endocytosis can be further divided in clathrin-coated vesicle-mediated endocytosis and non-coated vesicle-mediated internalization, the latter mechanism involves caveolae, cavaolae-like domains, and rafts.

In caveolae-mediated endocytosis cav1 represents the coat of the internalized caveolae vesicles. Furthermore, it has been demonstrated that a pathway independent from clathrin and caveolae exists. This endocytic process occurs when both clathrin and caveolae-mediated endocytosis is blocked. In this process the vesicles are larger than caveolae and smaller than coated pits.[45]

The major demonstration of caveolae-mediated endocytosis comes from studies of internalization of GPI-anchored proteins. The uptake of 5-methylentetraydrofolate is mediated by a GPI-anchored receptor.[46] Electron microscopy studies have shown that this receptor is also localized in endosomes, demonstrating that its internalization is not only due to potocytosis but is an endocytic event.[47] To demonstrate that caveolae are dynamic structures involved in endocytosis Simons and co-workers have developed an assay which used the GPI-anchored protein alkaline phosphatase (PLAP) as a marker of caveolae-mediated endocytosis.[48] Because of the enzymatic activity of this protein, it is possible to measure the decrease in the cell surface

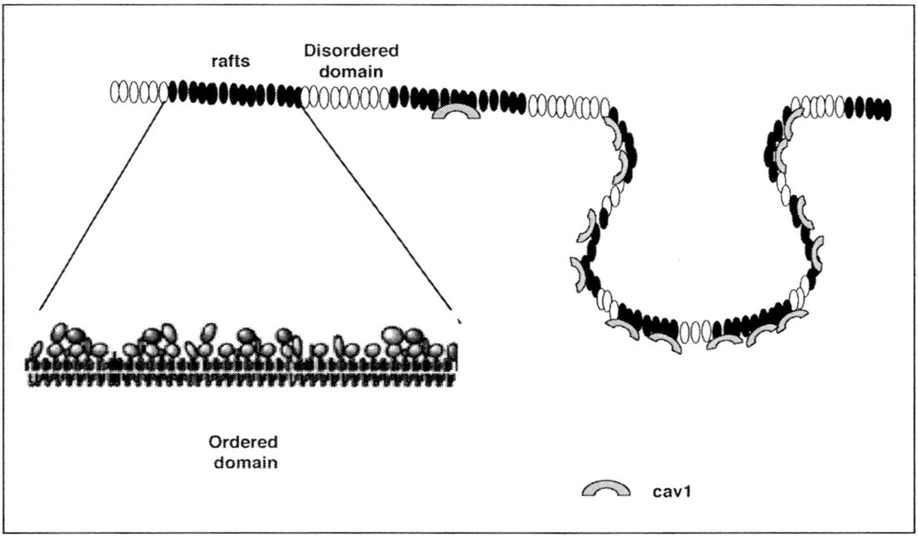

Figure 2. Schematic rapresentation of rafts and caveolae at the plasma membrane.
White color indicates the disordered domain, black represents the ordered domains (rafts). The position of caveolin1 (cav1) is indicated.

phosphatase activity following internalization. They have shown that in the absence of antibodies against PLAP, following incubation at 37° C there was no change in enzymatic activity at the cell surface. However following antibody-induced crosslinking of this protein a significant decrease of enzymatic activity occurred.[48]

Furthermore, PLAP and the folate receptors colocalize at the cell surface with the integral membrane protein stomatin, which is included in rafts. PLAP and stomatin have been found together in late endosomal-lysosomal vesicles, as shown by the fact that they colocalize with the lysosomal marker LAMP2.[49]

Internalization studies of cholera toxin following interaction with its receptor GM1, which is raft associated, have demonstrated that this toxin is internalized via a caveolae-mediated process clearly distinct from clathrin-mediated endocytosis.[48] However, this data is still debated because in neurons, which do not possess cav1, cholera toxin seems to be internalized via a raft-independent pathway.[50]

More recently the dynamics of caveolae in living cells has been studied using GFP-cav1 chimeras. During the endocytic process from the plasma membrane, fast moving vesicles can be observed. Cholesterol plays a key role in this mechanism, because depletion of cholesterol abolishes caveolae-mediated endocytosis.[51]

An important demonstration that rafts are involved in endocytosis comes from studies of lipid internalization.[52] Lipid analogues have been used to show that trafficking of lipids is dependent upon their physical properties. Lipids with long saturated hydrophobic tails are sorted differentially from those with short or unsaturated tails. The formation of lateral association of certain lipids induce a specific curvature in the membrane that enhances segregation of some membrane components. Thus, lipid composition plays a key role in the sorting of certain molecules in the endocytic pathway. Raft lipids are continuously endocytosed into early endosomes from which they could recycle to the plasma membrane directly, or indirectly following transport to the Golgi.[53] Moreover, it has been demonstrated that endocytic organelles can sort membrane components based on their preferencial association with specific lipids.[52]

Numerous studies suggest that rafts are also involved in the recycling pathways. Maxfield and co-workers[44] characterized the kinetics of internalization of GPI-anchored proteins in CHO cells using a fluorescent folate derivative bound to the GPI-anchored folate receptor. It was shown that it recycled slower than receptors that were not rafts associated. They demonstrated that internalized GPI proteins are found initially in the same compartment containing markers of receptor-mediated internalization (such as transferrin receptor), but the rate of their recycling to the surface is different. They also showed that separation of GPI proteins from transferrin receptor in early endosomes depends on the lipid composition of the membrane, in particular cholesterol levels seem to be responsible for the retention of GPI-anchored proteins in the endocytic compartment, because in cholesterol depleted cells they recycle with the same kinetics observed for the transferrin receptor.[44] Using fluorescent lipid probes it has been demonstrated that following internalization, these molecules are found in sorting endosomes and than quickly recycle on the cell surface.[54]

Caveolae-like mediated endocytosis occurs in cell types lacking cav1, such as lymphocytes. It has been demonstrated that cholera toxin enters lymphocytes via caveolae-like structures. In Jurkat T cells lymphoma depletion of cholesterol affects endocytosis of cholera toxin, suggesting that microdomains are important for its internalization.[55]

Because in caveolae-like structures the cav1 coat is absent it is difficult to discriminate between caveolae-like endocytosis and mechanisms independent from this pathway. Nevertheless, like caveolae, also caveolae-like structures are dependent on cholesterol for their formation. Thus, by perturbation of cholesterol homeostasis it has been possible to identify both caveolae- and clathrin-independent endocytic pathways. In fact, cholesterol depletion affects the formation of both these types of vesicles.

Rafts and Pathogens

Rafts have been implicated in the mechanism of action and internalization of different bacterial toxins, bacteria and virus. We will analyze different examples.

Toxins

In this section we will describe the role of rafts in the internalization and/or activity of bacterial toxins. Toxins can be classified in three groups dependent upon their mechanism of action. Group I toxins act at the cell surface where they induce transmembrane signaling. Group II toxins include pore forming toxins, which bind the surface receptors in an inactive form and, following activation by proteolytic cleavage, oligomerize into pores which disrupt the integrity of the target membrane. Finally, group III toxins, also called A-B type toxins, are formed by two different subunits, where the B subunit is responsible for the binding to the cell surface receptor and the A subunit exerts the cytotoxic activity.

Group I Toxins

LPS (bacterial lipopolysaccharide) is a member of the group 1 toxins, which exert their cytotoxicity by inducing a signaling cascade following binding to the host cell surface. LPS binds to a GPI-anchored protein, CD14,[56] which is enriched in rafts and which induces a MAPK signaling cascade that can also be regulated by rafts association/dissociation of some components (see ref. 6 for review).

Group II Toxins

A group II toxin, which has been clearly demonstrated to be dependent upon rafts for its activation and internalization is aerolysin, which is secreted by *Aeromonas hydrophila* of the *Vibrionacee* family. This bacterium is the aetiological agent of different infections, such as gastroenteritis and septicemia. The inactive soluble dimer precursor pro-aerolysin[57] binds to specific receptors present at the surface of target cells. Depending on the cell type, several receptors have been identified, all of which are GPI-anchored proteins.[58] Following binding to the receptors the protoxin is activated by proteolytic cleavage of the C-terminus followed by homo-oligomerization into ring-like amphipathic complexes, which insert into the membrane and form pores. Formation of these channels leads to loss of ions and small molecules from the cell[59] and to vacuolation of the endoplasmic reticulum.[59]

Binding of aerolysin to GPI-anchored receptors is important for its cytotoxicity. In fact T lymphocytes, which do not expose GPI-anchored proteins at their surface because of an inability to synthetize the anchor are much less sensitive to this toxin.[60] By immunofluorescence analysis it has been shown that the plasma membrane distribution of aerolysin is not homogeneous, and warming up the toxin-treated cells at 37° C increases the punctate staining of the toxin. This clusterizing upon warming up the cells is typical of raft-associated molecules.[58] Furthermore in immunofluorecence analysis antibody-cross linked aerolysin partially colocalized with Cav1, suggesting that this toxin clusters both in caveolae and in not-caveolar rafts.[58]

Aerolysin is able to oligomerize at low efficiency in solution in the absence of membranes. Neverthless, both oligomerization and channel formation occur at much lower toxin concentration when aerolysin is incubated with membranes.[58] Furthermore, by Triton X-100 extraction and sucrose gradient separation it has been shown that all forms of this toxin (proaerolysin, aerolysin, and its oligomers) are highly enriched in DIGs,[58] suggesting that rafts play a role in lowering the concentration of this toxin that is necessary to kill the cells.

Disruption of rafts by cholesterol depletion does not affect binding or processing of proaerolysin but reduces dramatically its ability to cluster upon binding with the receptor at the plasma membrane and, more importantly, prevents its oligomerization.[61] Thus, rafts could serve to locally concentrate the toxin and may be the site for its oligomerization. Neverthless, rafts do not have any effect on pore formation by aerolysin. It has been demonstrated that removal of cholesterol does not affect the sensitivity to this toxin.[62] Moreover, it has been shown that liposomes containing rafts are no more sensitive to aerolysin than raft-free liposomes.[62] Thus, because binding to the receptor is the rate-limiting step in aerolysin pathogenicity, it would appear that rafts association concentrates the receptor-toxin complexes at specific sites on the plasma membrane.

It is possible that, like aerolysin, other pore forming proteins need lipid rafts to function. For example, it has been demonstrated that in vitro *Vibrio cholera* cytolysin needs both cholesterol and sphingolipids to form pores.[63] Many strains of *V. cholerae* secrete a hemolytic toxin, *Vibrio cholerae* cytolysin (VCC), which is responsible for intestinal and extraintestinal infections. This toxin contains a signal peptide that is cleaved during secretion. The extracellular procytolysin is activated by a proteolytic cleavage of an N-terminal fragment, and can be further cleaved at the C-terminus to produce another active form.[63] This toxin first binds the plasma membrane as a monomer, then oligomerizes to form pores. Using an in vitro-system it has been shown that liposomes containing sphingolipids (in particular galactosylceramide) are more sensitive to this toxin.[63] Thus, VCC requires a particular lipid composition of the plasma membrane for infection. In fact, the susceptibility to this toxin increases in the presence of cholesterol and cholesterol-associated sphingolipids, especially ceramide.[63]

Moreover, the earthworm toxin lysenin binds sphingomyelin.[63] This protein has been demonstrated to be responsible for the hemolytic activity of the coelomic fluid from the earthworm *Eisenia fetida fetida* (*Anellida, Oligo chaeta, Lumbricidae*).[64] Various studies suggest that glycoproteins of the cell membrane act as receptors for this toxin, because preincubation with acetylated and methylated carbohydrates inhibits its activity.[64] The hemolytic activity of this toxin is strongly blocked when it is preincubated with sphingomyelin.[64] Using unilamellar liposomes as a model system it has been shown that specific sphingolipids are required for the binding and the activity of this toxin. Cholesterol alone does not induce binding of lysenin to liposomes, while complexes of sphingomyelin and cholesterol induce both binding and oligomerization of the toxin.[64]

The thiol-activated toxins, such as streptolysin O, lysteriolysin O, and perfringolysin O also require cholesterol to form channels in the membrane[65] and it seems that cholesterol is required for the conformational changes of these toxins that is necessary for their insertion into the membrane. Therefore, for all these toxins it seems that microdomains play a key role in the process of toxin oligomerization and pore formation.

VacA toxin is produced by the bacterium *Helicobacter pylori,* which is a Gram-negative bacterium that infects the gastric mucosa. Recently it has been shown that this toxin forms channels following oligomerization of membrane-bound monomers.[66] VacA is synthesized as 140 kDa protoxin and is further processed to form a 90 kDa protein which is released in the extracellular environment.[67] This toxin may be further cleaved to produce a 34-37 kDa N-terminal fragment and a 58 kDa C-terminal fragment, which remain associated after processing.[67]

Vac A binds to specific receptors at the cell surface, such as the EGF receptor.[68] It then becomes internalized and finally localizes in the endocytic-endosomal compartment in which it exerts its toxic activity, producing cytoplasmic vacuolar degeneration.[69] Using specific markers of the endocytic compartment it has been shown that the vacuoles produced by the toxin derive from late endosomal vesicles and it has been proposed that this toxin induces the formation of hybrid compartments between late endosomes and lysosomes.[70] The formation of pores plays a key role in cell vacuolation, because endocytosed VacA channels stimulate the turnover of endosomal V-ATPase by enhancing the permeability of endosomal membranes to anions.[66]

Several data suggest that internalization and activity of VacA toxin involve lipid rafts. Recently it has been demonstrated that internalization of VacA toxin in Hep2 cells is dependent upon GPI-anchored proteins. In fact, the sensitivity of these cells to this toxin is impaired by treating the cells with phosphatidylinositol-specific phospholipase C (PI-PLC), which removes GPI-anchored proteins from the cell surface.[71] Moreover, overexpression of proteins affecting clathrin-mediated endocytosis, such as a dominant negative form of Eps 15, and five tandem Src-homology-3 domains of intersectin, did not inhibit vacuolation induced by VacA.[71] Conversely, actin depolymerization and cholesterol depletion affect VacA activity.[71]

Group III Toxins

Cholera toxin (CT) is the best characterized group III toxin which seems to require rafts for its action. Following binding of the B subunit to its membrane-receptor, the ganglioside GM1, which is clustered in lipid rafts, the intact holotoxin is internalized. After its internalization, the toxin is processed to generate the active A1 peptide (CT-A1), an ADP-ribosyltransferase which is responsible for the transfer of ADP-ribose from NAD+ to the stimulatory G protein, Gs. Earlier electron microscopy studies have shown that CT is internalized through small non coated invaginations.[72] Later studies on the ultrastructural plasma membrane distribution of GM1[48] and on the colocalization of gold CT-B with cav1 have reinforced the idea that cholera toxin is internalized via caveolae. Using cell lines that do not express cav1 and caveolae, like CaCo2 cells and lymphocytes, Fishman and coworkers have shown that specific inhibition of clathrin-mediated endocytosis by the drug chloropromazine does not affect either toxin internalization, or its activity.[55] On the contrary, depletion of cholesterol by filipin interferes with both endocytosis and activity of this toxin.[55] Because the effect of filipin on cholera toxin activity was similar in A481 (human epidermis) and Jurkat T lymphoma cells, that respectively express high levels or no cav1, these results indicate that perturbation of membranes by cholesterol extraction rather than the presence of caveolae is responsible for inactivation of the toxin. Thus, according to these data both internalization and activation of cholera toxin are dependent upon rafts both in caveolae containing cells and in cells lacking cav1 and caveolae.

However, it has also been shown that after internalization cholera toxin colocalizes in tubulovesicle structures with molecules which enter the cell through coated pits.[73] Indeed Futerman and colleagues have shown that in hippocampal neurons cholera toxin is rafts associated at the cell surface, and that it is internalized via a raft-independent pathway.[50]

CNF1 (cytotoxic necrotizing factor 1) is also a type of A-B toxin produced by pathogenic *E. coli.* Recent work has shown that CNF1, like ricin, is internalized by both clathrin and caveolae-like independent pathways.[74] In Hep 2 cells a dominant negative dynamin II blocks the clathrin-mediated endocytosis of the transferrin receptors, but not CNF1 internalization. On the contrary, agents able to inhibit internalization of cholera toxin via caveolae have no effect on endocytosis of CNF1.[74]

Both the plant ricin and the bacterial shiga toxins are part of the A-B toxin family. The activity of these toxins requires their internalization, their transport to the Golgi apparatus and then retrograde transport to the endoplasmic reticulum.[75] Ricin and shiga toxins bind to glycolipids or glycoproteins on the plasma membrane of target cells. However, they can be internalized by mechanisms which are independent of rafts. Ricin is endocytosed by coated pits, but it is still internalized when this pathway is blocked. It has therefore been studied as a model to investigate clathrin-independent endocytosis.[76] The mechanism of internalization of ricin is also independent from caveolae-mediated endocytosis. In fact, although ricin is bound to glycosphingolipids, it can still be endocytosed when both clathrin- and caveolae-mediated endocytosis are inhibited by depletion of cholesterol.[77] On the contrary, shiga toxin, which is also bound to a glycolipid receptor, Gb3, is endocytosed mainly by clathrin-coated pits.[78]

Very recently the mechanism of internalization of the Anthrax Toxin has been elucidated.[103] The toxin is composed if three subunits; edema factor (EF; a calmodulin-dependent cyclase), lethal factor (LF; a metalloprotease that targets MAPK kinases), and protective antigen (PA; a protein of 83 kD which is able to bind the antrax receptor[104] at the plasma membrane). EF and LF are responsible for the toxin virulence. However they cannot exert their lethal effects in the absence of PA because they are unable to reach their cytosolic targets. Once bound to ATR, at the cell surface PA is cleaved into a 63-kD subunit which is able to oligomerize into an eptameric ring which binds EF and LF. Abrami et al (2003) demonstrated that toxin oligomerization clusters the antrax receptor into lipid rafts. Altering raft integrity prevents LF delivery to the cytoplasm suggesting that lipid rafts could be therapeutic targets for drugs against anthrax. Interestingly, although the regulation of anthrax toxin endocytosis relies on rafts, internalization of the receptor does not occur via caveolae but via a clathrin-dependent pathway.

Bacteria

The internalization of bacterial pathogens is driven by interactions between bacterial adhesion molecules and host receptors. The binding activates a signal transduction cascade in the host cells leading to stimulation of pre-existing endocytic pathways. The fate of internalized bacteria depend on which endocytic pathway is induced.

E. coli expressing the lectin FimH can survive inside the cell following phagocytosis.[79] FimH protein interacts with its receptor, CD48, on the host plasma membrane. CD48 is a GPI-anchored protein, thus its interaction with FimH leads to rafts-mediated endocytosis. Abrahm and co-workers clearly demonstrated that in BMMC macrophage cells endocytosis of FimH expressing *E. coli* occurs in caveolae.[80] BMMC cells express cav1, which colocalizes with the CD48 receptor both on the cell surface and in vesicles.[80] By confocal microscopy analysis it has been demonstrated that cav1 is localized around internalized FimH bacteria and that endocytosed bacteria colocalize with the rafts markers GM1 and cholesterol. In these cells pretreatment with the B subunit of cholera toxin inhibits specifically internalization of FimH-expressing bacteria, because this toxin clusters the caveolae following binding with its receptor GM1. Moreover, in these cells perturbation of rafts by cyclodestrin exerts a dose-dependent inhibition of internalization of FimH-expressing bacteria, while endocytosis of *E. coli* lacking FimH is unaffected by this drug.[80] Accordingly, following internalization FimH bacteria float to rafts fractions on sucrose density gradients, while extracellular bacteria are found in the pellet.[80] Furthermore, the discovery that also leukocytes express cav1 and contain caveolae supports a role for caveolae in bacterial internalization.[81]

The bacterium *Campylobacter jejuni* is also internalized via raft-mediated endocytosis. It has been proposed that entry of bacteria via rafts is the mechanism responsible for the intracellular survival of this pathogen.[82]

It is known that the luminal pH decreases along the degradative pathway, while it increase in the recycling vesicles. Recycling endosomes are less acidic than early endosomes because they lack vacuolar ATPase.[83] Caveolae-mediated endocytosis seems to follow the recycling pathway. Consequently, the bacteria can escape the degradative lysosomal compartment. In agreement with this hypothesis, it has been demonstrated that recycling endosomes are enriched in lipid rafts[83] and they are distinct from early endosomes.

Enveloped Viruses

Fusion of the virus envelope with the host cell membrane is a key event in infection. To infect a cell the virus binds a surface protein of the host and then its envelope fuses directly with the plasma membrane of the host cell. Alternatively, the virus enters the cell by receptor-mediated endocytosis and its envelope then fuses with the host cell endosome membrane.

Influenza

Influenza virus is an enveloped RNA virus containing single-stranded and segmented RNA of negative polarity. This virus enters the cell by direct fusion of its envelope with the plasma membrane of the host cell. The viral envelope glycoprotein of the influenza virus, HA (hemoagglutinin), was the first transmembrane protein to be shown to be rafts associated when expressed in eucaryotic cells.[84] In polarized epithelial cells influenza virus buds from the apical surface subsequent to the apical delivery of its envelope glycoprotein HA. Because rafts have been involved in apical protein sorting, HA targeting to the apical surface and budding of the virus from the apical side of epithelial cells favour a role of microdomains in these processes.[85]

It has been demonstrated that the envelope of fowl plague virus (FPV) from the influenza virus family contains detergent-insoluble complexes.[85] Following labeling with ^{3}H cholesterol, it has been shown that high amounts of cholesterol of the FPV virus envelope float on sucrose gradients, while the cholesterol present in the envelope of the vesicular stomatitis virus is less detergent-insoluble.[85] More importantly, HA is detergent-insoluble when exposed on the envelope of FPV, while it is mainly soluble when exposed on the envelope of a recombinant vesicular stomatitis virus expressing HA, suggesting that the lipid composition of the envelope is important for virus budding. Moreover, it has been demonstrated that palmitoylation of HA is necessary for its incorporation into virions.[86]

Budding of viruses from the infected cell requires two processes. First, both the viral structural proteins and the components of the envelope must be properly targeted individually or as subviral component to the plasma membrane. Second, all the viral components must interact with each other to initiate viral budding followed by the release of virions.

Several data suggest that HA plays a critical role to drive assembly and budding of the virus on specific sites of the plasma membrane. HA drives incorporation of the viral matrix protein M1 in detergent-resistant membranes. M1 mediates assembly of the virus because it is juxtaposed between the viral envelope and the nucleocapsid, therefore both assembly and budding of the virus occurs at specific plasma membrane domains.[86]

Mononegalovirales Measles Virus

The Mononegalovirales measles virus (MV) infection leads to an acute respiratory disease. Its RNA negative strand codes for six structural proteins: nucleoprotein (N), phosphoprotein (P), RNA polymerase (L), hemagglutinin (H), fusion protein (F), and matrix protein (M).[87] The F protein is synthesized as an inactive precursor, which is cleaved by host cell enzymes producing two active subunits linked by a disulfide bound. The ribonucleoparticle is composed by the RNA genome and by N, P, and L proteins. This particle is packaged into an envelope composed of H and F integral membrane glycoproteins plus the inner-membrane associated M protein. The virus enters the cell by direct fusion of its envelope with the plasma membrane of the host cell.

H protein mediates the binding with the receptor CD46 on human cells, while H and F proteins are necessary for the fusion between the virus and the host cell membrane. Following ribonucleoparticle assembly in the cytoplasm, it is directed to the plasma membrane to contact the envelope. M drives F and H proteins to the sites of virus assembly and subsequent budding. It has been demonstrated that MV proteins are raft-associated because they are detergent-insoluble.[88] Raft association of H and F occurs in the Golgi apparatus. M is synthetized on free ribosomes and associates with rafts independently of H and F.[88]

Importantly, MV buds from the apical surface in polarized MDCK cells[89] confirming the hypothesis that rafts association is important for apical sorting in these cells.[3] Thus, also in this case rafts compartmentalize the proteins forming the envelope and provide a specific site of virus assembly on the plasma membrane of infected cells.[88]

Human Immunodeficiency Virus Type 1

Human immunodeficiency virus type 1 (HIV 1) is a type C retrovirus responsible for AIDS. HIV 1 envelope glycoprotein (gp 160) is synthesized as a precursor, which is cleaved to form two subunits called the receptor binding gp120 and the transmembrane gp41. The gp120/gp41 complex is necessary for the binding of HIV 1 to its receptor CD4 on the host membrane. Binding of the gp120/gp41 complex to CD4 induces fusion of HIV1 envelope with the host membrane by recruitment of host chemokine receptors and induction of a conformational change of gp120. It has been demonstrated that the HIV 1 envelope excludes the CD45 host protein, despite its high concentration on the host plasma membrane.[90] As we have discussed in the Introduction, in T cells TCR-induced signaling occurs in rafts, which produce compartmentalization of specific signaling molecules and exclusion of others such as CD45 phosphatase. Thus, the initial observation that CD45 is excluded from the HIV envelope suggests that this virus buds from raft domains of the host cells membrane.[90] Furthermore it has been shown that HIV 1 incorporates some rafts-associated molecules, such as the ganglioside GM1 and the GPI-anchored proteins Thy-1, and CD59. These host cells-encoded proteins colocalize with viral proteins and by immunoprecipitation experiments it has been shown that anti-CD59 antibodies are able to immunoprecipitate HIV 1, while anti-CD45 antibodies do not capture the virus.[90] Furthermore, it has been demonstrated that gp41 as well as the Gag viral protein are resistant to TX 100 extraction.[90] Perturbation of rafts using β-cyclodestrin affects virus precipitation through gp41 antibodies, suggesting that rafts are involved in gp41 incorporation into the virus envelope.[90] In agreement with these data, inhibition of cholesterol synthesis decreases the production of virus from infected cells.[91]

Budding of this virus in rafts is an important step in its pathogenicity. In fact, both CD55 and CD59 that are rafts-associated inhibit steps in the complement pathway. Thus, budding in rafts represents an advantage for HIV 1, because it allows introduction of molecules capable of neutralizing complement in the virus envelope.[90]

Interestingly, removal of palmitoylation of gp160 affects its rafts association and its incorporation into HIV 1 envelope.[92] Low level incorporation of non-palmytoylated gp160 into the virus envelope results in a relevant decrement in HIV 1 infectivity.[92]

It has been shown that interaction of gp41 with the structural viral precursor $Pr55^{Gag}$ during HIV 1 assembly is important for the proper insertion of the envelope proteins in HIV 1 virions.[93] The cytoplasmic tail of gp41 contains an internalization signal. However, when $Pr55^{Gag}$ is coexpressed, internalization of gp41 is inhibited, suggesting that interaction between these two proteins prevents interaction of gp41 with the endocytic machinery.[94] It has been shown that interaction between these two proteins occurs in detergent-resistant fractions, suggesting that the formation of this complex in rafts is important for virus assembly because it prevents endocytosis of the envelope proteins.[95] All these data demonstrate that rafts drive proper incorporation of viral and host proteins into the HIV 1 envelope and that this specific incorporation plays a key role in HIV 1-mediated pathogenicity.

Interestingly it has been demonstrated that in epithelial cells budding of HIV 1 occurs from the basolateral surface of MDCK cells.[96] Previous studies suggest that rafts are involved in the sorting of proteins to the apical surface.[3,40,42] Neverthless, gp41 is rafts associated as well as other viral and host proteins important for HIV assembly, budding, and pathogenicity. Thus, it seems that the protein important for the proper site of budding of HIV is both rafts associated and sorted to the basolateral surface. This suggests that rafts association does not necessarily lead to apical delivery, as observed for PcP^{c} [97] other basolateral proteins in our laboratory (Pillich and Zurzolo, unpublished observations).

Simian Virus 40

Simian virus 40 (SV40) entry in host cells occurs by endocytosis and not, as with other viruses, following direct fusion of their envelope with the host plasma membrane. Some SV40 virus particles can be internalized via clathrin-mediated endocytosis. Nevertheless, the majority of the virus is endocytosed by caveolae leading to productive infection.[98] Interestingly, caveolae containing SV40 are targeted to the endoplasmic reticulum, rather than to the lysosomal compartment following a novel pathway that it has been recently characterized.[99] SV40 internalization occurs only 2 hours after its binding to the host membrane.[98] On the contrary, clathrin-mediated endocytosis of virus occurs a few minutes after binding to the membrane. Moreover, SV40 internalization is induced by an intracellular signaling cascade activated upon virus binding to the host plasma membrane.[100]

The following events are necessary for SV40 entry into the cell: SV40 binds to rafts on the cell surface, it then translocates to the caveolae invagination where the virus induces the signaling cascade, and this event finally induces virus entry into caveolae. SV40 induces up-regulation of c-myc.[100] Perturbation of rafts by nystatin affects c-myc upregulation upon SV40 infection, suggesting that rafts are necessary for the signaling induced by this virus.[101] Interestingly, some studies demonstrate that rafts play a key role in signal transduction (see Introduction). Perturbation of rafts did not affect SV40 binding to the host plasma membrane, but did block its internalization,[101] suggesting that this virus uses rafts mediated signal transduction to enter the cell. Furthermore, it has been demonstrated that cav1 and caveolae play a key role in SV40 infection.[37] Expression of caveolin mutants lacking the N-terminus of the protein inhibits internalization of SV40.[37] These mutants act as dominant negatives on the function of endogenous cav1 for virus entry into the cells, and they also block the H-ras signaling cascade.[37]

Unicellular Pathogens

Recently rafts association has also been examined in unicellular eukaryotes. Some differences have been detected between rafts from higher eukaryotes and unicellular organisms. Interestingly, in Trypanosomes none of the GPI-anchored proteins are rafts associated.[102] Nevertheless, a new *Trypanosoma brucei* protein that is rafts associated has been recently cloned.[102]

African trypanosomes are unicellular eukaryotes responsible for sleeping sickness in man. These parasites alternate between the mammalian host and the tsetse fly. This life cycle is characterized by the expression of two surface proteins called the variant surface protein (VSG) and procyclic acidic repetitive protein (PARP). The first one is expressed during mammalian infection and life in the blood cells, while the latter one is expressed during infection of the insect host. Recently a new protein has been identified (BARP) that is expressed only during infection of the bood cells. This protein is rafts associated and is exposed on the entire body of *T. brucei*, clusterized in rafts.[102] It is unknown whether rafts association of this protein is involved in the pathogenicity of this organism, but it is of interest that a cycle-specific protein of this pathogen is rafts associated.

Summary

It is clear from this breaf overview that several pathogens might use rafts to enhance their pathogenicity or the effectivenes of their toxins. Further studies will be necessary to clarify the mechanisms at the basis of these events. Similarly, it is rather evident that pathogens and toxins can be used as interesting and powerful tools to study the dynamics of rafts interaction and to understand the different funtions of these microdomains.

Addendum

In the last few years rafts and caveolae have been the subject of many conferences. For a more recent update on rafts identity, methods of detection and function see the report of the second joint Euresco Conference / EMBO workshop in a series on "Microdomains, Lipid Rafts and Caveolae" which will be published in 2003 (Zurzolo et al, EMBO Reports 2003).

References

1. Brown D, London E. Structure and function of sphingolipid- and cholesterol-rich membrane rafts. J Biol Chem 2000; 275:17221-17224.
2. Harder T, Simons J. Caveolae, DIGs, and the dynamics of sphingolipid-cholesterol microdomains. Curr Opin Cell Biol 1997; 9:534-542.
3. Simons K, Ikonen E. Functional rafts in cell membranes. Nature 1997; 387:569-572.
4. Brown R. Sphingolipid organization in biomembranes: What physical studies of model membranes reveal. J Cell Sci 1998; 111:1-9.
5. Rock P, Allietta M, Young W et al. Organization of glycosphingolipids in phosphatidylcholine bilayers: Use of antibody molecules and Fab fragments as morphological markers. Biochemistry 1990; 29:8484-8490.
6. Simons K, Toomre D. Lipid rafts and signal transduction. Nature Reviews 2000; 1:31-38.
7. Hooper N. Detergent-insoluble glycosphingolipid/cholesterol-rich membrane domains, lipid rafts and caveolae. Mol Membr Biol 1999; 16:145-156.
8. Parton R. Caveolae and caveolins. Curr Opin Cell Biol 1996; 8:542-548.
9. Scherer P, Okamoto T, Chun M et al. Identification, sequence and expression of caveolin-2 defines a caveolin gene family. Proc Nat Acad Sci USA 1996; 93:131-135.
10. Tang Z, Scherer P, Okamoto T et al. Molecular cloning of caveolin-3, a novel member of the caveolin gene family expressed predominantly in muscle. J Biol Chem 1996; 271:2255-2261.
11. Scherer P, Lewis R, Volonte D et al. Cell-type and tissue-specific expression of caveolin-2. Caveolins 1 and 2 co-localize and form a stable hetero-oligomeric complex in vivo. J Biol Chem 1997; 272:29337-29346.
12. Song K, Scherer P, Tang Z et al. Expression of caveolin-3 in skeletal, cardiac and smooth muscle cells. Caveolin-3 is a component of the sarcolemma and co-frationates with dystrophin and dystrophin-associated glycoproteins. J Biol Chem 1996; 271:15160-15165.
13. Monier S, Parton R, Vogel F et al. VIP21-caveolin, a membrane protein costituent of the caveolar coat, oligomerizes in vivo and in vitro. Mol Biol Cell 1995; 6:911-927.
14. Fra A, Williamson E, Simons K et al De novo formation of caveolae in lymphocytes by expression of VIP21-caveolin. Proc Natl Acad Sci USA 1995; 92:8655-8659.
15. Lipardi C, Mora R, Colomer V et al. Caveolin transfection results in caveolae formation but not apical sorting of glycosylphosphatidylinositol (GPI)-anchored proteins in epithelial cells. J Cell Biol 1998; 81:617-626.
16. Rothberg K, Heuser J, Donzell W et al. Caveolin, a protein component of caveolae membrane coats. Cell 1992; 68:673-682.
17. Stahlhut M, van Deurs B. Identification of filamin as a novel ligand for caveolin-1: Evidence for the organization of caveolin-1 associated membrane domains by the actin cytoskeleton. Mol Biol Cell 2000; 11:325-337.
18. Volonte D, Galbiati F, Li S et al. Flotillins/cavatellins are differentially expressed in cells and tussues and form hetero-oligo-meric complex with caveolins in vivo. Characterization and epitope-mapping of a novel flotillin-1 monoclonal antibody probe. J Biol Chem 1999; 274:12702-12709.
19. Snyers L, Umlauf E, Prohaska R. Oligomeric nature of the integral membrane protein stomatin. J Biol Chem 1998; 273:17221-17226.
20. Harder T, Gerke V. The annexin II2p112 complex is the major protein component of the Triton X-100-insoluble low-density fraction prepared from MDCK cells in the presence of Ca2+. Biochem Biophys Acta 1994; 1223:375-382.
21. Fiedler K, Lafont F, Parton R et al. Annexin XIIIb: A novel epithelial specific annaxin is implicated in vesicular traffic to the apical plasma membrane. J Cell Biol 1995; 128:1043-1053.
22. Harder T, Kellner R, Parton R et al. Specific release of membrane-bound annexin II and cortical cytoskeletal elements by sequestration of membrane cholesterol. Mol Biol Cell 1997; 8:533-545.

23. D Brown, J Rose. Sorting of GPI-anchored proteins to glycolipid-enriched membrane subdomains during transport to the apical cell surface. Cell 1992; 68:533-544.
24. Sheets E, Lee G, Simson R et al. Transient confinement of a glycophosphatidylinositol-anchored protein in the plasma membrane. Biochemistry 1997; 36:12449-12458.
25. Jacobson K, Dietrich C. Looking at lipid rafts? Trends Cell Biol 1999; 9:87-91.
26. Varma R, Mayor S. GPI-anchored proteins are organized in submicron domains at the cell surface. Nature 1998; 394:798-801.
27. Friedrichson T, Kurzchalia T. Microdomains of GPI-anchored proteins in living cells revealed by crosslinking. Nature 1998; 394:802-805.
28. Kenworthy A, Petranova N, Edidin M. High-resolution FRET microscopy of cholera toxin B-subunit and GPI-anchored proteins in cell plasma membranes. Mol Biol Cell 2000; 11:1645-1655.
29. Madore N, Smith K, Graham C et al. Functionally different GPI proteins are organized in different domains on the neuronal surface. EMBO J 1999; 18:6917-6926.
30. Roper K, Corbeil D, Huttner W. Retention of prominin in microvilli reveals distinct cholesterol-based lipid microdomains in the apical plasma membrane. Nature Cell Biol 2000; 2:582-592.
31. Melkonian K, Ostermeyer A, Chen J et al. Role of lipid modifications in targeting proteins to detergent-resistant membrane rafts. Many raft proteins are acylated, while few are prenylated. J Biol Chem 1999; 274:3910-3917.
32. Moran M, Miceli M. Engagement of GPI-linked Cd48 contributes to TCR signals and cytoskeletal reorganization: a role for lipid rafts in T cell activation. Immunity 1999; 9:787-796.
33. Brown D. The tyrosine kinase connection: How GPI-anchored proteins activate T cells. Curr Opin Immunol 1993; 5:349-354.
34. Janes P, Ley S, Magee A. Aggregation of lipid rafts accompanies signaling via the T cell antigen receptor. J Cell Biol 1999; 147:447-461.
35. Tansey M, Baloh R, Milbrandt J et al. GFRa-mediated localization of RET to lipid rafts is required for effective downstream signaling, differentiation, and neuronal survival. Neuron 2000; 25:611-623.
36. Paratcha G, Ledda F, Baars L et al. Released GFRalpha1 potentiates downstream signaling, neuronal survival, and differentiation via a novel mechanism of recruitment of c-Ret to lipid rafts. Neuron 2001; 29:171-184.
37. Roy S, Luetterforst R, Harding A et al. Dominant-negative caveolin inhibits H-ras function by dustrupting cholesterol-rich plasma membrane domains. Nature Cell Biol 1999; 1:98-104.
38. Prior I, Harding A, Yan J et al. GTP-dependent segregation of H-ras from lipid rafts is required for biological activity. Nature Cell Biol 2001; 3:368-375.
39. Brown D. Interactions between GPI-anchored proteins and membrane lipids. Trends Cell Biol 1992; 2:338-343.
40. Zurzolo C, Van't Hof W, van Meer G et al. Caveolin/VIP21 and glycosphingolipid clusters in the sorting of glycophosphatidylinositol-anchred protein in epithelial cells. EMBO J 1994; 13:42-53.
41. Danielsen E. Involvment of detergent-insoluble complexes in the intracellular transport of intestinal brush border enzymes. Biochemistry 1995; 34:1596-1605.
42. Lipardi C, Nitsch L, Zurzolo C. Detergent-insoluble glycophosphatidyl inositol-anchored proteins are apically sorted in Fischer Rat Thyroid cells, but interference with cholesterol or sphingolipids differentially affects detergent insolubility and apical sorting. Mol Biol Cell 2000; 11:531-542.
43. Martin-Belmonte F, Alonso M, Zhang X et al. Thyroglobulin is selected as luminal protein cargo for apical transport via detergent-resistant membranes in epithelial cells. J Biol Chem 2000; 275:41074-41081.
44. Mayor S, Sabharanjak S, Maxfield F. Cholesterol-dependent retention of GPI-anchored proteins in endosomes. EMBO J 1998; 17:4626-46387.
45. Lamaze C, Schmid S. The emergence of clathrin-independent pinocytic pathways. Curr Opin Cell Biol 1995; 7:573-580.
46. Anderson R. Potocytosis of small molecules and ions by caveolae. Trends Cell Biol 1993; 3:69-71.
47. Birn H, Selhub J, Christensen E. Cell fractionation and electron microscope studies of kidney folate-binding protein. Am J Physiol 1993; 264:C302-C310.
48. Parton R, Joggerst B, Simons K. Regulated internalization of caveolae. J Cell Biol 1994; 127:1199-1215.
49. Snyers L, Umlauf E, Prohaska R. Association of stomatin with lipid-protein complexes in the plasma membrane and endocytic compartment. E J Cell Biol 1999; 78:802-812.
50. Shogomori H, Futerman A. Cholera toxin is found in detergent-insoluble rafts/domains at the cell surface of hippocampal neurons but is internalized via a raft-independent mechanism. J Biol Chem 2001; 276:9182-9188.

51. Schnitzer J, Oh P, Pinney E et al. Filipin-sensitive caveolae-mediated transport in endothelium reduced transcytosis, scavenger endocytosis, and capillary permeability of select macromolecules. J Cell Biol 1994; 127:1217-1232.
52. Mukherjee S, Soe T, Maxfield F. Endocytic sorting of lipid analogues differing soley in the chemistry of their hydrophobic tails. J Cell Biol 1999; 144:1271-1284.
53. Mukherjee S, Maxfield F. Role of membrane organization and membrane domains in endocytic lipid trafficking. Traffic 2000; 1:203-211.
54. Hao M, Maxfield F. Characterization of rapid membrane internalization and recycling. J Biol Chem 2000; 275:15279-15286.
55. Orlandi P, Fishman P. Filipin-dependent inhibition of cholera toxin: evidence for toxin internalization and activation through caveolae-like domains. J Cell Biol 1998; 141:905-915.
56. Ulevitch R, Tobias P. Recognition of endotoxin by cells leading to transmembrane signaling. Curr Opin Immunol 1994; 6:125-130.
57. van der Goot F, Ausio J, Wong K et al. Dimerization stabilizes the pore-forming toxin aerolysin in solution. J Biol Chem 1993; 268:18272-18279.
58. Abrami L, Fivaz E, Glauser P et al. A pore-formin toxin interacts with a GPI-anchored protein and causes vacuolation of the endoplasmic reticulum. J Cell Biol 1998; 140:525-540.
59. Krause K, Fivaz M, Monod A et al. Aerolysin induces G-protein activation and Ca2+ release from intracellular stores in human granulocytes. J Biol Chem 1998; 273:18122-18129.
60. Brodsky R, Mukhina G, Nelson K et al. Resistance of paroxysmal nocturnal hemogobinuria cells to the glycophosphatidylinositol-binding toxin aerolysin. Blood 1999; 93:1749-1756.
61. Abrami L, Gisou F, van der Goot F. Plasma membrane microdomains act as concentration platforms to facilitate intoxication by aerolysin. J Cell Biol 1999; 147:175-184.
62. Nelson K, Buckley J. Channel formation by the glycosylphosphatidylinositol-anchored protein binding toxin aerolysin is not promoted by lipid rafts. J Biol Chem 2000; 275:19839-19843.
63. Zitzer A, Zitzer O, Bhakdi S et al. Oligomerization of Vibrio cholerae cytolysin yields a pentameric pore and has a dual specificity for cholesterol and sphingolipids in the target membrane. J Biol Chem 1999; 274:1375-1380.
64. Lange S, Nussler F, Krauschke E et al. Interaction of earthworm hemolysin with lipid membranes requires sphingolipids. J Biol Chem 1997; 272:20884-20892.
65. Bhakdi S, Bayley H, Valeva A et al. Staphylococcal alpha-toxin, streptolysin-O, and Escherichia coli hemolysin: Prototypes of pore-forming bacterial cytolysin. Arch Microbiol 1996; 165:73-79.
66. Szabo I, Brutsche S, Tombola F et al. Formation of anion-selective channels in the plasma membrane by the toxin VacA of Helicobacter pylori is required for its biological activity. EMBO J 1999; 18:5517-5527.
67. Cover T. The vacuolating cytotoxin of Helicobacter pylori. Mol Microbiol 1996; 20:241-246.
68. Seto K, Hayashi-Kuwabara Y, Yoneta T et al. Vacuolation induced by cytotoxin from Helicobacter pylori is mediated by EGF receptor in HeLa cells. FEBS Lett 1998; 431:347-350.
69. Ricci V, Sommi P, Fiocca R et al. Helicobacter pylori vacuolating toxin accumulates within the endosomal-vacuolar compartment of coltured gastri cells and potentiates the vecuolating activity of ammonia. J Pathol 1996; 183:453-459.
70. Molinari M, Galli C, Norais N et al. Vacuoles induced by Helicobacter pylori toxin contain both late endosomal and lysosomal markers. J Biol Chem 1997; 272:25339-25344.
71. Ricci V, Galmiche A, Doye A et al. High cell sensitivity to Helicobacter pylori VacA toxin depends on a GPI-anchored protein and is not blocked by inhibition of the clathrin-mediated pathway of endocytosis. Mol Biol Cell 2000; 11:3897-3909.
72. Montesano R, Roth J, Robert A et al. Non-coated membrane invaginations are involved in binding and internalization of cholera and tetanus toxins. Nature 1982; 296:651-653.
73. Tran D, Carpentier J, Sawano F et al. Ligands internalized through coated or noncoated invaginations follow a common intracellular pathway. Proc Natl Acad Sci USA 1987; 84:7957-7961.
74. Contamin S, Galmiche A, Doye A et al. The p21 Rho-activating Toxin Cytotoxic Necrotizing Factor 1 is endocytosed by a alathrin-independent mechanism and enters the cytosol by an acidic-dependent membrane translocation step. Mol Biol Cell 2000; 11:1775-1787.
75. Llorente A, Rapak A, Schmid S et al. Expression of mutant dynamin inhibits toxicity and transport of endocytosed ricin to the Golgi apparatus. J Cell Biol 1998; 140:553-563.
76. Sandvig K, van Deurs B. Endocytosis and intracellular transport of ricin:recent discoveries. FEBS Lett 1999; 452:67-70.
77. Rodal S, Skretting G, Garred O et al. Extraction of cholesterol with methyl-beta-cyclodextrin perturbs formation of clathrin-coated endocytic vesicles. Mol Biol Cell 1999; 10:961-974.
78. Sandvig K, van Deurs B. Endocytosis, intracellular transport and cytotoxic action of Shiga toxin and ricin. Physiol Rev 1996; 76:949-966.

79. Baorto D, Gao Z, Malaviya R et al. Survival of Fimh-expressing enterobacteria in macrophages relies on glycolipid traffic. Nature 1997; 389:636-639.
80. Shin J-S, Gao Z, Abraham S. Involvment of cellular caveolae in bacterial entry into mast cells. Science 2000; 289:785-788.
81. Kiss A, Geuze H. Caveolae can be alternative endocytotic structures in elicited macrophages. Eu J Cell Biol 1997; 73:19-27.
82. Mulvey M, Hultgren S. Bacterial spelunkers. Science 2000; 289:732-733.
83. Gagescu R, Demaurex N, Parton R et al. The recycling endosome of Madin-Darby Canine Kidney cells is a mildly acidic compartment rich in raft components. Mol Biol Cell 2000; 11:2775-2791.
84. Scheiffele P, Roth M, Simons K. Interaction of influenza virus haemagglutinin with sphingolipid-cholesterol membrane domains via its transmembrane domain. EMBO J 1997; 16:5501-5508.
85. Scheiffele P, Rietveld A, Wilk T et al. Influenza viruses select ordered lipid domain budding from the plasma membrane. J Biol Chem 1999; 274:2038-2044.
86. Zhang J, Pekosz A, Lamb R. Influenza virus assembly and lipid rafs microdomains: A role for the cytoplasmic tails of the spike glycoproteins. J Virol 2000; 74:4634-4644.
87. Horikami S, Moyer S. Structure, transcription, and replication of measles virus. Curr Top Microbiol Immunol 1995; 191:35-50.
88. Manie' S, Debreyne S, Vincent S et al. Measles virus structural components are enriched into lipid raft microdomains: A potential cellular location for virus assembly. J Virol 2000; 74:305-311.
89. Maisner A, Klenk H, Herrler G. Polarized budding of measles virus is not determined by viral surface glycoproteins. J Virol 1998; 72:5276-5278.
90. Nguyen D, Hildreth J. Evidence for budding of human immunodeficiency virus type 1 selectively from glycolipid-enriched membrane lipid rafts. J Virol 2000; 74:3264-3272.
91. Maziere J, Landureau J, Giral, Auclair, Fall, Lachgar A et al. Lovastatin inhibits HIV-1 expression in H9 human T lymphocytes cultured in cholesterol-poor medium. Biomed Pharmacother 1994; 48:63-67.
92. Rousso I, Mixon M, Chen B et al. Palmitoylation of the HIV-1 envelope glycoprotein is critical for viral infectivity. Proc Natl Acad Sci USA 2000; 97:13523-13525.
93. Freed E, Martin M. Virion incorporation of envelope glycoproteins with long but not short cytoplasmic tails is blocked by specific, single amino acid substitutions in the human immunodeficiency virus type 1 matrix. J Virol 1995; 69:1984-1989.
94. Egan M, Carruth L, Rowell J et al. Human immunodeficiency virus type 1 envelope protein endocytosis mediated by a highly conserved intrinsic internalization signal in the cytoplasmic domain of gp41 is suppressed in the presence of the Pr55gag precursor protein. J Virol 1996; 70:6547-6556.
95. Wyma D, Kotov A, Aiken C. Evidence for a stable interaction of gp41 with Pr55(Gag) in immature human immunodeficiency virus type 1 particles. J Virol 2000; 74:9381-9387.
96. Lodge R, Lalonde J-P, Lemay G et al. The membrane-proximal intracytoplasmic tyrosine residue of HIV-1 envelope glycoprotein is critical for basolateral targeting of viral budding in MDCK cells. EMBO J 1997; 16:695-705.
97. Sarnataro D, Paladino S, Campana V et al. PcP^c is sorted to the basolateral membrane of epithelial cells independently of its association with rafts. Traffic 2002; 3:810-821.
98. Anderson H, Chen Y, Norkin L. Bound simian virus 40 translocates to caveolin-enriched membrane domains and its entry is inhibited by drugs that selectively distrupt caveolae. Mol Biol Cell 1996; 7:1825-1834.
99. Pelkmans L, Kartenbeck J, Helenius A. Caveolar endocytosis of simian virus 40 reveals a new two-step vesicular-transport pathway to the ER. Nature Cell Biol 2001; 3:473-483.
100. Dangoria N, Breau W, Anderson H et al. Extracellular simian virus 40 induces an ERK/MAPK-independent signaling pathway that activates primary response genes and promotes virus entry. J Gen Virol 1996; 77:2173-2182.
101. Chen Y, Norkin L. Extracellular simian virus 40 transmits a signal that promotes virus enclosure within caveolae. Exp Cell Res 1999; 246:83-90.
102. Nolan D, Jackson D, Biggs M et al. Characterization of a novel alanine-rich protein located in surface microdomains in Trypanosoma brucei. J Biol Chem 2000; 275:4072-4080.
103. Abrami L, Liu S, Cosson P et al. Anthrax toxin triggers endocytosis of its receptor via a lipid raft-mediated clathrin-dependent process. J Cell Biol 2003; 160(3):321-328.
104. Bradley KA, Mogridge J, Mourez M et al. Identification of the cellular receptor for anthrax toxin. Nature 2001; 414(6860):225-229.

CHAPTER 4

Endosome-Phagosome Interactions in Pathogenesis

Carmen Alvarez-Dominguez, Carla Peña-Macarro and Amaya Prada-Delgado

Introduction

It is widely accepted now that parameters and factors regulating the secretory and endocytic pathways also govern the phagocytic routes. However, not until recently it was accepted that phagosomes may fuse with endosomes, independent on the nature of the phagocytic particle.[1] The initial models stated that phagosome maturation implied a direct transport towards a phagolysosome.[2] Nowadays, phagosome maturation is understood as a process that implies extensive interactions of phagosomes with lysosomes, endosomes as well as with Golgi-derived vesicles.[3]

Most of our knowledge on the phagocytic trafficking comes from studies with inert particles, either fixed *Staphylococcus aureus* or latex beads.[4,5] These models constituted important tools for analyzing the parameters regulating fusion of endosomes with phagosomes and also allowed for an extensive analysis of phagosomal composition. They also revealed that phagosome transformation involved multiple transient fusion and fission events, known as the "kiss and run" hypothesis.[5] According to this hypothesis phagosomes lose their features of plasma membrane/early endosomes decreasing markers such as the mannose (MnR) and transferrin (TfR) receptors, Rab5 or annexin. While they become late endosomes/lysosomes by enrichment of late markers such as, Rab7, Lamp-1 or cathepsin-D.[6,7] In contrast to inert particles, intracellular pathogens interfere with the normal process of maturation as a strategy to avoid the microbicidal mechanisms and elude the immune response. The development of new techniques for phagosome isolation[5] and the in vitro systems to analyze phagosome-endosome fusion[4] revealed the first rules for endosome fusion with phagosomes containing pathogenic agents. One of the most interesting information obtained with these systems, was that pathogens targeted their strategies to the endosomal trafficking regulators, Rab5 or Rab7.[8,9] These studies also elucidated the role of cytokines on phagosome biogenesis, interactions with other organelles and pathogen survival.[10-12] Moreover, pathogens and phagocytic models allowed for the description of novel antigen processing and presentation routes.[13] In this Chapter, we have integrated all this information relevant for pathogenesis: pathogen interference with intracellular trafficking, regulation of the phagocytic machinery, inactivation of the cell destructive machinery and the onset of an immune response.

Factors Involved in Phagosome-Endosome/Lysosome Fusion

Fusion of phagosomes with endosomes and lysosomes was first documented in vivo by loading the endocytic compartments with electrodense material and analyzing their delivery to phagosomes by electron microscopy. Moreover, when the electrodense material were gold particles, quantification analysis allowed to calculate the percentage of phagosome fusion with endosomes or lysosomes.

Intracellular Pathogens in Membrane Interactions and Vacuole Biogenesis, edited by Jean-Pierre Gorvel. ©2004 Eurekah.com and Kluwer Academic / Plenum Publishers.

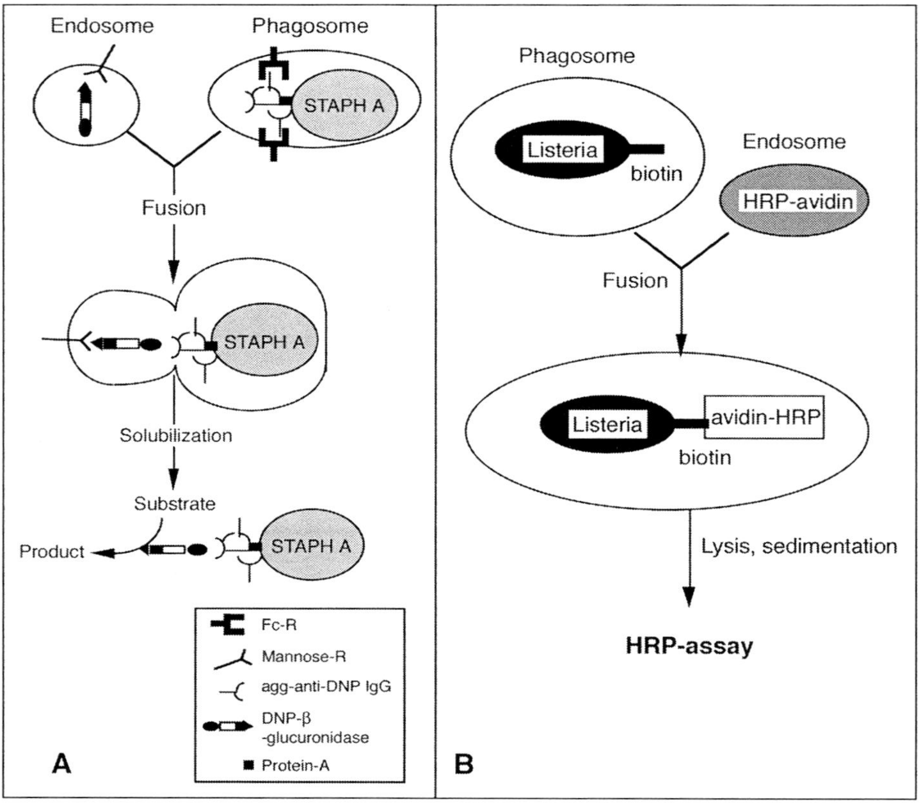

Figure 1. Schematic representation of the two models for phagosome-endosome fusion in vitro. Panel A, DNP-β-glucuronidase and Staphylococcus aureaus (STAPH A) particles coated with anti-DNP IgG antibodies are internalized separately via specific receptors (mannose receptor and FcR, respectively) by two different set of MØs. Phagosomes and endosomes are isolated and incubated together in fusion buffer supplemented with cytosol in the presence of DNP-BSA as scavenger to block unspecific binding. Phagosome-endosome fusion leads to association of the enzyme to the coated STAPH A particles. The sedimented particles after solubilization were assayed for β-glucuronidase activity. Panel B, biotinylated *Listeria* (dead or live) bacteria and HRP-avidin conjugated are internalized separately via specific receptors (*Listeria* receptor and mannose receptor, respectively) by two separate sets of MØs. Phagosomes and endosomes were isolated and incubated in fusion buffer supplemented with cytosol in the presence of avidin-insulin as scavenger. Phagosome-endosome fusion will allow for the association of the enzyme to the biotinylated bacteria. After membrane solubilization, bacteria are sedimented and analyzed for the presence of HRP onto them by developing the colorimetric reaction.

The first in vivo studies that described the interactions of phagosomes with endosomes[14] also revealed valuable information on phagosome maturation. This information included: (i) the dependence on the nature of the phagocytic particle for phagocytic processing. (ii) The observation that maturation arrest was determined by the apposition of the phagosomal membrane[15] or (iii) the function of transport vesicles that fuse with early endosomes on the recycling of proteins from the phagosomes.[16] However, these in vivo studies failed to establish the parameters regulating the fusion events. These parameters were unveiled with the development of in vitro cell-free systems for phagosome fusion and the new methods for isolation of fairly pure phagosomes and endosomes.

Phagosome-Endosome Fusion with Phagocytic Models

In the first cell-free in vitro system reported, the phagocytic probe consisted on fixed *Staphylococcus aureus* coated with mouse anti-DNP monoclonal antibody, while the endocytic probe was receptor-mediated internalized DNP-β-glucuronidase. These probes were internalized by different set of cells, isolated and mixed in an in vitro fusion assay described in Figure 1 (panel A). The parameters regulating phagosome fusion with endosomes, either containing inert particles (e.g., latex beads), dead or fixed microorganisms are summarized in Table 1. In brief, these fusion systems required energy (ATP), cytosolic and membrane-associated proteins, several GTP-binding proteins (Rab5,[8] heterotrimeric G-proteins,[17] members of the Rho family[18] or their downstream effectors) and N-ethylmaleimide (NEM) sensitive factors such as NSF.[8,17] Some of these parameters were also confirmed by other fusion methods that combined in vivo fusion events with in vitro analysis of latex beads phagosomes that received endocytic HRP. The system was developed later on, by analyzing the colorimetric reaction associated to the phagosomes.[5,19] Analysis of latex beads phagosomes indicated the importance of t-SNARES, synthaxins 2, 3 and 4, that were found onto phagosomes complexed with NSF and SNAP proteins.[20]

Another in vitro cell-free fusion assay in which the phagocytic probe consisted on biotinylated dead LM bacteria and the endocytic probe on avidin conjugated HRP[18] was described in Figure 1 (panel B). This assay confirmed those parameters previously reported with the other cell-free in vitro fusion assay. However, new information obtained with this assay indicated that Rab5 was the key regulator of early phagocytic fusion events, requiring the GTP-form of the protein but not GTP-hydrolysis. These parameters were similar to those reported for endosome-endosome fusion.[21,22] This assay also revealed that Rab5 association with phagosomes and function preceded and regulated NSF binding to phagosomes. The role of Rab5, ATP and NSF was confirmed by two recent reports using similar in vitro fusion assays as in Figure 1 (panel B). The first one used as the phagocytic probe avidin-conjugated latex beads and as the endocytic probe, biotinylated-HRP.[19] The second one used dead biotinylated-*Salmonella* phagosomes and avidin-HRP loaded endosomes.[23]

Phagosome-Endosome Fusion with Pathogens

The two pathogenic models for phagosome-endosome fusion reported to date: *Listeria monocytogenes* and *Salmonella typhimurium* used the in vitro system described in Figure 1B. Albeit the marked differences between both pathogens, both pathogenic fusion systems showed similarities on regulatory parameters[8,23] and differences with the dead-bacteria systems as indicated in Table 1. The system with live pathogens presented faster rates of fusion due to efficient recruitment of Rab5 and NSF and was ATP and cytosolic NSF independent. These differences between the live and dead bacteria systems suggested that live bacteria were driving their own fusion events and providing their energy source. Another explanation suggests that live bacteria phagosomes were predocked with all the machinery that allows fusion to proceed at a higher and more efficient manner.

Phagosome-Lysosome Fusion

In general, fusion of lysosomes with phagosomes is regulated by the same parameters as fusion with endosomes (Fig.1) independently of the system used in vivo, in vitro, in vivo-in vitro combination[8,12,19,23,25] or SLO-permeabilized cells.[26] The fusion system required ATP, cytosolic proteins, NSF and microtubules. Live pathogens such as *Salmonella* or *Listeria* presented very low percentages of fusion with lysosomes while their dead bacteria counterparts clearly fused with them. In vivo systems for phagosome-lysosome fusion and in vitro analysis indicated that the Rab regulator depended on the microorganism, Rab7 for dead *Salmonella*[9] and Rab5 for dead *Listeria* or latex beads phagosomes.[8,20,25] This latter Rab5 regulated phagosome-lysosome fusion was also reported for pathogenic *Listeria* phagosome-lysosome fusion on activated MØs.[12] It is possible that the size of the particles explained the discrepancies on the

Table 1. Differences between dead and live pathogens in phagosome-endosome fusion

Condition[b]	Relative Fusion of Phagosomes with[a] Dead-Pathogens	Live-Pathogens
control	++	++
- cytosol	-	-
+ trypsin	-	-
- ATP	-	++
+ NEM	-	++
+ anti-NSF	-	++
+ KCl wash	-	+/-
- cytosolic-Rab5	+/-	-
+ anti-Rab5	-	-
+ GDI	+/-	-
time required	20-25 min	5 min

[a]Isolated phagosomes containing biotinylated dead or live bacteria and isolated endosomes containing avidin-conjugated HRP were resuspended in complete fusion buffer supplemented with cytosolic proteins (1 mg/ml) otherwise indicated. Vesicles and cytosol were incubated 60 min. at 37°C to allow fusion. Membranes are lysed in the presence of avidin-insulin as scavenger and the HRP activity associated with the bacteria reflected the extent of fusion. Results indicated the relative fusion compared to controls that have the highest value (++) corresponded to 1.00 relative units (OD units/ mg of protein), (+) value corresponded to 0.5 relative units, (+/-) corresponded to 0.25 relative units and (-) to less than 0.1 relative units.
[b]Conditions tested were the following: control, 1 mg/ml cytosolic proteins;-ATP, an ATP-depleting system (5 mM glucose, 25 U/ml hexokinase) substituted the ATP-regenerating system;-cytosol, no cytosolic proteins are added; + trypsin, both set of vesicles were incubated (1 h, 4°C) with 10 µg/ml trypsin. To quench excess of trypsin, the samples were incubated (30 min. 4°C) with a trypsin inhibitors cockail; + NEM, vesicles and cytosol were incubated with 3 mM NEM (30 min. 4°C) and excess was quenched with 3 mM DTT; + anti-NSF, mAb 4A6 (0.1 µg/assay) was added to the fusion reaction; + anti-Rab5, mAb 4F11 (0.1 µg/assay) was added to the fusion reaction; + GDI, phagosomes were pretreated with purified GDI (6 µg/assay, 20 min, 30°C) in the presence of 1 mM GDP and untreated endosomes; + KCl wash, phagosomes were incubated with 0.5 M KCl (10 min. 4°C). Vesicles were sedimented to remove excess salt and resuspended in complete cytosol with ATP-depleting system;-cytosolic-rab5, 1 mg/ml cytosolic proteins were incubated with 1 µg of mAb 4F11 and protein A-sepharose beads (2 h, 4°C). Beads were sedimented and supernatants used in the fusion reaction; time required, fusion reactions were performed as in controls with 1 mg/ml cytosolic proteins. At different times (2 min-60 min) fusion reaction were stopped and HRP activity associated to bacteria measured. Results reflected the minimum time required to achieve a highest relative fusion value of 1.00 (++).

Rab protein regulators for these fusion events, as size influences the fate and processing of phagocytic particles.[15,27] Large particles as *Salmonella* (2-4 µm) may require Rab7;[9] while smaller particles as latex beads or *Listeria* (1 µm) required Rab5.[8,12,19,25]

Pathogen Interference with Trafficking and Immune Response

Although pathogen interference with intracellular trafficking may have the purpose to increase bacterial survival, it can also drive to evasion of the immune system. This immune evasion can involve inhibition of antigen processing and presentation, since the peptide-MHC-class II loading compartment coalesced in the endosomal/lysosomal routes with pathogen compartments. In brief, peptide-MHC class II loading is achieved after the following sequence of events that can be partially explained in Figure 2. First, MHC-class II molecules composed by α and β chains are assembled in the ER. Here, class II α/β heterodimers bind to the invariant chain (Ii). The complexes Ii-class II α/β heterodimers are transorted along the endocytic pathway where Ii is gradually degraded by endosomal protesases. This degradation generates a Ii-derived peptide: CLIP, which occupies the class II peptide-binding site. CLIP-class II complexes accumulate in a prelysosomal compartment (late endosomes, Le), the MHC-II loading compartment. Here, CLIP is removed by interaction of the class II complex with the human HLA-DM (known as H2-M in the murine system). DM stabilizes the empty form of class II molecules and promotes the exchange of an exogenous peptide for CLIP (Ag peptide in Fig. 2). Once an exogenous peptide is bound tightly by class II molecules, the occupied class II molecule acquires a stable conformation. Since Ii is dissociated from the class II complex, the peptide-class II complexed is transported to the cell surface. Molecular stability is detected biochemically by resistance to dissociation of α and β chains in the presence of SDS at room temperature. CLIP-loaded or empty class II molecules dissociate under these conditions (see class II unstable molecules in Fig. 2). As it can be expected there are several stages of this process where pathogens can interfere.[28] The first stage of interference can be located at the level of early endosomes or sorting endosomes (see Fig. 2) after Ii-class II complexes have been transported to this comparmtent. Further processes can be inhibited by pathogens that interfere with the maturation of early endosomes to late endosomes, such as *Listeria monocytogenes*. In an earlier level as the sorting of Ii-class II complexes in the TGN or also at this stage may be localized the *Chlamydia* interference. On a latter stage, at the level of DM stabilization; CLIP release and exchange by exogenous pepdides; processes that occur in a late endocytic compartment; might interfere *Mycobacterium, Salmonella* and *Helicobacter*. The *Coxiella* interference can be located a further stage since this pathogen compartment already contained class II-DM complexes. The pathogen interference occurs in a phagolysosome by increasing the class II-DM association rate, leading to empty unstable class II molecules. Moreover, this MHC-class II antigen processing inhibition may facilitate chronic infections by decreasing T cell responses to microbial antigens. Whatever the case, these interferences (e.g., intracellular trafficking and antigen processing and presentation) might be related. A summary of our point of view to integrate both interferences is shown in Figure 2. In general, it can be assumed that pathogens interfering with antigen processing and presentation should be responsible for chronic infections or at least, presenting chronic phases on their infection. Although the list of pathogens is probably very large, here we have focused exclusively on those ones whose intracellular targets are known. We have included in our description some pathogens as *Brucella* or *Legionella* that diverged from the intracellular trafficking routes but have not been reported to inhibit antigen processing. As well as *Yersinia*, a pathogen with a good knowledge on the intracellular interference mechanism and bacterial effectors, inhibit their own phagocytosis, and therefore it cannot be considered as intracellular interfering pathogens.

The phenomenon of antigen processing inhibition is not restricted to live bacteria but also to structural compounds present onto dead bacteria[29] also capable of interfering. This inhibition is especially relevant for MØs, while unappreciated for other APCs (e.g., B or dendritic cells).[30] Two chemical saccharide structures resulted especially inhibitory by their accumulation onto MØ lysosomes and reduction of lysosomes fusion capabilities: polysaccharides with (i) two anomeric carbon atoms of two interlinked sugars or with (ii) several sulfate groups per disaccharide repeating unit. The glycolipids from *Mycobacterium tuberculosis*, the polysialic acid capsule of *Neisseria meningitidis* and the peptidoglycan of *Listeria monocytogenes* belonged to the first group and to the latter group: *Mycobacterium* spp. and *Chlamydia trachomatis*.[31,32] In spite of the relevance of this chemical structure interference, we will describe here the interference caused by live pathogens.

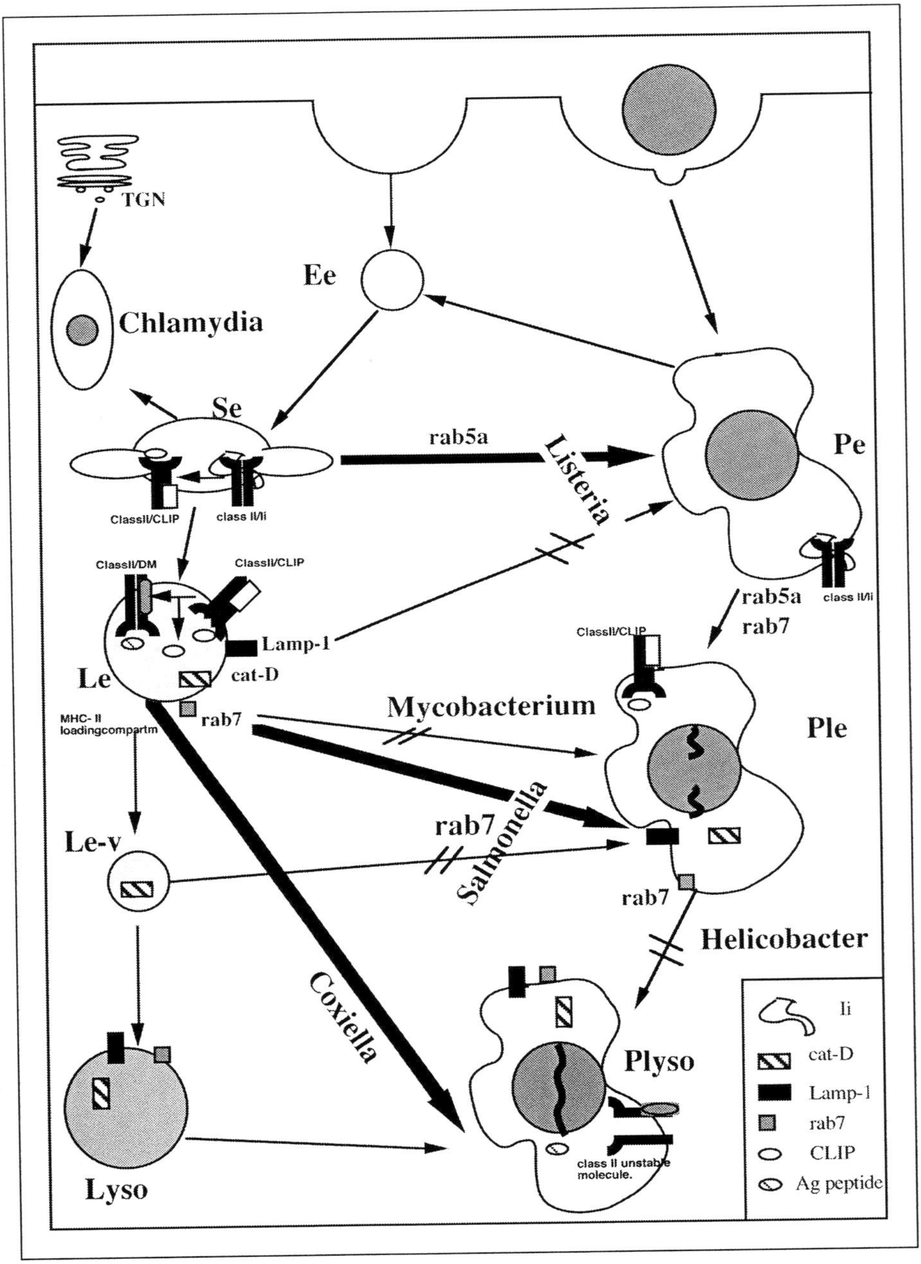

Figure 2. Please see legend on following page.

Figure 2. Pathogen interference with intracellular trafficking and antigen presentation. After phagosome formation, early phagosomes (Pe) may fuse with different endocytic vesicles such as early (Ee) or sorting endosomes (Se). Within these Se endosomes class II molecules complexed to the invariant chain, Ii, (class II/Ii) are degraded sequentially by endosomal proteases to form an Ii peptide, CLIP, which binds to the class II complex and prevents other peptides binding (class II/CLIP). From the Se compartment, *Chlamydia* vacuoles diverged to a specialized compartment interactive with the trans-Golgi network (TGN) and receiving sphingomyelin. This pathogen interferes with the sorting of Ii-class II complexes from the TGN to Se. The early endosomal trafficking regulator, Rab5a, regulates the fusion processes of Ee and Se with the Pe. Maturation of Pe implies interactions with late endocytic vesicles (Le or Le-v). Le is considered the MHC-II loading compartment. CLIP is released and exogenous peptide (stripped circles, called Ag peptide) bound to class II-complexes due to DM association with these complexes (class II/DM). The first interference corresponded to *Listeria* due to accumulation of Rab5a and upregulation of fusions with Se. Latter on Rab5a is locked and interactions with several Le-v are prevented, as those containing Lamp-1 or cathepsin-D (Cat-D). Nevertheless, some Le-v interactions are allowed, those containing Rab7. The interference with antigen processing should be located at the level of CLIP release and exogenous peptide loading. In a subsequent step from Pe to Ple will act also *Mycobacterium* that accumulated Rab5a but failed to localized Rab7 or cathepsin-D. However, some fusion with Le-v containing Lamp-1 may occur. Antigen processing inhibition is localized to a defective transport of immature class II complexes (class II/Ii or class II/CLIP). *Salmonella* interference can be localized in a latter step, this pathogen phagosomes contained transiently Rab5a, accumulated Rab7 and therefore lgps as Lamp-1, Lamp-2 or LAP. However, there is a selective inhibition to fuse with Le-v containing soluble lysosomal enzymes such as cathepsin-D. This strategy causes arrest of the phagosome progression to a Plyso, remaining in a specialized Ple compartment that inhibits antigen processing at the level of CLIP replacement by exogenous peptides. The putative bacterial proteins involved in the intracellular inteferences are: SopE and SipC. A subsequent stage of interference constituted *Helicobacter* inhibition, that also accumulates Rab7, H+-ATPase and lgps, while lacked soluble lysosomal enzymes as cathepsin-D. The function of Rab7 is inhibited by the pathogen that creates a swollen compartment arrested in a Ple compartment. This interference is controlled by the bacterial toxin, VacA, that is also involved in the antigen processing inhibition at the level of exogenous peptide loading. The last step is inhibited by *Coxiella* action. This pathogen promotes the maturation to a swollen Plyso compartment that interferes with antigen processing. The bacterial strategy consisted in increase the association of DM molecules with class II complexes, causing empty class II complexes that behave as unstable complexes (class II unstable molecules).

Brucella

Brucella is a facultative gram-negative intracellular bacterium that causes chronic infections. Its manifestation is presented as abortion and infertility in cattle and as undulant fever in humans. The bacterium remains in membrane-bound compartments in professional and non-professional phagocytes until the host cell dies. At first, *Brucella* phagosomes acquired endosomal markers, such as EEA1 but after transformation into an autophagosome (containing the autophagic marker MDC; the ER marker, sec61β and the late endosomal markers, lamp-1 and lamp-2) diverts from classical phagocytic routes and avoids fusion with endosomes and lysosomes. Then, the bacterium is translocated to a safe location, the ER, where it replicates.[33,34] In professional phagocytes, the pathway is different. Phagosomes delay fusion with resident lysosomes and with newly formed endosomes and lacked autophagosomal or ER markers.[35] There are no reports investigating its interference with antigen processing, therefore, we have excluded *Brucella* cycle from our draft on Figure 2. However, lack of reports does not necessarily means a lack on antigen processing interference.

Legionella pneumophila

Legionella pneumophila is a gram-negative bacterium, causative agent of Legionnaries disease. The host cells are professional phagocytes, including different alveolar MØs.[36]

After internalization by coiling phagocytosis, the bacterium is located in an autophagosome (markers were defined for *Brucella* autophagosome) that evades fusion with endosomes/lysosomes but interacts with other organelles. In summary, *Legionella* phagosomes lacked transferrin receptor, MHC-class II, β2-microglobuline, lamp-1, lamp-2 and cathepsin-D but contained 5'-nucleotidase or complement receptors (CR3). This autophagosome presents receptors to recruit Rab7, not requiring as a prerequisite acquisition of Rab5. Rab7 is lost later on upon transformation of the autophagosome into a replication vacuole, a ribosome-studded compartment.[36-39] Two genes encoded by a type-IV secretion system control the interference of *Legionella* with the classical maturation routes and promote survival: the *dot* and the *icm* genes. The strategy may imply factors transported by these genes to remodel the phagosome into a specialized organelle that recruits nutrients and additional membranes to support bacterial growth inside. The phagosomal features of this bacterium suggested interference with antigen processing and presentation or at least, meant that antigen processing takes place outside the phagosome.[40] As the case of *Brucella* there are no reports yet, therefore we have excluded from our draft on intracellular trafficking and antigen processing interference of Figure 2.

Coxiella

Coxiella burnetii is the etiological agent of Q fever.[41] Acute phase in humans is presented as an atypical pneumonia or a prolonged flu but persistent bacteria may result in a chronic phase characterized by circulating immune complexes inflamatory diseases. The intracellular cycle consisted in living and replicating within a low pH (4.7-4.8) phagolysosome, with characteristics of secondary lysosomes (cathepsin-D, acid phosphatase, Lamp-1 and Lamp-2 positive phagosomes). These phagolysosomes are not normal sized but swollen lysosomes causing anomalies in the triggering of the respiratory burst.[42] Two distinct morphological variants of the pathogen played a role in survival: a resistant to extracellular stress form that upon internalization transformed into a metabolically active form, sensitive to environmental changes, that delays phagolysosomal fusion.[43,44] The fact that this bacterium causes a chronic disease suggests an ability to evade a vigorous host immune response albeit residing in a late-endosome/lysosome. These phagosomes contained MHC class II and DM molecules, features of the MHC class II loading compartment. The bacterium interferes with antigen processing by enhancing the association between MHC class II and DM molecules that promotes unstable MHC class II molecules and prevent the normal loading of antigenic peptides.[45] A schematic representation is shown in Figure 2.

Chlamydia

Chlamydia is the causative agent of several significant human diseases (e.g., trachoma by *C. trachomatis*, sexually transmitted diseases or pneumonia by *C. pneumoniae*).[36] These bacteria undergo a biphasic developmental cycle within membrane-limited parasitophorous vacuoles, termed inclusions. The bacterium form that is internalized is the environmentally stable elementary body (EB) infectious form, which inside the cells differentiates into the reticulated bodies (RB) which then divide into progeny. Individual vacuoles containing EBs fuse to form a large perinuclear vacuole, the inclusion. The chamydial inclusion is distinct from phagolysosomes and does not display endocytic markers (MHC class II and DM negative vacuoles), being totally segregated from the endosomal-lysosomal pathway or recycling routes. The inclusion, however, fuse extensively with secretory vesicles from the TGN to recruit sphingomyelin,[36] maybe controlled by a specific set of Rabs.[43] *Chlamydia* interference with antigen processing was first observed by a decrease expression of MHC class II molecules in infected cells, recently described as a MHC class II molecules degradation by interfering with the sorting of these molecules in the TGN.[28,46]

Helicobacter

Helicobacter pylori is a gram-negative bacterium that colonizes the gastric epithelium of at least 50% of the world's population and can cause a chronic infection of the stomach mucous, a gastroduodenal ulcer with tumorogenic processes.[47,48]

Phagocytosis and intracellular trafficking of *Helicobacter pylori* is characterized by homotypic phagosome fusion that creates a stable megasome with multiple viable bacilli, key event for its survival.[47] This bacterium interferes with the function of Rab7 in late transport events, forming compartments endowed with the vacuolar ATPase and with membrane markers of both late endosomes and lysosomes (Rab7 and lgps positive but CI-M6PR negative phagosomes).[49] This interference is caused by a bacterial secreted toxin, VacA, reponsible for inhibition with the intracellular trafficking and with T cell epitopes presented by newly synthesized class II molecules (the invariant chain dependent pathway). Antigen presentation of epitopes by the class II molecules recycling pool (invariant chain independent pathway) was not affected.[50]

Listeria monocytogenes

Listeria monocytogenes is a gram-positive facultative intracellular bacterium that invades professional and non-professional phagocytes and are responsible for severe pathologies in immunocompromised people, newborns and pregnant women.[51] During the short (30 min) intraphagosomal period the bacterium creates a favorable environment for its survival characterized by two steps: (i) Alive bacterium avoids phagosome maturation by inactivation of Rab5a and inhibition of lysosomal interactions and lysosomal proteins recruitment (cathepsin-D, Limp-II or Lamp-1 negative phagosomes).[8,12] (ii) Combined secretion of listeriolysin and PI-PLC lyses the phagosomal membrane. This interference with the classical maturation routes also correlated with antigen processing inhibition. Structural components of the cell wall or secreted toxins such as listeriolysin may be responsible for the antigen processing interference. The probable stage of interference localizes to the coupling of antigenic peptides with MHC class II molecules.[28,52] In fact, the features of the LM phagosome as an early compartment that interferes with transport to late endosomal vesicles, may easily explain the inhibition described above on antigen processing. It is possible that LM antigen processing interference favors the development of the chronic phase of LM infection, especially in immunocompromised subjects (e.g., SCID mice or AIDS patients).

Mycobacterium

Mycobacterium spp. is an intracellular bacillus that enters and multiplies within monocytes and MØs. This bacillus-host interaction can lead to elimination of the pathogen, long-term containing of the bacilli (latent infection) or active disease (primary or reactivated tuberculosis).[53] Following phagocytosis, *Mycobacterium* spp. resides and multiplies in phagosomes that resist fusion with lysosomes and are only mildly acidified due to removal of the proton ATPase. The phagosome is arrested along the maturation routes by accumulating Rab5 and early markers and avoiding the presence of Rab7, inhibiting Rab7 function from early to late compartments. These phagosomes contain MHC-class II molecules, transferrin receptor and Lamp-1, but lacked Rab7 and M6PR[54,55] as described in Figure 2. The interference with antigen processing is reflected by a decrease on MHC-class II and DM molecules expression and related to defective trafficking of immature class II molecules from the TGN. This inhibition occurs for both types of T cell epitopes, those depending on newly synthesized MHC class II molecules and those depending on the recycling pool.[56] This impairment on antigen processing may facilitate a chronic infection maintained by the infected MØs as reservoir that reactivate after years and help to hide the infection from the T cell effector signaling machinery of the immune system.

Salmonella spp.

Salmonella spp. are gram-negative facultative intracellular bacteria that trigger their uptake in phagocytic and non-phagocytic cells by macropinocytosis.[3] *Salmonella* modulates the phagosomes to survive inside and replicate after a lag of several hours. Recent studies have demonstrated that *Salmonella* bypassed the normal endocytic route and persisted in a specialized compartment of low pH. Early markers are found but lost rapidly such as Rab5, TfR and EEA1 and the vacuole becomes enriched on lgps such as, Lamp-1, Lamp-2, LAP. However, these phagosomes lacked M6PR and soluble lysosomal proteins as cathepsin-D. The mechanism to accumulate lgps onto the vacuoles is regulated by the accumulation of Rab7 that triggers recruitment of transport vesicles coming from the late endosomes.[57] The intracellular trafficking of the bacterium inside MØs is slighty different. Fusion with lysosomes is avoided and fusion with endosomes was promoted by the accumulation of Rab5-NSF and the exclusion of Rab7.[23] *Salmonella* spp. harbor a especialized type III secretion system that allows for direct translocation of several bacterial proteins directly into the host cell. Two proteins known as SopE and SptP have been reported as the proteins responsible for the cytoskeleton rearrangements characteristic of this bacterium entry.[58,59] They act as effectors of Rho-exchange and GAP activities, respectively. However, SopE, stimulator of Rho-exchange activity which is secreted into the MØ cytosol, has also been found to bind Rab5, therefore maybe favoring the presence of Rab5-GTP onto these phagosomes.[23] Another type III secretion protein with a known effect on intracellular trafficking is SpiC, a protein secreted to the MØ cytosol that inhibits phagosome-lysosome fusion.[60] It is possible that the alteration on the phagosome exerted by *Salmonella* also interferes with MHC-class II antigen presentation.

Yersinia

The pathogenic *Yersiniae* include enteric pathogens and the causative agent of bubonic plague, *Yersinia pestis*. All the species are facultative intracellular pathogens but the bacterium does not grow intracellularly, except *Y. pestis* that enters and grows inside macrophages.[3,61] The pathogenicity of *Yersinia* results from its impressive ability to overcome the defenses of the mammalian host and to overwhelm it with massive growth.[62] *Yersinia* internalization occurs by a zipper-type mechanism in which bacterial surface proteins bind to host cell surface receptors,[53] and it uses the no-acidification of its phagosome as a way of survival inside the cell. *Y. pseudotuberculosis* phagosome has the same amount of vacuolar ATPase as other phagosomes, but there is reduced activity of this proton pump.[63]

Yersinia also uses a type III secretion system to inject bacterial proteins into eukaryotic cells integrated in the Yop virulon. There are three effectors of these virulon that exert a negative role on cytoskeleton dynamics and, thereby, contributing to the resistance of *Yersinia* to phagocytosis by macrophages. Yop H is specifically targeted to the focal complexes, Yop E disrupts actin filaments and Yop T exerts a dramatic depolymerizing effect on actin by modifying RhoA, the small GTPase that regulates the formation of stress fibers.[62] In difference to *Salmonella*, this bacterium uses these gene products to inactivate macrophages, disrupting phagocytosis. Therefore, *Yersinia* escapes antigen presentation by class II by avoiding internalization by the macrophage.

Cytokines and Phagosome-Endosome/Lysosome Interactions

MØs are professional phagocytes well armed for pathogen destruction. The most powerful armament against pathogens is triggered by several cytokines, a process known as macrophage activation. Thus, cytokines such as IFN-γ, TNF-α, IL-6, IL-1β, GM-CSF and G-CSF are critical in host resistance against bacterial infections. Among all these cytokines, IFN-γ is perhaps the best documented with a clear bactericidal ability. The bactericidal actions accompanying IFN-γ activation were: (i) phagosomal remodeling to confine pathogens to this environment and restrict in their growth (i.e., *Listeria* or *Mycobacterium*).[64-66] (ii) Assembly of the phagocyte NADPH oxidase to produce toxic molecules and (iii) induction of MHC class II molecules expression.

Regarding phagosome remodeling, few molecules mediated this bactericidal capability: (i) Nramp1, a MØ-restricted protein preferentially localized to late endosomes/lysosomes and involved in destruction of *Leishmania*, *Mycobacterium* and *Salmonella*.[67-69] (ii) IGTP, a recently described IFN-γ inducible protein critical for *Toxoplasma* clearance and (iii) Rab5a, a small GTPase that upon IFN-γ induction is involved in *Listeria monocytogenes* clearance.[12] All three of them may have a role in phagosome-endosome/lysosome fusion.[10,12,19,67] It is possible that, the IFN-γ inducible Rab5a small GTPase acquires novel functions upon cytokine stimulation. Not only fusion of phagosomes with lysosomes, but also the phagocyte NADPH oxidase assembly with production of toxic molecules due to rac2 activation and translocation to the phagosomes. The combination of these newly acquired rab5a functions confined the pathogen to the phagosomal environment and clearly compromised its viability.

Mycobacterium phagosomes, whose phagosome maturation is arrested[70] upon IFN-γ stimulation, advances these phagosomes to later stages of maturation, characterized by a lower pH, accumulation of vacuolar ATPases and fusion with lysosomes.[64]

The IFN-γ bactericidal action on the phagocyte NADPH oxidase assembly can be the target of some pathogens as *Coxiella brunetti* that diminished the respiratory burst accompanied by the swelling on the phagolysosomal vacuoles. This inhibition on the respiratory burst is probably caused by secreted bacterial proteins that caused a deficiency in the assembly of the phagocyte NADPH oxidase.[42]

Finally, although IFN-γ induction of MHC class II molecules cannot be categorized as a bactericidal action itself, it becomes the target of some pathogens not for survival purposes but for chronic phase establishment of their infections. For instance, *Mycobacterium* spp. reduced the expression of the MHC class II transactivator, CIITA that affects the MØ activation state. A similar effect was found for *Chlamydia*. This bacterium also inhibited MHC class II molecule expression[46] by degrading a downstream transcriptional factor, the USF-1. This factor acts after activation of the JAK/STAT pathways and is required for expression of the MHC class II transactivator, CIITA. Degradation of USF-1 by the bacterium resulted in diminished expression of CIITA and since CIITA is an obligate mediator for MHC class II gene transcription, degradation of USF-1 also suppressed MHC class II expression.

IFN-γ is not the exclusive cytokine that drives MØ activation, G-CSF is a MØ activating cytokine controlled by the expression of the transcriptional factor NF-IL-6. This transcriptional factor was shown to control clearly *Brucella abortus* infection and growth.[11] The mechanism involves: (i) endosome-phagosome fusion events putatively controlled by Rab5, (ii) translocation of respiratory burst (phox) elements present on endocytic compartments to the *Brucella* phagosomes and (iii) transformation of the phagosome into a hostile environment with reducing agents, acid hydrolase and radicals.

Future Applications

It is clear that the analysis of phagosome interactions with the endosomal/lysosomal compartment is critical for the pathogenesis of a great variety of parasites. The research performed in this area has provided not only with the parameters and regulators of these interactions but also have revealed a key information that can be considered as a new concept in cell biology: pathogens interfere with the fusion and transport machinery along the endosomal/lysosomal routes. It has been the extensive studies on this new area with different bacteria, what has revealed the regulators and more importantly, new functions for these regulators. On this regard, the recent studies with cytokines have been able to establish a correlation between cytokine signaling and phagosome-endosome or phagosome-lysosome interactions.[11,12,64-66] They also have revealed novel functions for endosomal regulators as Rab5a in the clearance of *Listeria*, the transport of lysosomal proteins and the activation and translocation of rac2 to the phagosomes with production of ROI listericidal molecules.[12] Another lysosomal protein, Nramp1, is also regulated by cytokines and involved in *Mycobacterium, Salmonella* and *Leishmania* clearance.[67-69] Also the GTPase, IGTP, is controlled by cytokines and critical for *Toxoplasma* destruction.[71] Nonetheless, exclusively Rab5a plays a role in controlling interactions with endocytic organelles as well as in pathogen elimination. The function of the other two molecules, Nramp1 and IGTP in the endosomal/lysosomal routes have not been studied yet. We believe this emerging area would reveal critical information for the innate immunity. Once bacterial molecules involved in these processes and trafficking targets would be unveiled, it will be possible to manipulate the system to increase the natural resistance against pathogens.

We also envision that those studies trying to characterize pathogen molecules responsible for intracellular trafficking interference as those reported for *Salmonella*,[23,58-60] *Legionella*,[39]

Helicobacter[49,50] and *Yersinia*[62] will be extended to other pathogens such as *Chlamydia, Coxiella, Listeria* or *Mycobacterium.* It is also possible that those bacterial intracellular trafficking interfering molecules will inhibit antigen processing as the case of VacA from *Helicobacter.* Common bacterial inhibitors for intracellular trafficking and antigen processing would constitute ideal molecules to incorporate on the vaccine design. The success of these vaccines would indicate those bacterial molecules that trigger the onset of acquired immunity. Moreover, those interfering molecules of common pathways may share structural analogy and even same intracellular targets. These latter molecules would be good candidates for designing live, recombinant, multimeric vaccines. On this regard some pathogens described in here have already been used on live recombinant vaccines such as *Listeria,*[72] *Salmonella* or *Mycobacterium.*[73]

References

1. Beron W, Alvarez-Dominguez, Mayorga LS et al. Membrane trafficking along the phagocytic pathway. Trends Cell Biol 1995; 5:100-104.
2. Silverstein SC, Greenberg S, DiVirgilio F et al. Phagocytosis. In: Paul WE, ed. Fundamental Immunology. New York: Raven Press, 1989:703-719.
3. Meresse S, Steele-Mortimer O, Moreno E et al. Controlling the maturation of pathogen-containing vacuoles: A matter of life and death. Nature Cell Biol 1999; 1:183-188.
4. Mayorga LS, Bertini F, Stahl PD. Fusion of newly formed phagosomes with endosomes in intact cells and in a cell-free system. J Biol Chem 1991; 266(10):6511-6517.
5. Desjardins M, Huber LA, Parton RG et al. Biogenesis of phagolysosomes proceeds through a sequential series of interactions with the endocytic apparatus. J Cell Biol 1994; 124(5):677-688.
6. Pitt A, Mayorga LS, Stahl PD et al. Alterations in the protein composition of maturing phagosomes. J Clin Invest 1992; 90(5):1978-1983.
7. Desjardins M, Celis JE, van Meer G et al. Molecular characterization of phagosomes. J Biol. Chem 1994; 269(51):32194-32200.
8. Alvarez-Dominguez C, Barbieri AM, Beron W et al. Phagocytosed live Listeria monocytogenes influences Rab5-regulated in vitro phagosome-endosome fusion. J Biol Chem 1996; 271(23):13834-13843.
9. Hashim S, Mukherjee K, Raje M et al. Live Salmonella modulate expression of rab proteins to persist in a specialized compartment and escape transport to lysosomes. J Biol Chem 2000; 275(21):16281-16288.
10. Alvarez-Dominguez C, Stahl PD. Interferon-gamma selectively induces Rab5a synthesis and processing in mononuclear cells. J Biol Chem 1998; 273(51):33901-33904.
11. Pizarro-Cerda J, Desjardins M, Moreno E et al. Modulation of endocytosis in nuclear factor IL-6(-/-) macrophages is responsible for a high susceptibility to intracellular bacterial infection. J Immunol 1999; 162(6):3519-3526.
12. Prada-Delgado A, Carrasco-Marin E, Bokoch G et al. IFN-γ listericidal action is mediated by novel rab5a functions at the phagosomal stage. J Cell Biol 2001; 276(22):19059-19065.
13. Harding CV. Phagocytic processing of antigens for presentation by MHC molecules. Trends Cell Biol 1995; 5:105-109.
14. Lang T, de Chastellier C, Ryter A et al. Endocytic membrane traffic with respect to phagosomes in macrophages infected with non-pathogenic bacteria: Phagosomal membrane acquires the same composition as lysosomal membrane. Eur J Cell Biol 1988; 46(1):39-50.
15. de Chastellier C, Thilo L. Phagosome maturation and fusion with lysosomes in relation to surface property and size of the phagocytic particle. Eur J Cell Biol 1997; 74(1):49-62.
16. Pitt A, Mayorga LS, Schwartz AL et al. Transport of phagosomal components to an endosomal compartment. J Biol Chem 1992; 267(1):126-132.
17. Beron W, Colombo MI, Mayorga LS et al. In vitro reconstitution of phagosome-endosome fusion: Evidence for regulation by heterotrimeric GTPases. Arch Biochem Biophys 1995; 317(2):337-342.
18. Castellano F, Montcourrier P, Chavrier P. Membrane recruitment of rac1 triggers phagocytosis. J Cell Sci 2000; 113(17):2955-2961.
19. Jahraus A, Tjelle TE, Berg T et al. In vitro fusion of phagosomes with different endocytic organelles from J774 macrophages. J Biol Chem 1998; 273(46):30379-30390.
20. Hackam DJ, Rotstein OD, Bennett M et al. Characterization and subcellular localization of target membrane soluble NSF attachment protein receptors (t-SNAREs) in macrophages. Syntaxins 2, 3, and 4 are present on phagosomal membranes. J Immunol 1996; 156(11):4377-4383.
21. Gorvel JP, Chavrier P, Zerial M et al. Rab5 controls early endosome fusion in vitro. Cell 1991; 64(5):915-925.

22. Barbieri MA, Li G, Colombo MI et al. Rab5, an early acting endosomal GTPase, supports in vitro endosome fusion without GTP hydrolysis. J Biol Chem 1994; 269(29):18720-18722.
23. Mukherjee K, Siddiqi SA, Hashim S et al. Live Salmonella recruits N-ethylmaleimide-sensitive fusion protein on phagosomal membrane and promotes fusion with early endosome. J Cell Biol 2000; 148(4):741-753.
24. Alvarez-Dominguez C, Roberts R, Stahl PD. Internalized Listeria monocytogenes modulates intracellular trafficking and delays maturation of the phagosome. J Cell Sci 1997; 110(6):731-743.
25. Alvarez-Dominguez C, Stahl PD. Increased expression of Rab5a correlates directly with accelerated maturation of Listeria monocytogenes phagosomes. J Biol Chem 1999; 274(17):11459-11462.
26. Funato K, Beron W, Yang CZ et al. Reconstitution of phagosome-lysosome fusion in streptolysin O-permeabilized cells. J Biol Chem 1997; 272(26):16147-16151.
27. Oh YK, Alpuche-Aranda C, Berthiaume E et al. Rapid and complete fusion of macrophage lysosomes with phagosomes containing Salmonella typhimurium. Infect Immun 1996; 64(9):3877-3883.
28. Brodsky FM, Lem L, Solache A et al. Human pathogen subversion of antigen presentation. 1999; Immunol Rev 168:199-215.
29. Leyva-Cobian F, Unanue ER. Intracellular interference with antigen presentation. J Immunol 1988; 141(5):1445-1450.
30. Gonzalez-Fernandez M, Carrasco-Marin E, Alvarez-Dominguez C et al. Inhibitory effects of thymus-independent type 2 antigens on MHC class II-restricted antigen presentation: Comparative analysis of carbohydrate structures and the antigen presenting cell. Cell Immunol 1997; 176(1):1-13.
31. Leyva-Cobian F, Outschoorn IM, Carrasco-Marin E et al. The consequences of the intracellular retention of pathogen-derived T-cell-independent antigens on protein presentation to T cells. Clin Immunol Immunopathol 1997; 85(1):1-15.
32. Harding CV, Unanue ER. Antigen processing and intracellular Ia. Possible roles of endocytosis and protein synthesis in Ia function. J Immunol 1989; 142(1):12-19.
33. Pizarro-Cerda J, Moreno E, Sanguedolce V et al. Virulent Brucella abortus prevents lysosome fusion and is distributed within autophagosome-like compartments. Infect Immun 1998; 66(5):2387-2392.
34. Pizarro-Cerda J, Meresse S, Parton RG et al. Brucella abortus transits through the autophagic pathway and replicates in the endoplasmic reticulum of nonprofessional phagocytes. Infect Immun 1998; 66(12):5711-5724.
35. Arenas GN, Staskevich AS, Aballay A et al. Intracellular trafficking of Brucella abortus in J774 macrophagues. Infect Immun 2000; 68(7):4255-4263.
36. Sinai AP, Joiner KA. Safe haven: The cell biology of nonfusogenic pathogen vacuoles. Annu Rev Microbiol 1997; 51:415-462.
37. Clemens DL, Lee B, Horwitz MA. Mycobacterium tuberculosis and Legionella pneumophila phagosomes arrested maturation despite acquisition of rab7. Infect Immun 2000; 68(9):5154-5166.
38. Clemens DL, Lee B, Horwitz MA. Deviant expression of rab5 on phagosomes containing the intracellular pathogens Mycobacterium tuberculosis and Legionella pneumophila is associated with altered phagosomal fate. Infect Immun 2000; 68(5):2671-2684.
39. Coers J, Monahan C, Roy CR. Modulation of phagosome biogenesis by Legionella pneumophila creates an organelle permissive for intracellular growth. Nature Cell Biol 1999; 1:451-453.
40. Ojcius DM, Gachelin G, Dautry-Varsat A. Presentation of antigens derived from microorganisms residing in host-cell vacuoles. Trends Microbiol 1996; 4:53-58.
41. Hackstadt T. The diverse habitats of obligate intracellular parasites. Curr Op Microbiol 1998; 1:82-87.
42. Baca OG, Akporiaye ET, Rowatt JD. Possible biochemical adaptations of Coxiella burnetti for survival within phagocytes: Effect of antibody. In: Leive L, Schlesinger D, eds. Microbiology. Washington D.C.: American Society for Microbiology, 1984:269-272.
43. Heinzen RA, Scidmore MA, Rockey DD et al. Differential intecaction with endocytic and exocytic pathways distinguish parasitophorous vacuoles of Coxiella burnetii and Chlamydia trachomatis. Infect Immun 1996; 64(3):796-809.
44. Howe D, Mallavia LP. Coxiella burnetii exhibits morphological change and delays phagolysosomal fusion after internalization by J774.1 cells. Infect Immun 2000; 68(7):3815-3821.
45. Lem L, Riethof DA, Scidmore M et al. Enhanced interaction of HLA-DM with HLA-DR in enlarged vacuoles of hereditary and infectious lysosomal diseases. J Immunol 1999; 162:523-532.
46. Zhong G, Fan T, Liu L. Chlamydia inhibits interferon-γ inducible major histocompatibility complex class II expression by degradation of upstream stimulatory factor 1. J Exp Med 1999; 189:1931-1937.
47. Allen LH, Schlesinger LS, Kang B. Virulent strains of Helicobacter pylori demonstrated delayed phagocytosis and stimulate homotypic fusion in macrophages. J Exp Med 2000; 191(1):115-127.

48. de Bernard M, Arico B, Papini E et al. Helicobacter pylori toxin Vac A induces vacuole formation by acting in the cell cytosol. Mol Microbiol 1997; 26(4):665-674.
49. Papini E, Satin B, Bucci C et al. The small GTPase binding protein rab7 is essential for cellular vacuolation induced by Helicobacter pylori cytotoxin. EMBO J 1997; 16:15-24.
50. Molinari M, Salio M, Galli C et al. Selective inhibition of Ii-dependent antigen presentation by Helicobacter pylori toxin VacA. J Exp Med 1998; 187:135-140.
51. Lorber B. Listeriosis. Clin Infect Dis 1997; 24:1-11.
52. Cluff CW, Garcia M, Ziegler HK. Intracellular hemolysin-producing Listeria monocytogenes stratins inhibit macrophage-mediated antigen processing. Infect Immun 1990; 58:3601.
53. Deretic V, Fratti RA. Mycobacterium tuberculosis phagosome. Mol Microbiol 1999; 31(6):1603-1609.
54. Via LE, Deretic D, Ulmer RJ et al. Arrest of mycobacterial phagosome maturation is caused by a block in vesicle fusion between stages controlled by rab5 and rab7. J Biol Chem 1997; 272(20):13326-13331.
55. Clemens DL, Horwitz MA. Characterization of the Mycobacterium tuberculosis phagosome and evidence that phagosomal maturation is inhibited. J Exp Med 1995; 181:257-270.
56. Hmama Z, Gabuthuler R, Jefferies WA et al. Attenuation of HLA-DR expression by mononuclear phagocytes infected with Mycobacterium tuberculosis is related to intracellular sequestration of immature class II heterodimers. J Immunol 1998; 161:4882-4893.
57. Meresse S, Steele-Mortimer O, Finlay BB et al. The rab7 GTPase controls the maturation of Salmonella typhimurium-containing vacuoles in HeLa cells. EMBO J 1999; 18(16):4394-4403.
58. Hardt WD, Chen LM, Schuebel KE et al. S. typhimurium encodes an activator of Rho GTPases that induces membrane ruffling and nuclear responses in host cells. Cell 1998; 93(5):815-826.
59. Fu Y, Galan JE. A salmonella protein antagonizes Rac-1 and Cdc42 to mediate host-cell recovery after bacterial invasion. Nature 1999; 401(6750):293-297.
60. Uchiya K, Barbieri MA, Funato K et al. A Salmonella virulence protein that inhibits cellular trafficking. EMBO J 1999; 18(14):3924-3933.
61. Falkow S, Isberg RR, Portnoy DA. The interaction of bacteria with mammalian cells. Annu Rev Cell Biol 1992; 8:333-363.
62. Cornellis GR. Molecular and cell biology aspects of plague. Proc Natl Acad Sci USA 2000; 97 (16):8778-8783.
63. Tsukano H, Kura F, Inoue S et al. Yersinia pseudotuberculosis blocks the phagosomal acidification of B10.A mouse macrophages through the inhibition of vacuolar H^+-ATPase activity. Microbial Pathogen 1999; 27:253-263.
64. Schaible UE, Sturgill-Koszycki S, Schlesinger PH et al. Cytokine activation leads to acidification and increases maturation of Mycobacterium avium-containing phagosomes in murine macrophages. J Immunol 1998; 160(3):1290-1296.
65. Via LE, Fratti RA, McFalone M et al. Effects of cytokines on mycobacterial phagosome maturation. J Cell Sci 1998; 111(7):897-905.
66. Portnoy DA, Schreiber RD, Connelly P et al. Gamma interferon limits access of Listeria monocytogenes to the macrophage cytoplasm. J Exp Med 1989; 170(6):2141-2146.
67. Searle S, Bright NA, Roach TI et al. Localisation of Nramp1 in macrophages: Modulation with activation and infection. J Cell Sci 1998; 111(19):2855-2866.
68. Gruenheid S, Gros P. Genetic susceptibility to intracellular infections: Nramp1, macrophage function and divalent cations transport. Curr Op Microbiol 2000; 3(1):43-48.
69. Vidal S, Gros P, Skamene E. Natural resistance to infection with intracellular parasites: molecular genetics identifies Nramp1 as the Bcg/Ity/Lsh locus. J Leukoc Biol 1995; 58(4):382-390.
70. Sturgill-Koszycki S, Schaible UE, Russell DG. Mycobacterium-containing phagosomes are accessible to early endosomes and reflect a transitional state in normal phagosome biogenesis. EMBO J 1996; 15(24):6960-6968.
71. Taylor GA, Collazo CM, Yap G et al. Pathogen specific loss of host resistance in mice lacking the IFN-gamma inducible gene IGTP. Proc Natl Acad Sci USA 2000; 97:751-755.
72. Jensen ER, Shen H, Wettstein FO et al. Recombinant Listeria monocytogenes as a live vaccine vehicle and a probe for studying cell-mediated immunity. Immunol Rev 1997; 158:147-157.
73. Kauffman SH, Hess J. Impact of intracellular location of and antigen display by intracellular bacteria: Implications for vaccine development. Immunol Lett 1999; 65(1-2):81-84.

CHAPTER 5

Macrophages:

Agents of Immunological Surveillance or Targets for Pathogens?

Rosângela P. da Silva, Sigrid Heinsbroek, Bongi Ntolosi and Siamon Gordon

Macrophages (MΦ) are professional phagocytes, capable of high endocytic rates and of specialised uptake mechanisms like phagocytosis and macropinocytosis.[1,2] Associated with these endocytic pathways are signalling cascades leading to the triggering of microbicidal activity, antigen processing/presentation and secretion of soluble mediators such as chemokines and cytokines.[3] Thus, MΦ represent one of the cornerstones of the immune system, linking the innate and acquired responses.

Competent defence against micro-organisms relies on appropriate activation of MΦ effector functions.[4,5] These processes also present pathogens with potential targets that can be exploited in order to favour their own survival.[6] During the evolutionary interplay between micro-organisms and these cells, both non-obligate and obligate intracellular pathogens have had to develop specific strategies to counteract endocytic and killing mechanisms and to subvert their ability to elicit an inflammatory response that can prime the immune response.

Pathogens like *Haemophilus,* staphylococci and streptococci have capsules that inhibit a MΦ phagocytic response by preventing contact. [7,8] Non-encapsulated micro-organisms (e.g., *Yersinia*) can also inhibit phagocytosis by producing toxins that interfere with actin polymerisation and tyrosine phosphorylation, disrupting the ensuing signal transduction pathways.[9] Others like *Candida albicans* (Fig.1), *Shigella, Listeria* and *Trypanosoma cruzi* are phagocytosed but have developed mechanisms for escaping from MΦ after uptake.[10-13]

Micro-organisms that require the intracellular milieu stow away in the endocytic pathway. They can induce their own uptake (e.g., *Salmonella*) and can disrupt the maturation of endocytic organelles that lead to lysosomal fusion, digestion and exocytosis (e.g., *Toxoplasma, Mycobacteria* and *Legionella*).[14-17] Pathogens have also developed strategies to avoid or resist microbicidal mechanisms and to disrupt cytokine secretion and signalling. Lastly, some pathogens like *Shigella, Yersinia* and *Salmonella* can induce apoptotic death of the host cell, including MΦ. Apoptosis after bacterial entry is a complex event that can elicit or suppress inflammatory responses and it is not clear whether it favours the host or the pathogen.

This Chapter will review MΦ specific endocytic receptors and effector mechanisms. These cellular processes are mediated by a series of cytosolic elements that, upon interaction with cytoplasmic domains of plasma membrane receptors, trigger a series of signalling and membrane trafficking cascades that ultimately lead to enhanced endocytic uptake, killing of micro-organisms and priming of the immune system. Due to their essential role in the anti-microbial response, they are often targeted by pathogens in their struggle to get ahead in the survival game. Where appropriate, we will briefly mention pathways targeted by pathogens, more detailed information on these can be found elsewhere in this book.

Intracellular Pathogens in Membrane Interactions and Vacuole Biogenesis, edited by Jean-Pierre Gorvel.

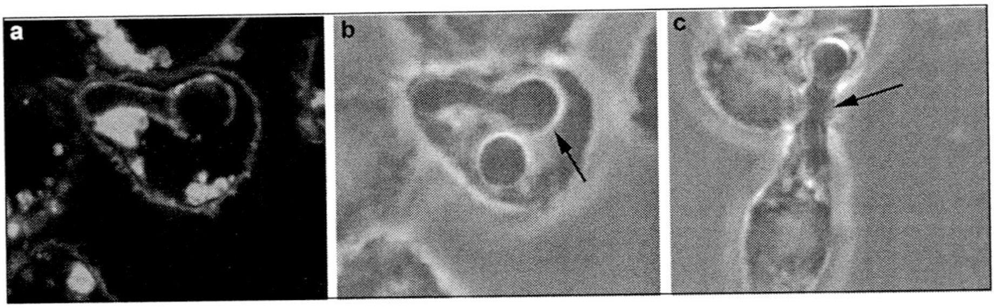

Figure 1.* Mϕ readily phagocytose the pathogenic yeast *Candida albicans*. The yeast recruits lysosomes to the periphery of the Mϕ and develops filaments within the Lysosomal Associated Membrane Protein (LAMP)-positive phagosome. Panels A and B show a filamentous form (arrow in panel B) growing within the LAMP+ phagosome. Panel A is a confocal micrograph of a RAW264.7 mouse M__ fluorescently labelled with rhodamine-phalloidin (red) and an anti-LAMP rat mAb plus a FITC-conjugated secondary Ab (green). Panel B is the correspondent confocal transmission image. Eventually filamental growth ruptures the phagosome and plasma membrane. Candida filamentous forms are invasive and can actively penetrate cells. Panel C is a transmission confocal micrograph of RAW264.7 cells showing a Candida filament (arrow) that, having ruptured the cell initially infected, is now invading an adjacent cell.
* These confocal micrographs were produced in the Sir William Dunn School of Pathology Imaging Suite, which is funded by a grant from The Wellcome Trust.

Macrophage Phagocytic Receptors

Central to phagocytosis by MΦ is the expression of a series of specific plasma membrane receptors that on the one hand greatly increase the range of particles able to interact with this cell, and on the other provide cytosolic sequences containing motifs that couple uptake to the effector response.[3,18] Further layers of complexity are added by the fact that interacting particles often engage more than one MΦ receptor and that these receptors can cooperate with one another.[19-21]

Most MΦ endocytic receptors are also "multi-tasking", having extracellular regions that contain domains capable of interacting with extracellular matrix and serum components, various endocytic ligands and phagocytic particles, including apoptotic cells and micro-organisms.[5,22-24] On the cytosolic face these receptors contain peptide sequences involved in the traditional clathrin-mediated endocytosis and also in responses as varied as binding, migration, macropinocytosis and phagocytosis, all accompanied by dramatic cytoskeletal reorganisation.[18,25] Phagocytic uptake can result from direct interaction with their targets, utilising structural determinants present on the surface of the particle (non-opsonic phagocytosis), or indirectly, by recognition of opsonins supplied by the host (opsonin-dependent phagocytosis). Table 1 shows a list of MΦ receptors reported to mediate phagocytic uptake.

Despite the complexity associated with different phagocytic mechanisms, a number of common features exist: particle internalization is initiated by the interaction of specific receptors on the surface of the phagocyte with ligands on the surface of the particle. This leads to the polymerisation of actin at the site of ingestion, and the internalisation of the particle via an actin-based mechanism. After internalization, actin is shed from the phagosome and the phagosome matures by a series of fusion and fission events with vesicles of the endocytic pathway, culminating in the formation of the mature phagolysosome (for reviews, see refs. 26 and 27). Since endosome-lysosome trafficking occurs primarily in association with microtubules, phagosome maturation requires the co-ordinated interaction of the actin- and tubulin-based cytoskeleton.[1]

During the process of phagocytosis, phagocytes convert from aerobic to anaerobic respiration. Although anaerobic glycolysis results in the formation of lactic acid, the consequent drop in pH within the phagosome is mainly due to translocation of the endosomal Na^+/H^+ ATPase.[28] This lowered pH may be partly responsible for the death and digestion of the phagocytosed particles because the hydrolytic lysosomal enzymes all have acid pH optima.

Table 1. Mϕ receptors involved in phagocytosis

Opsonin-Dependent Receptors	Opsonin-Independent Receptors
$Fc_\gamma R$ I, IIA, IIIA	Complement receptor 3 (CR3) *1
Low affinity IgE receptor (CD23, $Fc_\varepsilon RII$)	Macrophage mannose receptor
IgA receptor ($Fc_\alpha R$)	β-glucan receptor
High affinity IgE receptor ($Fc_\varepsilon RI$)	Macrophage scavenger receptors
Complement receptors 1 and 3 (CR1, CR3)	Phosphatidylserine receptor
Vitronectin receptor ($\alpha_v\beta3$)	CD14*2
	Receptor for Advanced Glycosylation End Products (RAGE)

*1 CR3 has an extra lectin-like domain that functions independently of opsonisation.
*2 CD14 ligand interaction can be facilitated by binding of LPS to LBP.

Phagocytic cells make use of toxic forms of oxygen to kill ingested microbial cells. Superoxide, formed by the reduction of O_2 by NADPH oxidase, is converted at the acidic pH of the phagosome by superoxide dismutase, to yield singlet oxygen and hydrogen peroxidase. The phagocytic enzyme myeloperoxidase forms hypochlorous acid (HOCl) from chloride ions and hydrogen peroxide (H_2O_2), and the HOCl reacts with a second molecule of H_2O_2 to yield additional singlet oxygen. Another toxic oxygen metabolite is NO^- which is produced by NO^- synthase on the cell-membrane. The combined action of these oxygen-dependent phagocyte enzymes forms sufficient levels of toxic oxygen compounds to kill ingested bacterial cells by oxidising cellular constituents.[29-31]

Although oxidative microbicidal mechanisms have received most attention recently and may even be predominant, oxygen-independent microbicidal mechanisms also exist. Mechanisms employed by MΦ include induction of indoleamine 2,3-dioxygenase to deplete intracellular tryptophan and sequestration of iron. Both elements are essential to pathogen survival. Phagocytes, particularly granulocytes also synthesise a host of potent antimicrobial peptides and enzymes (for a review, see ref. 32).

Opsonic Receptors

Fc Receptor-Mediated Phagocytosis

Fc receptors (FcRs) are opsonin-recognising receptors that recognise the Fc portion of immunoglobulins. Most of the current understanding of signalling pathways leading to phagocytosis by MΦ comes from studies of Fc receptor uptake. However some of these cascades have also been reported to play a role in uptake by other MΦ receptors (see below).

There are two major classes of FcRs: receptors that activate effector functions and receptors that inhibit them. MΦ FcRs that are known to mediate phagocytosis and fall within the activation class include Fc γRI, Fc ΦγRIIA and Fc γRIII(Ravetch JV, 1997). All three types are members of the Ig superfamily and are capable of independently triggering phagocytosis, although the contribution of Fc γRIIB to phagocytosis is presently unclear.[33]

Initiation of Signalling: The src and p72syk Kinases

Signalling is initiated by ligand binding and clustering of cell surface FcRs. Members of the src family of tyrosine kinases associate with FcRs and are probably responsible for the early phase of phosphorylation of tyrosine residues within a peptide sequence in the FcγR cytoplasmic tail or associated γchain known as immunoreceptor tyrosine-based activation motifs (ITAM).[1,34] This serves to recruit p72Syk, an SH2-domain-containing tyrosine kinase, which becomes activated to phosphorylate multiple substrates, including neighbouring ITAMs. P72Syk is recruited from cytosolic as well as plasma-membrane-associated pools. It is conceivable that a small number of pre-formed p72Syk-ITAM complexes exist in resting cells.[18] Src tyrosine kinases also trigger the phosphorylation of other substrates including Cbl, Grb2, Shc, Sos, CrkL and SLP76. These are adaptor proteins that form complexes involved in downstream responses such as NADPH oxidase and MAP kinase (MAPK) activation.[35-39] Cbl has been shown to be involved in the phagocytic response itself.[40]

The notion that phosphorylation of ITAM tyrosines is catalysed by src tyrosine kinases was first supported by pharmacological findings that herbimycin A, a tyrosine kinase inhibitor relatively specific for src family kinases, potently suppressed FcR mediated functions.[25,41] In addition, src family members were found to physically associate with resting FcR, and their catalytic activity was shown to be augmented by receptor aggregation[42,43] Targeted disruption of single or multiple src family genes provided further evidence supporting the role of src family kinases in FcR functions.[44,45] FcR-mediated phagocytosis by $Lyn^{-/-}Hck^{-/-}Fgr^{-/-}$ macrophages is considerably delayed, but still preserved. In contrast, FcR phagocytosis in $Syk^{-/-}$ macrophages is severely compromised.[44] Recent findings suggest that src family kinases are differentially involved in FcR-signalling and that selective kinases, including Lyn and Hck, catalyse FcR-mediated phagocytosis.[46]

Events parallel and downstream of p72Syk activation during phagocytosis are less well understood. However, it is clear that programmes of both actin assembly and membrane trafficking are needed to trigger cytoskeletal alterations, pseudopod extension and phagosomal closure. Activation of phosphoinositide-specific phospholipase C (PI-PLC), PI3-kinases (PI-3K), various protein kinase C (PKC) isoforms and of members of the Rho family of small GTPases are among the best documented events involved in these programmes.

Regulation of Membrane Trafficking and Endocytic Activity by PI-3K Activation and Generation of Phosphoinositides

Previous data have shown PI-3 kinase activation is necessary for pseudopodia extension and proper phagosomal closure during Fc γR-mediated uptake.[47-49] Moreover, bacterial entry into host cells was shown to depend on PI-3 kinase activation.[50] Recent studies exploring the complex changes in phosphoinositide metabolism and the identification of new protein-lipid interacting domains are beginning to explain the crucial role played by PI-3 kinases during phagocytic uptake.[51]

Sequential PI-3 kinase activation generates phosphoinositide 3,4,5-triphosphate [PI(3,4,5)P3] after the cleavage of PI(4,5)P2 into diacylglycerol and inositol 1,4,5 triphosphate. These phosphoinositides, generated at the cytosolic side of the plasma membrane, provide anchoring sites for several signalling and scaffolding elements containing plekstrin-homology domains(PH), epsin NH2-terminal homology domains(ENTH) and FYVE domains.[52,53] Cytosolic proteins containing these domains are involved in membrane trafficking during clathrin-mediated endocytosis, macropinocytosis and phagocytosis.[52] One particular example is early endosomal autoantigen 1 (EEA1), a FYVE domain containing protein that is also an effector of Rab 5. PI3K activity is required for EEA1 recruitment to early phagosomal membranes, where it then controls phagosomal maturation through its interaction with oligomeric membrane fusion complexes containing NSF, α SNAP and the SNARE Syntaxin 13. PI3K activation and EEA1 recruitment are targeted during Mycobacterial infection as a way of arresting phagosomal maturation.[54] Phosphoinositides generated by PI-PLC cleavage can activate PKC isoforms and induce calcium transients.[55]

Cross-linking of FcR during phagocytic uptake induces a marked reduction in lipid mobility at the phagocytic cup.[51] This may partly explain the selective accumulation of PI(4,5)P2 and PI(3,4,5)P3 residues at the cup and may constitute the molecular basis for the spatial restriction of signalling events preceding actin rearrangements . Finally, PI3 kinase activation is necessary for recruitment of amphiphysin IIm, which in turn recruits dynamin to the early phagosome.[56] An isoform of dynamin, dynaminIIa, has been shown to play an important role in phagocytosis.[57]

The Rho Family of Small GTPases Are the Main Regulators of Actin Assembly

Previous studies suggest PI3 kinases interact with the GTP-bound form of the small GTPase Rac.[58] The Rho family of small GTPases, of which Rac is a member, are the main regulators of actin polymerisation within cells.[59] Their differential activation has been linked to the generation of specific actin structures; Rho activation induces stress-fibre formation, Rac activation is associated with lamelipodia extension and membrane ruffling, and Cdc42 activation leads to filopodia formation.[60-62] Interestingly, Rac and Cdc42 activation have been shown to play an essential role in actin polymerisation during FcR phagocytosis.[63,64] It is also noteworthy that Rac-GTP can activate PI(4)P5 kinase, leading to the formation of PI(4,5)P2. This phosphoinositide can then not only be targeted by PLC, but also bind and release actin-binding proteins, leading to the formation of free actin ends and new polymerisation.[52] Perhaps not surprisingly, micro-organisms that use the intracellular environment to thrive in the host have developed very complex mechanisms to activate these small GTPases (e.g., *Salmonella* and *Shigella*). This leads to extensive membrane ruffling, characteristic of macropinocytic uptake, which then increases their chances of uptake by the host cell.[14]

The WASP and Arp2/3 Complex Proteins Are Effectors of the Rho GTPases in the Actin Assembly Pathway

Induction of actin polymerisation by the GTP-bound forms of the small Rho GTPases during phagocytosis leads to recruitment of the Arp2/3 complex. [65] This family of actin binding proteins had previously been shown to play a crucial role in dendritic nucleation of actin, which is the type of actin arrangement seen in lamellipodia and pseudopodia. Arp2/3 complex recruitment and activation is mediated by the interaction of members of the Wiskot-Aldrich Syndrome Protein (WASP)/ N-WASP/Scar family of proteins with the activated form of the small GTPase Cdc42. WASP proteins also bind to profilin, another actin-binding protein involved in filament elongation.[66]

There is abundant evidence implicating WASP and Arp2/3 complex proteins in the actin polymerisation which leads to phagocytic uptake. Mutations in these proteins produce defects in mobility and phagocytosis exhibited by phagocytes from patients suffering from Wiskot-Aldrich Syndrome.[67] Arp2/3 complex recruitment is required for both FcR and CR3 phagocytosis. Overexpression of a dominant-negative form of Scar (ScarWA) in COS cells has been shown to block Arp2/3 complex recruitment and phagocytosis via FcR. Phagocytosis via CR3 was also affected (see below). However, ScarWA overexpression in the J774.A1 MΦ cell line failed to block either Arp2/3 complex or F-actin recruitment and FcR phagocytic uptake remained intact.[68]

Recruitment and activation of WASP and Arp2/3 complex proteins has been reported as the mechanism employed by pathogens such as Listeria, Shigella and Vaccinia to induce actin polymerisation and propel themselves out of host cells. Enteropathogenic *Escherichia coli* (EPEC) have also developed mechanisms to activate WASP and Arp2/3 proteins leading to actin polymerisation and the formation of cellular structures known as pedestals. Pedestal formation pushes the plasma membrane and the attached bacterium away from the cells by distances of up to 10μm.[14]

Phagocytosis Is Coupled to Activation of Various PKC Isoforms

Another well documented event during FcR ligation and phagocytosis is the activation of several PKC isoforms.[69] PKC is a family of enzymes divided in three groups according to structure and co-factor requirement. Classical PKC isoforms α, βI, βII and γ require Ca^{2+}, diacylglycerol (DAG) and phosphatidylserine (PS) for activation. Novel PKC isoforms δ, ε, η and θ lack the Ca^{2+} requirement, but are activated by DAG and PS. The last group of PKCs, atypical isoforms ζ and ν λ, bind PS, but require neither Ca^{2+} nor DAG for activation (for a review see ref. 70). Of all these isoforms PKCs α, β, δ and ε have been shown to translocate to FcR membranes in RAW264.7 cells. However, FcR phagocytosis proceeds normally in the absence of Ca^{2+} signalling and in Ca^{2+}-depleted cells only PKCs δ and ε were able to translocate. Moreover phagocytosis was inhibited by general PKC inhibitors, but not by classical PKC-specific inhibitors, strongly suggesting a role for novel PKCs δ and ε in the FcR phagocytic response. PKC α activation seems to be required for other Ca^{2+}-dependent effector mechanisms such as the respiratory burst.[71] Another line of evidence shows that MARCKS, the major PKC substrate, is known to regulate actin structures at the plasma membrane. MARCKS is rapidly phosphorylated during particle uptake, and MARCKS is also recruited to the phagosome with kinetics similar to those of F-actin.[72] PKC inhibitors that prevent phagocytosis also block the accumulation of PKC, MARCKS, F-actin, and a number of other cytoskeletal proteins beneath bound zymosan.[73]

Signalling Restriction and the Phagocytic Cup

One of the most striking observations in the phagocytosis field is the spatial restriction of signalling and actin cytoskeletal reorganisation to plasma membrane segments engaged by the particle, giving rise to the formation of a phagocytic cup. One possible explanation for this fact is that following receptor clustering, as in the T cell receptor-mediated immunological synapse, the local concentration of p72Syk increases relative to local concentrations of protein tyrosine phosphatases, thus favouring the localised accumulation of phosphorylated substrates.[18] However, recent studies by the same group have failed to detect selective exclusion of phosphatases from FcR phagocytic cups. 3'-phosphatase PTEN and 5'-phosphatase SHIP1 were studied and both were observed to distribute throughout the cytoplasm with little plasma membrane association. Surprisingly, SHIP1 was greatly concentrated in FcR phagocytic cups. This raises the intriguing possibility that contrary to events in the immunological synapse, concentration of phosphorylated molecules by membrane domains engaged in phagocytic uptake depends on as yet unidentified, alternative mechanisms that may involve restricted lipid mobility at the phagocytic cup.[51]

Other MΦ Effector Functions Activated by FcR Engagement

Signalling pathways leading to FcR uptake are connected to cascades involved in triggering oxidative and non-oxidative microbicidal mechanisms and also in secretion of cytokines and other inflammatory mediators. The activation of Src kinases and ITAM phosphorylation are linked to phosphorylation of the adaptor proteins Cbl, Grb2, CrkL, NCK, SLP-76 and Shc. These proteins form adaptor complexes that translocate to sites of FcR crosslinking and after phosphorylation regulate the guanine nucleotide exchange factors "son of sevenless" (SOS) and C3G.[37]

These events are associated with the conversion of Ras-GDP to Ras-GTP which then catalyses the activation of Raf1, leading to MAPK activation. This in turn leads to activation of other cellular components including PLA2 and nuclear transcripts c-fos and c-jun. MAPK activation leads to Rap1a and Rac2 activation, resulting in NADPH oxidase activation and the release of superoxide anions.[35-39] MAPK p38 activation has also been recently implicated in regulation of early endocytic trafficking via the GDI-rab5 complex.[74] PLA2 activation is required for release of inflammatory mediators such as arachidonic acid. MAPK activation, including ERKs, p38 and JNK, phosphorylates substrates that are part of nuclear transcription complexes involved in pro-inflammatory cytokine transcription.[75] It is not known if any pathogens the MAPK pathway in MΦ. However activation of the MAPK pathway has been inplicated in the generation of microbicidal responses against avirulent *Mycobacterium tuberculosis* and the release of inflammatory cytokines by *Salmonella*-infected MΦ.[76,77] This raises the intriguing possibility that inhibition of MAPK pathways might be advantageous for pathogenic micro-organisms.

Complement Receptor (CRs)-Mediated Phagocytosis

Complement proteins, present in serum, opsonize micro-organisms for phagocytosis by complement receptors on macrophages. CR3, the iC3b binding receptor which is a member of the integrin receptor family (αMβ2), has an additional lectin domain recognising β-D-glucans and some mannose- and N-acetyl-glucosamine-containing oligosaccharides. This binding site can recognise soluble and particulate ligands in a complement-independent manner. This integrin also plays an important role in phagocyte adhesion and migration. Studies using blocking mAb and the generation of CR3 null mice have shown CR3 is necessary for recruitment of MΦ to sites of inflammation. Additionally, patients suffering from Leukocyte Adhesion Deficiency (LAD) lack β2 integrin expression and present profound immunological defects, some of which might be attributed to lack of CR3 expression.[78] Among microbial molecules thought to bind to this domain are LPS, leishmanial lipophosphoglycan, mycobacterial polysaccharides, *Klebsiella pneumoniae* acylpolygalactoside and yeast β-glucans and zymosan.[79-82] This broad ligand capacity places CR3 in the pattern-recognition receptor category in addition to its opsonic receptor status.

Phagocytosis via CR3 Requires Activation through a Pathway Involving the Rap1 GTPase

In contrast to the powerfully activating phagocytic response elicited by FcR, CR-mediated phagocytosis is a relatively quiet affair. FcRs are constitutively active for phagocytosis, but CR can only bind particles and need additional stimuli to enable its uptake capacity.[83] CR3 activation by phorbol esters, TNF-α, PAF and LPS is known to require activation of the Rap1 GTPase. Rap1 activation leads both to increased binding and increased ingestion of particles via CR3, suggesting that changes in conformation of binding epitopes and cytoskeletal organisation are involved in CR3 activation.[84]

FcR and CR3 Phagocytosis Are Mediated by Differential Actin Assembly

Another difference relates to the way FcR and CR engagement induces actin assembly. Although all types of phagocytosis require actin polymerisation at the site of ingestion, results of electron microscopy (EM) studies demonstrate that IgG- and complement-opsonised particles are internalised differently by MΦ. During FcR-mediated phagocytosis, veils of membrane rise above the cell surface and tightly surround the particle before drawing it into the body of the macrophage. FcR-mediated ingestion occurs by a zippering process, in which FcRs in the macrophage plasma membrane interact sequentially with IgG molecules distributed over the surface of the ingested particle and the cytoskeletal proteins are diffusely distributed on the phagosomes.[1,85]

By contrast, EM data indicate that CR-mediated phagocytosis is a more passive process. Complement-opsonised particles appear to sink into the cell with elaboration of small, if any, pseudopodia. Moreover, the phagosome membrane is less tightly apposed to complement-opsonised particles, with point-like contact separating regions of looser membrane. These point-like contact areas are enriched with a variety of cytoskeletal proteins. Formation of these foci is blocked by inhibitors of PKC, but not by inhibitors of protein tyrosine kinases, although tyrosine phosphorylation has been reported to increase the efficiency of phagocytosis.[1,83,86,87]

Differential Actin Assembly Is Associated with Differential Activation of Small Rho GTPases by CR3 and FcR

These observations are possibly associated with the differential activation of small Rho GTPases induced by these receptors, leading to separate actin assembly pathways. FcR-mediated phagocytosis induces Cdc 42 and Rac activation while CR3-mediated phagocytosis induces activation of Rho.[64] While Rac and Cdc42 activation can induce the formation of new actin filaments, Rho activation allows only reorganisation of pre-existing filaments.[88]

Members of both the WASP and Arp2/3 complex proteins are recruited to FcR and CR3 phagosomes. However, inhibition of Arp2/3 recruitment and activation using a dominant-negative construct of the Scar protein, ScarWA, reveals further subtle differences in signalling for actin assembly between the two receptors. In the MΦ cell line J774.A1, ScarWA expression blocks CR3-mediated phagocytosis only, leaving FcR-mediated uptake unaffected.[68]

Other Effector Functions: More Tales of GTPase Activation

Much less is known about downstream events after CR3 engagement. However it has long been known that CR3-mediated phagocytosis fails to induce oxidative microbicidal mechanisms and pro-inflammatory cytokine secretion, suggesting that other differences exist between CR3 and FcR signalling.[87,89] Once more the differential activation of small Rho GTPases may account for this CR3-FcR distinction: Rac2 is one of the NADPH oxidase subunits and CR3 ligation fails to activate Rac. Some new data have recently thrown more light on CR3 signalling pathways. Overexpression of R-Ras in J774.A1 MΦ was shown to induce CR3 activation in a Rap1-dependent manner. Interestingly, the R-Ras activation pathway in these MΦ does not involve Raf, MAPK,PI-3K, or Ral exchange factors.[90] Since these are the pathways on which MΦ functions such as oxidative burst and cytokine release are dependent on, identification of the R-Ras/Rap1 alternate pathway may partly explain CR3's poor effector function activation.

The relatively quiescent CR-mediated uptake has been exploited by several pathogens in order to enhance their survival prospects. *Leishmania major* is one example of a pathogen that developed mechanisms to activate complement, leading to C3 fixation and CR-binding, but at the same time shed the C5-9 pore-forming complex and thus avoid complement-mediated lysis.[80,91,92] *Leishmania* binding to MΦ fails to induce oxidative responses even after IFN γ stimulation.[93] This parasite can modulate cytokine secretion by MΦ Its entry into susceptible MΦ also blocks IL12 secretion while inducing TGFβ and IL10 secretion, but it is not clear whether CR binding is connected to this. Lack of IL12 secretion and the concomitant secretion of the de-activating cytokines TGFβ and IL10 block the development of a Th1 type immune response and is of paramount importance in determining protozoan parasite survival in the host.[93,94]

Other Opsonic Receptors

Not much is known about signalling via other MΦ opsonic receptors. It is noteworthy however, that one of these receptors, the vitronectin receptor ($\alpha_v\beta3$) is involved in a cooperative interaction with the scavenger receptor CD36 in binding and entry of apoptotic cells in MΦ. Although the role of $\alpha_v\beta3$ in phagocytic signalling has not been studied, another vitronectin receptor ($\alpha_v\beta5$), which is involved in apoptotic cell uptake by dendritic cells, has been reported to recruit the p130^{cas}-CrkII-dock180 complex, which in turn activates Rac1.[95]

Non-Opsonic Receptors

One of the primary challenges to the innate immune system is the discrimination of a large number of potential pathogens from self, utilising a restricted number of phagocytic receptors. This challenge has been met by the evolution of a variety of pattern-recognition receptors that recognise conserved motifs on pathogens known as pathogen-associated motif patterns (PAMPs) that are not found in higher eukaryotes. These motifs have essential roles in the biology of the invading agents, and are therefore not subjected to high mutation rates.[96] PAMPs include mannans and β-glucans in the yeast cell wall, formylated peptides in bacteria, and lipopolysaccarides and lipoteichoic acids on the surface of Gram negative and Gram positive bacteria, respectively. Most MΦ non-opsonic receptors are involved in interactions with PAMPs and are therefore essential for the development of effector mechanisms, antigen loading and the priming of the immune response.

The Macrophage Mannose Receptor

The MΦ mannose receptor (MMR) is a single chain lectin with a short cytoplasmic tail and an extracellular domain including 8 lectin-like carbohydrate-binding domains. The cytoplasmic tail is crucial to both the endocytic and phagocytic functions of the receptor, but little is known about the signals that lead to phagocytosis.[97]

MMR recognises glycosylated molecules with terminal mannose, fucose or N-acetylglucosamine moieties through its carbohydrate-recognition domains and sulphated N-acetyl-galactosamine structures through its cysteine-rich domain. It efficiently internalises ligands through the endocytic and phagocytic pathways. Recognition of glycoconjugates on the surface of micro-organisms by the MR is involved in phagocytosis of several pathogens such as *Candida albicans, Leishmania donovani, Mycobacterium tuberculosis* and *Pneumocystis carinii.*[98-100] Routing of soluble and particulate material via MMR leads to antigen presentation to T cells and this antigen loading mechanism has been widely used both in MΦ and dendritic cells.[101] Additionally, several studies support the idea that MR ligation is coupled to activation of effector functions, as it triggers the secretion of lysosomal enzymes, production of O_2^- and pro-inflammatory cytokines.[102-106]

Forms of endocytic uptake associated with the MMR include clathrin-dependent receptor-mediated endocytosis, macropinocytosis and phagocytosis. Another important characteristic of MMR-mediated uptake is the subsequent recycling of the receptor back to the plasma membrane after internalisation, rather than targeting to lysosomes where it would be degraded. Both the transmembrane and cytoplasmic domains of the receptor have been shown to modulate correct internalisation and sorting as shown by experiments using chimaeric constructs of the FcR and MMR. Recently a di-aromatic peptide motif, Tyr^{18}-Phe^{19}, has been shown to direct correct sorting of the MMR. This di-aromatic motif had previously been shown to be involved in sorting of the cation-dependent mannose 6-phosphate receptor, but cytosolic molecules interacting with it have not been identified.[107] Another peptide motif, FENTYL, has been described as essential for triggering of the phagocytic response, but not much else is known about downstream signalling events leading to particle uptake and other effector mechanisms.[97]

Finally, it should not be forgotten that besides interacting with material of microbial origin, the MMR is also able to interact with host-derived ligands such as lysosomal enzymes. This type of "scavenger-like" activity by the MMR still needs to be reconciled with the prominent role thought to be played by this receptor in innate immune responses. Possible explanations include differential MMR-interation with its ligands, cooperation with other_receptors and the inherent activation state of MR-expressing MΦ.[108]

β-glucan receptor

Another MΦ lectin receptor, the β-glucan receptor, has recently been shown to be identical to the dendritic cell receptor Dectin 1.[109,110] The β-glucan receptor has been shown to interact with yeast derived β-glucans and NIH3T3 fibroblasts transfected with this receptor were able to bind and phagocytose heat-killed *Candida albicans.* Importantly, the peptide sequence in the cytoplasmic tail of the β-glucan receptor was predicted to contain an ITAM motif, but further interactions with either src kinases or p72syk have not been studied.

Scavenger Receptors

Scavenger receptors (SRs) are membrane glycoproteins which are defined by their capacity to bind and internalise modified lipoprotein (Table 2). SRs are structurally diverse, but they can be distinguished from other lipoprotein receptors by their ligand-binding properties. Ligands for SRs include modified proteins and lipoproteins, polyribonucleotides and non-physiological compounds such as asbestos and silica. They also include sugars, phospholipids, lipopolysaccharide and lipoteichoic acid, which may act as ligands for phagocytosis. While all these ligands are macromolecular and polyanionic, the precise requirements for binding are not yet fully understood and not all polyanions binding to the receptors. Of the SR family, SR-AI/II and SR-AIII have been implicated in innate immune mechanisms and will be the focus in this section. CD36 has also been recognised as a receptor for malarial trophozoite-infected red cells, in addition to its thrombospondin and Oxidised-LDL binding function, but a connection to the immune response has not been established to date (for reviews on the SR family see refs. 111-113).

Table 2.

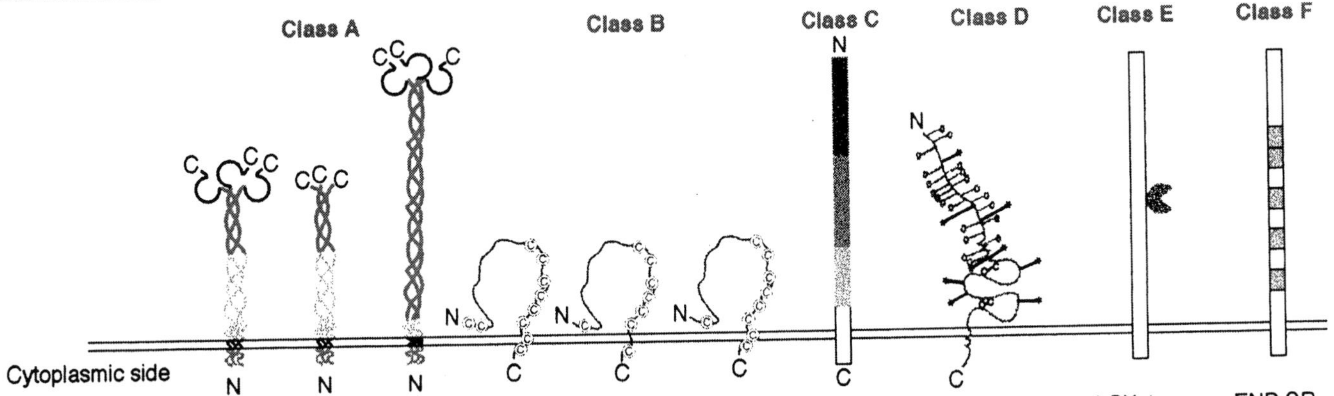

	SR-AI	SR-AII	MARCO	CD36	SR-BI	Croquemort	SR-CI	Macrosialin	LOX-1	END SR
Motif	Collagenous domain			C-terminal domain			Complement MAM domain	LAMP domain	C-type lectin	EGF domains
Ligands										
acLDL	+	+	+	+/–	+	?	+	?	?	+
oxLDL	+	+	?	+	+	?	?	+	+	+/–
LDL	–	–	?	–	+	?	–	?	–	–
Poly G/I	+	+	?	–	–	?	+	?	+	+
MBSA	+	+	?	+		?	?	?	?	+
Phosphatidylserine	–	–	?	+	+	?	?	?	?	?
Apoptotic cells	+	+	?	+	+	+	?	?	?	?

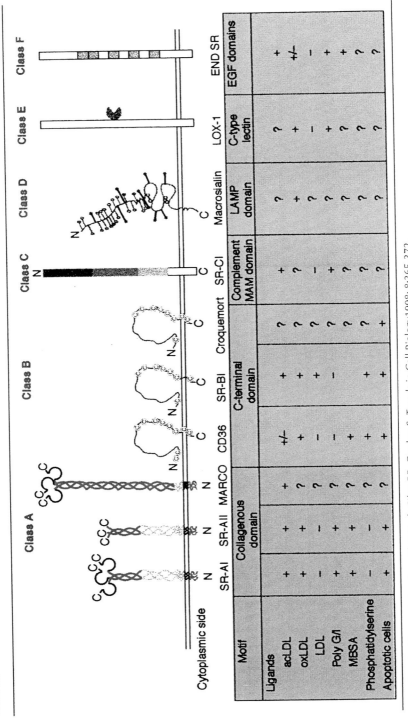

Reproduced with permission from Platt N, da Silva RP, Gordon S. Trends in Cell Biology1998; 8:365-372.

SR-AI/II

SR-AI/II is a 220kDa type II integral membrane protein expressed as a homotrimer of a ~77kDa subunit. Its extracellular region contains a spacer domain followed by an α-helical coiled-coil domain, implicated in trimer assembly and a collagenous domain, containing the only ligand binding site described to date.[114] SR-AI has an extra SR-specific cysteine-rich domain of no known function. Of the SR family, SR-AI/II have been reported to bind LPS, LTA and both gram-negative and gram-positive intact bacteria.[114-115] These receptors also contribute to MΦ adhesion.[116]

Their LPS/LTA binding activity raises the possibility that these receptors can function as pattern-recognition receptors and play a role in triggering effector mechanisms and priming of the immune response. However, their involvement in apoptotic cell uptake conflicts with this idea and as in the case of the MR more research is needed to ascertain receptor specificity in each response.[23]

In vivo and in vitro binding of *Listeria monocytogenes, Staphylococcus aureus* and heat-killed *Escherichia coli* to SR-A have been reported.[114] In vivo studies have used the SR-AI/AII$^{-/-}$ mouse to explore the role of this receptor in bacterial clearance and pathogenicity. SR-AI/AII$^{-/-}$ mice have been shown to be more susceptible to infection by *Listeria monocytogenes.* Impaired uptake activity by liver Kupffer cells and more extensive granuloma formation were also observed.[117] Infection of SR-AI/AII$^{-/-}$ mice with *Staphylococcus aureus* has shown decreased clearance of bacteria from the peritoneum and blood resulting in greater susceptibility of SR-AI/AII-deficient-mice to this pathogen.[118]

Finally experiments relating to SR-AI/II involvement in granulomatous infections has produced some contradictory data. BCG infection of SR-AI/AII-deficient mice have shown that these mice have no difference in susceptibility to this bacterium when compared to infection of control mice.[119] In contrast, analysis of granuloma formation after intravenous injection of *Corynebacterium parvum* revealed some early impairment of this response, although number of cells and granuloma size were normal after 14 days and the mice did not show increased susceptibility to infection.[120] The data highlighted above suggest that in some cases SR-AI/II play a role in the host protective immunity. However it is not clear if its contribution relates to LPS-triggered effector mechanisms, a response more typically associated with CD14 uptake and Toll receptors recruitment (see below), or to a possible scavenging role of toxic microbial components.

SR-AI/II phagocytic capacity and associated signalling cascades are not currently understood. There is some evidence that these receptors contribute to bacterial ingestion.[115] Additionally, transfection of the receptor in non-professional phagocytes induces binding and internalisation of apoptotic neutrophils.[23] Aged platelet ingestion has also been reported to involve SR-AI/II.[121] However a proper role for its cytoplasmic tail in triggering phagocytosis of bacteria or apoptotic cells has not yet been established. There is extensive receptor redundancy in ligand binding within the SR family and MΦ express a wealth of phagocytic receptors that could cooperate with SR-A to trigger phagocytosis. All studies to date employed complex particles capable of multiple receptor interactions and further study is needed to clarify SR-AI/II contribution to the phagocytic response.

Like other MΦ phagocytic receptors, SR-AI/II is also involved in clathrin-dependent receptor-mediated endocytosis. Their ability to endocytose large quantities of chemically-modified low-density lipoprotein (LDL) underlies the mechanism through which macrophages are converted into the foam cells that are characteristic of atherosclerotic plaques and vascular disease.[113]

The SR-A cytoplasmic domain does not contain the classical tyrosine-based peptide motif predicted to form a tight β turn and interact with clathrin adaptor proteins. Two studies have sought to identify internalisation motifs present in the SR-A cytoplasmic tail. A report analysing COS cell-transfected bovine SR-AI identified a VXFD motif essential to receptor internalization and plasma membrane localization. Another more recent study analyzing the murine receptor reported broadly similar conclusions. Mutation of Asp[25], the terminal amino-acid of the VXFD motif inhibited acetylated-LDL degradation.[122] Phosphorylation of Ser[21] and Ser[49] in the cytoplasmic tail has also been shown to interfere with receptor trafficking. Intriguingly, substitution of Ser[21] induced greater plasma membrane expression and acetylated-LDL degradation while mutation of Ser[49] reduced lipid uptake through reduced internalization.[123] Adaptor proteins interacting with these receptor trafficking motifs remain to be identified. Other signalling pathways reported as being activated by SR-AI/II binding of Acetylated and Oxidised LDL include MAPK, PLC-γ1, PI3K and PKC activation.[124,125] Endocytosis via SR-A has also been reported to recruit the src kinase Lyn. It is not clear, however, which functional aspects they contribute to and whether these responses are receptor-specific.

Uptake of modified antigens, via SR-AI/II lead to enhanced antigen presentation, indicating that besides its role in atherogenesis, endocytosis via SR-AI/II is related to antigenic stimulation of T cells.[126] Further identification of downstream cascades activated after endocytosis will undoubtedly clarify membrane trafficking pathways involved in antigen processing and presentation by MΦ.

As mentioned above, studies analysing effector mechanisms arrising from SR-AI/II ligation, particularly by LPS uptake or during phagocytic engagement, have been hampered by the ligand-binding promiscuity of the SR family and other MΦ receptors, making it very difficult to distinguish which receptors are contributing to each effect. One study, however, has shown that in contrast to CD14-/- mice, live BCG infection of SR-AI/AII-/- mice led to production of larger quantities of TNF-α and IL-6 and more susceptibility to endotoxic shock.[119] This suggests that signalling through these receptors reduces the release of pro-inflammatory cytokines but the possibility that SRAI/II can, in certain circumstances, activate effector functions needs to be further explored.

SR-AIII (MARCO)

MARCO is a 210kDa type II integral membrane protein, composed of homotrimers of a 55Kda subunit. Its primary structure is similar to SR-AI/II in its domain organisation, but in the MARCO molecule the spacer domain is immediately followed by a longer collagenous domain without the coiled-coil region. The C-terminal cysteine-rich domain is homologous to the comparable SR-AI region. One interesting distinction between SR-AI/II and MARCO relates to their tissue distribution. MARCO is a MΦ-restricted molecule expressed primarily in the marginal zone of the spleen and medullary cords of lymphnodes. SR-AI/II have a wider pattern of distribution, including MΦ in most organs and some endothelial cells.[127,128]

By far the most prominent feature of the MARCO literature is the data describing its ability to bind to bacterial cell components and intact bacteria such as Escherichia coli and Staphyloccocus aureus. Other polyanionic ligands such as Ficoll failed to bind to MARCO, but intriguingly, unlike the human homologue mouse MARCO was shown to bind to Acetylated-LDL.[129] Like SR-AI/II, MARCO has also been shown to bind to unopsonised environmental particles.[130]

Experiments employing cells transfected with intact and truncated forms of MARCO have shown that the binding site for bacterial cell wall components is located in the cysteine-rich domain and that only a small part of this domain, residues 432-442, is required for ligand interaction.[131]

LPS stimulation of J774.2 MΦ in vitro or in vivo LPS injection readily up-regulates MARCO expression. J774.2 cells stimulated with various pro-inflammatory cytokines failed to show increased MARCO expression, indicating that LPS itself is responsible for this effect. Six hours after LPS injection Kupffer cells in the liver were found to express MARCO. An even wider pattern of distribution was found after BCG infection or bacterial sepsis, including MΦ populations in liver, lung and spleen that do not normally express this receptor. Moreover, injection of some anti-MARCO monoclonal antibodies interfered with clearance of heat-killed bacteria in treated animals, further supporting a role for MARCO in antibacterial defence.[132]

The only report studying possible signalling events mediated by MARCO has shown other interesting data. Transfection of full-length MARCO into COS, CHO, NIH3T3, HeLa and 293 cell lines induce a dramatic change in cell morphology with active extension of filopodia and lamellipodia. These actin-based structures were most pronounced when cells were adhered to glass coverslips in the presence of fetal calf serum. Additionally, MARCO-expressing cells had a lower number of focal adhesions. Adhesion of cells to fibronectin, but not to laminin or vitronectin, and co-expression of dominant-negative construct of Rac1(V12N17), but not dominant-negativeCdc42, were shown to block MARCO's effect on actin assembly. These data indicate this receptor can signal to the actin cytoskeleton and induce actin assembly. They also suggest its expression may contribute to mophological changes observed in activated MΦ and that, analogous to SR-AI/II, MARCO is also involved in MΦ adhesion.[133] Further studies are necessary to clarify whether these actin remodelling events are related to any forms of endocytic uptake and whether they are associated with triggering of MΦ effector functions.

CD14 and the Toll-Like Receptor Family

Finally no Chapter on MΦ contributions to immune function is complete without mentioning CD14, its LPS-binding ability associated with recruitment of Toll receptors and potent signalling mechanisms.CD14 is a GPI-anchored protein primarily expressed by monocytes, but also expressed by some tissue MΦ notably liver Kupffer cells, and gingival fibroblasts. The plasma membrane associated receptor can be cleaved either by release from its GPI-anchor or by proteolytic cleavage, to form two separate soluble forms that can be detected in plasma, normal cerebrospinal fluid and urine from diseased patients. Expression of CD14 can be up-regulated by IFN-γ, TNF-α and some bacterial components, and down-modulated by Th2 cytokines.[134] Both membrane-bound and soluble forms of CD14 are able to bind LPS, but this interaction is facilitated by the serum LPS binding protein (LBP).[135] Although the best characterised CD14 interaction is that with LPS, this receptor has been reported to bind to Gram-positive bacterial components and even fungal antigens. CD14 ligation by LPS/sCD14 or LPS/LBP is followed by endocytic uptake and signalling through members of the Toll-like receptor family (TLR) which act as co-receptors for microbial components.[136,137] However, endocytosis of LPS by CD14 can be dissociated from the signalling response. Internalisation of LPS by CD14 depends on its aggregation state and can not be blocked by inhibitors of the signalling response such as co-expression of mutated, nonfunctional TLRs, generic inhibitors of LPS response or repeated stimulation.[138-140] Conversely, some antibodies and cytochalasin block LPS internalisation, but leave the signalling response intact.

Toll-Like Receptors

Toll was initially identified as important for *Drosophila* ontogenesis and antimicrobial resistance.[141] The recognition of sequence similarities between the cytoplasmic domains of Toll and the IL-1 receptor (the Toll-IL-1R module or TIR module) led to the discovery of a family of mammalian Toll-like molecules.[142] Another clue in the discovery of this remarkable example of evolutionary conservation between man and fly was the recognition of similarities between the fly Rel transcription factor Dorsal and the activation of the mammalian Rel transcription factor NF-κB by IL-1. Both are retained in the cytoplasm by proteins containing ankyrin repeats, Cactus and IκB, respectively. Dissociation of Dorsal and NF-κB from their respective inhibitors occurs after Toll/TLR/IL-1receptor ligation and leads to activation of homologous tyrosine kinases Pelle and IRAK, respectively (for reviews on Toll-like proteins see refs. 143 and 144) Another important event in the field was the discovery that the LPS-hyporesponsive phenotype in mice results from mutations in TLR4, which was then identified as a co-receptor for LPS.[145] Other TLR ligands include various microbial-derived molecules, such lipoteichoic acid (LTA), peptidoglycan (PGN), bacterial or mycobacterial lipoproteins, lipoarabinomannan and zymosan. TLR2 is activated primarily by PGN and lipoproteins, TLR4 is activated by LPS and LTA and bacterial unmethylated CpG DNA is necessary for TLR9 activation.[146-148]

The mechanisms underlying the TLR/CD14 interaction with microbial products are not fully understood. Since in the fly Toll binds to a proteolycally processed, cysteine-knot polypeptide, Spatzle, it was initially postulated in the mammalian cognate system that microbial compounds activate endogenous proteases that could catalyse other products, including CD14, that then interact with the TLR. However recent studies taking advantage of species-specific differences in the way Lipid A(the active moiety in LPS) triggers effector function have shown that a direct LPS-TLR4 interaction is the more likely scenario.[149,150] LipidIVa and *Rhodobacter sphaeroides* lipid A (RSLA) antagonises LPS responses in human cells, but act as agonists in hamster and murine cells. Remarkably Lipid A and RSLA antagonise the LPS response in hamster cells transfected with human TLR4 and human cells expressing hamster TLR4 can be activated by both compounds. Moreover, reconstitution of TLR4-null cells with murine TLR4 recontitutes the response, while human TLR4 transfection does not.

TLR signalling requires the recruitment of the MyD88 adaptor protein.[151] Recently another molecule, MD2, has been identified as necessary for TLR4 activation in MΦ. It is thought direct interaction of MD2with TLR4 stabilises the LPS/CD14/TLR complex. MD2 can also associate with TLR2 and undoubtely other accessory molecules are yet to be discovered.[152-154]

The TLR family of receptors has ten identified members so far, all containing the TIR domain in their cytoplasmic tails. Their extracellular regions contain two cysteine-rich motifs that flank the leucine-rich repeats, with the exception of TLR9, which contains only one cysteine-rich motif between the transmembrane region and leucine repeats.[155]

Signalling via all TLRs induces NF-κB activation and its nuclear translocation leading to transcriptional activation of pro-inflammatory cytokine genes.[155] TLR signalling depends on recruitment of the adaptor molecule MyD88, which then leads to recruitment of IRAK1.[151] The current hypothesis is that formation of the TLR/MyD88/IRAK complex leads to IRAK phosphorylation. Phosphorylated IRAK then activates TRAF6 leading to activation of the NF-kB signalosome. The latter is composed of the two kinases, IKKα and IKKβ and a regulatory subunit, IKKγ/NEMO. Activation of the signalosome proceeds after IKKβ phosphorylation leading to nuclear translocation of NF-κB.[156] Other signalling cascades, particularly the MAPK pathway, through kinases NIK and MEKK1 are thought to participate in TLR signalling to the signalosome.[157,158] Further involvement of the MAPKs in the LPS response is mediated by CD14 ligation, leading to activation of all three terminal MAPK (JNK, p38 and Erk1/2) and translocation of transcriptional factors AP1(c-jun+c-fos), ATF2 and ELK1.[134]

Some TLRs have been reported to co-operate with one another, thereby increasing their repertoire for ligand recognition.[159] Finally, with the exception of TLR9, all other TLRs translocate to phagosomal membranes and are now thought to be responsible for the potent pro-inflammatory responses triggered by bacterial ingestion.[147]

Conclusion

The last few years have brought tremendous advances in the understanding of triggering mechanisms and signalling cascades involved in the activation of MΦ effector function. Landmarks in the field were the identification of the mammalian TLR family and their fundamental role in the innate immunity, and the recent finding of yet another family of proteins, NOD/CARD, also involved in modulating inflammatory responses by MΦ and other cells (for a review see ref. 160). The discovery of these remarkably well conserved mechanisms of microbial recognition heralded a new era of interest in the innate immune responses and seem to promise yet more exciting results. Successful pathogens will no doubt have developed mechanisms, still to be identified, to counteract activation of these pathways.

Our knowledge of how phagocytic uptake proceeds, particularly via the FcR, has also increased substantially. The identification of small Rho GTPases and their effectors as the main regulators of actin assembly and the role of phosphoinositide metabolism in regulating endocytic responses and endosomal/phagosomal membrane trafficking are particular examples. For every new step identified, virulent micro-organisms seem to be ahead of the game and very elaborate mechanisms exploiting the actin assembly machinery and disrupting membrane trafficking have been revealed.

We still need to learn more about signalling mechanisms triggered by other non-opsonic MΦ receptors and their relationship with each effector mechanism. Interaction with complex particles leads to multiple engagement of MΦ receptors and at present it is not clear which interaction leads to activation of which pathway and what the specific underlying signalling is. This is particularly important considering that both host- and microbial-derived molecules are among potential ligands for these receptors and that their interaction is connected to functions as diverse as adhesion, migration, receptor-mediated endocytosis and phagocytosis.

Another area that needs consideration is the inherent activation state of the MΦ. Are all pathways triggered by receptor ligation active in inflammatory and resident tissue MΦ? Last but not least, we still need to develop a complete picture to explain the prodigious endocytic capacity of MΦ including specialised uptake mechanisms such as phagocytosis and macropinocytosis. Most of the signalling elements identified so far are expressed by a variety of other cells. Cells that express these elements in combination are able to phagocytose, but their phagocytic capacity pales when compared to MΦ. Certainly, part of the answer lies in the expression of specific phagocytic receptors by MΦ. However, every experimenter in the field knows that expression of these receptors in non-professional phagocytes does not reconstitute phagocytic capacity to levels observed in MΦ. Further research should help us understand what makes MΦ truly professional phagocytes.

References

1. Aderem A, Underhill DM. Mechanisms of phagocytosis in macrophages. Annu Rev Immunol 1999; 17:593-623.
2. Morrissette N, Gold E, Adrem A. The macrophage—A cell for all seasons. Trends Cell Biol 1999; 9:199-201.
3. Aderem A, Underhill DM. Heterogeneity in macrophage phagocytosis. Adv Cell Mol Biol Memb Organ 1999; 5:195-213.
4. Le Page C, Genin P, Baines MG et al. Interferon activation and innate immunity. Rev Immunogenet 2000; 2:374-386.
5. Peiser L, Gordon S. The function of scavenger receptors expressed by macrophages and their role in the regulation of inflammation. Microbes Infect 2001; 3:149-159.
6. Ramakrishnan L, Falkow S. Pathogen strategies: A hitchhiker's guide to the macrophage. Adv Cell Mol Biol Mem Organ 1999; 6:1-26.
7. Cross AS. The biological significance of bacterial encapsulation. Curr Top Microbiol Immunol 1990; 150:87-95.
8. Mims C, Playfair J, Roitt I et al. Natural defenses in action. In: Medical Microbiology. London: Mosby International Limited:111-118.
9. Cornelis GL. The Yersinia Yop virulon, abacterial system to subvert cells of the primary host defense. Folia Microbiologica 1998; 43:253-261.
10. Káposzta R, Maródi L, Hollinshead M et al. Rapid recruitment of late endosomes and lysosomes in mouse macrophages ingesting Candida albicans. J Cell Sci 1999; 112:3237-3248.
11. Sansonetti PJ, Ryter A, Clerc P et al. Multiplication of Shigella flexneri within HeLa cells: Lysis of the phagocytic vacuole and plasmid-mediated contact hemolysis. Infec Immun 1986; 51:461-469.
12. Portnoy DA, Jacks PS, Hinrichs DJ. Role of hemolysin for intracellular growth of Listeria monocytogenis. J Exp Med 1988; 167:1459-1471.
13. Rodríguez A, Andrews NW. Regulated exocytosis of lysosomes: A novel pathway revealed by the interaction of Trypanosoma cruzi with host cells. Adv Cell Mol Biol Mem Organ 1999; 6:281-296.
14. Frischknecht F, Way M. Surfing pathogens and the lessons learned for actin polymerisation. Trends Cell Biol 2001; 11:30-38.
15. Joiner KA, Fuhrman SA, Miettinen HM et al. Toxoplasma gondii: Fusion competence of the parasitophorous vacuole in Fc receptor transfected fibroblast. Science 1990; 249:641-646.
16. Pieters J. Entry and survival of pathogenic mycobacteria in macrophages. Microbes Infect 2001; 3:249-255.
17. Horwitz MA. The Legionnaires' disease bacterium (Legionella pneumophila) inhibits phagosome-lysosome fusion in humamn monocytes. J Exp Med 1983; 158:2108-2126.
18. Greenberg S. Diversity in phagocytic signalling. J Cell Sci 2001; 114:1029-1140.

19. Kraus JC, Poo H, Zue W et al. Reconstitution of antibody-dependent phagocytosis in in fibroblasts expressing Fc γ receptor IIIB and complement receptor type 3. J Immunol 1994; 153:1769-1777.
20. Zhou M, Brown EJ. CR3 (Mac-1, αMβ2, CD11b/CD18) and Fc γ RIII cooperate in generation of a neutrophil respiratory burst: requirement for Fc γ RII and tyrosine phosphorylation. J Cell Biol 1994; 125:1407-1416.
21. Keisuke K, Takeshita S, Kensuke M et al. Functional association of CD9 with the Fc γ receptors in macrophages. J Immunol 2001; 166:3256-3265.
22. Platt N, Haworth R, Da Silva RP et al. Scavenger receptors and phagocytosis of bacteria and apoptotic cells. Adv Cell Mol Biol Mem Organ 1999; 5:71-85.
23. Platt N, da Silva RP, Gordon S. Recognizing death: The phagocytosis of apoptotic cells. Trends Cell Biol 1998; 8:365-372.
24. Linehan SA, Martinez-Pomares L, Gordon S. Macrophage lectins in host defense. Microbes Infect 2000; 2:279-288.
25. Greenberg S. Signal transduction of phagocytosis. Trends Cell Biol 1995; 5:93-99.
26. Desjardin M. Biogenesis of phagolysosmes: "The kiss and run" hypothesis. Trends Cell Biol 1995; 5:183-186.
27. Tjelle TE, Løvdal T, Berg T. Phagosome dynamics and function. BioEssays 2000; 22:255-263.
28. Hackham DJ, Rotstein OD, Zhang WJ et al. Regulation of phagosomal acidification. Differential targetting of Na^+/H^+ exchangers, Na^+/K^+ ATPases and vacuolar-type H^+-ATPase. J Biol Chem 1997; 272:29810-29820.
29. Michael U, Nathan S, Nathan US. Antimicrobial mechanisms of macrophages. Adv Cell Mol Biol Mem Organ 1999; 5:407- 439.
30. Nauddeef W. The NADPH-dependent oxidase of phagocytes. Proceedings of the Association of American Physicians 1999; 111:373-382.
31. Vasquez-Torres A, Fang FC. Oxygen-dependent anti-Salmonella activity of macrophages. Trends Microbiol 2001; 9:29- 33.
32. Ganz T. Oxygen-independent microbicidal mechanisms of phagocytes. Proceedings of the Association of American Physicians 1999; 111:390-395.
33. Ravetch JV. Fc receptors. Curr Opin Immunol 1997; 9:121-125.
34. Ravetch JV. Fc receptors: Rubor redux. Cell 1994; 78:553-560.
35. Park RK, Liu Y, Durden DL. A role for Shc, Grb2 and Raf in Fc γ RI signal relay. J Biol Chem 1996; 271:13342-13348.
36. Park RK, Kiono WT, Liu Y et al. Cbl-Grb2 interaction in myeloid immunoreceptor tyrosine activation motif signaling. J Immunol 1998; 160:5018-5027.
37. Park RK, Erdreich-Epstein A, Liu M et al. High affinity IgG receptor activation of Src family kinases is required for modulation of the Shc-Grb2-Sos complex and the downstream activation of nicotinamide adenine dinucleotide phosphate (reduced) oxidase. J Immunol 1999; 163:6023-6034.
38. Erdreich-Epstein A, Liu M, Kant AM et al. Cbl functions downstream of Src kinases in Fc gamma RI signaling in primary human macrophages. J Leukoc Biol 1999; 65:523-534.
39. Karimi K, Lennartz MR. Mitogen activated protein kinase is activated during IgG-mediated phagocytosis, but is not required for target ingestion. Inflammation 1998; 22:67-82.
40. Sato N, Moo-Kyung K, Schreiber AD. Enhancement of Fc γ receptor–mediated phagocytosis by transforming mutants of Cbl. J Immunol 1999; 163:6123-6131.
41. Allan LAH, Aderem A. Mechanisms of phagocytosis, Curr Opin Immunol 1996; 8:36-40.
42. Ghazizadeh S, Bolen JB, Fleit HB. Physical and functional assosiation of Src-related protein tyrosine kinases with Fcγ RII in monocytic THP-1 cells. J Biol Chem 1994; 269:8878.
43. Salcedo TW, Kurosaki P, Kanakaraj J et al. (), Physical and functional association of p56lck with Fcγ RIIIA (CD 16) in natural killer cells. J Exp Med 1993; 177:1475.
44. Crowley MT, Costello PS, Fitzer-Attas CJ et al. A critical role for Syk in signal transduction and phagocytosis mediated by Fcγ receptors on macrophages, J Exp Med 1997; 186:1027-1039.
45. Nishizumi H, Horikawa K, Mlinaric-Rascan I et al. A double-edged kinase Lyn: A positive and negative regulator for antigen receptor mediated signals. J Exp Med 1998; 187:1343.
46. Suzuki T, Kono H, Hirose N et al. Differential involvement of Src family kinases in Fcγ receptor-mediated phagocytosis. J Immunol 2000; 165:473-482.
47. Araki N, Johnson MT, Swanson JA. A role for phophatidylinositide 3-kinase in the completion of macropinocytosis and phagocytosis by macrophages. J Cell Biol 1996; 135:1249-1260.
48. Ninomiya N, Hazeki K, Fukui Y et al. Involvement of phosphatidyl inositol 3 kinase in Fcgamma receptor signalling. J Biol Chem 1994; 269:22732-22737.
49. Cox D, Tseng CC, Bjekic G, Greenbersg SA. A requirement for phophatidylinositol 3-kinase in pseudopod extension. J Biol Chem 1999; 274:1240-1247.

50. Braun L, Ohayon H, Cossart P. The InlB protein of Listeria monocytogenes is sufficient to promote entry into mammalian cells. Mol Microbiol 1998; 27:1077-1087.
51. Marshall JG, Booth JW, Stambolic V et al. Restricted accumulation of phophatidylinositol 3-kinase products in a plasmalemmal subdomain during Fc receptor-mediated phagocytosis. J Cell Biol 2001; 153:1369-1380.
52. Simonsen A, Wurmser AE, Emr S et al. The role of phosphoinositides in membrane transport. Curr Op Cell Biol 2001; 13:485-492.
53. McPherson PS, Kay BK, Hussein N. Signaling on the endocytic pathway. Traffic 2001; 2:375-394.
54. Fratti RA, Backer JM, Gruenberg J et al. Role of phophatidylinositol 3-kinase and Rab 5 effectors in phagosomal biogenesis and mycobacterial phagosome maturation arrest. J Cell Biol 2001; 154:631-644.
55. Oancea E, Meyer T. Protein kinase C as a molecular machinery for decoding calcium and diacylglicerol signals. Cell 1998; 95:307-318.
56. Gold ES, Morrissette NS, Underhill DM et al. Amphiphysin IIm, a novel amphiphysin II isoform, is required for macrophage phagocytosis. Immunity. 2000; 12:285-292.
57. Gold ES, Underhill DM, Morrissette NS et al. Dynamin 2 is required for phagocytosis in macrophages.J Exp Med 1999; 190:1849-1856.
58. Hill KM, Huang Y, Yip SC et al. N-terminal domains of the class ia phosphoinositide 3-kinase regulatory subunit play a role in cytoskeletal but not mitogenic signaling. J Biol Chem 2001; 276:16374-16378.
59. Hall A. Rho GTPases and the actin cytoskeleton, Science 1998; 279:509-514.
60. Ridley A, Hall A. The small GTP-binding protein rho regulates the assembly of focal adhesions and actin stress fibers in response to growth factors. Cell 1992; 70:389-399.
61. Ridley A, Paterson HF, Johnston CL et al. The small GTP-binding protein rac regulates growth factor induced membrane ruffling. Cell 1992; 70:401-410.
62. Kozma R, Ahmed S, Beat A et al. The Ras-related protein Cdc42Hs and bradykinin promote formation of peripheral actin microspikes and filopodia in Swiss NIH3T3 fibroblasts. Mol Cell Biol 1995; 15:1942-1952.
63. Cox D, Chang P, Zhang Q et al. Requirements for both Rac1 and Cdc42 in membrane ruffling and phagocytosis in leukocytes, J Exp Med 1997; 186:1487-1494.
64. Caron E, Hall A. Indentification of two distinct mechanisms of phagocytosis controlled by different Rho GTPases. Science 1998; 282:1717-1721.
65. Machesky LM, Gould KL. The Arp2/3 complex: a multifunctional actin organiser. Curr Op Cell Biol 1999; 11:117-121.
66. Mullins RD. How WSP-family proteins and the Arp2/3 complex convert intracellular signals into cytoskeletal structures. Curr Op Cell Biol 2000; 12:91-96.
67. Symons M, Derry JM, Kerlak B et al. Wiskott-Aldrich syndrome protein, a novel effector for the GTPase Cdc43Hs, is implicated in actin polymerisation. Cell 1998; 84:723-734.
68. May R, Caron E, Hall A et al. Involvement of the Arp2/3 complex in phagocytosis mediated by Fc γR or CR3. Nature Cell Biology 2000; 00:1-3.
69. Zheleznyak A, Brown EJ. Immunoglobulin-mediated phagocytosis by human monocytes requires protein kinase C activation. Evidence for protein kinase C translocation to phagosomes. J Biol Chem 1992; 267:12042-12048.
70. Liu WS, Heckman CA. The sevenfold way of PKC regulation. Cell Signal 1998; 10:529-542.
71. Larsen EC, DiGennaro JA, Saito N et al. Differential requirement for classic and novel PKC Isoforms in respiratory burst and phagocytosis in RAW264.7 cells. J Immunol 2000; 165:2809-2817.
72. Aderem A.The MARCKS brothers: A family of protein kinase C substrates. Cell 1992; 71:713-716.
73. Allan LAH, Aderem A. A role for MARCKS, the alpha isozyme of protein kinase C and myosin I in zymosan phagocytosis by macrophages. J Exp Med 1995; 182:829-840.
74. Cavalli V, Vilbois F, Corti M et al. The stress induced MAP kinase p38 regulates endocytic trafficking via the GDI:Rab5 complex. Mol Cell 2001; 7:421-432.
75. Kyriakis, JM, Avruch J. Protein kinases cascades activated by stress and inflammatory cytokines. BioEssays 1996; 18:567
76. Perkvist N, Stendahl O. Activation of human neutrophils by Mycobacterium tuberculosis H37Ra involves the phospholipase c g 2, Shc adapter protein, and p38 mitogen-activated protein kinase. J Immunol 2000; 164:959-965.
77. Procyk KJ, Rippo MR, Testi R et al. Distinct mechanisms target stress and extracellular signal-activated kinase 1 and Jun N-terminal kinase during infection of macrophages with Salmonella. J Immunol 1999; 163:4924-4930.
78. Ehlers MRW. CR3: A general purpose adhesion recognition receptor essential for innate immunity. Microbes Infect 2000; 2:289-294.

79. Troesltra A, De Graaf Miltenburg LAM, Van Bommel T et al. Lipoplolysaccharide-coated erythrocytes activate human neutrophils via CD14 while subsequent binding is throgh CD11b/CD18. J Immunol 1999; 162:4220-4225.
80. Wilson ME, Pearson RD. Role of the CR3 and mannose receptors in the attachment and ingestion of Leishmania donovani by human mononuclear phagocytes, Infection and Immunity 1988; 56:363-369
81. Ehlers MRW, Daffé M. Interactions between Mycobacterium tuberculosis and the host cell: Are mycobacterial sugars the key? Trends Microbiol 1998; 6:328-335.
82. Isberg RR, VanNhieu GT. Binding and internalization of micro-organisms by integrin receptors. Trends Microbiol 1994; 2:10-14.
83. Kaplan G. Differences in the mode of phagocytosis with Fc and C3 receptors in macrophages, Scand J Immunol 1977; 6:797-807.
84. Caron E, Self AJ, Hall A. The GTPase Rap1 controlsfunctional activation of macrophage αMβ2 by LPS and other activators. Curr Biol 2000; 10:974-978.
85. Swanson JA, Baer SC. Phagocytosis by zippers and triggers. Trends Cell Biol 1995; 5:89-93.
86. Allan LAH, Aderem A. Molecular definition of distinct cytoskeletal structures involved in complement- and Fc receptor-mediated phagocytosis in macrophages. J Exp Med 1996; 184:627-637
87. Ravetch JV, Clynes RA. Divergent roles for Fc receptors and complement in vivo. Annu Rev Immunol 1998; 16:421-432.
88. Machesky LM, Hall A. Role of actin polymerisation and adhesion to extracellular matrix in Rac- and Rho-induced cytoskeletal organisation. J Cell Biol 1997; 138:913-926.
89. Wright SD, Silverstein SC. Receptors for C3b and C3bi promote phagocytosis but not the release of toxic oxygen from human phagocytes. J Exp Med 1983;158:2016-2023.
90. Self AJ, Caron E, Paterson HF, Hall A. Analysis of R-Ras signalling pathways. J Cell Sci 2001; 114:1357-1366.
91. Da Silva RP, Hall BF, Joiner KA et al. CR1, the C3b receptor, mediated binding of infective Leishmania major metacyclic promastigotes to human macrophages. J Immunol 1989; 143:617-622.
92. Puentes SM, Da Silva RP, Sacks DL et al. Serum resistance of metacyclic stage Leishmania major promastigotes is due to release of C5b-9. J Immunol 1990; 145:4311-4316.
93. Bogdan C, Rollinghoff M. The immune response to Leishmania: Mechanisms of parasite control and evasion. Int J Parasitol 1998; 28:121-134.
94. Bogdan C, Rollinghoff M. How do protozoan parasites survive inside macrophages? Parasitol Today 1999; 15:22-28.
95. Albert ML, Kim J-ll, Birge RB. αvβ5 integrin recruits the Crkll-Dock180-Rac1 complex for phagocytosis of apoptotic cells. Nature Cell Biol 2000; 2:899-905.
96. Medzhitov R, Janeway Jr CA. Innate Immunity: Impact of the adaptive immune response. Curr Op Immunol 1997; 9:4-9.
97. Ezekowitz RAB, Sastry K, Bailly P et al. Molecular characterization of the humal macrophage mannose receptor: Demonstration of a multiple carbohydrate recognition-like domains and phagocytosis of yeast in COS cells. J Exp Med 1990; 172:1785-1794.
98. Káposzta R, da Silva RP, Marodi L et al. Cellular mechanisms of phagocytosis of Candida by murine macrophages. Adv Cell Mol Biol Mem Organ 1999; 6:317-331.
99. Astarie-Dequeker C, N'Diaye AN, Le Cabec V et al. The mannose receptor mediated uptake of pathogenic and nonpathogenic Mycobacteria and bypasses bactericidal responses in human macrophages. Infect Immunity 1999; 67:469-477.
100. Ezekowitz RAB, Williams DJ, Koziel H et al. Uptake of Pneumocystis carinii mediated by the macrophage mannose receptor. Nature 1991; 351:155-158.
101. van Bergen J, Ossendorp F, Jordens R et al. Get into the groove! Targeting antigens to MHC class II. Immunol Rev 1999; 172:87-96.
102. Berton G, Gordon S. Modulation of macrophage mannosyl-specific receptors by cultivation on immobilized zymosan. Effects on superoxide-anion release and phagocytosis. Immunology 1983; 49:705-715.
103. Bodmer JL, Dean RT. Does the induction of macrophage lysosomal enzyme secretion by zymosan involve the mannose receptor. Biochem. Biophys Res Commun 1983; 113:192-198.
104. Oshumi Y, Lee YC. Mannose-receptor ligands stimulate the secretion of lysosomal enzymes from rabbit alveolar macrophages. J Biol Chem 1987; 262:7955-7962
105. Shibata Y, Metzger WJ, Myrvik QN. Chitin particle-induced cell-mediated immunity is inhibited by soluble mannan. Mannose receptor-mediated phagocytosis initiates IL-12 production. J Immunol 1997; 159:2462-2467.
106. Yamamoto Y, Klein TW, Friedman H. Involvement of the mannose receptor in cytokine interleukin-1β(IL-1β), IL-6, and granulocyte macrophage colony-stimulating factor responses, but not in

chemokine macrophage inflammatory protein 1b (MIP-1b), MIP-2, and KC responses, caused by attachment of Candida albicans to macrophages. Infect Immun 1997; 65:1077-1082

107. Schweizer A., Stahl PD, Rohrer J. A di-aromatic motif in the cytosolic tail of the mannose receptor mediates endosomal sorting. J Biol Chem 2000; 275:29694-29700.
108. Linehan SA, Martinez-Pomares L, da Silva RP et al. Endogenous ligands of carbohydrate recognition domains of the mannose receptor in murine macrophages, endothelial cells and secretory cells; potential relevance to inflammation and immunity. Eur J Immunol 2001; 31:1857-1866.
109. Czop JK, Kay J. Isolation and characterisation of b-glucan receptors on human mononuclear phagocytes. J Exp Med 1991; 173:1511-1520.
110. Brown G, Gordon S. A new receptor for β-glucans. Nature 2001; In press.
111. da Silva RP, Platt N, de Villiers WJS et al. Membrane molecules and macrophage endoytosis: scavenger receptor and macrosialin as markers of plasma-membrane and vacuolar functions. Biochem Soc Trans 1995; 24:220-224.
112. Platt N, Gordon S. Scavenger receptors: Diverse activities and promiscous binding of plyanionic ligands. Chem Biol 1998; 5:193-203.
113. Krieger M, Acton S, Ashkenas J et al. Molecular flypaper, host defense, and atherosclerosis. Structre, binding propreties and functions of macrophage scavenger receptors. J Biol Chem 1993; 263:4569-4572.
114. Platt N, Haworth R, Da Silva RP et al. Scavenger receptors and phagocytosis of bacteria and apoptotic cells. Adv Cell Mol Biol Mem Organs 1999; 5:71-85.
115. Peiser L, Gough PJ, Kodama T et al. Macrophage class A scavenger receptor-mediated phagocytosis of Escherichia coli: Role of cell heterogeneity, microbial strain, and culture conditions in vitro. Infectd Immunity 2000; 68:1953-1963.
116. Fraser I, Hughes D, Gordon S. Divalent cation-independent macrophage adhesion inhibited by monoclonal antibody to murine scavenger receptor. Nature 1993; 364:343-346.
117. Suzuki H, Kurihara Y, Takeya M et al. A role for macrophage sacvenger receptors in atherosclerosis and susceptibility to infection. Nature 1997; 386:292-296.
118. Thomas CA, Li Y, Kodama T et al. Protection from lethal gram-positive infection by macrophage scavenger receptor-dependent phagocytosis. J Exp Med 2000; 191:147-156.
119. Haworth R, Platt N, Keshav S et al. The macrophage scavenger receptor typeA is expressed by activated macrophages and protects the host against lethal endotoxic shock. J Exp Med 1997; 186:1419-1431.
120. Hagiwarea SI, Takeya M, Suzuki H et al. Role of macrophage sacvenger receptors in hepatic granuloma formation in mice. Am J Pathol 1999; 154:705-720.
121. Brown SB, Clarke M, Magowan L et al. Constitutive death of platelets leading to scavenger receptor-mediated phagocytosis. A caspase-independent cell clearance program. J Biol chem 2000; 275:5987-5996.
122. Morimoto Y, Wada Y, Hinagata J et al. VXFD in the cytoplasmic domain of macrophage scavenger receptors mediates their efficient internalisation and cell-expression. Biol Pharm Bull 1999; 22:1022-1266.
123. Fong LG, Le D. The processing of ligands by the class A scavenger receptor is dependent on signal information located in the cytoplasmic domain. J Biol chem 1999; 274:3608-3616.
124. Hsu HY, Hajjar DP, Khan KM et al. Ligand-binding to macrophage scavenger receptor A induces urokinase-type plasminogen activator expression by a protein kinase-dependent signaling pathway. J Biol Chem 1998; 273:1240-1246.
125. Falcone DJ, McCaffrey TA, Vergilio JA. Stimulation of macrophage urokinase expression by polyanions is protein kinase C-dependent and requires protein and RNA synthesis. J Biol Chem 1991; 266:22726-22732.
126. Abraham R, Choudhury A, Basu SK et al. Disruption of T cell tolerance by directing a self antigen to macrophage-specific scavenger receptors. J Immunol 1997; 158:4029-4035.
127. Elomaa O, Kangas M, Sahlberg C et al. Cloning of a novel bacteria-binding receptor structurally related to scavenger receptors and expressed in a subst of macrophages. Cell 1995; 80:603-609.
128. Kraal G, van der Laan LJW, Elomaa O et al. The macrophage receptor MARCO. Microbes Infect 2000; 2:313-316.
129. Elsourbagy NA, Xiaotong L, Terret J et al. Molecular characterisation of a human scavenger receptor, human MARCO. Eur J Biochem 2000; 267:919-926.
130. Palecanda A, Paulauskis J, Al-Mutari E et al. Role of the scavenger receptor MARCO in alveolar macrophage binding of unopsonised environmental particles. J Exp Med 1999; 189:1497-1506.
131. Elomaa O, Sankala M, Pikkrainen T et al. Structure of the human macrophage MARCO receptor and characterization of its bacteria-binding region. J Biol Chem 1998; 273:4530-4538.

132. van der Laan LJW, Döpp EA, Haworth R et al. Regulation and functional Involvement of macrophage scavenger receptor MARCO in clearance of bacteria in vivo. J Immunol 1999; 162:939-947.
133. Pikkrainen T, Bränström A, Tryggvason K. Expression of macrophage MARCO receptor induces formation of dendritic plasma membrane processes. J Biol Chem 1999; 274:10975-10982.
134. Landmann R, M Mllre B, Zimmerli W. CD14, new aspects of ligand and signal diversity. Microbes Infect 2000; 2:295-304.
135. Pugin J, Heumann D, Tomasz A et al. CD14 is a pattern-recognition receptor. Immunity 1994; 1:509-516.
136. Kirshning CJ, Wesche H, Merrill Ayres T et al. Human Toll-like receptor 2 confers responsiviness to bacterial lipopolysaccharide. J Exp Med 1998; 188:2091-2097.
137. Yang RB, Mark MR, Gray A et al. Toll-like receptor 2 mediates lipopolysaccharide-induced cellular signalling. Nature 1998; 395:284-288.
138. Kitchens RL, Munford RS. CD14-dependent internalisation of bacterial lipopolysaccharide (LPS) is strongly influenced by LPS aggregation but not by cellular responses. J Immunol 1998; 160:1920-1928.
139. Gegner JA, Ulevitch RJ, Tobias PS. Lipoploysaccharide (LPS) siganl transduction and clearance: Dual roles for LPS-binding protein and membrane CD14. J Biol Chem 1995; 270:5320-5325.
140. Pugin J, Kravchenko V, Kirklan, T et al. Glycosylphosphatidylinositol-anchored or integral membrane form of CD14. Infect Immunity 1998; 66:1174-1180.
141. Lemaitre B, Nicholas E, Michaut L et al. The dorso-ventral regulatory gene cassette spaetzle/Toll/Cactus controls the potent antifungal response in Drosophila adults. Cell 1996; 86:973-983.
142. Medzhitov R, Preston-Hurlburt P, Janeway CA. A human homologue of the Drosophila Toll protein signals activation of adaptive immunity. Nature 1997; 388:394-397.
143. Imler J-L, Hoffmann JA. Toll receptors in innate immunity. Trends Cell Biol 2001; 11:304-311.
144. Muzio M, Mantovani A. Toll-like receptors. Microbes and Immunity 2000; 2:251-255.
145. Poltorak A, He X, Smirnova I et al. Defective LPS signaling in C3H/HeJ and C57BL/10ScCr mice: Mutations in the Tlr4 gene. Science 1998; 282:2085-2088.
146. Ulevitch RJ, Tobias PS. Recognition of Gram-negative bacteria and endotoxin by the innate immune system. Curr Opin Immunol 1999; 11:19–22.
147. Underhill DM, Ozinsky A, Hajjar AM et al. The Toll-like receptor 2 is recruited to macrophage phagosomes and discriminates between pathogens. Nature 1999; 401:811–815.
148. Hemmi H, Takeuchi O, Kawai T et al. A Toll-like receptor recognizes bacterial DNA. Nature 2000; 408:740–745.
149. Lien E, Chow JC, Hawkins LD et al. Toll-like receptor 4 imparts ligand-specific recognition of bacterial lipopolysaccharide. J Clin Invest 2000; 105:497-504.
150. Poltorak A, Ricciardi-Castagnoli P, Citterio S et al. Physical contact between lipopolysaccharide and Toll-like receptor 4revealed by genetic complementation. Proc Natl Acad Sci USA 2000; 97:2163–2167.
151. Medzhitov R, Preston-Hurlburt P, Kopp E et al. MyD88 is an adaptor protein in the hToll/IL-1 receptor family signaling pathways. Mol Cell 1998; 2:253-258.
152. Shimazu R, Akashi S, Ogata H et al. MD-2, a molecule that confers lipopolysaccharide responsiveness on Toll-like receptor 4. J Exp Med 1999; 189:1777-1782.
153. Akashi S, Ogata H, Nagai Y et al. Cutting edge: Cell surface expression and lipopolysaccharide signaling via the toll-like receptor 4-MD-2 complex on mouse peritoneal macrophages. J Immunol 2000; 164:3471-3475.
154. Dziarski R, Wang Q, Miyake K et al. MD-2 enables Toll-like receptor 2 (TLR2)-mediated responses to lipopolysaccharide and enhances TLR2-mediated responses to Gram-positive and Gram-negative bacteria and their cell wall components. J Immunol 2001; 166:1938-1944.
155. Rock FL, Hardiman G, Timans JC et al. A family of human receptors structurally related to Drosophila Toll, Proc Natl Acad Sci USA 1997; 95:558-593.
156. Israel A. The IKK complex: An integrator of all signals that activate NF-κB? Trends Cell Biol 2000; 10:129-133.
157. Muzio M, Natoli G, Saccani S et al. The human Toll signaling pathway: Divergence of NF-κB and JNK/SAPK activation upstream of TRAF6. J Exp Med 1998; 187:2097-2101.
158. Wang Q, Dziarski R, Kirschning CJ et al. Micrococci and peptidoglycan activate TLR2—>MyD88—>IRAK—>TRAF—>NIK—>IKK—>NF-kappaB signal transduction pathway that induces transcription of interleukin-8. Infect Immunity 2001; 4:2270-2276.
159. Ozinsky A, Underhill DM, Fontenot JD et al. The repertoire for pattern recognition of pathogens by the innate immune system is defined by cooperation between toll-like receptors. Proc Natl Acad Sci USA 2000; 97:13766-13771.
160. Dangl JL, Jones JDG. Plant pathogens and integrated defence responses to infection. Nature 2001; 411:826-833.

Chapter 6

Intestinal Epithelial Cells:
A Route of Entry for Entero-Invasive Pathogens

Philippe J. Sansonetti and Guy Tran Van Nhieu

Summary

Bacterial pathogens have evolved a large number of strategies to colonise the intestinal epithelium. These may, however, be summarised in two major "pathovars": (i) adherent microorganisms that remain extracellular and bind to the apical pole of the intestinal epithelium, whereas (ii) invasive microorganisms have the capacity to penetrate into epithelial cells, thus defining a "pathovar" which itself encompasses a large array of different strategies. Recent recognition of the genetic bases of bacterial pathogenicity (i.e., cataloging of virulence genes located on plasmids, bacteriophages or pathogenicity islands) and analysis of the molecular cross talks that are established between the pathogen and its target cells have made sense of this diversity of interactions. The comparative strategies of pathogens that invade the intestinal epithelial barrier to conduct their pathogenic process are reviewed, with particular emphasis on bacteria such as *Shigella, Yersinia* and *Salmonella*. Each of these species represents a paradigm of interaction that is "dissected" with regard to its molecular and cellular characteristics of epithelial invasion and crossing. Particular interest is paid to the bacterial effectors and their eukaryotic cell targets. The signalling pathways induced by these invasive pathogens that alter the epithelial cell actin cytoskeleton will also be discussed as a recurrent theme during entry and dissemination into epithelial cells. Other signalling pathways alter the trafficking of cellular vesicles and induce changes in the intracellular compartment in which they reside, thus creating niches favourable to their survival and growth.

Functional Anatomy of the Intestinal Epithelial Barrier

The intestinal mucosa is a vast area covered by a single layer of epithelial cells that constitutes, in addition to its absorbtive and digestive properties, a very efficient barrier against the resident bacterial flora, particularly in the colon, but also against pathogens. Exclusion of the pathogens is not simply due to the physical presence of a continuous lining of cells that are tightly bound by junctions.[1] Epithelial cells also extend a thick array of microvilli on their apical surface which is prolonged by a thick layer of highly glycosylated, membrane-associated mucins forming the brush-border-associated glycocalyx.[2] It also responds to a multiplicity of factors such as the mucus layer that serves as a matrix entrapping bacteria and washing them away, according to intestinal peristaltism, while also concentrating a panoply of antibacterial factors: secreted IgA antibodies and antibacterial substances such as lactoferrin, lysozyme and cryptdins, a family of short hydrophobic antibacterial peptides produced by Paneth cells in intestinal crypts.[3] One should not forget, however, that the intestinal mucosa has also an immunological function. Effective immune surveillance against pathogens requires that the intestinal mucosa samples microbial antigens if not the microbes themselves. This supposes the presence of sites in the epithelium that are able to translocate those particulate antigens and microbes to the antigen presenting cells associated with the lymphoid follicles that are compo-

Intracellular Pathogens in Membrane Interactions and Vacuole Biogenesis, edited by Jean-Pierre Gorvel.

nents of the inductive arm of the mucosal immune system. These sites correspond to the follicle-associated epithelium (FAE) which barely represents 1/10 000 of the entire villous surface. They are characterised by the presence of M cells (M for microfold). These cells that are derived from regular villous epithelial cells,[4] lack microvilli, produce very little glycocalyx, and express intense endocytic activity that accounts for active translocation of particulate antigens to the underlying lymphoid tissue. These cells are often forming a large pocket containing a mononuclear cell, lymphocyte or macrophage. Differentiation of FAE into M cells is a complex process that involves close interaction between epithelial and lymphoid cells.[5]

Categories of Pathogens with Regard to Their Interaction with the Intestinal Epithelial Barrier

Intestinal pathogens can be classified into two major categories: those that adhere to the intestinal epithelium and those that invade the intestinal epithelium. To the former category belong pathogens such as *Vibrio cholerae* and enterotoxigenic *Escherichia coli* (ETEC). These microorganisms reach the apical pole of the intestinal lining and adhere through specific recognition of carbohydrates linked to glycoproteins or glycolipids of the epithelial cell membrane. From this extracellular site of colonisation which affects mostly the proximal small intestine, they produce toxins such as cholera toxin (CT) or thermolabile (LT) and thermostable (ST) toxins that act as pharmacological agents antagonising reabsorbtion of sodium and water and acting as agonists for secretion of chloride and water in the crypts, thus causing the diarrheal symptoms. To the latter category belong enteroinvasive pathogens such as *Shigella, Salmonella* and *Yersinia.* These are characterised by their capacity to invade the intestinal epithelium. From this site of invasion, several scenarios can be envisioned: *Shigella* remains essentially local, causing major inflammatory destruction of the colonic and rectal mucosa. *Yersinia* proceeds to loco-regional infection, involving the mesenteric lymph nodes draining the area of the gut that has been infected, in general the distal portion of the small intestine. *Salmonella* subsequently proceeds to systemic dissemination, such as in the case of typhoid fever, following infection of the small intestine by *Salmonella typhi.* It is clear that these different behaviours reflect genetic differences among these invasive pathogens that dictate a particular pattern of infection. In all cases, however, the way the microorganism interacts with the epithelial cells will have a significant impact on the pathogenic profile.

Cellular Routes of Invasion by Enteric Pathogens (M Cells vs. Villous Epithelial Cells vs. CD18-Mediated Pathways)

Recent contributions combining cell assay systems and in vivo models of intestinal invasion by bacteria have provided us with a complex picture. Three routes of invasion can currently be considered, as summarized in Figure 1 for *Salmonella.*[6]

M Cells of the FAE As a Route of Invasion

A variety of bacterial species, viruses and protozoans have been shown to translocate through the intestinal epithelium via M cells reviewed in refs. 7 and 8. Interestingly, these microbes take advantage of the physiological route of mucosal sampling of antigens to cross the epithelial barrier. This brings significant change in the classical concept of mucosal invasion involving primary translocation through the regular villous epithelium since the pathogen's primary and early "task" will be to survive the deleterious effect of phagocytic cells (i.e., resident macrophages) that prevail in these areas, instead of entering straight into epithelial cells via their apical pole, a process that appears difficult to achieve, even for these invasive microbes.[9] Invasion of the epithelium thereby becomes the secondary event in the chronology of mucosal invasion. On the other hand, this entry route is a strong incentive to use live attenuated vectors derived from these pathogens in order to optimise mucosal immunisation thanks to their natural targeting to the inductive sites of the intestinal immune system.[10]

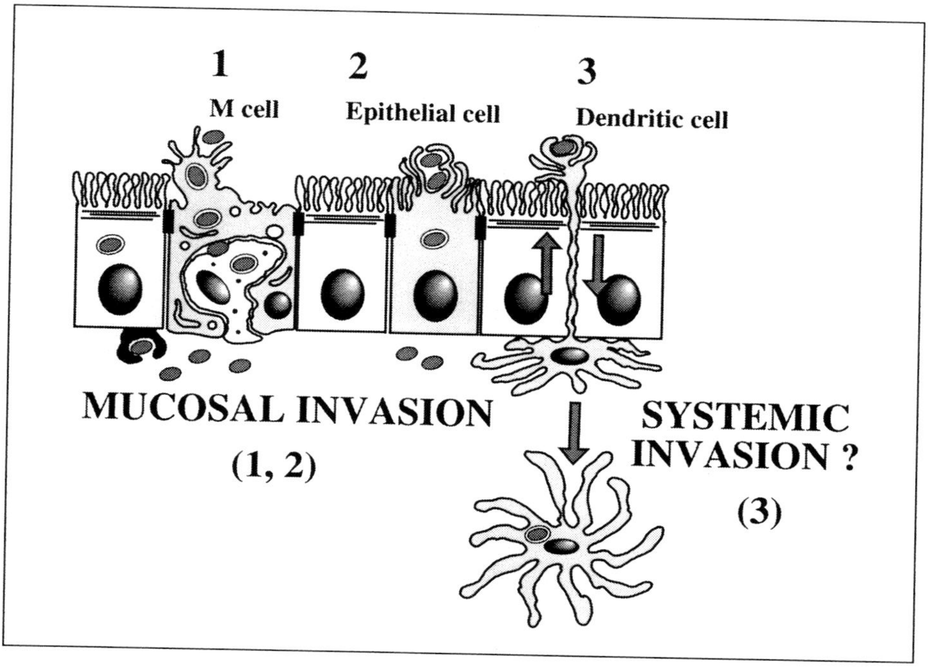

Figure 1. Three possible routes of crossing of the intestinal epithelial barrier as described in the case of *Salmonella*. (1) Translocation through M cells, (2) translocation through villous epithelial cells, (3) luminal capture by phagocytic cells (i.e., dendritic cells), also called the CD18-dependent route.

In the case of *Shigella* (Fig. 2), clinical observations have shown that prior to development of the full blown dysenteric syndrome, early inflammatory lesions of the colo-rectal mucosa often resemble aphtoid ulcers. Histopathological description of these lesions generally reveals the presence of a lymphoid follicle.[11] Experiments carried out in monkeys,[12] as well as in a rabbit model of infection of ligated intestinal loops,[13] have confirmed these clinical observations. In the latter model, bacteria were shown to selectively translocate through M cells.[14] So far, no specific adherence system mediating the interaction between *Shigella* and the luminal side of M cells has been identified. However, shigellae expressing an invasive phenotype (see below for description) are more efficient at translocating through M cells than their non-invasive counterpart, indicating that the invasive microbe is not translocated as an inert particle, but promotes its efficient internalization. Passage seems to occur rapidly, without lysis of the endocytic vacuole and cytotoxicity to the cell, bacteria then find themselves delivered into the intraepithelial pocket of the M cell where they are phagocytosed by the dendritic cells and resident macrophages prevailing in this dome area. Once phagocytosed, the survival strategy of *Shigella* is to cause rapid apoptosis of the macrophage, a process that has been described both in vitro and in vivo.[14,15] This is likely to allow these bacteria access to the baso-lateral side of epithelial cells where they can efficiently enter. Indeed, the molecular mechanism of macrophage killing by *Shigella* that involves activation of caspase 1[16] also initiates inflammation by causing maturation of two pro-inflammatory cytokines: IL-1β and IL-18.[17] This early inflammatory process leads to quick disruption of the epithelial barrier, thereby facilitating further *Shigella* invasion. In consequence, M cells appear to play a strategic role in *Shigella* invasion by facilitating its crossing of the epithelial barrier and by putting bacteria in a situation to trigger a signalling cascade that disrupts the intestinal epithelium. Further bacterial entry into epithe-

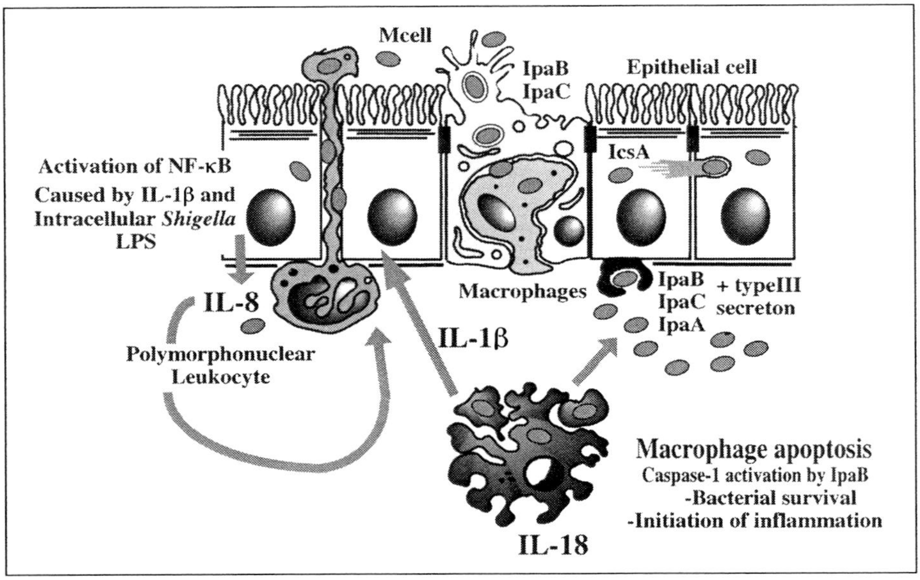

Figure 2. Crossing of the intestinal epithelium by *Shigella*.

lial cells and initiation of cell spread, as described below, achieve the characteristic local and strongly pro-inflammatory process of *Shigella* infection.

Yersinia enterocolitica cause diarrheal diseases, whereas *Yersinia pseudotuberculosis* usually causes mild or infra-clinical enteric symptoms, but may cause subsequent mesenteric lymphadenitis that sometimes leads to systemic diffusion. Like *Shigella*, these bacteria are primarily crossing the intestinal epithelium through the FAE (Fig. 3), particularly in the terminal ileum where mucosal lymphoid follicles are aggregated as Peyer's patches.[18] There is good evidence for the existence of a potent mechanism of *Yersinia* binding and internalization into M cells. Invasin, a 103 kDa, chromosomally-encoded outer membrane protein of *Y. pseudotuberculosis* binds β1 integrins that are expressed apically on M cells, unlike polarised villous epithelial cells in which they are expressed baso-laterally. Inv negative mutants still adhere to and invade M cells, but at a much lower level than the wild type strain and their colonisation potential for Peyer's patches is considerably reduced.[19] Binding of Invasin to β1 integrins is of high affinity and leads to internalization by a zippering process.[20] Other *Yersinia* surface proteins such as Ail, PsaA and YadA may also contribute to binding and internalization, thus accounting for the residual invasive activity of *inv* mutants.[21] As already mentioned for *Shigella*, once it has reached the dome of the follicle, *Yersinia* must survive attack by resident macrophages. In order to achieve this survival strategy, *Yersinia* express an antiphagocytic property that is caused by injection of several protein effectors disrupting the cytoskeletal assembly processes through a plasmid-encoded type III secreton.[22,23] Among these antiphagocytic effectors are YopH, YopT and YopE. YopH is a tyrosine phosphatase which rapidly dephosphorylates paxilin, $p130^{cas}$ and FAK, proteins that are involved in the assembly of cytoskeletal complexes involved in phagocytosis.[24] YopT provokes the depolymerisation of actin filaments by inducing redistribution of the RhoA GTPase which plays a major role in the formation of stress cables.[25] YopE is a potent cytotoxin expressing a GAP function that negatively regulates the function of the small GTPases of the Rho family involved in phagocytosis.[26] As a result, yersiniae remain essentially extracellular in infected Peyer's patches and mesenteric lymph nodes. This allows their extracellular survival as well as possible Inv-mediated entry into epithelial cells which is not fully abolished by the antiphagocytic mechanisms that apply to the macrophage.

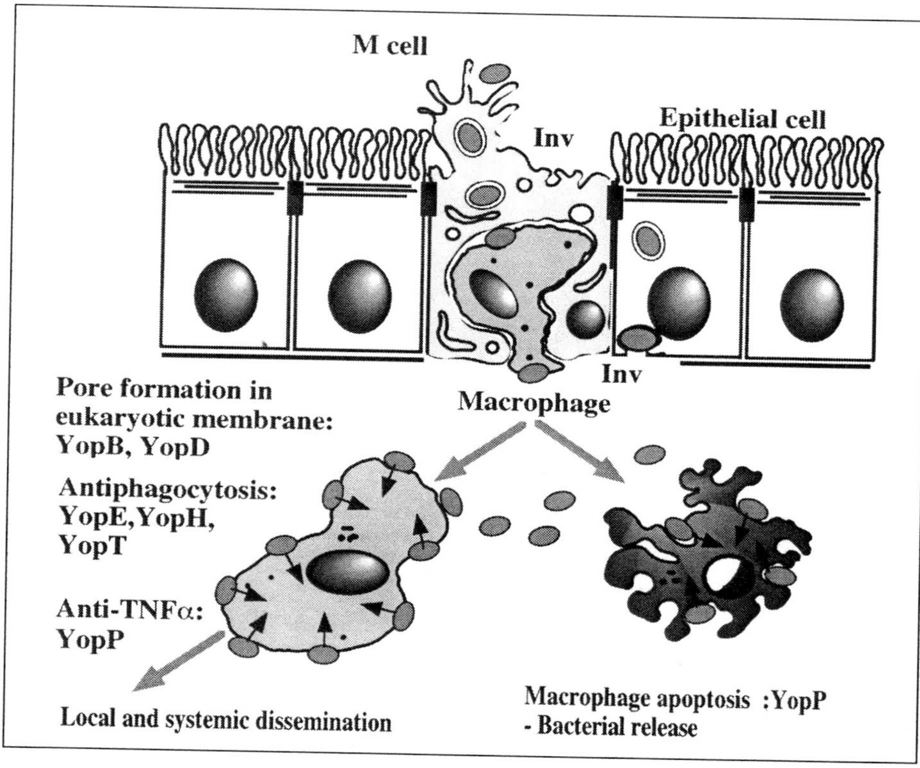

Figure 3. Crossing of the intestinal epithelium by *Yersinia*.

Following oral administration, *Salmonella typhimurium* crosses the epithelial barrier and causes systemic dissemination resulting in fatal septicaemia (Fig. 4). A close picture is seen in humans infected by *Salmonella typhi*. Following inoculation of mouse ligated intestinal loops, *S. typhimurium*, like *Shigella* and *Yersinia*, show clear tropism for M cells with epithelial translocation through the FAE.[27,28] *Salmonella* (and possibly other invasive pathogens) may, however, follow alternative routes of entry that will be considered later. Although similar conditions have not been tested, *S. typhimurium,* unlike *Shigella*, appear to be cytotoxic for M cells.[28,29] Long Lpf fimbriae mediate a somewhat specific adherence to murine M cells.[30] However, Lpf may be only one of several factors representing a redundant system of interaction. A carbohydrate containing galactose-β(1-3)-galactosamine has been shown to serve as a receptor for *S. typhimurium* on Caco-2 cells,[31] but its relevance in vivo is unknown. Adherence of *S. typhimurium* to M cells is followed by ruffling of the cell membrane and macropinocytosis reflecting cytoskeletal changes similar to those shown to occur in vitro in cultivated cells.[32] The invasion complex encoded by the *Salmonella* pathogenicity island 1 (SPI1) and additional effector proteins secreted through the type III secreton encoded by SPI1 are likely to contribute to invasion of M cells. SPI1 mutants are not cytotoxic for M cells and their capacity to cross the intestinal barrier is seriously impaired whereas their virulence remains intact following systemic infection.[33] Once it has reached the dome of lymphoid follicles, *Salmonella* adopts a strategy that is clearly different from those described for *Yersinia* and *Shigella*. Following its phagocytosis by resident macrophages[34] (and probably also dendritic cells), expression of SPI1 is associated with SipB-dependent apoptotic killing of the phagocytes via activation of caspase-1.[35] However, it appears that *Salmonella* have evolved a strategy of survival inside phagocytes, particu-

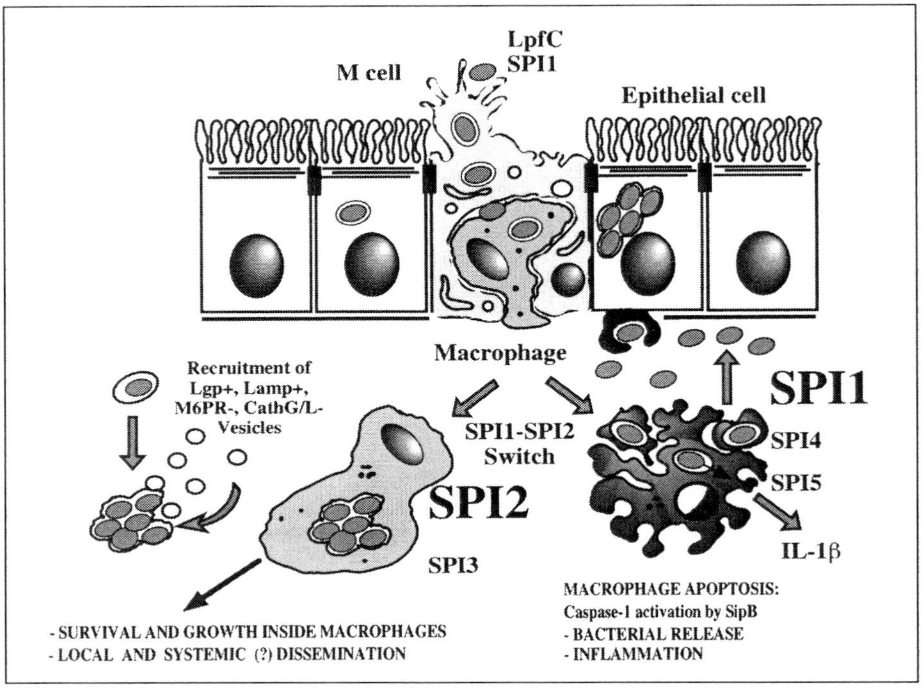

Figure 4. Crossing of the intestinal epithelium by *Salmonella*.

larly macrophages, which may facilitate its systemic dissemination. SPI2, another pathogenicity island encoding an alternative type III secreton and its dedicated effector proteins,[36] as well as a series of phoP/phoQ-regulated genes[37] appear essential for *Salmonella* survival and growth inside macrophages.

A Villous Epithelial Cell Route of Invasion?

Although M cells seem to account in a large part for initial crossing of the intestinal epithelium by enteroinvasive microorganisms, alternative routes of crossing are likely to exist based on both in vitro and in vivo evidence. Thirty five years ago, Takeuchi showed electron microscopic evidence for direct invasion of villous epithelial cells by *Salmonella* in vivo.[38] In our own studies with *Shigella*, we have repeatedly observed invasion of intestinal villi in the absence of Peyer's patch in the rabbit ligated model of *Shigella* infection.[39-41] However, invasion of the villous epithelium appears significantly later (i.e., 8h) than invasion of the FAE (i.e., 2-4h), the size of the bacterial inoculum is "gigantic" (i.e., $5x10^9$ cfu, compared to some 100 cfu required to cause the disease in humans and infection proceeds in a close system (i.e., ligated loop), thereby preventing washing out of bacteria by intestinal fluid. Although this model has appeared crucial to our understanding of the pathogenesis of shigellosis, the artificial conditions in which infection is carried out may overwhelm the otherwise efficient mechanisms of epithelial protection against apical invasion, thereby by-passing the key role of M cells. In addition, it has now become clear that intestinal epithelial cells respond to invasive pathogens by expressing pro-inflammatory cytokines and chemokines.[42]

It has also been shown that a consequence of this reprogramming of epithelial cells to produce pro-inflammatory molecules was leading to attraction and further trans-epithelial migration of polymorphonuclear leukocytes (PMN), thereby disrupting the permeability of the epithelium.[43] Preliminary evidence indicates that a direct interaction of *Salmonella* without internalization is sufficient to trigger transepithelial migration of PMNs.[44] It is likely that in *Salmonella* and *Shigella,* some of the proteins injected by the type III translocon trigger the activation of the pro-inflammatory transcription factors such as NF-κB and AP-1. In the case of *Shigella,* this process facilitates bacterial invasion of these epithelial cells via their baso-lateral pole that seems, in general, more permissive to bacterial entry.[39,45] It seems therefore that transepithelial signaling induced by these bacteria may facilitate epithelial invasion in areas that do not possess FAE and lymphoid structures.

A CD18-Dependent Route of Infection for Salmonella *and Other Invasive Bacteria ?*

Even though SPI1-deficient mutants of *S. typhimurium* are deficient in invading both M cells and villous epithelial cells, they conserve their capacity to disseminate extraintestinally and to kill infected mice.[46] In addition, Inv negative mutants of *Y. enterocolitica* are unable to colonise Peyer's patches and the corresponding mesenteric lymph nodes, but they still retain their capacity of systemic dissemination.[47] These observations have suggested that alternative routes to M cells and villous epithelial cells may exist for crossing the epithelial barrier. Recent evidence indicates that in the case of *S. typhimurium,* one such alternative route exists that permits systemic dissemination. This route involves CD18-expressing mononuclear phagocytes.[48] A SPI1 negative mutant of *S. typhimurium* has recently been shown to conserve residual ability to cross the murine intestinal epithelium and to reach the spleen after oral administration. Bacterial uptake appears to be mediated by dendritic cells which open the tight junctions and send dendrites on the luminal side of the epithelium where they sample the bacteria. These dendritic cells express tight junction proteins that may be involved in resealing of the epithelium, thereby preserving the impermeability of the epithelial barrier while moving transcellularly.[49]

Molecular and Cellular Mechanisms of Epithelial Cell Invasion

Beyond the complexities of the mechanisms that lead to epithelial translocation of enteroinvasive pathogens and their further local (i.e., *Shigella*), regional (i.e., *Yersinia*) and systemic (i.e., *Salmonella/Yersinia*) dissemination, it remains that entry of these bacteria into intestinal epithelial cells, particularly following the crossing of M cells, is at the centre of the process. Recent progress in understanding how these bacteria manipulate the cytoskeleton of these non-phagocytic cells to promote their own entry and further invasion are reviewed in this Chapter. Two major mechanisms of bacterial internalization into epithelial cells have been described.[50] The "zipper" process corresponds to tight envelopment of the bacterial body by the mammalian cell membrane that is determined by high-affinity binding of a bacterial surface protein with a transmembrane receptor molecule of the mammalian cell surface that is involved in cell adhesion, i.e., the *Yersinia* Invasin binding to integrins of the β1 family.[51] or the *Listeria monocytogenes* Internalin A binding to E-cadherin.[52] The "trigger" process corresponds to bacteria causing massive cytoskeletal changes in the mammalian cell underneath its site of interaction, very similar to those induced by delivery of a growth factor, thereby causing a ruffling process that internalises the bacterial body in a macropinocytic vacuole.[53]

Y. pseudotuberculosis, a paradigm of "zippering" entry. The major protein that allows entry into epithelial cells is Invasin (Inv), an outer membrane protein of 986 aa.[51] Inv interacts with β1 integrins (α3β1, α4β1, α5β1, α6β1 and αVβ1) which are involved in adherence of epithelial cells to the extracellular matrix. The C-terminal domain of Inv forms a compact domain, or "super domain", that interacts with integrins. In spite of its lack of an RGD motif that is characteristic of the binding site of matrix proteins to integrins, it is a competitive inhibitor of fibronectin binding to β1 integrins. Inv is able to oligomerise.[54] and to bind β1 integrins with an affinity constant that is much higher than that of fibronectin.[55] These two properties, clustering of integrins and high affinity binding, are thought to account for transition between a

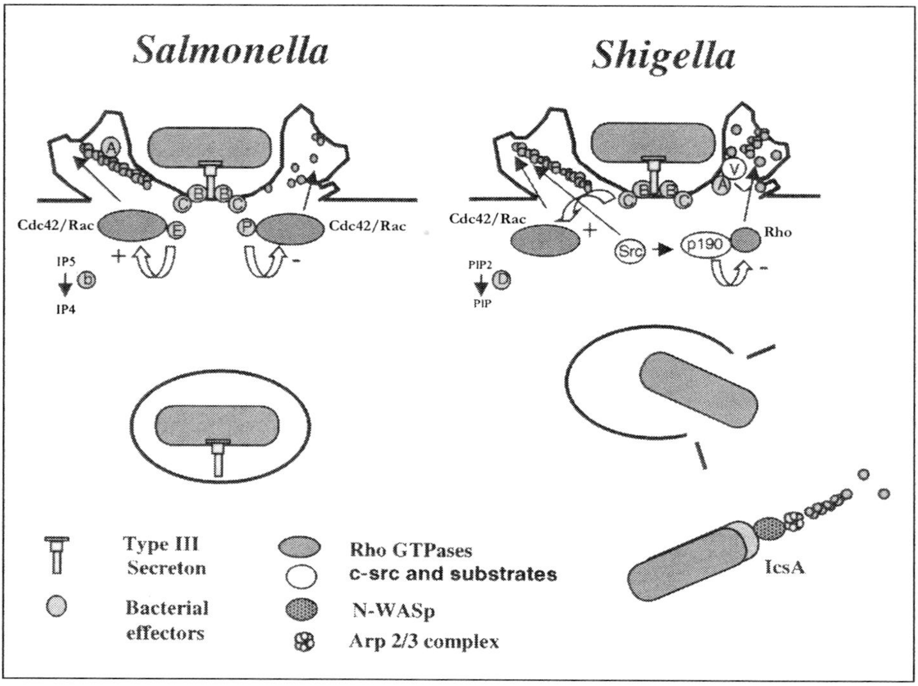

Figure 5. Comparative signalling during entry of *Salmonella* and *Shigella* into epithelial cells. The "trigger" paradigm.

process of physiological adherence to a process of internalization. The cytoplasmic domain of the β1 integrin transmits signals to the cell cytoskeleton that induce internalization, some residues being essential for signal transduction; this domain, in physiological conditions, interacts with components of focal complexes and adherence plaques. Inhibitors of tyrosine phosphorylation inhibit internalization of *Yersinia* and p125FAK, the focal adhesion kinase is involved, probably by allowing the negative regulation of adhesion structures that are incompatible with the bacterial entry process.[56] The cytoplasmic domain of the α chain is not required for internalization. Consistently, alterations of the β1 chain cytoplasmic domain that decrease focal adhesion association increase internalization. It is likely that loosening of these interactions may increase receptor motility in the membrane, thereby facilitating internalization.

Shigella and *Salmonella*: paradigms of "triggering" entry involving a type III secreton, but models of differential intracellular behaviour (Fig. 5). The entry process of *Shigella* and *Salmonella* finds its unity in both microorganisms expressing a type III secreton which has been visualised in both species: Mxi-Spa in *Shigella*[57] and Inv-Spa in *Salmonella*,[58] and has started to be characterised with regard to its protein components in both species.[58-60] These secretons are composed of a cytoplasmic bulb followed by a disk-like structure that spans the inner and outer membranes. A needle-like structure crosses the previous domains and extends outside the outer membrane with an average size of 60 nm. This needle structure is involved in the secretion of bacterial effectors that are released upon contact between the bacteria and their target cell. It is likely that the first secreted products participate in the formation of a pore in the eukaryotic cell membrane. IpaB and IpaC in the case of *Shigella*, SipB and SipC in the case of *Salmonella* have been shown to be part of this pore that accounts for the intracellular translocation of the other effectors.[57,61] Actin nucleation and its subsequent polymerisation is triggered by the signalling domain that is anchored into the eukaryotic membrane by the type III translocon. This leads to the formation of filopodes and lamellipodes that associate in a localised ruffle that internalises the bacterial body by a macropinocytic process.[62]

Effector Proteins and Signals Involved in Salmonella *Entry into Epithelial Cells*

Effectors have been identified that are delivered via the type III secreton Inv-Spa. Two homologous proteins, SopE1 and SopE2 act as exchange factors (GEF) for the small GTPases Cdc42 and Rac.[63] They catalyse the exchange of GDP for GTP on these small GTPases, thereby leading to a cascade of activation signals leading to actin polymerisation.[64] Activated, GTP-bound Cdc42 or Rac interact with proteins of the WASP family that, in turn, bind and activate Arp2/3, a complex of seven proteins that induces actin nucleation.[65] Alternatively, or in a coordinated manner, SipC, an effector protein encoded by SIP1 that inserts into the eukaryotic cell membrane may directly induce actin nucleation.[66] SipA, another product of SPI1, binds and stabilises actin filaments, thereby favoring the outward extension of cell protrusions.[67] SopB, an inositol-phosphatase is also transferred through the type III secreton, although its exact role in the entry process is unknown.[68] To complete the entry process and "repair" the major local cytoskeletal alterations induced during the early steps of entry, *Salmonella* induces actin depolymerisation through the translocation of another effector protein, SptP. This protein functions as a GAP (GTPase activating protein) that down regulates the function of Cdc42 and Rac by stimulating their GTPase activity, thus inducing a switch towards their inactive GDP-bound form.[69]

Effector Proteins and Signals Involved in Shigella *Entry into Epithelial Cells*

In spite of some homology between the *Salmonella* Sip effectors of entry and their Ipa counterparts in *Shigella*, it appears that their entry mechanisms into epithelial cells show significant differences. First, the sequence of the *Shigella* virulence plasmid that is necessary and sufficient to promote entry into epithelial cells has not shown any gene homologue to *sop*E or *spt*P.[70] Three effector proteins secreted by the type III secretory apparatus of *Shigella* have been shown to induce the signals required for the cytoskeletal rearrangements causing bacterial entry. IpaC, which is a component of the pore allowing the translocation of effector proteins is also involved in triggering actin polymerisation, probably through its C-terminal domain exposed into the host cell cytoplasm.[71] The mechanism by which IpaC operates is still uncharacterized. IpaA binds to vinculin, a focal adhesion protein that plays a key role in the anchoring of actin filaments to the plasma membrane.[72] High-affinity binding of IpaA to the N-terminal head of vinculin triggers its unfolding thereby allowing F-actin-binding to its C-terminus. This leads to the formation of the so-called actin cup, a focal adhesion-like structure that seems essential to carry out *Shigella* entry.[73] Surprisingly, the IpaA-vinculin complex also induces actin depolymerisation.[74] This activity is likely to participate in the transition from filopodial to lamellipodial structures causing major extensions off the cell surface in the actin cup structure and in the final resorbtion of the entry focus after entry has occured. IpgD, another secreted effector protein has a phosphatidyl-inositol phosphatase activity that may account for the relaxation of the membrane-cytoskeletal association, but a detailed account of the actual function of this protein is still required.[75] The proto-oncogen c-src that is recruited at the entry site regulates the time-wise organisation of the signalling process induced by the *Shigella* effector. Depending on the stage of the entry process, c-src either enhances the polymerisation of actin filaments at the early stage, or down regulates it at a later stage.[76,77]

Differential Intracellular Behavior of Salmonella *and* Shigella *Dictates Differential Pathogenic Properties*

Once intracellular, *Shigella* and *Salmonella* adopt different behaviours, as *Salmonella* remain entrapped inside a vacuolar compartment, whereas *Shigella* disrupt their endocytic vacuole and escape into the cytoplasm.[78] By polymerising actin at one bacterial extremity, *Shigella* has the capacity to move intracellularly and from cell to cell. This allows very efficient epithelial colonisation that is a key feature of *Shigella* pathogenesis as mutants that have lost this capacity are severely impaired in their virulence, both in vitro and in vivo[79-81] including in human volunteers.[82] Intracellular motility of *Shigella* is caused by the polar expression of an outer membrane protein of 1102 aa, IcsA, that like the entry effectors, is encoded by the virulence plasmid of *Shigella*.[80] The N-terminal portion of IcsA, through a series of glycine-rich repeats (GRRs), binds to N-WASP (Egile et al, manuscript in preparation)[83] and activates this protein by causing its unfolding in a way that makes its C-terminal domain (i.e., the VCA domain). The VCA domain then becomes available for recruitment and binding of the Arp2/3 complex, thereby causing actin nucleation and polymerisation.[84] In consequence, this tripartite complex (i.e., IcsA, N-WASP and Arp2/3) appears necessary and sufficient to cause actin nucleation/ polymerisation and thereby to promote bacterial motility in the cytoplasm at the speed at which actin can polymerise. Bacterial bodies may therefore move at a speed of 10 microns per minute and even sometimes faster. This mechanism is similar to the one that promotes intracellular motility of *L. monocytogenes*, although in this case the Arp2/3 complex binds directly onto ActA, the functional analogue of IcsA in this species.[85] *Shigella* uses this capacity for intracellular movement to spread from cell to cell in the context of a differenciated, polarised epithelial lining. In addition to motility related to IcsA-mediated actin polymerisation, cell to cell spread also involves engagement by *Shigella* of components of the intermediate junction.[86] It seems that the protrusion formed by the spreading microorganism is actively endocytosed by the adjacent cell in a process that requires activation of myosin II .[87] Following internalization of the protrusion, the two membrane layers are lysed by a mechanism that requires secretion of the pore-forming IpaB and IpaC proteins.[88-90] Current research is aimed at further defining molecular details of this cell to cell spreading process. Altogether, this particular behavior of *Shigella* appears to have two major consequences that bear heavily on the pathogenic profile of the disease: it tends to force the progression of infection into a scheme of predominant epithelial invasion, thus explaining that shigellosis, in spite of its potential severity, remains essentially a mucosal infection. As recent evidence indicates that *Shigella*, once intracellular, induce strong and sustained activation of the pro-inflammatory transcription factor NF-κB, large epithelial colonisation achieved by the intercellular spreading phenotype of *Shigella* that remain protected from the intervention of phagocytic cells, may account for the massive inflammatory destruction of mucosal tissues which is characteristic of bacillary dysentery.[91]

A majority of studies have dealt with the intracellular behavior of *Salmonella* inside macrophages (as reviewed in ref. 92). *Salmonella*-containing vacuoles, in HeLa cells, were initially shown to form two populations: one that appeared separated from the endocytic route and one that followed a "classical" phagocytic pathway.[93] Both compartments may be prone to *Salmonella* survival and intracellular growth. However, live pathogenic *Salmonella* are now currently associated with an atypical compartment that acquires some lysosomal markers such as Lgp or Lamp, but does not acquire mannose-6-phosphate receptors and cathepsin D and L which are markers of late maturation towards terminal lysosomes.[94] These data indicate that *Salmonella* is able to interrupt the maturation of its compartment in order to achieve its intracellular survival and growth strategy. Rab-7 may control addition of the membranous material constituting this compartment by recruiting and fusing vesicles that are rich in Lpg and poor in cathepsins.[94] This compartment is also characterised by its capacity to induce the formation of tubular structures in HeLa cells.[95] The physiological relevance of the occurrence of these structures is not yet clear, although they appear to be characteristic of virulent strains. It has recently become obvious that the SPI2 pathogenicity island of *Salmonella* is essential for controlling this particular maturation process. A large part of the research that has been carried out so far in non-polarized epithelial cells now needs to be considered in the perspective of polarity as a major parameter in the context of bacterial translocation through the epithelial barrier.

Conclusion

Entero-invasive bacterial pathogens have provided ample opportunities, thanks to a multidisciplinary approach integrating molecular genetics, cell biology, biochemistry, immunology and experimental medicine, to decipher the signalling processes and their consequences on the dynamic changes in the cell cytoskeleton, on the trafficking of cell compartments, on the mechanisms of cell survival and death that characterise the complex process by which these microbes subvert, disrupt, invade and eventually destroy barriers such as the intestinal epithelium. As shown in this review, these pathogenic processes are much more complex than simple apical entry followed by epithelial translocation and delivery of the invasive microorganism to sub-epithelial tissues. More than bacterial-epithelial cell translocation, one is dealing with a complex array of sequential interactions involving not only the bacteria, but also various cell populations such as the epithelial cells themselves and phagocytes. Entero-invasive bacteria actually provide us with invaluable information on the co-ordination of the innate immune response of the host to pathogens at the very early stage of the infectious process.

Acknowledgements

Work in PMM is supported by the Howard Hughes Medical Institute. PS and GTVN would like to thank their colleagues of PMM for work and discussion in the field and Colette Jacquemin for editing this manuscript.

References

1. Madara JL, Nash S, Moore R et al. Structure and function of the intestinal epithelial barrier in health and disease. Monogr Pathol 1990; 31:306-324.
2. Maury J, Nicoletti C, Guzzo-Chambraud I et al. The filamentous brush border glycocalyx, a mucin-like marker of enterocyte hyperpolarisation. Eur J Biochem 1995; 228:323-331.
3. Sanderson IR, Walker WA. Mucosal barrier, an overview. In: Ogra PL, Lamm ME, Bienenstock J, Mestecky J, Strober W, McGhee JR, eds. Mucosal Immunology. 2nd Ed. San Diego: San Diego Academic Press, 1999:5-17.
4. Neutra MR. M cells in antigen sampling. Curr Top Microbiol Immunol 1999; 236:17-32.
5. Kerneis S, Bogdanova A, Kraehenbuhl JP et al. Conversion by Peyer'patch lymphocytes of human enterocytes into M cells that transport bacteria. Science 1996; 277:949-952.
6. Vasquez-Torrez A, Fang F. Cellular routes of invasion by enteropathogens. Curr Opin Microbiol 2000; 3:54-59.
7. Owen RL. Uptake and transport of intestinal macromolecules and microorganisms by M cells and Peyer's patches—A historical and personal perspective. Sem Immunol 1999; 11:157-163.
8. Sansonetti P, Phalipon A. M cells as ports of entry for enteroinvasive pathogens: Mechanisms of interaction, consequences for the disease process. Sem Immunol 1999; 11:193-203.
9. Mounier J, Vasselon T, Hellio R et al. Shigella flexneri enters human colonic Caco-2 cells through the baso-lateral pole. Infect Immun 1992; 60:237-248.
10. Phalipon A, Sansonetti PJ. Live attenuated Shigella flexneri mutants as vaccine candidates against shigellosis and vectors for antigen delivery. Biologicals 1995; 23:125-134.
11. Mathan MM, Mathan VI. Intestinal manifestations of invasive diarrheas and their diagnosis. Rev Infect Dis 1991; 13(Suppl.4):S314-S318.
12. Sansonetti PJ, Arondel J, Fontaine A et al. OmpB (osmo-regulation) and icsA (cell to cell spread) mutants of Shigella flexneri. Evaluation as vaccine candidates. Probes to study the pathogenesis of shigellosis. 1991; Vaccine 9:416-422.
13. Wassef JS, Keren DF, Mailloux JL. Role of M cells in initial antigen uptake and in ulcer formation in the rabbit intestinal loop model of shigellosis. Infect Immun 1989; 57:858-863.
14. Sansonetti PJ, Arondel J, Cantey RJ et al. Infection of rabbit Peyer's patches by Shigella flexneri: Effect of adhesive and invasive bacterial phenotypes on follicular-associated epithelium. Infect Immun 1996; 64:2752-2764.
15. Zychlinsky A, Prévost MC, Sansonetti PJ. Shigella flexneri induces apoptosis in infected macrophages. Nature 1992; 358:167-169.

16. Hilbi H, Moss JE, Hersh D et al. Shigella-induced apoptosis is dependent on caspase-1 which binds to IpaB. J Biol Chem 1998; 273:32895-32900.
17. Sansonetti PJ, Phalipon A, Arondel J et al. Caspase-1 activation of IL1β and IL-18 are essential for Shigella flexneri induced inflammation. Immunity 2000; 12(5):581-590.
18. Grutzkau A, Hanski C, Hanhn H et al. Involvement of M cell in the bacterial invasion of Peyer's patches: a common mechanism shared by Yersinia enterocolitica and other enteroinvasive bacteria. Gut 1990; 31:1011-1015.
19. Clark MA, Hirst BH, Jepson A. M-cell surface β1 integrin expression and invasin-mediated targeting of Yersinia pseudotuberculosis to mouse Peyer's patch M cells. Infect Immun 1998; 66:1237-1243.
20. Isberg RR. Uptake of enteropathogenic Yersinia by mammalian cells. Curr Top Microbiol Immunol 1996; 209:1-24.
21. Marra A, Isberg RR. Invasin-dependant and invasin-independant pathways for translocation of Yersinia pseudotuberculosis across the Peyer's patch intestinal epithelium. Infect Immun 1997; 65:3412-3421.
22. Fällmann M, Persson C, Wolf-Watz H. Yersinia proteins that target host cell signalling pathways. J Clin Invest 1997; 99:1153-1157.
23. Cornelis G. The Yersinia deadly kiss. J Bacteriol 1998; 180:5495-5504.
24. Persson C, Carballeira N, Wolf-Watz H et al. The PTPase YopH inhibits uptake of Yersinia, tyrosine phosphorylation of $p130^{cas}$ and FAK, and the associated accumulation of these proteins in peripheral focal adhesions. EMBO J 1997; 16:2307-2318.
25. Zumbihl R, Aepfelbacher M, Andor A et al. The cytotoxin YopT of Yersinia enterocolitica induces modifications and cellular redistribution of the small GTP-binding protein RhoA. J Biol Chem 1999; 274:29289-29293.
26. Black D, Bliska J. The RhoGAP activity of the Yersinia pseudotuberculosis cytotoxin YopE is required for antiphagocytosis function and virulence. Mol Microbiol 2000; 37:515-527.
27. Clark MA, Jepson MA, Simmons NL et al. Preferential interaction of Salmonella typhimurium with mouse Peyer's patch M cells. Res Microbiol 1994; 145:543-552.
28. Jones BD, Ghori N, Falkow S. Salmonella typhimurium initiates murine infection by penetrating and destroying the specialised epithelial M cells of Peyer's patches. J Exp Med 1994; 180:15-23.
29. Kohbata S, Yokobata H, Yabuuchi E. Cytopathogenic effect of Salmonella typhi GIFU10007 on M cells of murine ileal Peyer's patches in ligated ileal loops: An ultrastructural study. Microbiol Immunol 1986; 30:1225-1237.
30. Baumler AJ, Tsolis RM, Heffron F. The lpf fimbrial operon mediates adhesion of Salmonella typhimurium to murine Peyer's patch. Proc Natl Acad Sci USA 1996; 93:279-283.
31. Giannasca KT, Giannasca PJ, Neutra MR. Role of apical membrane glycoconjugates in adherence of Salmonella typhimurium to intestinal epithelial cells. Infect Immun 1996; 64:135-145.
32. Finlay BB, Falkow S. Salmonella interaction with polarised human intestinal Caco-2 epithelial cells. J Infect Dis 1990; 162:1096-1106.
33. Penheiter KL, Mathur N, Giles D et al. Non-invasive Salmonella typhimurium mutants are avirulent because of an inhability to enter and destroy M cells of ileal Peyer's patches. Molec Microbiol 1997; 24:697-709.
34. Hopkins SA, Niedergang F, Corthesy-Theulaz IE et al. A recombinant Salmonella typhimurium vaccine strain is taken up and survives within murins Peyer's patch dendritic cells. Cellular Microbiol 2000; 2:59-68.
35. Hersh D, Monack DM, Smith MR et al. The Salmonella invasin SipB induces macrophage apoptosis by binding to Caspase-1. Proc Natl Acad Sci USA 1999; 96:2396-2401.
36. Shea J, Hensel M, Gleeson C et al. Identification of a virulence locus encoding a second type III secretion system in Salmonella typhimurium. Proc Natl Acad Sci USA 1996; 93:2593-2597.
37. Miller SI. PhoP/PhoQ: Macrophage-specific modulator of Salmonella virulence? Mol Microbiol 1991; 5:2073-2078.
38. Takeuchi A. Electron microscope studies of experimental Salmonella infection. Am J Pathol 1966; 50:109-136.
39. Perdomo JJ, Cavaillon JM, Huerre M et al. Acute inflammation causes epithelial invasion and mucosal destruction in experimental shigellosis. J Exp Med 1994; 180:1307-1319.
40. Sansonetti PJ, Arondel J, Cavaillon JM et al. Role of IL-1 in the pathogenesis of experimental shigellosis. J Clin Invest 1995; 96:884-892.
41. Sansonetti PJ, Arondel J, Cantey RJ et al. Infection of rabbit Peyer's patches by Shigella flexneri: Effect of adhesive or invasive bacterial phenotypes on follicular-associated epithelium. Infect Immun 1996; 64:2752-2764.
42. Jung HC, Eckmann I, Yang SK et al. A distinct array of pro-inflammatory cytokines is expressed in human colon epithelial cells in response to bacterial invasion. J Clin Invest 1995; 95:55-65.

43. McCormick BA, Miller SI, Carnes D et al. Transepithelial signaling to neutrophils by Salmonellae: A nouvel virulence mechanism for gastroenteritis. Infect Immun 1995; 63:2302-2309.
44. Gewirtz AT, Siber AM, Madara JL. Orchestration of neutrophil movement by intestinal epithelial cells in response to Salmonella typhimurium can be uncoupled from bacterial internalization. Infect Immun 1999; 67:608-617.
45. Perdomo JJ, Gounon P, Sansonetti PJ. Polymorphonuclear leukocyte transmigration promotes invasion of colonic epithelial monolayer by Shigella flexneri. J Clin Invest 1994; 93:633-643.
46. Everest P, Ketley J, Hardy S et al. Evaluation of Salmonella typhimurium mutants in a model of experimental gastroenteritis. Infect Immun 1999; 67:2815-2821.
47. Pepe J, Miller VL. Yersinia enterocolitica invasin: A primary role in the initiation of infection. Proc Natl Acad Sci USA 1993; 90:6373-6377.
48. Vasquez-Torres A, Jones-Carson J, Baumler AJ et al. Extraintestinal dissemination of Salmonella via CD18-expressing phagocytes. Nature 1999; 401:804-808.
49. Rescigno M, Urbano M, Valzasina B et al. Dendritic cells express tight junction proteins and penetrate gut epithelial monolayers to sample bacteria. Nature Immunol 2001; 2:361-367.
50. Isberg RR. Discrimination between intracellular uptake and surface adhesion of bacterial pathogens. Science 1991; 252:934-938.
51. Isberg RR, Barnes P. Subversion of integrins by enteropathogenic Yersinia. J Cell Science 2001; 114:21-28.
52. Mengaud J, Ohayon H, Gounon P et al. E-cadherin is the receptor for internalin, a surface protein required for entry of Listeria monocytogenes into epithelial cells. Cell 1996; 84:924-932.
53. Finlay BB, Cossart P. Exploitation of mammalian host cell functions by bacterial pathogens. Science 1997; 276:718-725.
54. Dersch P, Isberg RR. A region of the Yersinia pseudotuberculosis invasin protein enhances integrin-mediated uptake into mammalian cells and promotes self-association. EMBO J 2000; 19:2008-2014.
55. Tran Van Nhieu G, Isberg RR. The Yersinia pseudotuberculosis invasin protein and human fibronectin bind to mutually exclusive sites on the $\alpha5\beta1$ integrin receptor. J Biol Chem 1991; 266:24367-24375.
56. Alrutz M, Isberg RR. Involvement of focal adhesion kinase in invasin-mediated uptake. Proc Natl Acad Sci USA 1998; 95:13658-13663.
57. Blocker A, Gounon P, Larquet E et al. Role of Shigella's type III secretion system in insertion of IpaB and IpaC into the host membrane. J Cell Biol 1999; 147:683-693.
58. Kubori T, Matsushima Y, Nakamura D et al. Supramolecular structure of the Salmonella typhimurium type III protein secretion system. Science 1998; 280:602-605.
59. Tamano K, Aizawa AC, Katayama E et al. Supramolecular structure of the Shigella type III secretion machinery: the needle part is changeable in length and is essential for delivery of effectors. EMBO J 2000; 19:3876-3887.
60. Blocker A, Jouihri N, Larquet E et al. Structure and composition of the Shigella flexneri "needle complex" a part of its type III secreton. Mol Microbiol 2001; 39: 652-663.
61. De Geyter C, Wattiez R, Sansonetti PJ et al. Characterization of the interaction of IpaB and IpaD, proteins required for entry of Shigella flexneri into epithelial cells, wiht a lipid membrane. Eur J Biochem 2000; 267:5769-5776.
62. Adam T, Arpin M, Prévost MC et al. Cytoskeletal rearrangements and the functional role of T-plastin during entry of Shigella flexneri into HeLa cells. J Cell Biol 1995; 129:367-381.
63. Galan J, Zhou D. Striking the balance: modulation of the actin cytoskeleton by Salmonella. Proc Natl Acad Sci USA 2000; 97:8754-8761.
64. Hall A. Rho GTPases and the actin cytoskeleton. Science 1998; 279:509-514.
65. Welch MD. The world according to Arp: regulation of actin nucleation by the Arp2/3 complex. Trends Cell Biol 1999; 9:423-427.
66. Hayward R, Koronakis V. Direct nucleation and bundling of actin by the SipC protein of invasive Salmonella. EMBO J 1999; 18:4926-4934.
67. Zhou D, Mooseker MS, Galan JE. Role of the Salmonella typhimurium actin-binding protein SipA in bacterial internalization. Science 1999; 283:2092-2095.
68. Zhou D, Hernandez L, Shears S et al. A Salmonella inositol phosphatase acts in conjunction with other bacterial effectors to promote host cell actin cytoskeleton rearrangements and bacterial internalization. Mol Microbiol 2001; 39:248-259.
69. Fu Y, Galan JE. A Salmonella protein antagonises Rac-1 and Cdc42 to mediate host cell recovery after bacterial invasion. Nature 1999; 401:293-297.
70. Buchreiser C, Glaser P, Rusniok C et al. The virulence plasmid pWR100 and the repertoire of proteins secreted by the type III secretion apparatus of Shigella flexneri. Mol Microbiol 2000; 38:1-14.

71. Tran Van Nhieu G, Caron E, Hall A et al. IpaC determines filopodia formation during Shigella entry into epithelial cells. EMBO J 1999; 18:3249-3262.
72. Tran Van Nhieu G, Ben Ze'ev A, Sansonetti PJ. Modulation of bacterial entry in epithelial cells by association between vinculin and the Shigella IpaA invasin. EMBO J 1997; 16:2717-2729.
73. Bourdet-Sicard R, Rudiger M, Sansonetti PJ et al. Vinculin is unfolded by the Shigella protein IpaA, and the complex promotes F-actin depolymerization. EMBO J 1999; 18 :5853-5862.
74. Bourdet-Sicard R, Egile C, Sansonetti PJ et al. Diversion of cytoskeletal process by Shigella during invasion of epithelial cells. Microbes & Infection 2000; 2:813-819.
75. Niebuhr K, Jouihri N, Allaoui, A et al. IpgD, a protein secreted by the type III secretion machinery of Shigella flexneri, is chaperoned by IpgE and implicated in entry focus formation. Mol Microbiol 2000; 38:8-19.
76. Skoudy A, Mounier J, Aruffo A et al. CD44 binds to the Shigella IpaB protein and participates in bacterial invasion of epithelial cells. Cell Microbiol 2000; 2:19-23.
77. Dumenil G, Sansonetti PJ, Tran Van Nhieu G. Src tyrosine kinase activity down-regulates Rho-dependent responses during Shigella entry into epithelial cells and stress fiber formation. J Cell Sciences 2000; 113:71-80.
78. Sansonetti PJ, Ryter A, Clerc P et al. Multiplication of Shigella flexneri within HeLa cells: lysis of the phagocytic vacuole and plasmid-mediated contact hemolysis. Infect Immun 1986; 51:461-469.
79. Makino S, Sasakawa C, Kamata K et al. A genetic determinant required for continuous reinfection of adjacent cells on large plasmid in S. flexneri 2a. Cell 1986; 46:551-555.
80. Bernardini ML, Mounier J, d'Hauteville H et al. Identification of icsA, a plasmid locus of Shigella flexneri which governs bacterial intra- and intercellular spread through interaction with F-actin. Proc Natl Acad Sci USA 1989; 86:3867-3871.
81. Sansonetti PJ, Arondel J. Construction and evaluation of a double mutant of Shigella flexneri as a candidate for oral vaccination against shigellosis. Vaccine, 1989; 7:443-450.
82. Coster T, Hoge CW, Van De Verg LL et al. Vaccination against shigellosis with attenuated Shigella flexneri 2a strain. Infect.Immun. 1999; 67:3437-3443.
83. Miki H, Suetsugu S, Takenawa T. WAVE, a novel WASP-family protein involved in actin reorganization induced by Rac. EMBO J 1998; 17:6932-6941.
84. Egile C, Loisel TP, Laurent V et al. Activation of the CDC42 effector N-WASP by the Shigella icsA protein promotes actin nucleation by Arp2/3 complex and bacterial actin-ased motility. J Cell Biol 1999; 146:1319-1332.
85. Cossart P. Actin-based motility of pathogens: the Arp2/3 complex is a central player. Cell Microbiol 2000; 2:195-205.
86. Sansonetti PJ, Mounier J, Prévost MC et al. Cadherin expression is required for the spread of Shigella flexneri between epithelial cells. Cell 1994; 76:829-839.
87. Rathman M, De Lanerolle M, Ohayon H et al. Myosin light chain kinase plays an essential role in S. flexneri dissemination. J Cell Sciences 2000; 113: 3375-3386.
88. Page AL, Ohayon H, Gounon P et al. Role of IpaB and IpaC in the intercellular dissemination of Shigella flexneri. Cellular Microbiology 1999; 1:183-193.
89. Schuch R, Sandlin RC, Maurelli AT. A system for identifying post-invasion functions of invasion genes: requirements for the Mxi-Spa type III secretion pathway of Shigella flexneri in intercellular dissemination. Mol Microbiol 1999; 34:675-689.
90. Rathman M, Jouirhi N, Allaoui A et al. The development of a FACS-based strategy for the isolation of Shigella flexneri mutants that are deficient in intercellular spread. Molecular Microbiol 2000; 35:974-990.
91. Sansonetti PJ. Rupture, invasion and inflammatory destruction of the intestinal barrier by Shigella, making sense of prokaryote-eukaryote cross-talk. FEMS Microbiology Reviews 2001; 25:3-14.
92. Haas A. Reprogramming the phagocytic pathway—Intracellular pathogens and their vacuoles (Review). Mol Memb Biol 1998; 15:103-121.
93. Garcia del Portillo F, Finlay BB. Targetting of Salmonella typhimurium to vesicles containing lysosomal membranes glycoproteins bypasses compartments with mannose-6-phosphate receptors. J Cell Biol 1995; 129:81-97.
94. Méresse S, Steele-Mortimer O, Finlay BB et al. The Rab-7 GTPase controls the maturation of Salmonella typhimurium-containing vacuoles in HeLa cells. EMBO J 1999; 18:4394-4403.
95. Garcia del Portillo F, Zwick MB, Leung KY et al. Salmonella induces the formation of filamentous structures containing lysosomal membrane glycoproteins in epithelial cells. Proc Natl Acad Sci USA 1993; 90:10544-10548.

CHAPTER 7

Life and Death of *Brucella* within Cells

Edgardo Moreno and Javier Pizarro-Cerdá

Introduction

Brucellosis is a contagious bacterial disease of animals and humans. It is caused by organisms of the genus *Brucella*, composed of at least seven species displaying different affinities for host mammals. *Brucella abortus* is a parasite of bovines, *Brucella melitensis* of caprines and ovines, *Brucella suis* of swine and reindeer, *Brucella canis* of canines, *Brucella ovis* of male ovines, *Brucella neotomae* of wood rats and *Brucella* maris of marine mammals.[1,2] In humans the syndrome is severe, and without treatment the disease may lead to a fatal outcome. Traditionally, *Brucella* organisms have been defined as facultative intracellular pathogens because, being bacteria that multiply in bacteriological media, they also possess the faculty of replicating within cells.[1] However, the truth seems to prevail in the opposite side of this asseveration: *Brucella* organisms are pathogens which ultimate goal is to propagate in their preferred niche, the cell. In this environment, *Brucella* parasites survive and divide within membrane bound compartments of professional and non-professional phagocytes,[3-5] with preference for cells of the reticuloendothelial system and reproductive organs.[6] Despite the fact that the first member of the genus *Brucella* was discovered more than one hundred years ago, the intracellular life cycle and virulence mechanisms of these pathogens are just being unveiled (Fig. 1). The present work is an attempt to critically revise the available data on *Brucella* intracellular pathogenesis, under the horizon of a new discipline that has been called cellular microbiology.

Brucella Is Translocated by M Cells

The primary route of *Brucella* infection are the cells of mucosal surfaces.[6] In bovines, *B. abortus* laying in the ileum are ingested through a zipper-like internalisation mechanism by M cells but not by enterocytes.[7] In these phagocytic cells, the ingested bacterial organisms are located, as single entities or in small clusters, within transient vacuoles that do not show signs of fusion with lysosomes. The phagocytic vacuolar membrane does not associate with ribosomes, and some of the *Brucella*-containing compartments demonstrate coingested opaque material. The engulfed *Brucella* are eventually translocated by M cells to the gastrointestinal-associated lymphatic space, where they are taken up by macrophages and neutrophils.[7] The numbers of intracellular *Brucella* in M cells decrease with time after bacterial inoculation, indicating that microorganisms are steadily translocated. Alternatively, some ingested *Brucella* may be killed within these cells; although this needs to be demonstrated. After penetrating mucosal layers, both virulent and vaccine *B. abortus* S19 strains induce an inflammatory response in the submucosa.[8]

Brucella Binds and Penetrates Host Cells

Sequences coding for putative proteins similar to those necessary to build pili or flagella, which serve as cell adherence molecules in other bacteria,[9] are present in the *Brucella* genome (GeneBank AF019251).[10] In spite of this, none of these structures have been observed in

Intracellular Pathogens in Membrane Interactions and Vacuole Biogenesis, edited by Jean-Pierre Gorvel.

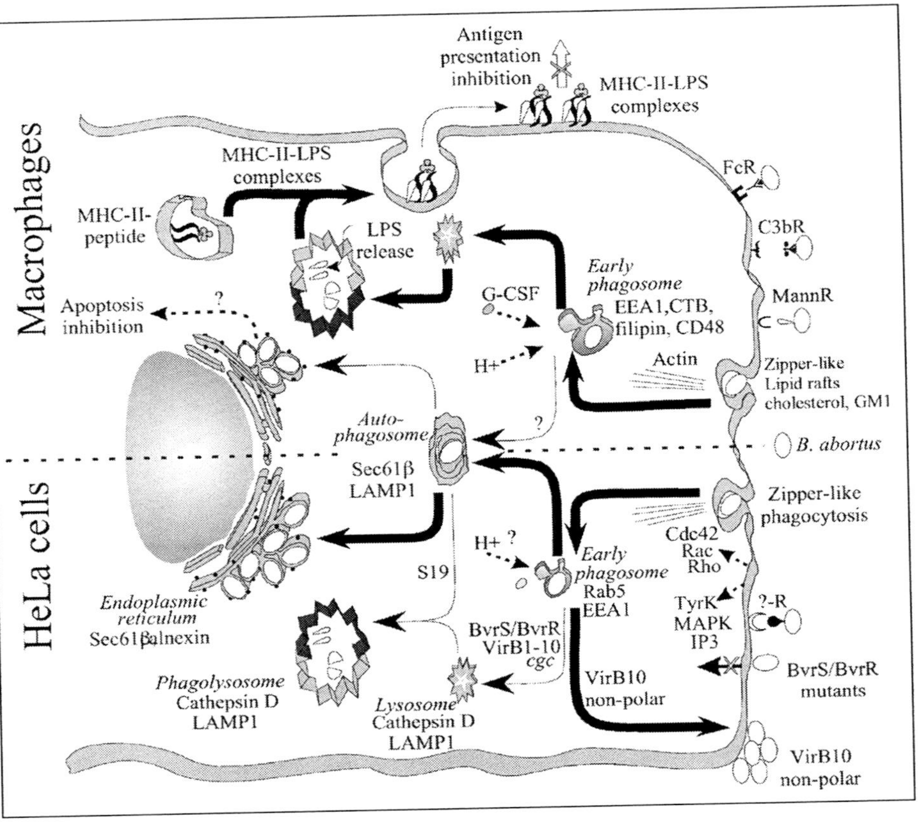

Figure 1. Schematic model of *B. abortus* invasion and intracellular trafficking in macrophages and epithelial HeLa cells. *Brucella* organisms bind to discrete sites of cells via receptor molecules (FcR, C3bR, MannR), some of which are not known (?-R), and penetrate by a zipper-like phagocytosis with moderate recruitment of actin filaments, activation of small GTPases (Cdc42, Rac, Rho) and signals mediated by second messengers (Tyr-K, MAP-K, IP3). The ingested bacterium, is initially found within early phagocytic compartments which may be acidify (H+) by acquisition of specific proton pumps (broken arrows). In macrophages, G-CSF is involved in promoting fusions between endocytic vacuoles and early phagosomes. In non-professional phagocytic epithelial cells, most of the ingested virulent brucellae reach the ER by the autophagocytic route (tick arrows) while only a few bacteria are digested within phagolysosomes (thin arrows). On the contrary, in macrophages, most of the virulent brucellae are destroyed within phagolysosomes (thick arrows), while only a few reach the ER (thin arrows). In macrophages, the released LPS converge with MHC-II compartments and the complexes MHC-II-LPS are transported to the cell membrane, inhibiting antigen presentation (crossed white arrow). Some mutants (BvrS/BvrR) are defective in penetration (crossed arrow), others are incapable to avoid fusion of early phagosomes with lysosomes (BvrS/BvrR, VirB1-VirB10, cgs), while others transit from autophagosomes to phagolysosomes (S19) or from early phagosomes to the cell membrane (non-polar VirB10). Signals for apoptosis inhibition may be released from the *Brucella* replicating niche (broken arrow).

Brucella cells, precluding (at least for the moment) their expression and participation as attachment molecules in this genus. However, the involvement of non-fibrillar adhesins remains open, since sequences compatible with adhesive protein molecules have been found in *B. abortus*.[11]

Various studies have reported that non-virulent smooth and rough *Brucella* mutants bind more readily and in larger amounts to the surface of professional and non-professional phagocytes than virulent strains.[12-14] However, there is not a strict direct correlation between the expression of O-chain and native hapten (NH) polysaccharides and binding to cells.[15,16] For instance, the virulent smooth *Brucella* and the "natural" rough *B. ovis* and *B. canis* (not expressing O-chain or NH) bind in very low numbers to HeLa cells. In contrast the transposon-defective perosamine synthetase rough *B. abortus* (*per*A mutant) not expressing O-chain or NH molecules, and the spontaneous rough *Brucella* mutants, expressing small amounts of lipid-bound and soluble NH in the envelopes, bind to these cells in high numbers.[15,16] It seems that at least in cultured cells, physiologically "healthy" *Brucella* have the tendency to attach in low numbers to epithelial cells.

During infection with rough *Brucella* mutants, the host cell plasma membrane, is frequently ticker in the region directly adjacent to the bacteria than during infection with smooth bacteria, occasionally forming coated-pits.[12] These observations, may not reflect the real physiological interactions between virulent smooth organisms and their host cells, since these cellular structures have not been observed during phagocytosis of smooth *Brucella*. In contrast, increased adherence displayed by many mutant *Brucella* seems to be the result of altered outer membrane properties which expose "new" sites for attaching in a non-physiological manner.[12-18] It is known that polysaccharide moieties in the outer membrane may hide hydrophobic and ionic charges on the surface of the bacteria that obstruct membrane domains capable of participating in the nonspecific interactions between *Brucella* and the host cells.[19] In addition, the absence of surface O-chain and NH exposes core sugars to the external surface, and gives more access to outer membrane proteins, which may serve as ligand sites.[20,21]

Invasion of Professional Phagocytes

Brucella organisms bind and penetrate more efficiently to professional than to non-professional phagocytes.[14] Still, the number of *Brucella* associated to professional phagocytes is considerably lower (from one to four-fold less) than the numbers observed with other intracellular *Proteobacteria*, such as *Salmonella, Shigella* or *Legionella*. Opsonized *Brucella* attaches and is more readily ingested by phagocytes than non-opsonized bacteria. This phenomenon is more clearly observed with activated than with non-activated professional phagocytes.[22-26] As in the case of *Mycobacterium*,[27] it may be that highly opsonized *Brucella* could take advantage of phagocytosis by Fc or complement receptors to invade cells expressing these receptors. However, opsonization seems to negatively affect the rate of survival and multiplication of *Brucella* within phagocytes, suggesting that Fc and complement receptor mediated phagocytosis rather work in favor of host cells.[24,28,29] *Brucella* lipopolysaccharide (LPS) and NH do not activate the alternative complement pathway,[30,31] Therefore, these polysaccharides do not serve as direct attachment of C3b on the bacterial surface. The fact that non-opsonized *Brucella* organisms bind, penetrate and reproduce within phagocytes from non-immunized animals,[14,32-34] indicates the existence of a receptor-ligand mechanism independent from the Fc and complement receptors.

Exclusion of negative charged molecules on the site of *Brucella* attachment has been observed on the macrophage membrane, suggesting specific interactions between both cell types (see addendum).[35] *Brucella* organisms (either opsonized or non-opsonized) are rapidly internalized by murine macrophages and human monocytes through a zipper-like mechanism, after inducing a moderate recruitment of actin filaments generating marked structural changes at the cell surface.[34,35] Penetration of *Brucella* to bovine mononuclear phagocytes is inhibited by bacterial cell envelops, antibody against the a chain of the MAC-1 integrin (CD11b), O-chain polysaccharide and denatured IgG, regardless if the cells originated from naturally resistant or susceptible cattle to brucellosis. In addition, fibronectin, mannan and antibodies against C3 also inhibit the penetration of non-opsonized bacteria to phagocytes from brucellosis resistant cattle.[32] Since the O-chain of LPS and mannan inhibit the binding of *Brucella* to bovine macrophages, it is likely that lectin-like receptors on the surface of phagocytic cells participate in the *Brucella* uptake. This idea is also supported by the observation that the strong binding of *Brucella* onto murine B-lymphocytes is inhibited by alpha methyl mannosamine and LPS.[36]

Invasion of Non-Professional Phagocytes

In HeLa cells, *B. abortus* attaches to cellular extensions that are compatible with adhesion plaques and between cell to cell contacts (Guzmán-Verri et al, see addendum) and penetrates host cells by a zipper-like phagocytosis mechanism.[37-39] Some of these observations have been performed in intoxicated monolayers with *Clostridium difficile* toxins B (TcdB) and BF (TcdBF) infected with *Brucella* (Guzmán-Verri C, et al, see addendum). TcdB intoxicated cells retract their body, leaving cellular spikes attached to the substrate. In contrast, TcdBF induces cell rounding and retraction of cellular spikes from the substratum. *Brucella* organisms attach to the cellular spikes of TcdB treated cells. On the contrary, in TcdBF intoxicated cells, *Brucella* organisms bind in reduced numbers, and when binding is observed, the bacteria locate on the cellular body.

Uptake of killed or live *Brucella* by Vero cells is suppressed by inhibitors of energy metabolism (iodoacetate and dinitrophenol), inhibitors of receptor-mediated endocytosis (monodansylcadaverine, amantadine and methylamine) and repressors of endosomal acidification (chloroquine, ammonium chloride and monensin). These drugs are capable of inhibiting penetration when added before the bacterial inoculum, but not when added after the inoculation period, suggesting that the infection process occurs via specific receptor molecules and requires energy input from the host cell.[40] It has been proposed that $\alpha5\beta1$ integrin mediates the adhesion of rough mutant *B. abortus* RB51 on bovine trophoblasts.[41] However this observation has not been reproduced in HeLa cells in spite of the highly adherent nature of mutant rough *Brucella* to cells and inert substrates.[16]

The participation of cytoskeleton, second messengers and GTPases in the internalization of *B. abortus* to HeLa cells has been recently investigated (Guzmán-Verri C, et al, see addendum). Phalloidin staining has revealed a modest recruitment of actin filaments in the site of *Brucella* attachment. Inhibition of actin filaments by drugs,[16,40] or by different clostridial toxins (TcdB, TcdBF. TcdA, TcsLT), which functionally modify the actin cytoskeleton through interaction with small GTPases of the Rho family, hampers internalization but not binding to cells (Guzmán-Verri C, see addendum). Infection is also inhibited by chemicals and toxins that increase the levels of cyclic-AMP (dibutyl-cyclic-AMP and *Vibrio cholerae* enterotoxin) but it is stimulated by toxins and chemicals that increase the levels of cyclic-GMP (*Escherichia coli* enterotoxin A and dibutyl-cyclic-GMP). This suggests an inverse relationship between these two second-messengers during *Brucella* internalization.[40] Similarly, wortmannin (reduces the level of inositol triphospate [IP_3]), considerably inhibits the internalization of *Brucella* to HeLa cells, suggesting involvement of the IP_3-kinase during this process (Guzmán-Verri C, see addendum). Since the level of cyclic GMP usually increases when the inositol phospholipid pathway is activated, it is likely that binding of *Brucella* to cells stimulates the generation of IP_3 via phospholipase C activation. Other cellular kinases, such as tyrosine kinases and MAP kinases seem to be required for physiological internalization to HeLa cells, since inhibition of these enzymes hampers bacterial penetration (Guzmán-Verri C, see addendum).

B. abortus attaches in larger numbers and is internalized more efficiently after intoxication of HeLa cells with the cytotoxic necrotizing factor (CNF) which deamidates the small GTPases Rho, Rac, and Cdc42, inducing ruffles and stress fiber formation (Guzmán-Verri C, et al, see addendum). In these cells, *Brucella* organisms bind to the ruffles as well as to discrete sites in the cell membrane. Moreover, HeLa cells transfected with dominant negative variants of Rho, Rac and Cdc42 proteins, are considerably less infected than control cells. On the contrary, the positive counterparts of these small GTPases expressed in HeLa cells, stimulate binding and penetration of *Brucella* with direct recruitment of Cdc42. Microtubule depolymerizing agents such as nocodazole and colchicine, partially reduce the internalization but not the replication of *Brucella* in HeLa cells (Guzmán-Verri C, see addendum).[40] This strategy for invading cells differs from those employed by other intracellular pathogens such as *Salmonella* and *Shigella*. Despite the obvious differences between *Brucella* and *Listeria*, both types of pathogens seem to employ similar strategies for invading cells, inducing a zipper-like internalization mechanism.[42]

Survival within Polymorphonuclear Neutrophils

Neutrophils are the primary line of defense during *Brucella* invasion. These cells are capable if ingesting and digesting *Brucella* faster and more efficiently than other cells.[43-45] After incubation with neutrophils, virulent *Brucella* organisms are steadily destroyed. From 50% to 80% of the ingested bacteria are killed within 2 to 5 hours.[44,45] This microbicidal activity seems to be more efficiently performed when bacteria are opsonized.[26,46] Once ingested, the bacteria are found within vacuoles, some of which already demonstrate fusions with lysosome-like granules. At later times, fusions of azurophilic granules increase in most but not all *Brucella* containing vacuoles. These events proceed without extensive neutrophil degranulation.[44] Exudates from lesions, milk and blood from infected animals demonstrate *Brucella* within resident neutrophils with no signs of replication.[7,47] It seems that some intracellular bacteria are capable of resisting for several hours or even for days the microbicidal assault displayed by neutrophils, which in time may release live bacteria on sites where it could invade more temperate host cells.[7]

Life within Non-Professional Phagocytes

As stated before, even at a multiplicity of infection of 100-500 bacteria per cell, the initial number of virulent *Brucella* per infected epithelial Vero or HeLa cell is low: one or two bacteria observed per infected cell.[3,12,13,48] Even if the bacterial inoculum is augmented several folds, the rate of infection per cell remains low, suggesting that not all cells are permissive to infection or, alternatively, that not all bacteria obtained from bacteriological cultures are able to attach to these cells. Despite this, once the virulent brucellae bind to host cells, the penetration efficiency is close to 100%.[14] Immediately after their internalization, *Brucella* organisms localize within single membrane compartments generally containing only one bacterium. During the first hours (from 1 to 5), no clear signs of intracellular bacterial replication is observed, and the rate of cellular infection remains relatively constant for the next 5 to 8 hours. After 24 to 48 hours, the number of bacteria, estimated as colony-forming units in a gentamicin survival assay, increases several logs from the initial inoculum, indicating extensive bacterial replication within the infected cells. Although many cells may possess large numbers of intracellular bacteria, the cell monolayers do not show any sign of cytopatic effects and the individual cells remain attached to the matrix. A small proportion of cells contains a low number of *Brucella* with no signs of bacterial degradation, while only a restricted sub-population of cells demonstrates bacterial degradation within phagolysosome-like compartments. The small number of bacteria within cells at late infection time periods is not the result of new infections, since normally low concentrations of gentamicin (0.5 mg/ml) are able to control extracellular bacteria that can be released from damaged cells.[3,48] Alternatively, the low number of live bacteria within host cell vacuoles after 48 hours, may be a reminiscence of chronic infections in animals. Multivesicular bodies, compatibles with autophagosomes are frequently present in infected cells but not in control monolayers.[3,12,13]

Under the electron microscope, the cell cytoplasm is filled with *Brucella* within ribosome-lined cisternae, compatible with rough endoplasmic reticulum (ER).[12,13] In the most heavily infected cells, bacteria are also seen in the perinuclear space. Although, the nuclear membrane is constrained due to the large number of brucellae surrounding the nucleus, none of them invade this organelle. These heavily parasited cells do not look apoptotic or present signs of necrosis. Moreover, dividing cells with the cytoplasm packed with bacteria are frequently observed (see addendum). Nevertheless, beyond 48 hours mechanical cellular rupture proceeds as consequence of *Brucella* overgrowth. The antibiotics in the tissue culture medium kill the freed bacteria, generating a rapid decrease in the number of colony forming units in the gentamicin survival assay.[48]

Very similar events to those described in Vero and HeLa cells at 48 hours after infection have been described in trophoblasts from experimentally infected animals.[37,38] Most if not all the *Brucella* are within the ER, and despite being filled with bacteria, infected trophoblasts do not show signs of degeneration, and the cellular junctions look normal. Matching to what has been observed in cultured monolayers, bacteria remain within the ER compartments of individual cells, with no signs of horizontal movement from cell to cell. Indeed, non-infected cells are bound together with heavily infected trophoblasts, demonstrating that even in vivo, *Brucella* organisms remain within vacuoles of their host cells until released. The trophoblast nucleus is never parasited although it is constrained by bacteria that lay very close or within cisternae that collide with the nuclear membrane. Similar to epithelial cell monolayers, at later times the infected trophoblasts seem to fracture, due to the large number of intracellular bacteria that eventually are released to the lumen.

Transient Interaction with the Early Endosomal Network

During the first minutes after invasion, both the virulent *B. abortus* strain 2308 as well as the attenuated strain 19 interact with an intracellular compartment related to the early endosomal network. This is confirmed by the presence of markers such as the transferrin receptor, the small GTP-binding protein rab5, or the early endosomal antigen 1 (EEA1) in the *Brucella*-containing compartments.[3] Several of the above-cited markers have been also observed in early phagosomes of other intracellular pathogens such as *Salmonella typhimurium, Leishmania donovani, Mycobacterium tuberculosis* and *Listeria monocytogenes*.[49-52] All these parasites use very different strategies to associate and penetrate host cells, suggesting that there is an invariable minimal cellular machinery regularly required to accomplish the internalization steps of an external agent. This seems to be more important when it is the host cell that performs an active role in the process: a different phenomenon could be expected in the case of the *Brucella* close relative *Bartonella bacilliforms*, or the protozoan *Toxoplasma gondii* and *Trypanosoma cruzi*, in which are the motile forces of the pathogens themselves that drive the invasion processes.[53-55]

The association of *B. abortus* with the early endocytic network is transient, since after 10 minutes of internalization, the number of *Brucella*-containing compartments labeled either with rab5 or EEA1 decreases significantly, and no labeling is detected with these markers after 30 min postinoculation.[16] The integrity of the early endosomal system is relevant to the subsequent normal trafficking of *B. abortus* in host cells. For instance, in the cell line NIH3T3 rab5Q79L, in which the activated form of rab5 (bound to GTP) is expressed, an important fraction of the internalized parasites are unable to escape from the early *Brucella*-containing compartments supporting bacterial replication within giant vesicles labeled with rab5.[16] However, after 48 h of infection, *B. abortus* proliferation is attenuated in NIH3T3 rab5Q79L cells in comparison to the wild type NIH3T3 counterparts (see addendum).[16] The augmentation of the endocytic activity of the mutant cells could be responsible for an increase in the delivery of gentamicin to intracellular compartments: indeed, an important fraction of the intracellular brucellae remains associated with the early endosomal network, exposing the bacteria to the bactericidal activity of the antibiotic within this compartment. Alternatively, this unnatural generated compartment may not be suitable for the adequate delivery of the necessary nutrients for intracellular bacterial replication. Vacuole acidification seems to be required for intracellular bacterial replication since chloroquine, ammonium chloride and monensin, all substances that inhibit endosomal acidification, are capable of reducing the number of intracellular bacteria at early times but not at later times after infection.[40] Acidification step without the acquisition of lysosomal markers may be necessary for the activation of virulence genes as it occurs with other parasites.[56,57]

Association with the Autophagocitic Machinery

After their transit through early phagosomes, neither the virulent *B. abortus* strain 2308 nor the attenuated strain 19 interact with the late endosomal network at 30 min postinvasion.[3] In contrast, latex beads or dead bacteria-containing phagosomes interact transiently with late endocytic compartments, characterized by the presence of the small GTP-binding protein rab7 or the mannose 6-phosphate receptors. This result is confirmed after infection of NIH3T3 rab7Q67L cells (in which rab7 is expressed in its GTP-bound form) with *B. abortus* 2308. In this mutant cell line, bacterial replication is similar as in wild-type cells (see addendum).[16]

Following these initial steps, the *Brucella*-containing compartment is transformed gradually and after 1 h of internalization, both virulent 2308 and attenuated 19 strains are present in an intracellular multimembranous compartment decorated with the lysosomal-associated membrane protein-(LAMP) 1, but devoid of the luminal lysosomal hydrolase cathepsin D. This finding supports previous propositions in the sense that virulent *B. abortus* inhibits the fusion of its phagosome with lysosomal compartments.[58] Several criteria permitted the identification of this late *Brucella*-containing compartment as an autophagosome. First, the multimembranous nature of LAMP-1 positive/cathepsin D negative vacuole is highly reminiscent of autophagosomes. Second, this compartment is labeled by the marker monodansylcadaverine, known to accumulate in autophagosomal bodies. Third, the ER marker Sec61β is present in this *Brucella*-containing vacuole, attesting to an ER-related origin of this compartment. Fourth, modulation of the autophagocytic process regulates the intracellular fate of the internalized brucellae.[3,48]

The presence of LAMP-1 in the late *Brucella*-containing compartment could be explained by a direct delivery of this molecule from the Golgi complex to the maturing autophagosomes. It is interesting to note that LAMP-1 has been widely associated with pathogen-containing compartments, such as the vacuoles of *S. typhimurium, L. donovani*[49,50] and *L. monocytogenes* (Pizarro-Cerdá J, unpublished results). This molecule may be present in these compartments accidentally as a bystander, an outcome resulting only from the different trafficking pathways followed by this molecule, transported in certain cases to the plasma membrane before being delivered to the lysosomes.[59] However, the actual function of the LAMP family of glycoproteins has not been clearly defined.[60] It would be interesting to determine if this molecule actually plays an active role that could be relevant to the intracellular survival of all the intracellular pathogens described above. The association of an intracellular pathogen with the autophagic pathway is not unique to *B. abortus*, and it has also been observed in the case of *Legionella pneumophila*.[61]

How these bacteria are able to interact with the autophagic cascade is still unknown. An interaction between early endocytic compartments and autophagic vacuoles has already been detected,[62] indicating that a physical connection could exist between early *Brucella*-containing compartment and autophagosomes. Several scenarios could then be conceived in order to explain the transfer of the pathogen from one intracellular compartment to the other. First, there could be a fusion between the *Brucella*-containing compartment and an already formed autophagic vacuole. However, how the bacteria are finally found within the luminal space of a multimembranous compartment could not be directly explained by this hypothesis. A second, but highly improbable possibility, is the escape of *B. abortus* from the *Brucella*-containing compartment to the cytoplasmic space, where the bacteria could be captured by nascent autophagosomes, but free brucellae are seldom observed in the cytosol of infected cells. A third possibility would be that the autophagosomal vacuoles are formed by invagination of the ER membranes around *Brucella*-containing compartments. Nevertheless, the absence of other ER markers such as BiP or ribophorin in the *Brucella*-containing compartment[3] would contradict this hypothesis. A modification of this alternative would be that only specialized regions of the ER, devoid of BiP or ribophorin, are involved in autophagosome formation.

Replication within the Endoplasmic Reticulum (ER)

In contrast to what is observed with attenuated and killed bacteria, most of the intracellular virulent *Brucella*-containing phagosomes, loses the LAMP-1 labeling and never acquire lysosomal markers in non-professional phagocytes.[3,48] However, this final *Brucella*-containing compartment retains the Sec61β labeling and acquires other markers of the ER such as the protein disulfide isomerase and calnexin.[3] The morphology of the final *Brucella*-containing compartment also differs from that of the autophagosomal stage: only a single membrane is detected around the replicating brucellae, and their intracellular location correspond to the perinuclear area of the infected cells.[3,48] All these data suggest that the virulent *B. abortus* transits from autophagosomes to the ER of host cells, where actual bacterial multiplication occurs, confirming previous ultrastructural studies in trophoblasts and other mammalian cell lines.[37,13] Additional evidence confirms the nature of the final niche of *B. abortus* replication in host cells. First, treatment of infected cells with brefeldin A, that normally induces the reorganization of the Golgi complex around the ER, induces the colocalization of Golgi markers around the *Brucella*-containing compartment.[3] Second, the treatment of infected cells with proaerolysin, a drug from *Aeromonas hydrophyla* that induces vacuolization of the ER in target cells, induces vacuolation of *Brucella*-containing compartment.[3] Treatment of target cells with proaerolysin before *Brucella* inoculation impairs the bacterial replication process and induces the degradation of virulent strain 2308[16] suggesting that the integrity of the structure of the ER is indispensable for the appropriate multiplication of *B. abortus*.

The possible benefits involved in the association of intracellular pathogens with the host-cell ER have not been characterized yet. Besides *B. abortus, L. pneumophila, T. gondii*, and even simian virus 40 multiply in this intracellular environment, revealing a convergent evolution path in non-related organisms.[61,63,64] In addition of being a strategy for avoiding lysosomal fusion during the final steps of intracellular invasion, association of *B. abortus* with the host ER could be a means of obtaining metabolites synthesized or translocated to this compartment. The strategy would be to take advantage of the biosynthetic enzymes, protein-conducting channels or peptide pores to increase the local nutrient supply,[65] fulfilling the complex nutritional requirements for the bacterial growth. A possible prediction of this model would be that the cellular stock of short-lived molecules would decrease in *B. abortus*-infected cells due to the shortage of amino acids and peptides or to the blocking of the biosynthetic process of new proteins in the ER. However, heavily infected cells show no decrease in stock levels of short-lived molecules such as LAMP-1 and LAMP-2 (Pizarro-Cerdá J, unpublished results).

Are Trophoblasts a Source of Growth Factors?

In the pregnant animal, *Brucella* preferentially replicates in placental trophoblasts during the middle and late stages of gestation, only after these cells actively secret steroids. The reason for this affinity and the process leading to abortion after midgestation are not known. Infected trophoblasts produce cortisol, a steroidal hormone not normally generated by the placenta.[66] The increased levels of prostaglandin F2α and decreased production of progesterone, coupled with augmented synthesis of estrogens and cortisol in infected trophoblasts at mid and late stages of gestation, are identical to the changes observed at the initiation of parturition in non-infected cattle. It may be that trophoblast infection induces synthesis of steroids and modifies the metabolism of prostaglandin precursors such as arachidonic acid, because these hormones may be used as growth factors by the *Brucella*. The increased hydrophobicity of the *Brucella* organisms outer membrane,[67] together with its preference for replicating within the ER of these cells,[37,38] may represent an evolutionary a adaptation for extracting hormones readily available within the cistern of this organelle.

Life Cycle within Macrophages

During the first two hours of *Brucella* infection in macrophages, the bacteria localize within single phagosomes with no signs of microbial degradation.[16,22,68] The number of ingested *Brucella* depends upon the inoculum size, opsonization and macrophage activation.[34] At these early time points, fusions between lysosomes-like compartments and some *Brucella* containing phagosomes can be detected, while a few intracellular bacteria localize within multimembrane compartments. After 12 to 15 hours of ingestion, the number of *Brucella* decreases in one to two logs, and more than one bacteria per vacuole are frequently seen.[24,33,69-71] An important proportion of the *Brucella*-containing compartments have fused with lysosomal granules and many of the bacteria are degraded, although a few of them appear intact, generally surrounded by a single vacuolar membrane.[22,25,72] At later times (from 15 to 24 h), the number of intracellular bacteria increases. While some macrophages are capable to control the infection, others become an adequate substrate for *Brucella* replication.[72,73] Still, a proportion of bacteria are digested and bacterial debris are evident in activated macrophages. From 24 to 48 hours the number of intracellular *Brucella* per infected cell raises until the intracellular space of the phagocytic cells is filled with parasitic bacteria.[16,68,69,73] Similar to what has been described for non-professional phagocytes, the nuclear membrane looks constrained, due to the large number of microorganisms surrounding the nucleus, however, none of the bacteria invade this organelle. These heavily parasited macrophages do not look apoptotic, vacuolated nor present signs of necrosis. Nevertheless, beyond 48 hours cellular rupture seems to proceed as consequence of *Brucella* overgrowth, and release of the bacteria occurs.

Trafficking within Macrophages

It has been proposed that phagosome acidification dependent of proton ATPases, occurring at early (1 h) but not at later times (after 7 h), is essential for bacterial replication.[71] Other investigators have observed that acidification of phagosomes take place at early (0.5 to 2 h) as well as later (20 h) times.[22] An acidic environment necessary for activating genes for pursuing the intracellular life cycle may be required by *Brucella* organisms.[74] Even though, lowering the pH of the phagosome containing *Brucella* may be important for preparing the replicating environment for this pathogen, this may not be sufficient, since phagosome containing alive bacteria acidifies at similar rate as those containing killed organisms.

The biogenesis of phagolysosomes and the ability to degrade invading microorganisms involve a regulated series of interactions between phagosomes and endocytic organelles.[9,75,76] During the first stages after infection (from 0.5 to 2 h), an important proportion of bacterial-containing phagosomes seem to be reluctant to fuse with newly internalized vesicles[16,68] At ten minutes after inoculation, *Brucella* organisms are transiently detected in phagosomes, characterized by the presence of early endosomal antigen- (EEA) 1. At one hour postinoculation, bacteria are located within a compartment positive for LAMP but negative for mannose-6-phosphate receptor and cathepsin D protein, indicating that virulent *Brucella* avoids fusion with late endosomes and lysosomes.[16,68] At later times (from 12 to 24 h), fusions between newly internalized vesicles and *Brucella* containing vacuoles are common events. However, the lysosomal markers LAMP and cathepsin D are excluded from vacuoles containing replicating *Brucella*,[16] indicating that some bacteria have actively avoided the constitutive degradative pathway commonly followed by inert particles.[75] Mature compartments containing alive replicating *Brucella* (after 24 h) are devoid of late endosomal marker mannose-6-phosphate receptor and lysosomal proteins (LAMP, and cathepsin D) but are positive for the ER marker Sec61β, suggesting that at least some bacteria have reached the same replicating niche (the ER) as in non-professional phagocytes.[16,68] Other investigators have proposed that in macrophages, *Brucella* organisms do not transit through autophagosomes and that the replicating niche in these cells resemble phagolysosomes.[22] These investigators found that very few compartments containing *Brucella* colocalize with specific markers for autophagosomes and ER. At this point it is not possible to reconcile these two positions.

In murine knockout macrophages deficient in fusion events, both virulent 2308 and attenuated S19 *B. abortus* strains, replicate more readily than in normal macrophages.[68] Restoration of vesicle fusiogenicity re-establishes the microbicidal activity, indicating that fusion of vacuoles containing *Brucella* with newly internalized vesicles are necessary for controlling intracellular *Brucella* replication. Therefore, it seems that non-fusiogenic vacuoles containing *Brucella*, are more likely to reach the final intracellular replicating niche that those vacuoles that fuse with lysosome-like compartments. Electron microscopy of infected macrophages reveals that the replicating pattern displayed by *Brucella* within macrophages at later times[69] is reminiscent of the replicating pattern within the ER of trophoblasts and epithelial cells.[13,37]

Cellular Functions during Infection

Transit of virulent *Brucella* from autophagosomes to the ER requires a cellular machinery that has not been identified. In CNF intoxicated cells, the intracellular trafficking to the ER via the autophagocytic route, remains intact, in spite that the actin cytoskeleton is functionally arrested (Chaves-Olarte, et al, see addendum). Similarly, segregation of chromosomes and nuclei assemble are not defective, while the formation of contractile ring for cell cytokinesis, mediated by a actin-myosin structures, is impaired in these CNF intoxicated cells. Colchicine added during the infecting period, has a modest effect in internalization, although it reduces the number of cells with intracellular replicating bacteria and has a profound effect in the morphology of the parasited cells, generating polymicronuclei.[40] It is possible that microtubule but not actin structures are involved in the retrograde transport of *Brucella*-containing compartment to the ER. The participation of molecular motors such as dynein, which normally promote the retrograde motion of membrane-bound vesicles through microtubules, is a likelihood mechanism that must be considered during the intracellular biogenesis of *Brucella*.

Proteins from the coat protein I (COPI) complex, known to participate in the anterograde transport of vesicles in the Golgi apparatus,[77] have recently been implicated in the retrograde transport of vesicles to the ER.[78] The presence of several subunits of the COPI complex in compartments associated with the endocytic cascade[79] suggests that these molecules could establish a link between endocytic compartments, autophagosomal vacuoles and the ER, a link that is necessary for the intracellular trafficking of *B. abortus*. Molecules from the SNARE family or small GTP-binding proteins of the rab family could also be implicated in the retrograde transport of the *Brucella*-containing compartments. Recently, it has been demonstrated that autophagocytosis could be blocked by GTP-gammaS (a non-hydrolizable analogue of GTP), suggesting that this process requires GTP-binding proteins.[80] It is interesting to note that the anterograde transport of vesicles from the ER is dependent on the small GTP-binding protein Sarpl[81] and the COPI-like complex, COPII.[82] Similar molecules could be implicated in autophagosomal formation, and could be under *B. abortus* control in order to induce its retrograde transport to the ER.

Brucella Virulence Mechanisms

Mechanisms of Entry into Host Cells

Only efficiency of invasion and intracellular replication, but not adherence to cell surface, positively correlates with *Brucella* virulence (see section 3). As stated before, a number of natural or laboratory engineered non-virulent rough and smooth mutants bind cells in a larger proportion than wild type *Brucella*.[14-16] Mutations in *bvr*S or *bvr*R genes of a two regulatory system hamper the penetration of smooth *B. abortus* in HeLa cells and considerably reduce the efficient invasion of murine macrophages, without affecting the bacterial binding.[14] One of the genes (*bvr*S) codes for a sensor protein (BvrS) displaying histidine kinase activity, while the second (*bvr*R) codes for a regulatory protein (BrvR) that promotes or inhibits transcription. Dysfunction in the *brv*R or *brv*S impairs the bacterial invasion, intracellular trafficking and virulence. Knock-out of both *bvr* genes seems to be lethal (O'Callaghan D, personal communication). The *bvr* mutants, although smooth, have the tendency to accumulate on the surface of host cells. Rough *per*A mutant, deficient only in the expression of surface O-chain and NH binds to cells in larger numbers (unpublished results); however, a rough *bvrS* mutant (double mutant), binds to cells in the same proportion as its precursor *bvrS* strain (unpublished results). Therefore, the exposed elements on the surface of rough *Brucella* necessary for adherence to cells, seem to be absent in the *bvr*S mutant.

Comparative studies revealed that *bvrS*/*bvr*R mutants are deficient in at least two sets of outer membrane group 3 proteins (Guzmán-Verri C, et al, see addendum). One of them corresponds to the Omp25 family, while the other is close related to the *Rhizobium* outer membrane protein RopB. Transcriptional analysis of the group 3 Omp RNAm reveals deficient expression of these proteins in both mutants indicating that their absence from the outer membrane is the result of a regulated translation. The role of Omp25 in virulence is further suggested by the fact that disruption of the *omp*25 gene attenuates *Brucella*.[83] Functional and structural analysis indicate that the core and lipid A, but not the O-chain of the LPS, have subtle but detectable changes with respect to the wild type molecule. In this respect, the BrvR/BrvS two-component regulatory system may be similar to other regulatory systems described in gram negative intracellular bacteria.[84] Although the BvrR/BvrS has low homology with PhoP/PhoQ, it is interesting that, among other characteristics, the PhoP/PhoQ activates the PmrA-PmrB system, which in turn, regulates the lipid A structure. The overall conclusion is that multiple genes, several of them important for the expression of outer membrane proteins and LPS synthesis, may be under the control of *bvr*R/*bvrS*, and that these elements are important for virulence.

Outer Membrane versus Bactericidal Substances

The most conspicuous structural defect that render *Brucella* organisms avirulent is the absence of the O-chain and the concomitant absence of the related NH polysaccharide molecules: in other words, a defect that results from the dissociation from smooth to rough phenotype.[12-17,43,85-87] In general, smooth *Brucella* are more resistant than rough strains to the killing action of bactericidal substances of phagocytes,[45,86] indicating that the O-chain and NH participate in this phenomenon. The role of LPS in the permeability properties and in resistance to bactericidal substances has been definitively established in a series of experiments involving *Brucella* rough strains and the construction of LPS chimeras.[15,85,86,88] When the heterologous LPS inserted in the outer membrane of susceptible bacteria corresponds to the less sensitive smooth *B. abortus*, the chimeras are more resistant to bactericidal cationic molecules; in contrast, when LPS is from the more sensitive bacteria, the chimeras are more susceptible to the action of bactericidal peptides. There is a direct correlation between the amount of heterologous smooth LPS on the surface of chimeric cells and sensitivity to bactericidal substances.[85] Although this particular resistance to bactericidal molecules may be related to the core and lipid A structures, there is a contribution of the O-chain and the outer membrane-associated NH, as suggested by the difference in susceptibility between the rough and smooth *Brucella* strains.[85,86] It is worth noting that the resistance of *Brucella* cells to cationic molecules is more conspicuous when the smooth LPS is integrated in its native outer membrane. This is demonstrated by the fact that *Brucella* smooth LPS micelles are partially permeabilized by the action of bactericidal peptides while *Brucella* cells are not.[85]

The O-chain and the NHs are only two of several factors necessary for virulence.[89] Dysfunction of the BrvR-BrvS two regulatory system generates an altered outer membrane, and in consequence an increased susceptibility to bactericidal cationic substances and surfactants. This defect is partially restored by inserting wild type LPS in the outer membrane of the *bvr* mutants (unpublished results), reinforcing the idea that in addition to LPS, other outer membrane molecules are also important for this function. Since bactericidal cationic peptides are molecules devoted to cell defense against pathogenic parasites,[90] it is likely that some of the outer membrane features relevant for resistance to lysosomal substances are also under the control of the BvrR-BvrS system.

Neutralization of Oxygen and Nitrogen Intermediates

Brucella organisms produce phosphomonoesterases, high concentration of periplasmic medium size class II cytochrome c, superoxide dismutases (Sod) and catalase, all proteins which may be involved in the protection of the bacteria against free hydrogen peroxide and superoxide radicals generated by phagocytic cells.[91-93] In contrast to some intracellular parasites such as

Legionella or *Leishmania, Brucella* phosphomonoesterase does not block the production of superoxide anion, does not hydrolyze phosphatidylinositol diphosphate or IP_3 molecules.[92] A collection of *Brucella* mutants not expressing Cu/Zn-SodC or catalase proteins have been generated on the expectation that these defects would increase the sensitivity of the mutants to bactericidal action of oxidative intermediates produced during the respiratory burst.[91,93] It has been demonstrated that the expression of these two enzymes concomitantly increase after exposure of the bacteria to the presence of peroxide or superoxide ions,[91] suggesting at first glance an adaptive response to the oxidative conditions of lysosomes. In spite of their increased sensitivity to hydrogen peroxide,[91] the catalase-deficient mutants, replicate at the same rate as wild type bacteria in mice, indicating that, at least in this host animal, catalase *per se* does not play a significant role in virulence.[94] Similarly, *Brucella* SodC deficient mutants exhibit virulence, establishing chronic infections in mice.[95] Similar to what has been observed with the catalase and SodC mutants, the urease defective mutants do not display attenuation and resemble the parental pathogenic strains in virulence assays.[94,96] However, urease still may play a role during the initial steps of host invasion in the penetration of the gastric mucous membrane.[97]

Role of Stress Proteins

Significant changes in the pattern of *Brucella* proteins expressed during intracellular growth have been recorded.[98] In addition to variations in the expression level of 73 proteins, repression of 50 proteins and induction of 24 new proteins occurs during growth of *B. abortus* within macrophages.[98] Acid and oxidative conditions as well as nutritional and heat stresses induce the synthesis of "new" bacterial proteins. However, the quantity of these "new" molecules produced in vitro is not equivalent to the amount expressed within macrophages.[99,100]

*Brucella htr*A, codes for the periplasmic heat shock-induced serine protease HtrA, likely to participate in the degradation of damaged proteins. Resembling *Salmonella* and *Yersinia htr*A mutants, *Brucella htr*A mutants demonstrate higher sensitive to oxidative substances, reduced survival in neutrophils and defective replication in macrophages.[101-103] The expression of *htrA* is regulated by a complex set of signal transduction pathways,[104] including a sigma factor (RpoE) and an anti-sigma factor (RseA), a two-component regulatory system (CpxRA) and two phosphoprotein phosphatases (PrpA and PrpB). Comparable to other *Brucella* stress proteins, such as RecA, which mutation reduces multiplication in mice, *htrA* mutants establish chronic infections suggesting residual virulence.[93,103,105] In goats, htrA *Brucella* mutants display a more attenuated phenotype.[101]

Regulated expression of *dna*K under oxidative stress seems to be used by bacterial pathogens to stand the respiratory burst of phagocytes.[106] Insertional inactivation of *Brucella dna*K and *dna*J, coding for the stress molecular chaperone DnaK and DnaJ respectively, have led to the conclusion that DnaK but not DnaJ is required for growth at 37 °C.[107] Experiments performed with both mutants at 30 °C, demonstrate that the *Brucella dna*K mutant survived but failed to multiply within phagocytes, whereas the parent strain and the *dna*J mutant multiplied normally. *B. suis* null mutants for ClpATPase chaperonine behave similarly to the wild type strain, indicating that ClpA by itself is dispensable for intracellular growth.[108] The role of other potential *Brucella* virulent genes such as those coding for the heat shock protein GroEL[109] have not been investigated in detail. In *Haemophilus ducreyi* the GroEL protein is associated to the bacterial surface, suggesting a possible involvement in the attachment to host cells.[110]

It has been proposed that intracellular *Brucella* organisms enter into a period of starvation that favor their resistance to oxidative conditions found within vacuoles.[111-113] *B. abortus* mutants of the RNA chaperone host factor-1 (HF-1) do not replicate in macrophages, but initially multiply in mice. It is likely that this stress protein, normally required for stationary growth, is also necessary during intracellular replication. The HF-1 protein participates, among several pathways, in the regulation of the sigma factor RpoS, required for maintenance of the stationary phase growth rate.[112] As a consequence, bacterial deficient *hfq* mutants, display an impaired stress response and problems for adapting to a stationary phase growth. These mutants, impaired for long-term survival under nutrient deprivation, also demonstrate growth stage- and medium-dependent sensitivity to hydrogen peroxide and a decreased capacity to resist acidic environments, conditions likely to be found intracellularly. *Brucella lon* mutants are impaired in their capacity to resist hydrogen peroxide, puromycin and display reduced survival in macrophages and significant attenuation in mice during the initial periods, but not in later times. The ATPase dependent Lon protease is one of the principal enzymes involved in the turnover of stress-damaged proteins. The level of transcription of this protein increases in response to several environmental stresses. It has been proposed that Lon function, as stress response protease, is required in *Brucella* during the initial stages of infection, but it is not essential for the establishment and maintenance of chronic infections in the host.[113]

Inhibition of Lysosome-Phagosome Function

In contrast to killed bacteria, alive *Brucella* escapes from the constitutive endocytic route by making its compartment non-fusiogenic with lysosomes, indicating that active bacterial mechanism working intracellularly are obligated in this process. Cells treated with killed *Brucella,* harbor a small number of intact bacteria together with bacterial degradation products scattered throughout the cytoplasm. At later stages cathepsin D, a well-known marker for lysosomes, colocalizes with vacuoles containing killed *B. abortus* and bacterial degradation products, attesting that phagosomes have fused with lysosomes. In contrast, intracellular virulent *Brucella* organisms seldom colocalize with cathepsin D, indicating that virulent bacteria avoid fusion with lysosomes.[48]

In vitro fusion experiments between vacuoles containing *Brucella* and lysosomes isolated from macrophages, showed that while compartments containing killed bacteria fuse with lysosomes, vacuoles containing live *Brucella* do not fuse with these organelles.[114] Intracellular *Brucella* organisms do not hamper the fusion of latex beads with lysosomes, suggesting that the key virulent factor is not the general inhibition of lysosome-phagosome fusion in cells, but rather the active modification of the *Brucella*-containing compartment, making this organelle non-fusiogenic with lysosomes.

It has been found that phagosomes containing live *Mycobacterium* retain a host protein (TACO) that prevents delivery of lysosomes into the parasite-containing phagosome.[115] Similarly, it has been proposed that *Leishmania* parasite inhibits phagolysosome biogenesis through insertion of a lipophosphoglycan into the phagosome membrane precluding by this manner the fusion of the phagosome with lysosomes.[116] Indirect evidence for inhibition of neutrophil "degranulation" by *B. abortus* has been obtained from studies of the effects of live or heat-killed organisms and bacterial extracts. *B. abortus* extracts apparently devoid of enzymes and LPS but containing 5'-guanosine monophosphate and adenine inhibited neutrophil degranulation.[117] The same molecules seem to inhibit the myeloperoxidase-hydrogen peroxide-halide activity by specifically hampering degranulation of peroxidase-positive polymorphonuclear granules.[118] Similarly, studies conducted with *B. abortus* extracts were capable of inhibiting phagosome-lysosome fusion in macrophages.[58] It has been observed that smooth *Brucella* LPS enhanced the intracellular survival of rough mutant bacteria in bovine neutrophils.[119] Despite this, some authors have proposed that LPS does not participate in the inhibition phagosome-lysosome fusion.[58,120] The above investigations, although valuable, do not explain the fact that only alive virulent *Brucella* is capable of hampering lysosomal fusion.[3,16,48,114]

Lipopolysaccharide Fails to Stimulate Lysosomal Activity

Several studies have cited the *Brucella* LPS molecule as a virulence factor.[85,86,89] However, in contrast to what has been proposed for other LPSs, the virulence of *Brucella* LPS is not expressed in the classical form of an endotoxin active molecule. For instance, it is known that LPSs from most gram negatives are able to activate phagocytic cells by stimulating the respiratory burst, inducing the production of bactericidal nitrogen intermediates and the generation of active cytokines.[121-123] All these mechanisms contribute to the destruction of intracellular bacteria and promote the fusion of the ingested bacteria with lysosomes.[112,124] *Brucella* LPS practically does not stimulate the oxidative burst or stimulate the release of lysozyme in phagocytic cells, and fails to generate significant amounts of bactericidal reactive nitrogen intermediates.[69,70,89] Furthermore, *Brucella* LPS induces very little release of interferon gamma (IFNγ) or tumor necrosis factor (TNF), substances which are known to enhance lysosomal fusion and the microbicidal mechanisms of phagocytic cells.[125,126] In this respect, the low biological activity of *Brucella* LPS may be envisioned as an advantage of *Brucella* parasites to adapt to intracellular life. Within this context, the *Brucella* LPS is considered a virulence factor.[85,89,127]

Iron Capture and Bactericidal Activity

B. abortus organisms secrete siderophores necessary for chelating iron.[128-130] This iron capturing molecules may compete for intracellular iron within the macrophage and inhibit iron mediated bactericidal killing systems.[129] Iron loaded macrophages have enhanced capability to kill or prevent the replication of intracellular *B. abortus*.[131] This effect was demonstrated with opsonized or non-opsonized bacteria, and with attenuated or virulent strains of *B. abortus*. The augmented bactericidal activity in the presence of iron may be indirectly mediated by the Heber-Weiss reaction.[131] In this system, hydroxyl radicals are generated from hydrogen peroxide in the presence of Fe^{++}, which is oxidized to Fe^{+++}. Generation of more hydroxyl radicals and Fe^{+++} results from the reaction between superoxide and Fe^{++}. The fact that the bactericidal action could be blocked with hydroxyl scavengers, supports this general idea. It is likely that in activated macrophages the concentration of superoxide ions and hydrogen peroxide increase within the lysosomes; thus, providing the substrates for the iron-catalyzed Heber-Weiss reaction. Similar phenomena have been observed with other Gram negative and Gram positive bacteria.

B. abortus siderophore 2,3 dihdroxybenzic acid protects killing of *Brucella* by activated macrophages during the first 12 h and increases the number of intracellular brucellae recovered after 48 h infection.[129] Mutations in the gene chorismate synthase *aro*C, necessary for chorismate synthesis which itself is required for the synthesis of para-amino benzoic acid, a precursor of 2,3 dihdroxybenzic acid, reduces *Brucella* virulence, supporting the role of this siderophore during intracellular survival.[132,133] Therefore, *Brucella* siderophores, capable to chelate intracellular iron necessary for the generation of bactericidal active radicals, may work as virulent factors when defined in the right context. However, the fact that the *Brucella* bacterioferritin deletion mutant replicates within phagocytes similarly to the parent strain,[128] controverts the "virulent" role that some iron capture molecules may have. Furthermore, exogenous iron does not enhance the replication rate as it occurs with *Listeria*,[134] suggesting that *Brucella* is capable of replicating under conditions of low iron.

Controlling the Endocytic Pathway

Up to now, only a reduced number of *Brucella* genes necessary for controlling the intracellular trafficking within cells has been demonstrated. *Brucella bvr* mutants forced to penetrate into CNF-treated HeLa cells, follow the endocytic degradative pathway (Chaves-Olarte, et al, see addendum), indicating that this two component regulatory system is also necessary for controlling vacuole maturation. Similarly, cyclic glucans in plant pathogenic bacteria are essential factors for parasitism.[135] *Brucella* species possess a periplasmic cyclic-1,2-glucan which is not osmotically regulated, in contrast to what happen in *Agrobacterium* or *Rhizobium* parasites.[136] *Brucella cgs* mutants, unable to produce cyclic glucan, are avirulent in mice and incapable of avoiding fusion with lysosomes (Gorvel JP, personal communication).[137] Since *Brucella cgs* mutants adhere and invade cells similarly to the wild type bacteria, it is likely that cyclic glucans are necessary for intracellular trafficking and not for attachment and penetration.

Recently, it has been described a genetic system (*virB*) encoding for a type IV secretion machinery,[10,138] that is also involved in regulating the intracellular trafficking of *Brucella* (see addendum).[139] Type IV secretion systems are complex organelles composed of several proteins, some of which concomitantly span the inner and the outer membrane. These systems are specialized in transferring molecules (proteins or DNA) from the internal to the external milieu of the bacterial cell.[140] The *Brucella* type IV secretion system composed of 13 open reading frames, possesses similarity with the VirB complex of other cell associated bacteria such as *Agrobacterium, Legionella* and *Rickettsia*. Polar mutations in the *vir*B1 and *vir*B10 abolish the ability of *Brucella* to replicate in mice and cells.[139] Non-polar mutation in *vir*B8 and *vir*B10, coding for internal membrane proteins generates attenuated bacteria. Mutations in *vir*B12 and *vir*B13 do not demonstrate defects (Comerci D, personal communication).[10]

Mutations in *vir*B genes do not have effect in the attachment and internalization of *Brucella* to cells. *Brucella* organisms harboring either polar or non-polar mutations in *vir*B10 are capable to penetrate cells, as the wild type *Brucella*, localizing in LAMP-1 positive compartments at early times of infection. However, after this period, *vir*B10 polar mutants are sorted to degradative compartments positive for lysosomal markers, while non-polar *vir*B10 mutants remain within compartments devoid of lysosomal or ER proteins, but retaining early endosomal markers. After 12 hours, a large proportion of intracellular non-polar *vir*B10 mutant bacteria, are recycled to the cell surface. Once outside, the bacteria seem to replicate on the cell surface. These results indicate that although the *vir*B genes are not required for attachment or invasion, they are necessary for regulating the intracellular trafficking from early endosomes to the ER. In addition, the VirB10 product seems to be essential for preventing the fusion of *Brucella* containing vacuoles with lysosomes. Therefore, the absence VirB11 would avoid the correct assembling of the type IV secretion machinery necessary for secreting bacterial substances required for avoiding lysosomal fusion. It follows that the absence of VirB10, will allow a defective but partially functional secretory system competent for controlling lysosomal fusion, but incapable of releasing substances necessary for promoting retrograde *Brucella* trafficking to the ER.

The putative *vir*B10 and *vir*B11 genes seem to encode for an internal transmembrane protein of unknown function (structural?) and for a cytoplasmic or inner membrane protein that has a conserved Walker A NTP binding motif, respectively. Gene reporter analysis has revealed that the *Brucella* VirB system activates during the first 12 h after internalization, then the expression of the system diminishes, corresponding to the replicating time in the ER (Comerci D, personal communication). Therefore, it seems that the type IV secretion machinery is required for controlling the brief trafficking of *Brucella* through the endocytic network until the bacterium reaches its replicating niche within the ER. Once the bacterium has reached the ER, the secretion apparatus may be turned off.

Trafficking from the endocytic network to the ER via autophagosomes, not only requires live virulent bacteria, but it also needs a physiologically normal endocytic machinery. For instance, in cells expressing an activated form of rab5 (bound to GTP), an important fraction of the internalized parasites are unable to escape from the early *Brucella*-containing compartment (see addendum).[16] Although, a fraction of bacteria are capable to reach the ER in these mutant cells, unexpectedly a few of *Brucella* escape from these "early" giant vesicles to the cytoplasm (Gorvel JP, personal communication). These events suggest that *Brucella* is capable of adapting to different intracellular environments by controlling its own intracellular trafficking by direct interaction with its host vacuole.

Escaping from Autophagosomes

The first step in the maturation of autophagosomes is the acquisition of lysosomal membrane-associated proteins by the nascent autophagosomes. Then, acidification of the maturating compartment occurs by inclusion of the H^+-ATPase, and finally delivery of acid hydrolases allows the degradation of intravacuolar isolated cytoplasmic materials.[141] In fact, it has been shown that nocodazole treatment causes the accumulation of acidic autophagosomes that lack acid hydrolases, supporting the concept that vacuole acidification and acquisition of hydrolytic enzymes are separate events.[142] The presence of LAMP-1 and LAMP-2 but the absence of cathepsin D, at 2 h postinoculation in *B. abortus* containing phagosomes, also supports the model of a stepwise maturation of autophagic vacuoles. Whether *B. abortus*-containing autophagosomes are able to acquire the H^+-ATPase, remains to be established. *Brucella* phagosomes seem to acidify rapidly after infection in murine J774 macrophages.[22,71] This acidification step seems to be independent of phagosome-lysosome fusion. In Vero cells, repressors of endosome acidification such as chloroquine and ammonium chloride, reduce the number of *Brucella* with respect to controls.[40] Adjustment of intravacuolar pH has been shown as essential for the activation of virulence genes in certain intracellular parasites. Parasitophorous vacuoles of *Leishmania amazonensis* maintain an acidic pH in infected macrophages.[56] New sets of proteins are synthesized by *S. typhimurium* upon infection of host cells.[54] This latter bacteria induces reduction of phagosome acidification in order to activate the virulence genes of the *PhoP/PhoQ* complex.[74] *B. abortus* is able to synthesize a new set of proteins during macrophage infection,[98,99] suggesting that intracellular conditions (one of which could be intravacuolar pH) may contribute to the activation of genes necessary for trafficking from autophagosomes to the ER. The critical difference between S19 and S2308 seems to be the inhibition of autophagosome maturation by the virulent strain. The reduced pathogenicity of S19 may lay in the incapacity of this strain to respond to environmental stimuli present in the autophagosome (acidification, for example) that could activate virulence genes for the expression of important proteins for remodeling the autophagosome.

Intracellular Nutrient Uptake

The overall higher hydrophobicity of *Brucella* cell envelopes and the close association among the macromolecules of the outer membrane are the factors implicated in the selective penetration of nutrients inside the bacterial cell.[67,143] It has been demonstrated that the absence of a barrier to hydrophobic substances is linked to the structure of the *Brucella* lipid A.[85] The advantages of possessing an hydrophobic envelope for intracellular bacteria, has been further stressed by the finding that LPS molecules become highly hydrophobic during *Rhizobium* bacteroid development.[144] The net result of this structural change in the LPS is that intracellular bacteroids harbor a more hydrophobic outer membranes than the free-living rhizobiae. This adaptive condition seems to promote the exchange of nutrients and favor intracellular life of the bacteroids. Obvious comparisons between the intracellular life style of *Brucella* and *Rhizobium* emerge, since these two bacteria are phylogenetically close relatives.[127]

The importance of cell permeability in *Brucella* was first noticed by Gerhardt et al, in 1950 (reviewed in ref. 145). These authors found that the rate of oxidation of glutamate was slower in virulent *B. abortus* than in attenuated strains. Subsequent work showed that this was due to a reduce cell permeability for the substrate.[145] The hydrophobic pathway rather than metabolic differences,[67] accounts for the effect that sexual hormones may have in the metabolic rate displayed by *Brucella*.[146] Penetration of hydrophobic siderophore chelates through the outer membrane could explain why iron induced outer membrane proteins are not observed in *B. abortus*, under conditions in which they are manifested in other bacteria.[130] In fact, the moderately hydrophobic iron chelator bipyridyl is readily taken up by *Brucella* (López-Goñi I, and Moriyón I, personal communication).

As in other Gram negatives, several porins are present in the outer membrane of *Brucella*, which may selectively acquire nutrients released within its replicating compartments. It has been proposed that some porin molecules may be specifically expressed during intracellular parasitism.[147,148] For instance, Omp2a, not expressed in vitro, when cloned in *Escherichia coli*, it increases the permeability to maltodextrins, reinforcing the idea that this protein may function during intracellular growth as specific translocator for molecules delivered within the replicating niche.

Auxotrophic and Cell Cycle Genes during Intracellular Life

Many of the mutations described in this section, may hamper the adequate extraction of nutrients from the replicating niche. It is likely that these genes are devoted to the regulation of vital functions not directly involved with virulence, but with more general aspects of the *Brucella* physiology. Mutants displaying reduced virulence and/or reduced intracellular survival within macrophages have been identified by signature-tagged transposon mutagenesis.[149] Several of these attenuated mutants carry auxotrophic defects such as those necessary for leucine, arginine or aromatic acid biosynthesis, while others carry deficiencies in the synthesis chorismate for the generation of para-aminobenzoic acid necessary for quinone synthesis, 2,3 dihydrobenzoic acid and folic acid. Mutations in genes involved in the glucose metabolism, such as those coding for phosphogluocose isomerase gene, also attenuate *Brucella*, probably due to several pleiotropic defects, including the synthesis of cell wall peptidoglycan. Similarly, transposon insertion in the *gpt*-like gene coding for hypoxanthine-guanine biosynthesis attenuates *Brucella* probably due to alterations in the nuclotide biosynthesis. In support of this are the *B. melitensis pur*E (purine auxotrophic) mutants that display reduced growth in macrophages.[150,151] The *pur*E mutation has minimal effect on internalization, but effectively blocks intracellular replication. The *pur*E mutation may cause bacterial death, simply as result of starvation.

Other mutants, also identified by signature-tagged mutagenesis have defects in regulatory systems, such as LysR transcriptional regulator, which is part of a positive regulator system of virulence genes in several bacteria[149] Mutants in the expression of NtrY protein, a sensor of an Ntr-related regulator, are weakly attenuated probably due to a pleiotropic negative effect on the *ntr* regulon.[152] Similar to other bacteria, mutations in genes involved in glutamine metabolism reduce the ability of *Brucella* to replicate in macrophages.[149]

The *Brucella ccr*M gene codifies for a CcrM DNA methyltransferase that catalyzes the methylation of the adenine in sequences GanTC. This gene performs important functions during cell division and therefore it is essential for viability.[153,154] Increase in the *ccr*M copy number, in addition to alter the morphology of the bacterial cells, attenuates *Brucella*, indicating that controlled cell cycle and bacterial division are necessary for intracellular survival. Mutations in the *bac*A gene which codes for a putative cytoplasmic membrane transport protein render *B. abortus* avirulent.[155] *Brucella bac*A mutants display altered transport of molecules, and then may have problems for replicating within cells.

Maintaining the Host Cell Alive

Some parasites may promote programmed cell death while others are prone to prevent it, prolong cell life or are even capable to stimulate replication of their host cells. In the case of *Proteobacteria* of the alpha subdivision such as *Rickettsia, Agrobacterium, Rhizobium* and *Brucella*, as well as other intracellular parasites that causes chronic infections (*Mycobacterium* and *Chlamydia*), all prevent cell death. After *Brucella* inoculation, heavily infected trophoblasts, epithelial cells or macrophages do not display signs of necrosis or apoptosis.[3,12,13,38,69] DNA synthesis, microtubule spindle formation, chromosome migration, karyokinesis and cytokinesis are not inhibited by intracellular *Brucella* (Chaves-Olate, et al, see addendum). As consequence, dividing cells filled with brucellae are frequently observed in vivo an in vitro.[12,13,40] In CNF treated HeLa cells, cytokinesis is inhibited due to paralysis of actin filaments without affecting nuclear division. When these cells are infected with *Brucella*, karyokinesis proceeds without signs of degeneration, despite the large number of intracellular *Brucella* within the ER (Chaves-Olarte, et al, see addendum). Since cycloheximide does not inhibit intracellular bacterial replication, it is feasible to propose that de novo host protein synthesis is not required during parasitism.[40]

Live but not killed *B. suis* prevents programmed cell death of infected human monocytes.[33] This suggests that infection protects host cells from several cytotoxic activities generated during the immune response. Since both, invaded and non-invaded cells are guarded against apoptosis, it has been suggested that this protective mechanism is mediated through soluble substances released during bacterial infection. Apoptosis inhibition was independent of LPS and requires the overexpression of the *A1* gene by infected cells, a member of the *bcl-2* family involved in the survival of blood forming cells.

In contrast to other LPSs, *Brucella* LPS practically does not display endotoxicity and is a poor inducer of cytotoxic mediators.[30,89,125] This property shared by other intracellular animal pathogens such as *Rickettsia, Legionella* and *Bartonella*, may be envisioned as an evolutive advantage of *Brucella* parasites for adaptation to intracellular life.[39,85,89] In this sense it is not surprising that the lipid As of the intracellular parasites *L. pneumophila* and *B. abortus* show low endotoxicity and reduced ability to stimulate cells.[30,89,125,156] This property may be idiosyncratic and useful for intracellular bacteria.

Modulation of the Immune Response

Once inside macrophages, pathogens might diminish or abrogate their antigen presentation capacity, thus reducing the T cell-mediated immune responses. LPSs from different bacteria have been shown to modulate the immune responses in several systems.[157-160] *B. abortus* LPS molecules accumulate inside lysosomal compartments and associate with MHC-II proteins in antigen presenting cells.[161-163] The intracellular LPS, which remains for long periods without being degraded, is exported to the cell surface where it forms stable macrodomains.[163] Once inside macrophages, *B. abortus* LPS impairs the MHC-II presentation pathway, but not MHC-I presentation of foreign peptide antigens. This impairment results neither from a deficient uptake or catabolism of the native antigen nor from a reduced MHC-II surface expression, reduced number of B7 membranous costimulatory molecules or by defective alpha/beta dimmer formation. In addition, this inhibitory effect is not due to a direct suppressive action of LPS on T cells, independent of macrophages.[163]

Brucella LPS macrodomains at the macrophage plasma membrane are highly enriched in MHC-II molecules, suggesting that the LPS-MHC-II macrodomains may impair the appropriate recognition of protein peptide-MHC-II complexes by CD4+T cells.[163] The presence of MHC-II-LPS macrodomains do not prevent the binding of antigen peptides into the groove of MHC-II molecules. Therefore, the LPS-induced interference on MHC-II antigen presentation is likely to occur distal to intracellular events leading to the meeting of antigenic peptides and MHC-II molecules. LPS molecules embedded in the membrane of MHC-II-positive compartments may interact with already peptide-MHC-II forming ternary complexes, which then recycle to the plasma membrane. In one model, the LPS O-chain, facing the external milieu, could prevent the accessibility of MHC-II complexes to their specific T cell receptor. Another model is related to a superantigen-like function in T cell activation. It is known that superantigens modify the geometry of TCR-peptide/MHC-II complexes which may be less critical for T cell activation than certain other factors, in particular those involved in the stability of the resulting complex.[164,165] The serial triggering[166-168] and kinetics proofreading models[169] of T cell activation suggest that the short half lives of TCR-peptide/MHC-II complexes are required for efficient T cell stimulation. The presence of LPS-MHC-II macrodomains in macrophages has been detected even 60 days after infection, thus highlighting the remarkable stability of these surface LPS macrodomains. In this model, LPS could down-regulate T cell responses by stabilizing the MHC-II/peptide complexes at the cell surface of antigen. Consequently, in contrast to super antigens, LPS would be less efficient at triggering T cells because they form TCR-LPS-MHC-II complexes with a very long half-life.

The in vitro inhibition of the immune response correlates with that observed in vivo upon infection by *Brucella*. It is worth noting that, chronic brucellosis is accompanied by a general immunosuppression that can be revealed by using an IL-2 detection system.[170,171] Infected macrophages may exert a negative feedback, diminishing lymphocyte proliferation in response to *Brucella* antigens.[172,173] It has been proposed that chronically infected macrophages may fail to act as targets of T cells and may down-regulate T lymphocytes function.[4] Since LPS is released from bacteria inside host cells and that LPS is not degraded by peritoneal macrophages,[162] it can hypothesize that the *Brucella*-induced immunosuppression could be attributed to bacteria-associated or bacteria-released LPS. In this respect, *Brucella* LPS may play a central role in the immunosuppression observed upon brucellosis infection and may account for the presence of anergic T cells in infected patients.[174]

Experiments have demonstrated that *Brucella* is capable to infect small lymphocytes in the cortical zone of lymph nodes.[151] The infected lymphocytes harboring intracellular bacteria within vacuolar compartments with no signs of degradation look normal. At the present time, the implications of lymphocyte parasitism by *Brucella* are not known. The fact that *Brucella* LPS complexes with MHC-II in these cells,[161] opens the possibility that LPS released by intracellular *Brucella* could modulate antigen presentation in lymphocytes as it does in macrophages.[163]

Control of the Intracellular Infection

In spite of the role played by antibodies and cytotoxic T cells in murine brucellosis,[23,175,176] it is generally accepted that the ultimate brucellicidal activity in most animals is mediated by a type-I immune response.[4,5,177] Therefore, a robust respiratory burst, a powerful production of microbicidal nitrogen intermediates and a more efficient phagosome-lysosome fusion performed by the activated macrophages are needed to kill intracellular *Brucella*.[24,178]

Several studies have highlighted the role of IFNγ, TNFα, IL-6, IL-1β, IL-10, IL-12, GM-CSF and G-CSF in host resistance against bacterial infections.[4,5,179-183] Particularly, during *Brucella* infection IFNγ, TNFα, IL-2 and IL-10, IL-12 seem to control the intracellular growth of *Brucella* strains within macrophages, whereas IL-1α, IL-4, IL-6 and GM-CSF do not have consistent effect.[69,179,184,185] Among the various cytokines, IFNγ is the most relevant for generating macrophages with strong brucellicidal activity. Moreover, IL-2, IL-10 and IL-12, cytokines that influence the acquired cellular resistance and specifically contribute to control the brucellae multiplication, seem to work via the IFNγ-dependent pathway. The role of TNFα in brucellosis is not completely clear. In contrast to what has been observed in murine macrophages, *Brucella* strains do not induced TNFα in human macrophages upon infection, a effect that is promoted by killed organisms but not by opsonization.[28] Pre-treatment of human macrophages with exogenous TNFα, significantly inhibits the rate of *Brucella* intracellular replication. Live *Brucella* is capable to inhibit the production of TNFα. On the other hand, TNFα may no be essential for the induction of acquired cellular resistance but it is likely to directly activate effector cells for limiting the multiplication of intracellular *Brucella*.[184-187] It is known that *Brucella* LPS is a poor inducer of cytokine production and an inefficient activator of phagocytic cells.[89,125,178] Moreover, it seems that CD14 molecule is a not a receptor for *Brucella* LPS (unpublished results), precluding the activation of macrophages by this mechanism. In this respect it is unlikely that macrophage activation and secretion of IL-12 occurs via the LPS molecule.

It has been demonstrated that bactericidal activity of activated macrophages and expression of macrophage specific cytokines depend upon the expression of NF-IL6.[183,188-195] Upon activation of NF-IL6 knockout macrophages by IFNγ, induction of transcription of TNFα, IL-6, IL-1β, GM-CSF, M-CSF, IL-10, IL-12 is comparable to that observed in normal mice.[183] Strikingly, no induction of G-CSF expression is observed in NF-IL6 knockout macrophages being this defect restricted to macrophages and fibroblasts.[183] NF-IL6 knockout mice display a high susceptibility to *Salmonella* and *Listeria* infections, suggesting that NF-IL6 plays a role in controlling intracellular parasites.[183] Attenuated *B. abortus* S19 is capable to replicate in NF-IL6 knockout murine macrophages. The level of multiplication is comparable to that observed in normal macrophages infected with pathogenic brucellae strains.[68] The role of NF-IL6 in the inhibition of intracellular bacterial replication is related to its control of endocytosis and membrane fusion events between endosomes and *Brucella*-containing phagosomes. Addition of G-CSF, restores both endocytosis and the morphology of endosomes, together with the bactericidal activity. During *Brucella* infection, it has been observed that endocytosis but not recycling is affected in NF-IL6 knockout macrophages,[68] suggesting that G-CSF promotes fusion events in early phagosomes.

In NF-IL6 knockout macrophages, NO synthetase is not impaired.[183] However, the production of reactive oxygen intermediates is lower in NF-IL6 than that in wild-type macrophages,[183] suggesting that NF-IL6 may control the expression of other elements of the respiratory burst. Indeed, it is known that G-CSF enhances the respiratory burst in phagocytes.[196] NO synthetase was found to be associated to intracellular membrane vesicles different from lysosomes and peroxisomes in murine macrophages.[197] These vesicles could translocate to *Brucella*-containing phagosomes in normal macrophages and be hampered in NF-IL6-deficient macrophages, due to the lack of fusion between endosomes and phagosomes. Therefore, one hypothesis is that G-CSF, by completely restoring endosome-phagosome fusion, allows elements of the respiratory burst, present in endocytic compartments, to reach the *Brucella*-containing phagosomes and thus to partially reestablish the bactericidal activity of macrophages.[68] Under these conditions, attenuated *Brucella* could be targeted to lysosomes and killed, whereas pathogenic bacteria could still replicate but in a lesser extent than in resting macrophages.[198]

Concluding Remarks

M cells, neutrophils, non-activated macrophages from newly infected hosts, activated macrophages from infected animals and non-professional phagocytes, all serve different purposes during the course of *Brucella* infection. While the translocation of ingested *Brucella* occurs through M cells, the first line of defense against *Brucella* invasion is secured by neutrophils.[7] Although neutrophils, are not the preferred niche for *Brucella* replication, some of the ingested bacteria are capable to withstand destruction inside this leukocytes.[23,43] In turn, this event may favor the spreading of the parasite to other tissues.[6,7] In the second line of defense are the macrophages, which similar to neutrophils, could destroy an important proportion of the ingested *Brucella*. Macrophages may also serve as substrate for *Brucella* replication as well as vehicles for brucellae transportation to other tissues. In the pregnant animal, *Brucella* organisms invade the erythrophagocytic trophoblasts, which are the preferred replicating host cells and the site from which the bacteria spreads to the fetus.[37,47] Generally, immune individuals control brucellosis via activation of the macrophagic system. Depending upon the animal species, humoral response may serve as an important aid for phagocytosis and for deviating the intracellular route of the ingested bacteria to destructive compartments. To this we must add, that in spite of the close phylogenetic relationship shown by the different *Brucella* species,[200] there are some species and strains more pathogenic than ohers.[1,26,201,202]

The idiosyncratic functions played by the various host cells, certainly mark the differences observed in the intracellular trafficking of *Brucella*. Professional phagocytes are more prone to destroy *Brucella* organisms and control their intracellular replication than non-professional phagocytes. For instance, when *Brucella* strains of low and high virulence are compared, it is clear that bacteria displaying low virulence, do not withstand the very powerful destructive machinery of professional phagocytes, while several of the same strains, are capable of replicating within non-professional phagocytic cells[12,13,25] Similarly, defective macrophages in fusion events are more permissive cells, allowing attenuated S19 to multiply.[68] Moreover, while in non-professional phagocytes most of the internalized bacteria transit within intracellular compartments that do not fuse with late endosomes and lysosomes, in macrophages, a relatively large proportion of the phagosomes fuse with acidic vacuoles and lysosome-like compartments.[22] In non-professional phagocytes most of the ingested *Brucella* organisms transit from early endosomes to autophagosomes, and then to the ER, while in macrophages just a small proportion of *Brucella*-containing compartments seem to follow this autophagocytic route to the ER.

In contrast to the proposition that in both, macrophages and non-professional phagocytes, *Brucella* organisms follow similar routes for reaching the ER,[3,16,68] other investigators have suggested that in macrophages, *Brucella* delays maturation of phagosomes into phagolysosomes.[22] Although, at this point these two postures can not be reconciled, it is clear that from the pure microscopically point of view a parallelism between replicating *Brucella* within the intracellular space of macrophages and non-professional phagocytes exist at later times. As it is the case during *Listeria* infection, at the end, it may be that in macrophages, most of the *Brucella* organisms are destroyed within lysosomes, while only a few bacteria arrive to the ER through the autophagocytic route. However, once these few bacteria are in the ER, they are capable of replicating inside this compartment which is non-fusiogenic with lysosomes. In non-professional phagocytes the events are reversed: only a few internalized *Brucella* are destroyed within lysosomes, while most of the *Brucella* arrive to the ER trafficking through the safer autophagocytic route (Fig. 1).

Obviously, the most economical pathway for the bacteria would be to use the same molecular machinery for intracellular trafficking and for subtracting resources from the different host cells, rather than to possess different molecular strategies for each cell type. An example of this, can be seen in *Legionella* species, bacteria that seem to use the same strategy to parasitize their free living host amoebae and the resident macrophages located in the lung of infected individuals.[199] Thinking in this direction, the phagolysosome route and the autophagosome route establish different constrains than must be surpassed. Similarly, the phagolysosomes and ER, perform different functions and display a different set of resources for the intracellular bacteria. Therefore, it seems unlikely that two mechanisms to adapt equally well to two very different intracellular environments have evolved in *Brucella*. However, the paths and rules of evolution are intricate and very often they do not follow our anthropological conceptions. In this respect, the question is open, and in spite of our favoritism for the autophagocytic route and replication within the ER in both, professional a non-professional phagocytes, we must pay considerable attention to alternative intracellular routes proposed for phagocytic cells.

Based in several investigations,[127,203,204] we demonstrated, more than one decade ago, the close phylogenetic relationship between *Brucella* and members of the alpha-2 *Proteobacteria*, and proposed a model to explain the evolution of *Brucella* parasitism from a common free-living ancestor.[127,204] More recently, we have adjusted our proposition and validated our hypothesis[39,205,206] taking into consideration comparative studies carried out by us[14,207] and by others.[208,209] These studies have contributed to understand the biology of this pathogen, and in the identification of various factors necessary for *Brucella* parasitism. For instance, the existence of well known genetic systems in *Rhizobium, Sinorhizobium, Agrobacterium* and *Phyllobacterium,* which are among the closest relatives of *Brucella,* have served to carry out comparative studies that have concluded in the description of potential virulence genes and to put forward several hypothesis on the virulence mechanisms.[10,11,139,155] The *bvr*S/*bvr*R two component regulatory system, the *vir*B type IV secretions machinery, the periplasmic cyclic glucan and the microcines transporter *bac*A, are genes and products all related to virulence, which were first described in some of these pathogens and endosymbionts.

The fine adjustments between *Brucella* parasites and their host cells are the result of a prolonged and intimate association between both parties, in the understanding that the evolutionary process does not operate by inventing but rather by reinventing on the grounds of preexisting structures. For instance, the periplasmic domains (involved in environmental sensing) of the sensory proteins (ChvG, ExoS and BvrS) of the two regulatory systems necessary for bacterial parasitism of *Agrobacterium, Sinorhizobium* and *Brucella* show less similarity than other protein domains, implying that they were derived for sensing different stimuli.[14] However, it is remarkable that the two intracellular bacteria (*S. meliloti* and *B. abortus*) are more similar in this region than the pericellular one (*A. tumefaciens*). Likewise, the VirB systems of *Brucella* and *Agrobacterium,* although similar in many respects, they function in different cellular environments, one delivering Ti DNA to govern its host cell from the external surface of the cytoplasmic membrane, and the other to transfer molecules inside the phagosome for controlling its intracellular trafficking. Although the absence of cyclic glucan render *Brucella, Agrobacterium* and *Rhizobium* incompetent to parasitize, in *Brucella* this oligosaccharide is not osmoregulated[136] and captures lipids (Moriyón I, personal communication), implying that this periplasmic molecule plays a different role in each of these parasitic bacteria.

Cell associated alpha-2 *Proteobacteria* share many structural characteristics in their outer membranes, several of which have been regarded as important for invading animal and plant cells (see Chapter 14). For instance, the closest relative of *Brucella*, the human opportunistic *Ochrobactrum*, is sensitive to bactericidal cationic peptides, in spite of having almost identical lipid As and sharing many outer membrane physical and chemical properties.[207] As in other alpha-2 *Proteobacteria*, the core oligosaccharide of *Ochrobactrum* possess negative charged galacturonic acid, sugar which accounts for this sensitivity. Being the resistance to bactericidal cationic peptides, an essential property related to *Brucella* parasitism, their absence in *Brucella* LPS is conspicuous and likely to be a key structural evolutionary variation.

Recently, it has been determined that the LPSs of *Rhizobium* intracellular bacteroids become highly hydrophobic, in contrast to the LPS of free living bacteria.[144] It was suggested that the switch from hydrophilic LPS to a predominantly hydrophobic molecule is the result of an adaptive response for changing environments. In the case of *Brucella* the highly hydrophobic outer membrane seems to be well adapted to intracellular life.[143,204-207] This pattern, offers a starting point on which the strong hydrophobicity of the lipid A of many alpha-2 *Proteobacteria* is complemented in *Brucella* by modification of the LPS core and by the horizontal acquisition of N-formylperosamine genes for the synthesis of the hydrophobic O-chain and NH. The significance of these adaptive evolution of *Brucella* to its intracellular niche has been carefully reviewed (see Chapter 14), by comparing the LPSs of *Legionella* and *Brucella*, both intracellular pathogens which have arrived from the extremes of different phylogenies to a very similar solutions.

Local conditions may determine the plasticity and size of the genome during evolution,[205,206] as well as the plasticity of the outer membrane.[144,209] In contrast to other parasitic bacteria, most of the virulent genes identified in *Brucella* so far, have been received vertically from a common alpha-2 *Proteobacteria* ancestor, as revealed by the extensive amelioration of the sequences. One conspicuous exception may be the horizontal acquisition of genes necessary for the synthesis of the O-chain and NH polysaccharides.[210] These genes were probably acquired exclusively by the *Brucella* ancestor, since none of the other members of the alpha-2 *Proteobacteria* seem to synthesize this sugar, not even the closest phylogenetic relative of *Brucella*.[209] The incorporation of N-formylperosamine polymers on the framework of a quinovosamine-containing core and lipid A basic structures (shared by many other alpha-2 *Proteobacteria*), could have been a crucial step in the pathogenicity of *Brucella*. To this we must add the changes that generated a core-lipid A possessing low endotoxic and biological potency, that also conferred resistance to bactericidal substances and proportioned the hydrophobic properties of the *Brucella* outer membrane.

A recent analysis on the 20% of the *B. abortus* genome has revealed that an important proportion of the putative genes are related to alpha *Proteobacteria*, to secretion systems and to virulent genes of parasitic bacteria.[211] No doubt that in the near future more sequences will be available, many of which will be related to virulent genes of plant and animal bacterial pathogens (see addendum). Although sequencing studies are essential, at this point, it is not enough to inject mice and infect cells with organisms that have been mutated in potential virulent genes, but rather to assign specific functions and to understand the relevance that these sequences possess for *Brucella* parasitism. In addition, we need to overcome the concept that *Brucella* virulence is related to a single or a few conditions but rather to a complexity of factors that efficiently interact for parasitism. All these refined evolutionary adaptations on the "top" of established ancestral genes and structures, have generated a group of exquisite pathogenic organisms, perfectly adapted to parasitize their hosts cells. Under this perspective, it may well be that the most risky and uncomfortable environment for members of the genus *Brucella* is the extracellular milieu, from which these bacteria must escape in order to survive and persist as intracellular pathogens from one generation to another.

Addendum

During the publication of this review article, relevant data discussed here have been documented in specialized journals. The mechanism of *Brucella* internalization and intracellular trafficking in non-professional phagocytes (Guzmán-Verri et al. J Biol Chem 2001; 276:44435; Chaves-Olarte et al. Cell Microbiol 2002; 4:683) and in macrophages (Naroeni and Porte. Infect Immun 2002; 70:1640; Watarai et al. Cell Micrbiol 2002; 46:341) described in several sections of this review have been published as indicated. Functional regulation of BvrS/BvrR and VirB systems involved in *Brucella* internalization and intracellular trafficking have also been documented (Guzmán-Verri et al. Proc Natl Acad Sci USA 2002; 99:112375; Boschiroli et al. Proc Natl Acad Sci USA 2002; 99:1544). Publications and commentaries on the complete genome sequence analysis of *B. melitensis* (DelVecchio et al. Proc Natl Acad Sci USA 2002; 99:443; Moreno and Moriyón. Proc Natl Acad Sci USA 2002; 99:1) and *B. suis* (paulsen et al. Proc Natl Acad Sci USA 2002; 99:13148) have appeared, confirming several of the assertions of this article.

Acknowledgments

This work was supported by research contract ICA4-CT-1999-10001 from the European Community, RTD project NOVELTARGET-VACCINES and MICIT/CONICIT of Costa Rica.

References

1. Corbel MJ, Brinley-Morgan WJ. Genus Brucella Meyer and Shaw. In: Krieg NR, Holt JC, eds. Bergey's Manual of Systematic Bacteriology. Vol. 1. Baltimore: The Williams Wilkins Co., 1984:377-388.
2. Jahans KL, Foster G, Broughton ES. The characterization of Brucella strains isolated from marine mammals. Vet Microbiol 1997; 57(2):373-382.
3. Pizarro-Cerdá J, Meresse S, Parton RG et al. Brucella abortus transits through the autophagic pathway and replicates in the endoplasmic reticulum of nonprofessional phagocytes. Infect Immun 1998; 66(12):5711-5724.
4. Baldwin CL, Winter AJ. Macrophages and Brucella. Immunol Ser 1994; 60:363-380.
5. Liautard JP, Gross A, Dornand J et al. Interactions between professional phagocytes and Brucella spp. Microbiologia 1996; 12(2):197-206.
6. Enright FM. The pathogenesis and pathobiology of Brucella infection in domestic animals. Antigens of Brucella. In: Nielsen K, Duncan B, eds. Animal Brucellosis. Boca Raton: CRC Press, Inc, 1990:301-320.
7. Ackermann MR, Cheville NF Deyoe BL. Bovine ileal dome lymphoepithelial cells: Endocytosis and transport of Brucella abortus strain 19. Vet Pathol 1988; 25(1):38-35.
8. Cheville NF, Jensen AE, Halling SM et al. Bacterial survival, lymph node changes, and immunologic responses of cattle vaccinated with standard and mutant strains of Brucella abortus. Am. J Vet Res 1992; 53(10):1881-1888.
9. Finlay B, Falkow S. Common themes in microbial pathogenicity revised. Microbiol Mole Biol Rev 1997; 61(2):136-169.
10. O'Callaghan D, Cazevielle C, Allardet-Servent A et al. A homologue of Agrobacterium tumefaciens VirB and Bordetella pertussis Ptl type IV systems is essential for intracellular survival of Brucella suis. Mol Microbiol 1999; 33(6):1210-1220.
11. Ugalde RA. Intracellular lifestyle of Brucella spp. Common genes with other animal pathogens, plant pathogens, and endosymbionts. Microbes Infect 1999; 1(14):1211-1219.
12. Detilleux PG, Deyoe BL, Cheville NF. Entry and intracellular localization of Brucella spp. in Vero cells: Fluorescence and electron microscopy. Vet Pathol 1990; 27(5):317-328.
13. Detilleux PG, Deyoe BL, Cheville NF. Penetration and intracellular growth of Brucella abortus in non-phagocytic cells in vitro. Infect Immun 1990; 58(7):2320-2328.
14. Sola-Landa A, Pizarro-Cerdá J, Grilló JM et al. A two component regulatory system playing a critical role in plant pathogens and endosymbionts is present in Brucella abortus and controls cell invasion and virulence. Mol. Microbiol. 1998; 29(1):5265-5273.
15. Freer E, Pizarro-Cerda J, Weintraub A et al. The outer membrane of Brucella ovis shows increased permeability to hydrophobic probes and is more susceptible to cationic peptides than are the outer membranes of mutant rough Brucella abortus strains. Infect Immun 1999; 67(11):6181-6186.

16. Pizarro-Cerdá J. Traffic intracellulaire et survie de Brucella abortus dans les phagocytes professionnels et non professionnels. Doctoral Thesis, University de la Mediterranee Aix-Marseille II, Faculte des Sciences de Luminy, Marseille-Luminy, France, 1998:156-168.
17. Allen CA, Adams LG, Ficht TA. Transposon-derived Brucella abortus rough mutants are attenuated and exhibit reduced intracellular survival. Infect Immun 1998; 66(3):1008-1016.
18. Corbeil LB, Blau K, Inzana TJ et al. Killing of Brucella abortus by bovine serum. Infect Immun 1988; 56(12):3251-3261.
19. Aragón V, Díaz R, Moreno E, Moriyón I. Characterization of Brucella abortus and Brucella melitensis native haptens as outer membrane O-type polysaccharides independent from the smooth lipopolysaccharide. J Bacteriol 1995; 178(4):1070-1079.
20. Bowden RA, Cloeckaert A, Zygmunt M et al. Surface exposure of outer membrane protein and lipopolysaccharide epitopes in Brucella species studied by enzyme-linked immunosorbent assay and flow cytometry. Infect Immun 1995; 63(10):3945-3952.
21. Cloeckaert A, de-Wergifosse P, Dubray G. Identification of seven surface-exposed Brucella outer membrane proteins by use of monoclonal antibodies: Immunogold labeling for electron microscopy and enzyme-linked immunosorbent assays. Infect Immun 1990; 58(12):3980-3987.
22. Arenas GN, Staskevich AS, Aballay A. Intracellular trafficking of Brucella abortus in mice. Infect Immun 2000; 68(7):4255-4263.
23. Eze MO, Yuan L, Crawford RM et al. Effects of opsonization and gamma interferon on growth of Brucella melitensis 16M in mouse peritoneal macrophages in vitro. Infect Immun 2000; 68(1):257-263.
24. Gross A, Spiesser S, Terraza A et al. Expression and bactericidal activity of nitric oxide synthase in Brucella suis infected murine macrophages. Infect Immun 1998; 66(4):1309-1316.
25. Harmon BG, Adams LG, Frey M. Survival of rough and smooth strains of Brucella abortus in bovine mammary gland macrophages. Am J Vet Res 1988; 49(7):1092-1097.
26. Young EJ, Borchert M, Kretzer FL et al. Phagocytosis and killing of Brucella by human polymorphonuclear leukocytes. J Infect Dis 1985; 151(4):682-690.
27. Schorey, JS, Carrol MC, Brown EJ. A macrophage invasion mechanism of pathogenic mycobacteria. Science 1997; 277(5329):1091-1093.
28. Caron E, Liautard JP, Kohler S. Differentiated U937 cells exhibit increased bactericidal activity upon LPS activation and discriminate between virulent and avirulent. Listeria and Brucella species. J Leuk Biol 1994; 56(2):174-181.
29. Harmon BG, Adams LG, Templeton JW et al. Macrophage function in mammary glands of Brucella abortus-infected cows and cows that resisted infection after inoculation of Brucella abortus. Am J Vet Res 1989; 50(4):459-465.
30. Moreno E, Berman DT, Boettcher LA. Biological activities of Brucella abortus lipopolysaccharides. Infect Immun 1981; 31(1):362-370.
31. Hoffmann EM, Houle JJ. Failure of Brucella abortus lipopolysaccharide(LPS) to activate the alternative pathway of complement. Vet Immunol Immunopathol 1984; 5(1):65-68.
32. Campbell GA, Adams LG, Sowa BA. Mechanism of binding of Brucella abortus to mononuclear phagocytes from cows naturally resistant or susceptible to brucellosis. Vet Immunol Immunopathol 1994; 41(3-4):295-306.
33. Gross A, Terraza A, Ouahrani-Bettache S et al. In vitro Brucella suis infection prevents the programmed cell death of human monocytic cells. Infect Immun 2000; 68(3):342-351.
34. Kuzumawati A, Cazevieille C, Porte F et al. Early events and implication of F-actin and annexin I associated structures in the pathogenic uptake of Brucella suis by J-774A.1 murine cell line and human monocytes. Microb Pathog 2000; 28(6):343-352.
35. Gay B, Mauss H, Sanchez-Teff S. Aspect ultrastructuraux de la phagocytose in vivo et in vitro de Brucella par les macrophages du péritoine de la souris Ann Immunol 1981; 132 D :299-313.
36. Lee CM, Mayer EP, Molnar J et al. The mechanism of natural binding of bacteria to human lymphocyte subpopulations. J Clin Lab Immunol 1983; 11(2):87-94.
37. Anderson TD, Cheville NF. Ultrastructural morphometric analysis of Brucella abortus-infected trophoblasts in experimental placentitis. Bacterial replication occurs in rough endoplasmic reticulum. Am J Pathol 1986; 124(2):226-237.
38. Anderson TD, Cheville NF, Meador VP. Pathogenesis of placentitis in the goat inoculated with Brucella abortus II. Ultrastructural studies. Vet Pathol 1986; 23(3):227-239.
39. Pizarro-Cerdá J, Moreno E, Gorvel JP. Brucella abortus invasion and survival within professional and nonprofessional phagocytes. Microbes Infect 1999; 6(2):201-232.
40. Detilleux PG, Deyoe BL, Cheville NF. Effect on endocytic and metabolic inhibitors on the internalization and intracellular growth of Brucella abortus in Vero cells. Am J Vet Res 1991; 52(10):1658-1664.

41. Bress D, Steadham E, Stevens M et al. $\alpha5\beta1$ integrin mediates Brucella abortus strain RB51 adhesion on bovine trophoblastic cells. Vet Pathol 1996; 33 :615.
42. Finlay B, Cossart P. Exploitation of mammalian host cell functions by bacterial pathogens. Science 1997; 276(5313):934-938.
43. Kreutzer DL, Deyfus LA, Robertson DC. Interactions of polymorphonuclear leukocytes with smooth and rough strains of Brucella abortus. Infect Immun 1979; 23(3):737-742.
44. Riley LK, Robertson DC. Ingestion an intracellular survival of Brucella abortus in human and bovine polymorphonuclear leukocytes. Infect Immun 1984; 46(1):231-236.
45. Riley LK, Robertson DC. Brucellacidal activity of human and bovine polymorphonuclear leukocyte granule extracts against smooth and rough strains of Brucella abortus. Infect Immun 1984; 46(1):224-230.
46. Canning PC, Deyoe BL, Roth JA. Opsonin-dependent stimulation of bovine neutrophil oxidative. metabolism by Brucella abortus. Am J Vet Res 1988; 49(2):160-163.
47. Tobias L, Cordes DO, Schurig GG. Placental pathology of the pregnant mouse inoculated with Brucella abortus strain 2308. Vet Pathol 1993; 30(2):119-129.
48. Pizarro-Cerdá J, Moreno E, Sanguedolce V et al. Virulent Brucella abortus prevents lysosome fusion and is distributed within autophagosome-like compartments. Infect Immun 1998; 66(5):2387-2392.
49. Steele-Mortimer O, Méresse S, Gorvel JP et al. Biogenesis of Salmonella typhimurium-containing vacuoles in epithelial cells involves interactions with the early endocytic pathway. Cell Microbiol. 1999; 1(1):33-49.
50. Scianimanico S, Desrosiers M, Dermine JF et al. Impaired recruitment of the small GTPase rab7 correlates with the inhibition of phagosome maturation by Leishmania donovani promastigotes. Cell Microbiol 1999; 1(1):19-32.
51. Via LE, Deretic D, Ulmer RJ et al. Arrest of mycobacterial phagosome maturation is caused by a block in vesicle fusion between stages. controlled by rab5 and rab7. J Biol Chem 1997; 272(20):13326-13331.
52. Alvárez-Domínguez C, Stahl P. Increased expression of rab5a correlates directly with accelerated maturation of Listeria monocytogenes phagosomes. J Biol Chem 1999; 274(17):11459-11462.
53. Dobrowolski JM, Sibley LD. Toxoplasma invasion of mammalian cells is powered by the actin cytoskeleton. Cell 1996; 84(6) 933-939.
54. Tardieux I, Webster P, Ravesloot J et al. Lysosome recruitment and fusion are early events required for trypanosome invasion of mammalian cells. Cell 1992; 71(7):lll7-l130.
55. Scherer DC, DeBuron-Connors I, Minnick MF. Characterization of Bartonella bacilliformis flagella and effect of antiflagellin antibodies on invasion of human erythrocytes. Infect Immun 1993; 61(12):4962-4971.
56. Antoine JC, Prina E, Jouane C et al. Parasitophorous vacuoles of Leishmania amazonensis-infected macrophages maintain an acidic pH. Infect Immun 1990; 58(3):779-787.
57. Buchmeier NA, Heffron F. Induction of Salmonella stress proteins upon infection of macrophages. Science 1990; 248(4956):730-732.
58. Frenchick PJ, Markam RJF, Cochrane AH. Inhibition of phagosome-lysosome fusion in macrophages by soluble extracts of virulent Brucella abortus. Am J Vet Res 1985; 46(2):332-335.
59. Hunziker W, Geuze HJ. Intracellular trafficking of lysosomal membrane proteins. BioEssays 1995; 18(3):379-388.
60. Andrejewski N, Punnonen EL, Guhde G et al. Normal lysosomal morphology and function in LAMP-1-deficient mice. J Biol Chem 1999; 274(18):12692-12701.
61. Swanson MS, Isberg RR, Association of Legionella pneumophila with the macrophage endoplasmic reticulum. Infect. Immun. 1995; 63(9):3609- 3620.
62. Liou W, Geuze HJ, Geelen MJH et al. The autophagic and endocytic pathways converge at the nascent autophagic vacuoles. J Cell Biol 1997; 136(1):61-70.
63. Sinai AP, Webster P, Joiner KA. Association of host cell endoplasmic reticulum and mitochondria with the Toxoplasma gondii parasitophorous vacuole membrane: A high affinity interaction. J Cell Sci 1997; ll0(pt17):2117-2128.
64. Stang E, Kartenbeck J, Parton RG. Major histocompatibility complex class I molecules mediate association of SV40 with caveolae. Mol Biol Cell 1997; 8(1):47-57.
65. Sinai AP, Joiner KA, The cell biology of non-fusogenic pathogen vacuoles. Annu Rev Microbiol 1997; 51 :415-462.
66. Enright FM, Samartino L. Mechanisms of abortion in Brucella abortus infected cattle. Proceedings of the 98th annual meeting of the United States Animal Health Association, Richmond, Virginia. 1994:88-95.

67. Martinez-de-Tejada G, Moriyon I. The outer membranes of Brucella spp. are not barriers to hydrophobic permeants. J Bacteriol 1993; 175(16):5273-5275.
68. Pizarro-Cerdá J, Desjardins M, Moreno E et al. Modulation of endocytosis in nuclear factor IL-6(-/-) macrophages is responsible for a high susceptibility to intracellular bacterial infection. J Immunol 1999; 162(6):3519-3526.
69. Jiang X, Baldwin CL. Effects of cytokines on the intracellular growth of Brucella abortus. Infect Immun 1993; 61(1):124-134.
70. Jiang X, Leonard B, Benson R et al. Macrophage control of Brucella abortus by oxygen intermediates and nitric oxide. Cell Immunol 1993; 151(2):309-319.
71. Porte F, Liautard JP, Kohler S. Early acidification of phagosomes containing Brucella suis is essential for intracellular survival in murine macrophages. Infect Immun 1999; 67(8):4041-4047.
72. Gay B, Mauss H, Sanchez-Teff S. Identification of fibronectins in peritoneal macrophages during the phagocytosis of Brucella. An immunocytochemical study by electron microscopy. Virchows Arch B-Cell Pathol 1986; 52(2):169-176.
73. Pomales-Lebron A, Stinebring WR. Intracellular multiplication of Brucella abortus in normal and immune mononuclear phagocytes. Proc Soc Exp Biol Med 1957; 94(2):78-83.
74. Alpuche-Aranda CM, Swanson JA, Loomis WP et al. Salmonella typhimurium activates virulence gene transcription within acidified macrophage phagosomes. Proc Natl Acad Sci USA 1992; 89(21):10079-10083.
75. Desjardins M, Huber LA, Parton RG et al. Biogenesis of phagolysosomes proceeds though a sequential series of interactions with the endocytic apparatus. J Cell Biol 1994; 124(5):677-688.
76. Méresse S, Steele-Mortimer O, Moreno E et al. Controlling the maturation of pathogen-containing vacuoles:matter of life and death. Nature Cell Biol 1999; 1(7):E183-E188.
77. Orci L, Glick BS, Rothman JE. A new type of coated vesicular carrier that appears not to contain clathrin:its possible role in protein transport within the Golgi stack. Cell 1986; 46(2):171-184.
78. Orci L, Stamnes M, Ravazzola M et al. Bidirectional transport by distinct populations of COPI-coated vesicles. Cell 1997; 90(2):335-349.
79. Whitney JA, Gomez M, Sheff D et al. Cytoplasmic coat proteins involved in endosome function. Cell 1995; 83(5):703-713.
80. Kadowaki M, Venerando R, Miotto G et al. De novo autophagic vacuole formation in hematocytes permeabilized by Staphylococcus aureus alfa-toxin. J Biol Chem 1994; 269(5):3703-3710.
81. D'Enfert C, Wuestehube LJ, Lila T et al. Sec12b-dependent membrane binding of the small GTP-binding protein Sarpl promotes formation of transport vesicles from the endoplasmic reticulum. J Cell Biol 1991; 114(4):663-670.
82. Kuehn MJ, Schekman R. COPII and secretory cargo capture into transport vesicles. Curr Opin Cell Biol 1997; 9(4):477-483.
83. Elzer P, Edmonds M, Cloeckaert A. Brucella mutants lacking the outer membrane protein Omp25. Abstract 77. Brucellosis 2000. Nimes, France. 2000:79-80.
84. Gunn JS, Miller SI. PhoP-PhoQ activates transcription of pmrAB, encoding a two-component regulatory system involved in Salmonella typhimurium antimicrobial peptide resistance. J Bacteriol 1996; 178(23):6857-6864.
85. Freer E, Moreno E, Moriyón I et al. Brucella-Salmonella lipopolysaccharide chimeras are less permeable to hydrophobic probes and more sensitive to cationic peptides and EDTA than are their native Brucella spp. counterparts. J Bacteriol 1996; 178(20):5867-5876.
86. Martínez-de-Tejada, GM, Pizarro J, Moreno E et al. The outer membranes of Brucella spp. are resistant to bactericidal cationic peptides. Infect Immun.1995; 63(8):3054-3061.
87. Stevens MG, Olsen SC, Pugh GW Jr et al. Immune and pathologic responses in mice infected with Brucella abortus 19, RB51, or 2308. Infect Immun 1994; 62(8):3206-3212.
88. Páramo L, Lomonte B, Pizarro-Cerdá J et al. Bactericidal activity of Lys49 and Asp49 myotoxic phospholipases A2 from Bothrops asper snake venom. Synthetic Lys49 myotoxin II-(115-129)-peptide identifies its bactericidal region. Eur J Biochem 1998; 253(2):452-461.
89. Rasool O, Freer E, Moreno E et al. Effect of Brucella abortus lipopolysaccharides on the oxidative metabolism and enzyme release of neutrophils. Infect Immun 1992; 60(4):4-7.
90. Vaara M. Agents that increase the permeability of the outer membrane. Microbiol Rev 1992; 56(3):395-411.
91. Kim JA, Sha Z, Mayfield JE. Regulation of Brucella abortus catalase. Infect Immun 2000; 68(7):3681-3866.
92. Saha AK, Mukhopadhyay NK, Dowling JN et al. Characterization of a phosphomonoesterase from Brucella abortus. Infect Immun 1990; 58(5):1153-1158.

93. Tatum FM, Detilleux PG, Sacks JM et al. Construction of Cu-Zn superoxide dismutase deletion mutants of Brucella abortus: Analysis of survival in vitro in epithelial and phagocytic cells and in vivo in mice. Infect Immun 1992; 60(7):2863-2869.
94. Grilló MJ. Brucelosis experimental en ratones:control de calidad de vacunas y estudio de factores de virulencia en Brucella. PhD. Thesis, University of Zaragoza, Spain. 1997
95. Latimer E, Simmers J, Sriranganathan N et al. Brucella abortus deficient in copper/zinc superoxide dismutase is virulent in BALB/c mice. Microbiol Pathogen 1992; 12(2):105-113.
96. Jubier-Maurin V, Rodrigue A, Ouahrani-Bettache S et al. Identification of the nik gene cluster of Brucella suis: Regulation and contribution to urease activity. J Bacteriol 2001; 183(2):426-434.
97. Sangari FJ, Díaz J, Seoane A et al. The role of urease in Brucella infection. Abstract 97. Brucellosis 2000. Nimes, France. 2000:79-80.
98. Rafie-Kolpin M, Essenberg RC, Wyckoff JH. Identification and comparison of macrophage-induced proteins and proteins induced under various stress conditions in Brucella abortus. Infect Immun 1996; 64(12):5274-5283.
99. Lin J, Fitch TA. Protein synthesis in Brucella abortus induced during macrophage infection. Infect Immun 1995; 63(4):1409-1414.
100. Teixeira-Gomes AP, Cloeckaert A, Zygnunt MS. Characterization of heat, oxidative and acid stress responses in Brucella melitensis. Infect Immun 2000; 68(5):2954-2961.
101. Elzer PH, Phillips RW, Robertoson GT et al. The HtrA stress response protease contributes to resistance of Brucella abortus to killing by murine phagocytes. Infect Immun 1996; 64(11):4838-4841.
102. Phillips RW, Elzer PH, Robertson GT et al. A Brucella melitensis high-temperature-requirement A(HtrA) deletion mutant is attenuated in goats and protects against abortion. Res Vet 1997; 63(2):165-167.
103. Phillips RW, Elzer PH, Roop-II RM. A Brucella melitensis high temperature requirement A(HtrA) deletion mutant demonstrates a stress response defective phenotype in vitro and transient attenuation in the BALB/c mouse model. Microb Pathog 1995; 19(5):227-234.
104. Pallen, MJ, Wren BW. The HtrA of serine proteases. Mol Microbiol 1997; 26(2):209-221.
105. Tatum FM, Morfitt DC, Halling SM. Construction of a Brucella abortus RecA mutant and its survival in mice. Microb Pathog 1993; 14(3):177-185.
106. Caron E, Cellier M, Liautard JP et al. Complementation of a DnaK-deficient Escherichia coli strain with dnaK/dnaJ operon of Brucella ovis reduces the rate of initial intracellular killing within the monocytic cell line U937. FEMS Microbiol Lett 1994; 120(3):335-340.
107. Kohler S, Teyssier J, Cloeckaert A et al. Participation of the molecular chaperone DnaK in intracellular growth of Brucella suis within U937-derived phagocytes. Mol Microbiol 1996; 20(4):701-712.
108. Ekaza E, Gulloteau L, Teyssier J et al. Functional analysis of the ClpATPase ClpA of Brucella suis, and persistence of knockout mutant in BALB/c mice. Microbiology 2000; 46(Pt7):1605-1616.
109. Lin J, LG, Ficht TA. Immunological response of the Brucella abortus GroEL homologue. Infect Immun 1996; 64(10):4396-4400.
110. Frisk A, Ison CA, Lagergård T. GroEL heat shock protein of Haemophilus ducreyi: Association with cell surface and capacity to bind to eukaryotic cells. Infect Immun 1998; 66(3):1252-1257.
111. Alcantara B, Gee J, Roop-II RM, Characterization of Brucella abortus mutants defective in survival under conditions of starvation and stress. Abstract 100. Brucellosis 2000. Nimes, France. 2000:94.
112. Robertson GT, Roop-II RM. The Brucella abortus host factor I(HF-I) contributes to the stress resistance during stationary phase and is a major determinant of virulence in mice. Molec Microbiol 1999; 34(4):690-700.
113. Robertson GT, Kovach ME, Allen CA et al. The Brucella abortus Lon functions as a generalized stress response protease and is required for wild-type virulence in BALB/c mice. Molec Microbiol 2000; 35(3):577-588.
114. Naroeni A, Jouy N, Ouahrani-Bettache S et al. Brucella suis-impaired specific recognition of phagosome by lysosomes die to phagosomal membrane modifications. Infect Immun 2001; 69(1):486-493.
115. Ferrari G, Langen H, Naito M et al. A coat protein on phagosomes involved in the intracellular survival of Mycobacteria. Cell 1999; 97(4):435-447.
116. Desjardins M, Descoteaux, Inhibition of phagolysosomal biogenesis by Leishmania lipophosphoglycan. J Exp Med 1997; 185(12):2061-2068.
117. Canning PC, Roth JA, Deyoe BL. Release of 5'-guanosine monophosphate and adenine by Brucella abortus and their role in the intracellular survival of the bacteria. J Infect Dis 1986; 154(3):464-470.

118. Bertram TA, Canning PC, Roth JA. Preferential inhibition of primary granule release from bovine neutrophils by a Brucella abortus extract. Infect Immun 1986; 52(1):285-292.
119. Soto L, Rojas X, Alonso O. Estudio in vitro del efecto de fracciones de pared celular de Brucella sobre la actividad de leucocitos polimorfonucleares bovinos. Arch Med Vet 1991; 23(1):27-33.
120. Kreutzer DL, Robertson DC. Surface macromolecules and virulence in intracellular parasitism: Comparison of cell envelope components of smooth and rough strains of Brucella abortus. Infect Immun 1979; 23(3):819-828.
121. Brade H, Brade L, Rietschel ET. Structure-activity relationships of bacterial lipopolysaccharide (endotoxins). Zbl Bakt Hyg 1988; A268(2):151-179.
122. Cline MJ, Melmon KL, Davies WC et al. Mechanism of endotoxin interaction with human leukocytes. Brit J Haematol 1968; 15(6):539-547.
123. Kelly NM, Young L, Cross AS. Differential induction of tumor necrosis factor expressing rough and smooth lipopolysaccharide phenotypes. Infect Immun 1991; 59(12):4491-4496.
124. Silverstein SC, Steinberg TH. Host defense against bacterial and fungal infections. In: Davies DB, Dulbecco R, Eisen HN, Ginsberg HS, eds. Microbiology, 4th ed. Philadelphia: JB Lippincott Co., 1990:485-505.
125. Goldstein J, Hoffman T, Frasch C et al. Lipopolysaccharide(LPS) from Brucella abortus is less toxic than that from Escherichia coli, suggesting the possible use of B. abortus or LPS from B. abortus as carrier in vaccines. Infect Immun 1992; 60(4):1385-1389.
126. Keleti G, Feingold DS, Yongner JS. Interferon induction in mice by lipopolysaccharide from Brucella abortus. Infect Immun 1974; 10(1):182-283.
127. Moreno E, Stackebrandt E, Dorsch M et al. Brucella abortus 16S rRNA and lipid A reveal a phylogenetic relationship with members of the alpha-2 subdivision of the class Proteobacteria. Proc Natl Acad Sci USA 1996; 172(7):3569-3576.
128. Denoel PA, Crawford RM, Zygmunt MS et al. Survival of bacterioferritin deletion mutant of Brucella melitensis 16M in human monocyte-derived macrophages. Infect Immun 1997; 65(10):4337-4340.
129. Leonard BA, López-Goñi I, Baldwin CL. Brucella abortus siderophore 2,3-dihidroxybenzoic acid protects brucellae from killing by macrophages. Vet Res 1997; 28(1):87-92.
130. Lopez-Goñi, Moriyón I, Neilands JB. Identification of 2,3. dihidroxybenzoic acid as Brucella abortus siderophore. Infect Immun 1992; 60(11):4496-4503.
131. Jiang X, Baldwin CL. Iron augments macrophage-mediated killing of Brucella abortus alone and in conjunction with interferon-gamma. Cell Immunol 1993; 148(2):397-407.
132. Foulongne V, Walravens K, Bourg G et al. Aromatic compound-dependent Brucella suis is attenuated in both cultured cells and mouse models. Infect Immun 2001; 69(1):547-550.
133. Hong P, Tsolis R, Ficht TA. Identification if genes for chronic persistence of Brucella abortus in mice. Infect Immun 2000; 68(7):4102-4107.
134. Alford CE, King TE Jr, Campell PA. Role of transferrin, transferrin receptors, and iron in macrophage listericidal activity. J Exp Med 1991; 174(2):459-466.
135. Breedveld MW, Miller KJ. Cyclic β-glucans of members of the family Rhizobiceae. Microbiol Rev 1994; 58(2):145-161.
136. Briones G, Inon de Iannino N, Steinberg M et al. Periplasmic cyclic 1,2-beta-glucan in Brucella spp. is not osmoregulated. Microbiology 1997; 143(Pt4):1115-1124.
137. Iñón-de-Iannino N, Briones G, Tolmasky M et al. Molecular Characterization of cgs, the Brucella abortus cyclic b(1-2) glucan synthetase. Genetic complementation of Rhizobium meliloti ndvB and Agrobacterium tumefaciens chvB mutants. J Bacteriol 1998; 180(17):4392-4400.
138. Sieira R, Comerci DJ, Sanchez DO et al. A homologue of an operon required for DNA transfer in Agrobacterium is required in Brucella abortus for virulence and intracellular multiplication. J Bacteriol 2000; 182(17):4849-4855.
139. Comerci DJ, Martinez-Lorenzo MJ, Sieira R et al. Essential role of the VirB machinery in the maturation of the Brucella abortus-containing vacuole. Cell Microbiol 2001; (in press).
140. Christie PJ, Covacci A. Bacterial type IV secretion systems; DNA conjugation machines adapted to export virulent factors. In Cossart P, Bouquet P, Normark S, Rino Rappuoli(eds). Cellular Microbiology. American Society for Microbiology, Washington D.C. 2000:265-273.
141. Dunn WA. Autophagy and related mechanisms of lysosome-mediated protein degradation. Trends Cell Biol 1994; 4 :139-143.
142. Aplin A, Jasionowski T, Tuttle DL et al. Cytoskeletal elements are required for the formation and maturation of autophagic vacuoles. J Cell Physiol 1992; 152(3):458-466.
143. Cherwonogrodzky JW, Dubray G, Moreno E et al. Antigens of Brucella. In: Nielsen K, Duncan JR, eds. Animal Brucellosis. Boca Ratón: CRC Press Inc., 1990:19-64.

144. Kannenberg EL, Carlson RW. Lipid A and O-chain modifications causes Rhizobium lipopolysaccharides to become hydrophobic during bacteroid development. Mol Microbiol 2001; 39(2):379-391.
145. Dasinger BL, Wilson JB. Glutamate metabolism in Brucella abortus strains of low and high virulence. J Bacteriol 1962; 84(3):911-915.
146. Meyer ME. Evolution and taxonomy in the genus Brucella: Steroid hormone induction of filterable forms with altered characteristics after reversion. Am J Vet Res 1976; 37(2):207-210.
147. Marquis H, Ficht TA. The omp2 gene locus of Brucella abortus encodes two homologous membrane proteins with properties characteristic of porins. Infect Immun 1993; 61(9):3785-3790.
148. Ficht TA, Bearden SW, Sowa BA et al. DNA sequence and expression of the 36-kilodalton outer membrane protein gene of Brucella abortus. Infect Immun 1989; 57(11):3281-3291.
149. Foulongne V, Bourg G, Cazevieille C et al. Identification of Brucella suis genes affecting intracellular survival in an in vitro human macrophage infection model by signature-tagged transposon mutagenesis. Infect Immun 2000; 63(3):1297-1303.
150. Drazek ES, Houng HH, Crawford RM et al. Deletion of purE attenuates Brucella melitensis 16M for growth in human monocyte-derived macrophages. Infect Immun 1995; 63(9):3297-3301.
151. Cheville NF, Olsen SC, Jensen AE et al. Bacterial persistence and immunity of goats vaccinated with purE deletion mutant or the parental 16M strain of Brucella melitensis. Infect Immun 1996; 64(7):241-2439.
152. Dorrell N, Guigue-Talet P, Spencer S et al. Investigation into the role of the response regulator NtrC in the metabolism and virulence of Brucella suis. Microb Pathog 1999; 27(1):1-11.
153. Robertson GT, Reisenauer A, Wright R et al. The Brucella abortus CcrM DNA methyltransferase is essential for viability, and its overexpression attenuates intracellular replication in macrophages. J Bacteriol 2000; 182(12):3482-3489.
154. Wright R, Stephens C, Shapiro L. The CcrM DNA methyltransferase is widespread in the alpha subdivision of Proteobacteria, and its essential functions are conserved in Rhizobium meliloti and Caulobacter crescentus. J Bacteriol 1997; 179(18):5869-5877.
155. LeVier K, Phillips RW, Grippe VK et al. Similar requirements of plant symbiont and mammalian pathogen for prolonged intracellular survival. Science 2000; 287(5462):2492-2493.
156. Zähringer U, Knirel, Lindner B, Helbig JH et al. The lipopolysaccharide of Legionella pneumophila serogroup 1(strain Philadelphia 1): Chemical structure and biological significance. Prog Clin Res 1995; 392:113-139.
157. Cella M, Engering A, Pinet V et al. Inflammatory stimuli induce accumulation of MHC class II complexes on dendritic cells. Nature 1997; 388(6644):782-785.
158. Knolle PA, Germann T, Treichel U et al. Endotoxin down-regulates T cell activation by antigen-presenting liver sinusoidal endothelial cells. J Immunol 1999; 162(3):1401-1407.
159. Krieger JI, Grammer SF, Grey HM et al. Antigen presentation by splenic B cells: Resting B cells are ineffective, whereas activated B cells are effective accessory cells for T cell responses. J Immunol 1985; 135(5):2937-2945.
160. Uchiyama T, Kamagata Y, Yoshioka M. Mechanism of lipopolysaccharide-induced immunosuppression: Immunological activity of B cell subsets responding to T-dependent or T independent antigens in lipopolysaccharide-preinjected mice. Infect Immun 1984; 45(2):367-371.
161. Forestier C, Moreno E, Méresse S et al. Interaction of Brucella abortus lipopolysaccharide with major histocompatibility complex class-II molecules in B lymphocytes. Infect Immun 1999; 67(8):4048-4054.
162. Forestier C, Moreno E, Pizarro-Cerda J et al. Lysosomal accumulation and recycling of LPS to cell surface of murine macrophages, an in vitro and in vivo study. J Immunol 1999; 162(11):6784-6791.
163. Forestier C, Deleuil F, Lapaque N et al. Brucella abortus lipopolysaccharide in murine peritoneal macrophages acts as a down-regulator of T cell activation J Immunol 2000; 165(1):5202-5210.
164. Andersen PS, Lavoie PM, Sekaly RP et al. Role of the T cell receptor alpha chain in stabilizing TCR-superantigen-MHC class II complexes Immunity 1999; 10(4):473-483.
165. Kersh GJ, Kersh EN, Fremont DH et al. High and low-potency ligands with similar affinities for the TCR: The importance of kinetics in TCR signaling. Immunity 1998; 9(6):817-826.
166. Itoh Y, Hemmer B, Martin R et al. Serial TCR engagement and down-modulation by peptide:MHC molecule ligands:relationship to the quality of individual TCR signaling events. J Immunol 1999; 162(4):2073-2080.
167. Valitutti S, Muller S, Cella M et al. Serial triggering of many T-cell receptors by a few peptide-MHC complexes. Nature 1995; 375(6527):148-151.
168. Viola A, Lanzavecchia A. T cell activation determined by T cell receptor number and tunable thresholds Science 1996; 273(5271):104-106.
169. Rabinowitz JD, Beeson C, Lyons DS et al. Kinetic discrimination in T-cell activation. Proc Natl Acad Sci 1996; 93(4):1401-1405.

170. Zhang J. A study on the role of immunosuppression in the pathogenesis of brucellosis. Acta Academiae Medicinae Sinicae 1992; 14(3):168-172.
171. Zhang J, Gao B, Cun C et al. Immunosuppression in murine brucellosis. Chin Med Sci J 1993; 8(3):134-138.
172. Cheers C, Pavlov H, Riglar C et al. Macrophage activation during experimental murine brucellosis. III. Do macrophages exert feedback control during brucellosis? Cell Immunol 1980; 49(2):168-175.
173. Riglar C, Cheers C. Macrophage activation during experimental murine brucellosis. II. Inhibition of in vitro lymphocyte proliferation by Brucella-activated macrophages. Cell Immunol 1980; 49(1):154-167.
174. Renoux M, Renoux G. Brucellosis, immunodepression, and levamisole. Lancet 1977; 1(8007):372.
175. Araya LN. Elzer PH,. Rowe GE, Enright FM et al. Temporal development of protective cell-mediated and humoral immunity in BALB/c mice infected with Brucella abortus. J Immunol 1989; 143(10):3330-3337.
176. Araya LN, Winter AJ. Comparative protection of mice against virulent and attenuated strains of Brucella abortus by passive transfer of immune T cells or serum. Infect Immun 1990; 58(1):254-256.
177. Oliveira SC, Harms JS, Rech EL et al. The role of T cell subsets and cytokines in the regulation of intracellular bacterial infection. Brazil J Med Biol Res 31; 31(1):77-84.
178. López-Urrutia L, Alonso A, Nieto ML et al. Lipopolysaccharide of Brucella abortus and Brucella melitensis induce oxide synthesis in rat peritoneal macrophages. Infect Immun 2000; 68(3):1740-1745.
179. Fernandez-Lago L, Monte M, Chordi A.. Endogenous gamma interferon and interleukin-10 in Brucella abortus 2308 infected mice. FEMS Immunol Med Microbiol 1996; 15(2-3):109-114.
180. Flesch IE, Hess JH, Oswald IP et al. Growth inhibition of Mycobacterium bovis by IFN-alpha stimulated macrophages: Regulation by endogenous tumor necrosis factor-alpha and by IL-10. Internat Immunol 1994; 6(5):693-700.
181. Flynn JL, Goldstein MM, Chan J et al. Tumor necrosis factor-alpha is required in the protective immune response against Mycobacterium tuberculosis in mice. Immunity 1995; 2(6):561-572.
182. Sarmento A, Appelberg R. Involvement of reactive oxygen intermediates in tumor necrosis factor alpha-dependent bacteriostasis of Mycobacterium avium. Infect Immun 1996; 64(8):3224-3230.
183. Tanaka T, Akira S, Yoshida K et al. Targeted disruption of the NF-IL6 gene discloses its essential role in bacteria killing and tumor cytotoxicity by macrophages. Cell 1995; 80(2):353-361.
184. Ottones F, Liautard J, Gross A et al. Activation of human V gamma9delta2 T cells by a Brucella suis non-petide fraction impairs bacterial intracellular multiplication in monocytic infected cells. Immunology 2000; 100(2):252-258.
185. Zhan Y, Cheers C. Differential induction of macrophage-derived cytokines by live and dead intracellular bacteria in vitro. Infect Immun 1995; 63(2):720-723.
186. Zhan Y, Liu Z, Cheers C. Tumor necrosis factor alpha and interleukin-12 contribute to resistance to intracellular bacterium Brucella abortus by different mechanisms. Infect Immun 1996; 64(7):2782-2786.
187. Huang LY, Krieg AM, Eller N et al. Induction and regulation of Th1-inducing cytokines by bacterial DNA, lipopolysaccharide, and heat-inactivated bacteria. Infect Immun 1999; 26(12):6257-6263.
188. Natsuka S, Akira S, Nishio Y et al. Macrophage differentiation specific expression of NF-IL6, a transcription factor for IL-6. Blood 1992; 79(2):460-466.
189. Akira S, Isshiki H, Sugita T et al. A nuclear factor for IL-6 expression(NF-IL6) is a member of a C/EBP family. EMBO J 1990; 9(6):1897-1906.
190. Cao Z, Umek RM, McKnight SL. Regulated expression of three C/EBP isoforms during adipose conversion of 3T3-L1 cells. Genes Dev 1991; 5(9):1538-1552.
191. Chang CJ, Chen TT, Lei HY et al. Molecular cloning of a transcription factor, AGP/EBP, that belongs to members of the C/EBP family. Mol Cell Biol 1990; 10(12):6642-6653.
192. Descombes P, Chojkier M, Lichtsteiner S et al. LAP, a novel member of the C/EBP gene family, encodes a liver-enriched transcriptional activator protein. Gene Dev 1990; 4(9):1541-1551.
193. Katz S, Kowentz-Leutz E, Müller C et al. The NF-M transcription factor is related to C/EBP? and plays a role in signal transduction, differentiation and leukemogenesis of avian myelomonocytic cells. EMBO J 1993; 12(4):1321-1332.
194. Lowenstein CJ, Alley EW, Raval P et al. Macrophage nitric oxide synthase gene:two upstream regions mediate induction by interferon gamma and lipopolysaccharide. Proc Natl Acad Sci USA 1993; 90(20):9730-9734.
195. Scott LM, Civin CI, Rorth P et al. A novel temporal expression pattern of three C/EBP family members in differentiating myelomonocytic cells. Blood 1992; 80(7):1725-1735.
196. Yuan L, Inoue S, Saito Y et al. An evaluation of the effects of cytokines on intracellular oxidative production in normal neutrophils by flow cytometry. Exp Cell Res 1993; 209(2):375-381.

197. Vodovotz Y, Russell D, Xie QW et al. Vesicle membrane association of nitric oxide synthase in primary mouse macrophages. J Immunol 1995; 154(6):2914-2925.
198. Jones SM, Winter AJ. Survival of virulent and attenuated strains of Brucella abortus in normal and gamma interferon-activated murine peritoneal macrophages. Infect Immun 1992; 60(7):3011-3014.
199. Barker J, Lambert PA, Brown MR. Influence of intra-amoebic and other growth conditions on the surface properties of Legionella pneumophila. Infec Immun 1993; 61(8):3503-3510.
200. Verger J, Grimont F, Grimont PAD et al. Brucella, a monospecific genus as shown by deoxyribonucleic acid hybridization. Int J Syst Bacteriol 1985; 35(2):292-295.
201. Flores-Castro R, Baer GM. Brucellosis (Brucella melitensis). Zoonotic implications. In: Steele JH, ed. Handbook Series in Zoonoses. Boca Raton: CRC Press, Inc., 1979:195-211.
202. Young, EJ, Gomez CI, Yawn DH et al. Comparison of Brucella abortus and Brucella melitensis infections in mice and their effect on acquired cellular resistance Infect Immun 1979; 26(2):680-685.
203. De-Ley J, Mannheim W, Segers P et al. Ribosomal ribonucleic acid cistron similarities and taxonomic neighborhood of Brucella and CDC Group Vd. Int J Syst Bacteriol 1987; 37(1):35-42.
204. Moreno E. Brucella evolution. In: Plommet M, ed. Prevention of Brucellosis in Mediterranean Countries. International Centre for Advanced Mediterranean Agronomic Studies. Wageningen: Pudoc Scientific Publishers, 1992:198-218.
205. Moreno E. In search of bacterial species definition. Rev Biol Trop 1997; 45(2):753-771.
206. Moreno E. Genome evolution within the alpha Proteobacteria: Why do some bacteria not possess plasmids and others exhibit more than one different chromosome? FEMS Microbiol Rev 1998; 22(4):255-275.
207. Velasco J, Bengoechea JA. Brandenburg K et al. Brucella and its closest phylogenetic relative Ochrobactrum differ in outer membrane permeability and cationic peptide resistance. Infect Immun 2000; 68(6):3210-3218..
208. Jumas-Bilak E, Milchaux-Charachon S, Bourg GD et al. Unconventional genomic organization in the alpha subgroup of Proteobacteria. J Bacteriol 1998; 180(10):2749-2755.
209. Velasco J, Romero C, López-Goñi I et al. Evaluation of the relatedness of Brucella spp. And Ochrobactrum antrophi and description of Ochrobactrum intermedium spp. Nov., a new species with close relationship to Brucella spp. Inter J Syst Bacteriol 1998; 48(P3):759-768.
210. Godfroid F, Taminiau B, Danese I et al. Identification of the perosamine synthetase gene of Brucella melitensis 16M and involvement of lipopolysaccharide O side chain in Brucella survival in mice and in macrophages. Infect Immun 1998; 66(11):5485-5493.
211. Sánchez DO, Zandomeni RO, Cravero S et al. Gene discovery through sequencing of Brucella abortus. Infect Immun 2001; 69(2):865-868.

CHAPTER 8

Biogenesis of *Salmonella*-Containing Vacuoles in Eukaryotic Cells

Olivia Steele-Mortimer and Stéphane Méresse

Introduction

Salmonella enterica are facultative gram-negative intracellular pathogens. Although over two hundred closely related serovars have been identified, three serovars are most commonly associated with human disease. *S. enterica* serovar Typhi (*S. typhi*) causes systemic infection (typhoid fever) in humans and, according to the Centers for Disease Control and Prevention, is responsible annually for over 600,000 deaths worldwide. Non-typhoidal *Salmonella*, in particular *S. typhimurium* and *S. enteritidis*, are the leading cause of gastrointestinal disease in the developed world. *S. typhimurium* also causes a systemic, typhoid-like, infection in mice and consequently much of our understanding of Salmonellosis has been derived from this system. In particular it is apparent that the ability to invade and survive within both epithelial cells and macrophages are essential features of pathogenesis. In this Chapter we will discuss the biogenesis of the *Salmonella*-containing vacuole (SCV) in both cell types and the roles of bacterial and host cell factors that have been identified so far.

Life in a Vacuole

Like other intracellular bacteria *Salmonella* are internalized into a membrane bound vacuole or phagosome. However, while some bacteria, such as *Listeria*, escape from the vacuole and survive and replicate within the cytosol of the host cell[1] *Salmonella*, and others, including *Brucella*, *Chlamydia* and *Legionella* remain inside the vacuole. As *S. typhimurium* is unable to replicate in macrophage cytosol, the maintenance of a vacuolar membrane is crucial for virulence.[2] Life within a vacuole presents a number of challenges including avoidance of degradation by the host cell and the procurement of nutrients. It is now clear that the success of these intracellular pathogens lies in their unique and varied abilities to control their intravacuolar environment.

Virulence Factors

Most *Salmonella* virulence genes are clustered together in "pathogenicity islands" on the bacterial chromosome. *Salmonella* have five pathogenicity islands (SPI-1, 2, 3, 4, and 5) which have apparently been acquired by horizontal transfer from an unidentified source. Encoded on SPI-1 and SPI-2 are two distinct but structurally similar type III secretion systems (TTSS-1 and TTSS-2 also refereed to as Inv/Spa and Spi/Ssa). These secretion systems are evolutionarily conserved among several gram-negative plant and animal pathogens and are used to translocate effector proteins directly into host cells.[3] TTSS-1 is required for invasion of epithelial cells[4] and has been most extensively studied. It encodes for approximately 30 proteins including the translocated effectors and structural components of the secretion machinery as well as regulatory proteins and chaperones for the effectors. In addition, several effectors translocated by the

Intracellular Pathogens in Membrane Interactions and Vacuole Biogenesis, edited by Jean-Pierre Gorvel.

SPI-1 TTSS are encoded elsewhere on the genome.[5-9] SPI-2 encodes TTSS-2 in addition to a two component regulatory system and proteins of unknown function including putative Spi/Ssa effectors. TTSS-2 is required for growth within macrophages and for systemic infection in the host.[10-12] SPI-2 expression is induced in intracellular bacteria and can be induced in vitro by conditions that mimic the intravacuolar environment.[13,14] Therefore it is likely that TTSS2 effectors play an important role in the control of vacuole maturation.

In addition to SPI-1 and -2 most *Salmonella* serovars harbor virulence plasmids that are important for systemic infection. The size of these plasmids varies even though a highly conserved operon of 5 genes (spvRABCD) has been found in all serotypes. The functions of virulence plasmid encoded proteins remain obscure although they do appear to be involved in survival within macrophages.

The Intracellular Environment

Following oral ingestion *Salmonella* must cross the intestinal epithelium in order to establish infection. The major portal of entry is apparently M cells, although in vitro *Salmonella* have been found to be able to invade all cell types tested.[15,16] However, an alternative transport strategy from the gastrointestinal tract to the bloodstream by CD18+ phagocytes has been described.[17,18] Almost all the information we have on epithelial cell invasion has been obtained using cultured epithelial cells. Invasion is dependent on the induction of actin rearrangement which results in membrane ruffling on the cell surface similar to that induced by growth factors or activated oncogenes.[19,20] This dramatic process is brought about by the co-operative activity of several translocated SPI-1 TTSS effectors. Following bacterial internalization the membrane ruffles disappear and the actin cytoskeleton reverts to its original state.

Once *Salmonella* have crossed the intestinal epithelium M cells become necrotic, begin to die and are ultimately removed from the epithelium. After crossing the intestinal mucosa, *Salmonella* first encounter dendritic cells that are enriched in Peyer's patches and that could play an essential role in triggering an immune response.[21] *Salmonella* colonize Peyer's patches[16,22] and trigger the recruitment of macrophages in response to the release of the proinflammatory chemokines IL-8 by infected enterocytes. Infected macrophages then spread via the blood to the liver and spleen where *Salmonella* are later found intracellularly in large quantities.[23]

Although other cells types, including hepatocytes and neutrophils,[24-26] play significant roles in pathogenesis the biogenesis of the SCV has been best studied in cultured epithelial cells and macrophages. It is significant that intracellular survival, as well as invasion, of these cell types presents very different obstacles for the pathogen to overcome. Macrophages are a particularly hostile environment for intracellular bacteria such as *Salmonella* and the bacteria have consequently developed different mechanisms for survival in each cell type, which is reflected in the biogenesis of the SCV.

Vacuole Biogenesis in HeLa Cells

Following internalization there is a lag phase of 2-4 hours, which precedes the initiation of bacterial replication. During this time the bacteria modulate their environment and this can be followed by analysis of the protein composition of the SCV (Fig. 1). Regulation of vacuole biogenesis is initiated before bacteria are fully internalized. Only a subset of cell-surface proteins, such as the class I MHC molecule, are co-internalized with the bacteria.[27] Following internalization the SCV membrane rapidly acquires proteins which are characteristic of the early endocytic pathway, in particular the early endosome specific proteins rab5, EEA1 and the transferrin receptor.[28,29] These proteins are rapidly lost from the maturing SCV and simultaneously proteins associated with later stages of the endocytic pathway are acquired. Intriguingly this process also demonstrates selectivity. Thus, while lysosomal glycoproteins (Lgp: lamp1, lamp2, cd63) and the vacuolar ATPase (v-ATPase) are acquired within 60min of invasion, other endocytic marker proteins such as the mannose 6-phosphate receptor-a late endosome marker- and soluble lysosomal enzymes are excluded.[28,30] It has been demonstrated that selective

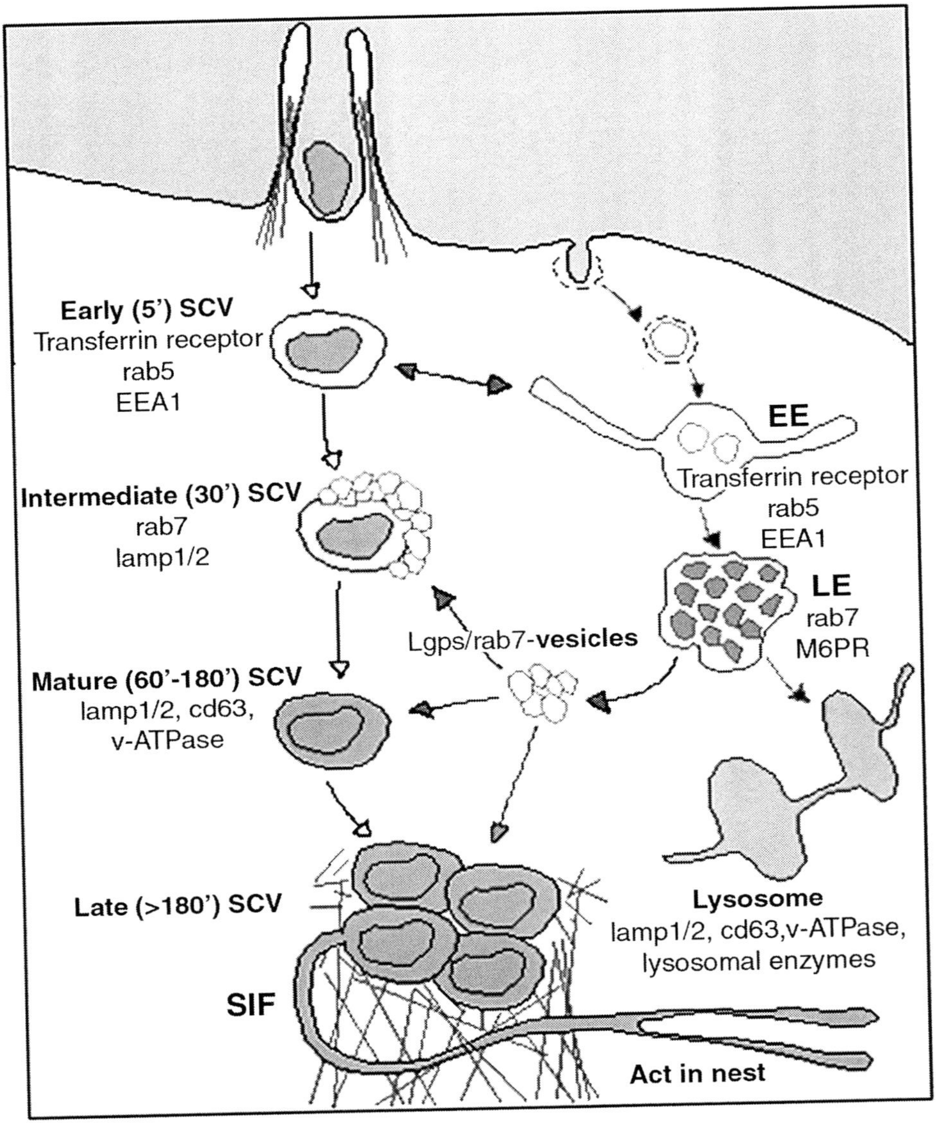

Figure 1. Intracellular pathway of *Salmonella*.
After internalization, the bacterium is found in a nascent SCV interacting with early endosomes (EE), probably by a 'kiss and run' mechanism, acquiring EEA1, rab5 and the transferrin receptor. Then, the SCV matures to an intermediate stage characterized by the accumulation of Lgps/rab7-vesicles likely arising from late endosomes (LE). This leads to the formation of a mature SCV surrounded by an actin meshwork in which the bacterium starts to replicate after a lag phase of 2-3 hours. Replication is concomitant with the formation of *Salmonella*-induced filaments (SIF).

delivery of Lgp is not due to direct fusion with lysosomes but via so far uncharacterized rab7- and Lgp-containing vesicles. These contain very low amounts of the lysosomal enzyme cathepsin D and may be intermediate cargo components of the late endocytic pathway.[31] As reported recently the SCV acquires Lgp from the endocytic recycling pool.[29]

Co-incident with the onset of intracellular replication are two spectacular modifications of the bacterial environment. The first is the formation of an F-actin meshwork attached to the basal side of the infected cell in which SCVs are nested.[32] Inhibition of intracellular replication has been observed in the presence of actin-depolymerization agents. The second is the appearance of membrane tubules in the host epithelial cell which appear to interconnect individual SCVs.[33] The membranes of these tubules, termed *Salmonella*-induced filaments or SIF, contain large amounts of the lysosomal glycoproteins and indeed may acquire almost all detectable cellular Lgp. While the morphology of these structures is reminiscent of the tubular lysosomes which have been described in certain cell types[34] they are not readily accessible to internalized markers and contain very low levels of the lysosomal enzymes which are characteristic of those structures.[33] In contrast to the dynamic rapidly elongating and shrinking tubular lysosomes SIFs appear to be essentially stable structures (Méresse et al, unpublished observations).

Vacuole Biogenesis in Macrophages

The characterization of the SCV in macrophages remains controversial and a consensus image of the maturation process has not yet emerged. Analysis in macrophages is complicated by the heterogeneity of the cell activation stage of macrophagic cell lines (Raw 264.7 and J774.2) and bone marrow- and peritoneal-derived macrophages. Some studies have shown that SCV acidifies and eventually merges with lysosomes[35,36] while other suggest that *Salmonella* inhibits the fusion between SCVs and lysosomes.[37] More recent studies[38] and our results rather show that the maturation of SCV in HeLa cells and in primary or cultivated macrophages are pretty similar. In particular, vacuoles maturing in macrophages exclude the mannose 6-phosphate receptor and the lysosomal enzymes. Although SIFs have only been detected in non-phagocytic cells the bacterial effector protein, SifA, which is required for their appearance is essential for the maintenance of vacuole integrity in macrophages (see below).[2]

The mechanism by which phagocytes kill *Salmonella* implicates the NADPH oxidase.[39] People deficient in this enzymatic activity are susceptible to salmonellosis.[40] *Salmonella* apparently decrease the effectiveness of the NADPH oxidase by preventing its delivery to the SCV by a SPI-2 dependent mechanism.[41,42]

SCV Biogenesis in Other Cell Types

Interactions between pathogens and skin cells have not been extensively studied. Yet, this organ has constant interactions with the external medium and must possess very efficient defense mechanisms. The intracellular fate of *Salmonella* has been recently analyzed in various skin-derived cell lines and in primary culture of human melanocytes.[43] In melanoma cells, the two Lgps cd63 and lamp1 are acquired with very different kinetics suggesting that SCV maturation involves both lysosomal and melanosomal compartments. A very striking event in skin-derived cells is the arrest of bacterial septation giving rise to elongated non invasive bacteria, which are eventually released from the host cell. This original feature may provide the basis for a new and efficient host defense mechanism against *Salmonella* infection.

Dendritic cells play a major role in the capture and presentation of antigens in peripheral sites of pathogen entry, including the skin and the intestine. Recent studies have tackled the relationship between *Salmonella* and dendritic cells. In a ligated intestinal loop model, *Salmonella typhimurium* is found in the dendritic cells of the Peyer's patches few hours after infection indicating their role in infectious processes.[44] These cells have been show to take up bacteria directly in the gut lumen through enterocytes while preserving the integrity of the intestinal barrier.[18] TTSS-1 effectors, which mediate the triggered invasion of non phagocytic cells, are not required for the entry of *Salmonella typhimurium* in dendritic cells. Intracellular bacteria do not replicate but survive in a uncharacterized membrane compartment essentially devoid of late endosomal/lysosomal marker and not related to MHC class II compartment.[44,45]

Acidification of the SCV

In macrophages phagosome acidification is considered a major mechanism for killing intracellular organisms due to downstream effects such as increased phagosome-lysosome fusion,[46] free radical formation[47] and activation of lysosomal acid hydrolases.[48] Some intracellular pathogens such as mycobacterium successfully block phagosome acidification[49], but this does not appear to be a mechanism utilized by Salmonella. On the contrary, SCV acidification has been shown to be essential for *Salmonella* survival and can occur within 20-30 min following internalization into cultured murine macrophages.[50] Acidification is dependent on the host cell v-ATPase, which is responsible for acidification of endocytic organelles and is presumably necessary for the induction of gene expression necessary for intracellular survival. More recently in a study in which cultured epithelial cells and macrophages were compared it was found that v-ATPase activity is not required for intracellular survival in epithelial cells or in some cultured macrophages.[51] This dichotomy is probably a reflection of the complexity of the system and certainly does not preclude a role for low pH in induction of genes required for intracellular survival. Indeed Salmonella are able to regulate gene expression in response to many environmental signals that are encountered at all stages of infection including the intracellular environment of different cell types.[11,52-61] Furthermore, *Salmonella* have at least two acid tolerance responses that are induced by exposure to low pH and which involve differential regulation of a number of genes including essential virulence genes.[57,59,62]

Bacterial Effectors Involved in SCV Biogenesis

SCV maturation clearly diverges from the endocytic/phagocytic pathways. This specific process, which is necessary for the establishment of an intracellular niche fully competent for bacterial replication is under the control of virulent *Salmonella*. Factors encoded by a virulence plasmid, within pathogenicity islands and spread through the bacterial chromosome are required for virulence in infected animals (Table 1).

SPI-1 TTSS-1 Effectors

SPI-1 effectors have been conclusively shown to be essential for bacterial invasion of epithelial cells.[63] At this time only one SPI-1 effector has been implicated in vacuole biogenesis. SopE is encoded within a cluster of genes from a cryptic P2-like prophage and is translocated by TTSS-1.[6] SopE is a guanine-nucleotide exchange factor for the mammalian Rho GTPase Cdc42 and rac and as such is considered one of the key players involved in membrane ruffling and bacterial internalization in non-phagocytic cells.[64] Although SopE is absent from most *S. typhimurium* strains a highly homologous protein, SopE2, (69% sequence identity) is present in all strains. Evidence suggesting that SopE may have a role in vacuole biogenesis came in a recent study in which the characteristics of fusion between endosomes and SCVs isolated from macrophages was investigated. Cell free fusion assays indicated that N-ethylmaleimide-sensitive fusion protein (NSF) and the small GTP-binding protein rab5 are recruited to the SCV membrane and that this recruitment is dependent on the presence of live *Salmonella* in the vacuole. Analysis of bacterial proteins able to bind to GTP-Rab5 in vitro identified SopE while excluding several other effector proteins. Based on this finding it has been proposed that, by selectively recruiting GTP-bound rab5 to the SCV membrane, *S. typhimurium* can induce the translocation of other components of the fusion machinery, which are essential for vacuole biogenesis (Fig. 2).[65,66] However it is well admitted that SPI-1 does not play a key role in SCV maturation in macrophages. Indeed, SCV diverges from the phagocytic pathway regardless of the mechanism of entry (SPI-1-mediated trigger mechanism versus phagocytosis).[38] A possible explanation is that SopE would rather play a role during the onset of epithelial cell invasion.

The recruitment of rab7- and Lpg-enriched vesicles that mediate maturation of SCVs in epithelial cells is unique to invasive *Salmonella*[31] and such vesicles have not been observed accompanying maturation events of inert particles- or SPI-1-deficient *Salmonella*-containing compartments. This suggests the involvement of TTSS-1 bacterial effectors that remain to be identified (Fig. 2).

Table 1. Virulence factors involved in SCV biogenesis

Gene	Encodes	Function	Location	Refs.
envZ	regulator	•EnvZ-OmpR is a two component regulatory system, which via ssrA/ ssrB can transcriptionally regulate SPI-2.	in OmpB operon	79-82
ompR	regulator	•see *envZ*		79-82
sifA	TTSS-2 effector	•Vacuole membrane integrity •SIF formation	in potABCD operon at 27 cs	2, 68, 72
sopE	TTSS-1 effector	•endosome fusion	cryptic P2-like prophage at 61 cs	6, 65
spiC	TTSS-2 effector	•inhibition of eukaryotic membrane trafficking •SIF formation	SPI-2	67, 72
unidentified	TTSS-2 effector	•prevent the trafficking of NADPH complex to the SCV		41
spvR	regulator	•regulation of virulence plasmid gene expression	virulence plasmid	72
sseF	TTSS-2 effector	•SIF formation	SPI-2	72
sseG	TTSS-2 effector	•SIF formation	SPI-2	72
unidentified	TTSS-2 effector	•Formation of an actin meshwork around SCVs		32

NB : This list does not include SPI-2 structural proteins

SPI-2 TTSS Effectors

The SPI-2 secretion system is required for bacterial proliferation in macrophages[10-12] and its effectors likely control key processes of the SCV maturation. Considerable effort is currently being directed towards addressing the role of individual TTSS-2 effectors in vacuole biogenesis.

SpiC was the first SPI-2 secreted effector protein to be identified.[67] A SpiC mutant is unable to survive within macrophages and is highly attenuated in mice although invasion of epithelial cells in unaffected. When expressed in uninfected mammalian cells SpiC alters the trafficking of transferrin receptor and, in a cell-free system, purified SpiC inhibits endosome fusion.[67] Based on these findings it has been proposed that SpiC may inactivate a cytosolic factor normally required for endocytic membrane trafficking in host cells (Fig. 2).

A competitive screen for virulence gene interactions during systemic infections of mice identified SifA as essential for maintenance of SCV membrane integrity in macrophages.[2] SifA, so called because it is essential for the formation of SIFs in epithelial cells, is not encoded on SPI-2 but has limited homology with other secreted effectors. SifA is not the only effector

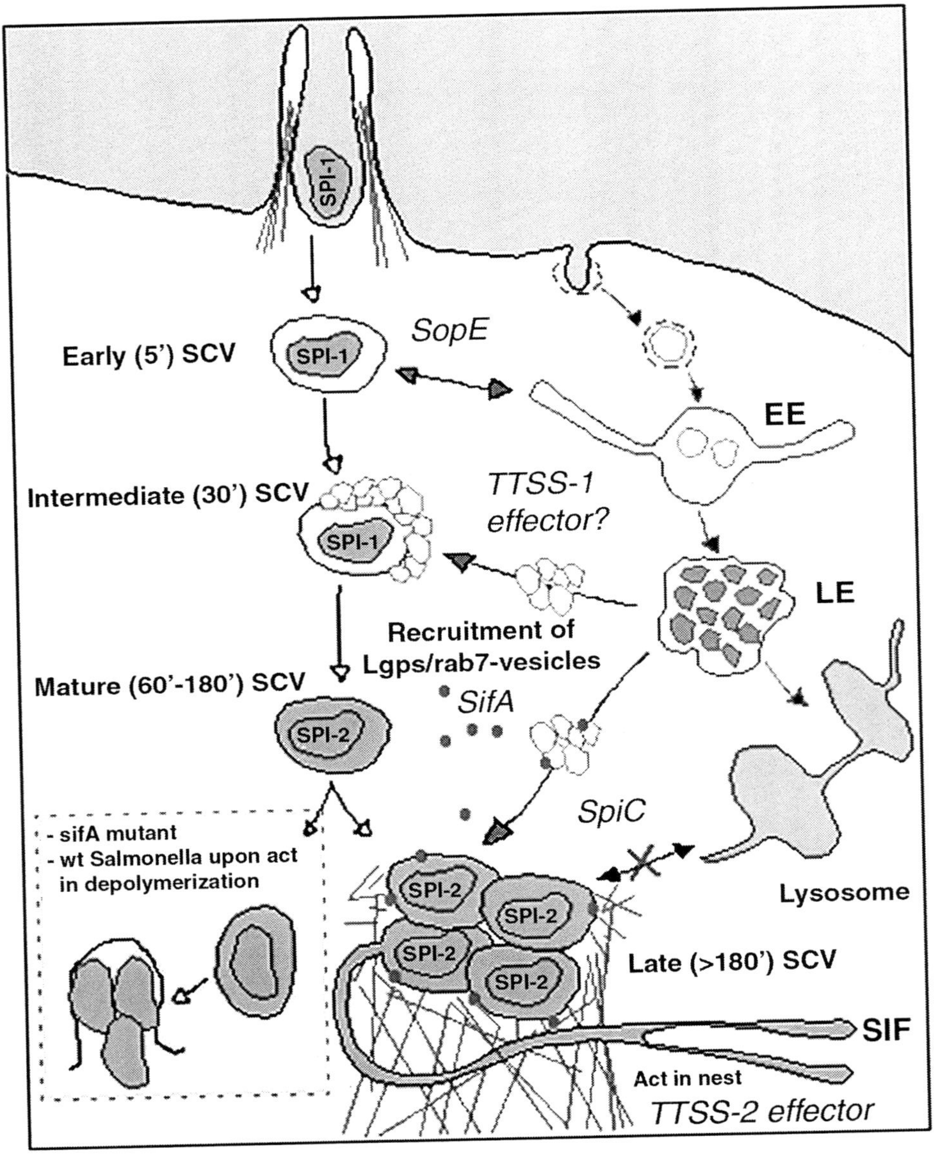

Figure 2. *Salmonella* effectors involved in SCV maturation.
TTSS-1 is responsible for the translocation of SopE, which in addition to its crucial role in the modification of the actin cytoskeleton regulates early endosome-nascent SCV interactions. It is likely that the recruitment of Lgps/rab7-vesicles is also under the control of a so far unidentified TTSS-1 effector. SpiC and SifA are two TTSS-2 effectors. SpiC inhibits the fusion between the *Salmonella*-replication niche and lysosomes. SifA controls the maintenance of the SCV and is necessary for the formation of SIFs. The formation of an actin meshwork in which SCVs are nested is under the control of an unidentified TTSS-2 effector. In the absence of SifA insufficient membrane is recruited to enclose the replicating bacteria, which become exposed to the cytosol. The same loss of vacuole is observed in the presence of actin depolymerisation agents.

required for SIF formation, thus it appears that SCV biogenesis is brought about by the co-operative effects of TTSS-2 effectors in a sophisticated process involving multiple interactions between pathogen and host cell.

SifA, although originally identified as a virulence factor essential for the formation of SIFs in epithelial cells,[68] is also necessary for replication in macrophages.[2, 69] A competitive screen for virulence gene showed that the *ompR,* SPI-2 and *sifA* loci interact during systemic infections of mice. Although encoded outside SPI-2, SifA is induced intracellularly and its expression regulated by the SPI-2 regulator *ssrA*. SifA in cultured epithelial cell restores the formation of SIFs upon infection with a SifA⁻ mutant.[2] These data indicate that SifA is a TTSS-2 effector. The intracellular fate of the SifA⁻ mutant is characterized by a breakdown of the vacuolar membrane resulting in release of mutant bacteria in the cytosol (Fig. 2).[2] Therefore, the integrity of a vacuolar membrane around *Salmonella* requires a functional *sifA* gene. For intracellular pathogens that replicate within a membrane-bound vacuole, there must be a progressive net increase in the surface area of the vacuolar membrane. SifA is thus a likely candidate to control directed membrane fusion events (Fig. 2). According to this model in the absence of SifA insufficient membrane would be recruited to enclose the replicating bacteria, which would become exposed to the cytosol. This conclusion is supported by the vacuolization of the Lgp-positive compartment in SifA-transfected HeLa cells.[69] The host target of SifA remains unidentified.

The formation of a meshwork of actin surrounding intracellular Salmonella requires a functional TTSS-2. Actin depolymerization agents cause the release of the bacterium into the cytosol of macrophages. This effect is specific as the membrane enclosing a SPI-2 mutant is insensitive to actin depolymerisation. Although reminiscent to the SifA mutant phenotype, this TTSS-2 effector is not involved in actin assembly around SCV.[32]

Both the maintenance of a vacuolar membrane, which is crucial for virulence,[2] and the formation of SIFs are likely to require complex spatial and temporal expression of TTSS-2 effectors, most of which remain to be identified or functionally characterized. For example, several TTSS-2 effectors have been identified on the basis of N-terminal similarities.[70,71] Two putative TTSS-2 effectors, SseF and SseG, have been identified as essential for SIF formation in a high-throughput screening of *S. typhimurium* mutants in epithelial cells.[72] Also, SsrA/B, a two component regulator encoded within SPI-2 controls the expression of at least 10 genes outside SPI-2 among which TTSS-2 are likely to be found.[73] Finally, the exclusion of NADPH oxidase from the SCV that prevents oxidative killing in infected phagocytes is dependent on a functional SPI-2.[41]

Other Virulence Factors

A role for virulence plasmid-encoded protein on the biogenesis of SCV has not been demonstrated. However recent data support such a possible role. Using a screen for mu*dj* transposon mutant negative for the formation of SIF, the virulence plasmid encoded SpvR has been identified.[72] SpvR regulates the *spvABCD* operon. Formation of SIFs was restored in a plasmid-cured strain, suggesting that SpvR may negatively regulate a virulence plasmid-encoded factor that inhibits the formation of membrane tubules. Among spvR regulated genes is *spvB,* the expression of which is induced intracellularly in macrophages and epithelial cells.[74] Interestingly the *spvB* gene encodes a mono-ADP-ribosyltransferase and actin was identified as its main target.[75-77] SpvB-expression results in a disorganization of the host cell cytoskeleton and cytopathic effects develop over 24 hours following infection in cells in which bacteria highly replicated.[78] Still it remains to understand how this toxin reaches the host cytosol.

Conclusions

Life within a vacuole presents a number of challenges including avoidance of degradation by the host cell and the procurement of nutrients. It is now clear that the success of virulent *Salmonella* lies in its unique and varied abilities to control its vacuolar environment. The strategy of these bacteria resides in the injection of effector proteins from the vacuole into the cytosol.

In addition to the discovery of new effectors, a challenge for the coming years will be the identification of interacting host partners.

Acknowledgments

We thank Emmanuel Boucrot, Jörg Deiwick, Thomas Henry and Nicolas Lapaque for reading the manuscript.

References

1. de Chastellier C, Berche P. Fate of Listeria monocytogenes in murine macrophages: Evidence for simultaneous killing and survival of intracellular bacteria. Infect Immun 1994; 62(2):543-553.
2. Beuzon CR, Meresse S, Unsworth KE et al. Salmonella maintains the integrity of its intracellular vacuole through the action of SifA. EMBO J 2000; 19(13):3235-3249.
3. Hueck CJ. Type III protein secretion systems in bacterial pathogens of animals and plants. Microbiol Mol Biol Rev 1998; 62(2):379-433.
4. Galan JE, Curtiss R III. Cloning and molecular characterization of genes whose products allow Salmonella typhimurium to penetrate tissue culture cells. Proc Natl Acad Sci USA 1989; 86(16):6383-6387.
5. Hardt WD, Urlaub H, Galan JE. A substrate of the centisome 63 type III protein secretion system of Salmonella typhimurium is encoded by a cryptic bacteriophage. Proc Natl Acad Sci USA 1998; 95(5):2574-2579.
6. Mirold S, Rabsch W, Rohde M et al. Isolation of a temperate bacteriophage encoding the type III effector protein SopE from an epidemic Salmonella typhimurium strain. Proc Natl Acad Sci USA 1999; 96(17):9845-9850.
7. Stender S, Friebel A, Linder S et al. Identification of SopE2 from Salmonella typhimurium, a conserved guanine nucleotide exchange factor for Cdc42 of the host cell [In Process Citation]. Mol Microbiol 2000; 36(6):1206-1221.
8. Galyov EE, Wood MW, Rosqvist R et al. A secreted effector protein of Salmonella dublin is translocated into eukaryotic cells and mediates inflammation and fluid secretion in infected ileal mucosa. Mol Microbiol 1997; 25(5):903-912.
9. Pfeifer CG, Marcus SL, Steele-Mortimer O et al. Salmonella typhimurium virulence genes are induced upon bacterial invasion into phagocytic and nonphagocytic cells. Infect Immun 1999; 67(11):5690-5698.
10. Hensel M, Shea JE, Waterman SR et al. Genes encoding putative effector proteins of the type III secretion system of Salmonella pathogenicity island 2 are required for bacterial virulence and proliferation in macrophages. Mol Microbiol 1998; 30(1):163-174.
11. Cirillo DM, Valdivia RH, Monack DM et al. Macrophage-dependent induction of the Salmonella pathogenicity island 2 type III secretion system and its role in intracellular survival. Mol Microbiol 1998; 30(1):175-188.
12. Ochman H, Soncini FC, Solomon F et al. Identification of a pathogenicity island required for Salmonella survival in host cells. Proc Natl Acad Sci USA 1996; 93(15):7800-7804.
13. Deiwick J, Hensel M. Regulation of virulence genes by environmental signals in Salmonella typhimurium. Electrophoresis 1999; 20(4-5):813-817.
14. Beuzon CR, Banks G, Deiwick J et al. pH-dependent secretion of SseB, a product of the SPI-2 type III secretion system of Salmonella typhimurium. Mol Microbiol 1999; 33(4):806-816.
15. Clark MA, Jepson MA, Simmons NL et al. Preferential interaction of Salmonella typhimurium with mouse Peyer's patch M cells. Res Microbiol 1994; 145(7):543-552.
16. Jones BD, Ghori N, Falkow S. Salmonella typhimurium initiates murine infection by penetrating and destroying the specialized epithelial M cells of the Peyer's patches. J Exp Med 1994; 180(1):15-23.
17. Vazquez-Torres A, Jones-Carson J, Baumler AJ et al. Extraintestinal dissemination of Salmonella by CD18-expressing phagocytes. Nature 1999; 401(6755):804-808.
18. Rescigno M, Urbano M, Valzasina B et al. Dendritic cells express tight junction proteins and penetrate gut epithelial monolayers to sample bacteria. Nat Immunol 2001; 2(4):361-367.
19. Galan JE, Pace J, Hayman MJ. Involvement of the epidermal growth factor receptor in the invasion of cultured mammalian cells by Salmonella typhimurium. Nature 1992; 357(6379):588-589.
20. Francis CL, Ryan TA, Jones BD et al. Ruffles induced by Salmonella and other stimuli direct macropinocytosis of bacteria. Nature 1993; 364(6438):639-642.
21. Hopkins SA, Niedergang F, Corthesy-Theulaz IE et al. A recombinant Salmonella typhimurium vaccine strain is taken up and survives within murine Peyer's patch dendritic cells. Cell Micro 2000; 2(1):59-68.

22. Jensen VB, Harty JT, Jones BD. Interactions of the invasive pathogens Salmonella typhimurium, Listeria monocytogenes, and Shigella flexneri with M cells and murine Peyer's patches. Infect Immun 1998; 66(8):3758-3766.
23. Richter-Dahlfors A, Buchan AMJ, Finlay BB. Murine salmonellosis studied by confocal microscopy: Salmonella typhimurium resides intracellularly inside macrophages and exerts a cytotoxic effect on phagocytes in vivo. J Exp Med 1997; 186(4):569-580.
24. Conlan JW, North RJ. Early pathogenesis of infection in the liver with the facultative intracellular bacteria Listeria monocytogenes, Francisella tularensis, and Salmonella typhimurium involves lysis of infected hepatocytes by leukocytes. Infect Immun 1992; 60(12):5164-5171.
25. Dunlap NE, Benjamin WH Jr, Briles DE. The intracellular nature of Salmonella infection during the early stages of mouse typhoid. Immunol Ser 1994; 60:303-312.
26. Lin FR, Wang XM, Hsu HS et al. Electron microscopic studies on the location of bacterial proliferation in the liver in murine salmonellosis. Br J Exp Pathol 1987; 68(4):539-550.
27. Garcia-del Portillo F, Pucciarelli MG, Jefferies WA et al. Salmonella typhimurium induces selective aggregation and internalization of host cell surface proteins during invasion of epithelial cells. J Cell Sci 1994; 107(Pt 7):2005-2020.
28. Steele-Mortimer O, Meresse S, Gorvel JP et al. Biogenesis of Salmonella typhimurium-containing vacuoles in epithelial cells involves interactions with the early endocytic pathway. Cell Microbiol 1999; 1(1):33-49.
29. Baldeon ME, Ceresa BP, Casanova JE. Expression of constitutively active Rab5 uncouples maturation of the Salmonella-containing vacuole from intracellular replication. Cell Microbiol 2001; 3(7):473-486.
30. Garcia-del Portillo F, Finlay BB. Targeting of Salmonella typhimurium to vesicles containing lysosomal membrane glycoproteins bypasses compartments with mannose 6-phosphate receptors. J Cell Biol 1995; 129(1):81-97.
31. Meresse S, Steele-Mortimer O, Finlay BB et al. The rab7 GTPase controls the maturation of Salmonella typhimurium-containing vacuoles in HeLa cells. Embo J 1999; 18(16):4394-4403.
32. Meresse S, Unsworth KE, Habermann A et al. Remodelling of the actin cytoskeleton is essential for replication of intravacuolar Salmonella. Cell Microbiol 2001; 3(8):567-577.
33. Garcia-del Portillo F, Zwick MB, Leung KY et al. Intracellular replication of Salmonella within epithelial cells is associated with filamentous structures containing lysosomal membrane glycoproteins. Infect Agents Dis 1993; 2(4):227-231.
34. Swanson J, Bushnell A, Silverstein SC. Tubular lysosome morphology and distribution within macrophages depend on the integrity of cytoplasmic microtubules. Proc Natl Acad Sci USA 1987; 84(7):1921-1925.
35. Alpuche Aranda CM, Swanson JA, Loomis WP et al. Salmonella typhimurium activates virulence gene transcription within acidified macrophage phagosomes. Proc Natl Acad Sci USA 1992; 89(21):10079-10083.
36. Oh YK, Alpuche-Aranda C, Berthiaume E et al. Rapid and complete fusion of macrophage lysosomes with phagosomes containing Salmonella typhimurium. Infect Immun 1996; 64(9):3877-3883.
37. Buchmeier NA, Heffron F. Inhibition of macrophage phagosome-lysosome fusion by Salmonella typhimurium. Infect Immun 1991; 59(7):2232-2238.
38. Rathman M, Barker LP, Falkow S. The unique trafficking pattern of Salmonella typhimurium-containing phagosomes in murine macrophages is independent of the mechanism of bacterial entry. Infect Immun 1997; 65(4):1475-1485.
39. Vazquez-Torres A, Jones-Carson J, Mastroeni P et al. Antimicrobial actions of the NADPH phagocyte oxidase and inducible nitric oxide synthase in experimental salmonellosis. I. Effects on microbial killing by activated peritoneal macrophages in vitro. J Exp Med 2000; 192(2):227-236.
40. Mouy R, Fischer A, Vilmer E et al. Incidence, severity, and prevention of infections in chronic granulomatous disease. J Pediatr 1989; 114(4 Pt 1):555-560.
41. Vazquez-Torres A, Xu Y, Jones-Carson J et al. Salmonella pathogenicity island 2-dependent evasion of the phagocyte NADPH oxidase. Science 2000; 287(5458):1655-1658.
42. Gallois A, Klein JR, Allen LA et al. Salmonella pathogenicity island 2-encoded type III secretion system mediates exclusion of NADPH oxidase assembly from the phagosomal membrane. J Immunol 2001; 166(9):5741-5748.
43. Martinez-Lorenzo MJ, Meresse S, de Chastellier C et al. Unusual intracellular trafficking of Salmonella typhimurium in human melanoma cells. Cell Microbiol 2001; 3(6):407-416.
44. Niedergang F, Sirard JC, Blanc CT et al. Entry and survival of Salmonella typhimurium in dendritic cells and presentation of recombinant antigens do not require macrophage-specific virulence factors. Proc Natl Acad Sci USA 2000; 97(26):14650-14655.

45. Garcia-Del Portillo F, Jungnitz H, Rohde M et al. Interaction of Salmonella enterica serotype Typhimurium with dendritic cells is defined by targeting to compartments lacking lysosomal membrane glycoproteins. Infect Immun 2000; 68(5):2985-2991.
46. McNeil PL, Tanasugarn L, Meigs JB et al. Acidification of phagosomes is initiated before lysosomal enzyme activity is detected. J Cell Biol 1983; 97(3):692-702.
47. Catterall JR, Sharma SD, Remington JS. Oxygen-independent killing by alveolar macrophages. J Exp Med 1986; 163(5):1113-1131.
48. Coffey JW, De Duve C. Digestive activity of lysosomes. I. The digestion of proteins by extracts of rat liver lysosomes. J Biol Chem 1968; 243(12):3255-3263.
49. Sturgill-Koszycki S, Schlesinger PH, Chakraborty P et al. Lack of acidification in Mycobacterium phagosomes produced by exclusion of the vesicular proton-ATPase [see comments] [published erratum appears in Science 1994 Mar 11;263(5152):1359]. Science 1994; 263(5147):678-681.
50. Rathman M, Sjaastad MD, Falkow S. Acidification of phagosomes containing Salmonella typhimurium in murine macrophages. Infect Immun 1996; 64(7):2765-2773.
51. Steele-Mortimer O, St-Louis M, Olivier M et al. Vacuole acidification is not required for survival of Salmonella enterica serovar typhimurium within cultured macrophages and epithelial cells. Infect Immun 2000; 68(9):5401-5404.
52. Heithoff DM, Conner CP, Hentschel U et al. Coordinate intracellular expression of Salmonella genes induced during infection. J Bacteriol 1999; 181(3):799-807.
53. Chen CY, Eckmann L, Libby SJ et al. Expression of Salmonella typhimurium rpoS and rpoS-dependent genes in the intracellular environment of eukaryotic cells. Infect Immun 1996; 64(11):4739-4743.
54. Valdivia RH, Falkow S. Bacterial genetics by flow cytometry: Rapid isolation of Salmonella typhimurium acid-inducible promoters by differential fluorescence induction. Mol Microbiol 1996; 22(2):367-378.
55. Janssen R, Verjans GM, Kusters JG et al. Induction of the phoE promoter upon invasion of Salmonella typhimurium into eukaryotic cells. Microb Pathog 1995; 19(4):193-201.
56. Deiwick J, Nikolaus T, Erdogan S et al. Environmental regulation of Salmonella pathogenicity island 2 gene expression. Mol Microbiol 1999; 31(6):1759-1773.
57. Bang IS, Kim BH, Foster JW et al. OmpR regulates the stationary-phase acid tolerance response of Salmonella enterica serovar typhimurium. J Bacteriol 2000; 182(8):2245-2252.
58. Walker SL, Sojka M, Dibb-Fuller M et al. Effect of pH, temperature and surface contact on the elaboration of fimbriae and flagella by Salmonella serotype Enteritidis. J Med Microbiol 1999; 48(3):253-261.
59. Foster JW, Moreno M. Inducible acid tolerance mechanisms in enteric bacteria. Novartis Found Symp 1999; 221:55-69.
60. Daefler S. Type III secretion by Salmonella typhimurium does not require contact with a eukaryotic host. Mol Microbiol 1999; 31(1):45-51.
61. Prouty AM, Gunn JS. Salmonella enterica serovar typhimurium invasion is repressed in the presence of bile. Infect Immun 2000; 68(12):6763-6769.
62. Garcia-del Portillo F, Foster JW, Finlay BB. Role of acid tolerance response genes in Salmonella typhimurium virulence. Infect Immun 1993; 61(10):4489-4492.
63. Galan JE. Interaction of Salmonella with host cells through the centisome 63 type III secretion system. Curr Opin Microbiol 1999; 2(1):46-50.
64. Galan JE, Zhou D. Striking a balance: modulation of the actin cytoskeleton by Salmonella. Proc Natl Acad Sci USA 2000; 97(16):8754-8761.
65. Mukherjee K, Siddiqi SA, Hashim S et al. Live Salmonella recruits N-ethylmaleimide-sensitive fusion protein on phagosomal membrane and promotes fusion with early endosome. J Cell Biol 2000; 148(4):741-753.
66. Mukherjee K, Parashuraman S, Raje M et al. SopE acts as a Rab5-specific nucleotide exchange factor and recruits non-prenylated Rab5 on Salmonella-containing phagosomes to promote fusion with early endosomes. J Biol Chem 2001;2 76(26):23607-23615.
67. Uchiya K, Barbieri MA, Funato K et al. A Salmonella virulence protein that inhibits cellular trafficking. EMBO J 1999; 18(14):3924-3933.
68. Stein MA, Leung KY, Zwick M et al. Identification of a Salmonella virulence gene required for formation of filamentous structures containing lysosomal membrane glycoproteins within epithelial cells. Mol Microbiol 1996; 20(1):151-164.
69. Brumell JH, Rosenberger CM, Gotto GT et alB. SifA permits survival and replication of Salmonella typhimurium in murine macrophages. Cellular Microbiology 2001; 2(2):74-84.
70. Brumell JH, Marcus SL, Finlay BB. N-terminal conservation of putative type III secreted effectors of Salmonella typhimurium. Mol Microbiol 2000; 36(3):773-774.

71. Miao EA, Miller SI. A conserved amino acid sequence directing intracellular type III secretion by Salmonella typhimurium. Proc Natl Acad Sci USA 2000; 97(13):7539-7544.
72. Guy RL, Gonias LA, Stein MA. Aggregation of host endosomes by Salmonella requires SPI2 translocation of SseFG and involves SpvR and the fms-aroE intragenic region. Mol Microbiol 2000; 37(6):1417-1435.
73. Worley MJ, Ching KH, Heffron F. Salmonella SsrB activates a global regulon of horizontally acquired genes. Mol Microbiol 2000; 36(3):749-761.
74. Fierer J, Eckmann L, Fang F et al. Expression of the Salmonella virulence plasmid gene spvB in cultured macrophages and nonphagocytic cells. Infect Immun 1993; 61(12):5231-5236.
75. Tezcan-Merdol D, Nyman T, Lindberg U et al. Actin is ADP-ribosylated by the Salmonella enterica virulence-associated protein SpvB. Mol Microbiol 2001; 39(3):606-619.
76. Otto H, Tezcan-Merdol D, Girisch R et al. The spvB gene-product of the Salmonella enterica virulence plasmid is a mono(ADP-ribosyl)transferase. Mol Microbiol 2000; 37(5):1106-1115.
77. Lesnick ML, Reiner NE, Fierer J et al. The Salmonella spvB virulence gene encodes an enzyme that ADP-ribosylates actin and destabilizes the cytoskeleton of eukaryotic cells. Mol Microbiol 2001; 39(6):1464-1470.
78. Libby SJ, Lesnick M, Hasegawa P et al. The Salmonella virulence plasmid spv genes are required for cytopathology in human monocyte-derived macrophages. Cell Microbiol 2000; 2(1):49-58.
79. Liljestrom P, Laamanen I, Palva ET. Structure and expression of the ompB operon, the regulatory locus for the outer membrane porin regulon in Salmonella typhimurium LT-2. J Mol Biol 1988; 201(4):663-673.
80. Mills SD, Ruschkowski SR, Stein MA et al. Trafficking of porin-deficient Salmonella typhimurium mutants inside HeLa cells: ompR and envZ mutants are defective for the formation of Salmonella-induced filaments. Infect Immun 1998; 66(4):1806-1811.
81. Lee AK, Detweiler CS, Falkow S. OmpR regulates the two-component system SsrA-ssrB in Salmonella pathogenicity island 2. J Bacteriol 2000; 182(3):771-781.
82. Worley MJ, Ching KH, Heffron F. Salmonella SsrB activates a global regulon of horizontally acquired genes. Mol Microbiol 2000; 36(3):749-761.

CHAPTER 9

Evasion of Phagosome Lysosome Fusion and Establishment of a Replicative Organelle by the Intracellular Pathogen *Legionella pneumophila*

Craig R. Roy and Jonathan C. Kagan

Abstract

Most human pathogens that survive and proliferate in normally sterile environments have evolved specialized virulence determinants that facilitate evasion of host defense mechanisms. Microbial strategies to evade host immunity are often geared towards professional phagocytes. This is because macrophages are always on the lookout for bacteria that have breached physical barriers to gain access to a privileged site. When foreign invaders are encountered, they are usually engulfed by macrophages and then destroyed when the phagosomes in which they reside are delivered to lysosomes. To avoid destruction, many microbial pathogens modulate vesicle trafficking in eukaryotic host cells in order to prevent phagosome lysosome fusion. Although many microbial pathogens have the ability to alter phagosome trafficking, it is unclear how most of them accomplish this feat. *Legionella pneumophila* are bacteria that have the ability to alter maturation of the endocytic vacuole in which they reside initially, allowing them to establish an organelle within phagocytic host cells that supports replication. Rather than fusing sequentially with early endosomes, late endosomes, then lysosomes; phagosomes containing *L. pneumophila* will associate rapidly with smooth vesicles and are remodeled into unique organelles decorated with ribosomes. Genetic analysis has revealed that a type IV-related secretion system is required by *L. pneumophila* to control phagosome biogenesis. It is believed that *L. pneumophila* use this specialized secretion system during uptake to inject proteins into the host cell. Proteins delivered into the host cell by *L. pneumophila* would then act on host factors that regulate vesicle trafficking. In theory, *L. pneumophila* could prevent phagosome maturation by either inhibiting the function of host factors required for trafficking of endocytic vesicles or they could promote rapid remodeling of the endocytic vacuole through subverting factors used for biogenesis of other cellular organelles. Recent evidence suggests that *L. pneumophila* may employ a combination of both strategies. Although the effector proteins being injected into host cells have not been identified, we propose that *L. pneumophila* use a type-IV related secretion system to deliver one set of effectors that will inhibit endocytic maturation momentarily, and then intracellular bacteria remodel their endocytic vacuole using a second set of effectors that subvert host factors used for biosynthetic transport. This allows *L. pneumophila* to create a specialized organelle that shares many features with the host cell endoplasmic reticulum.

Intracellular Pathogens in Membrane Interactions and Vacuole Biogenesis, edited by Jean-Pierre Gorvel.

The *L. pneumophila dot/icm* Genes Encode a Type IVB Transporter Required for Host Cell Pathogenesis

Legionella pneumophila are gram negative bacteria that can replicate inside of eukaryotic phagocytes.[1] When *L. pneumophila* gain access to the human lung, they can grow inside alveolar macrophages, which can result in a severe pneumonia called Legionnaires disease.[2-4] After uptake by phagocytes, phagosomes containing *L. pneumophila* avoid fusion with lysosomes[5] and are found intimately associated with host endoplasmic reticulum (ER).[6] The ability to replicate inside of eukaryotic host cells is clearly of great importance to *L. pneumophila*. In nature, *L. pneumophila* replicate inside of fresh water protozoan hosts,[7] and mutants of *L. pneumophila* that are unable to replicate intracellularly are avirulent in animal models of disease.[8-10]

Unlike a number of other intracellular pathogens, *L. pneumophila* can be cultured extracellularly on standard laboratory media[11] and is relatively easy to manipulate genetically.[12] These features have made *L. pneumophila* an excellent model organism for dissecting the molecular mechanisms that enable microbial pathogens to survive and proliferate within professional phagocytes. Towards this end several groups have independently conducted genetic screens to identify *L. pneumophila* mutants defective for host cell parasitism.[10,13-17] From these studies, it is clear that two general types of intracellular growth mutants can be isolated. The first category includes mutants that retain the ability to create a specialized phagosome that evades fusion with lysosomes. Thus, the intracellular growth defect exhibited by these mutants likely results from a genetic lesion that affects replication after formation of a protected niche. Mutants that fall into this category include those that are defective for iron utilization,[18] amino acid biosynthesis,[19] nucleotide biosynthesis[19] and chromosome replication.[20]

The key to understanding how *L. pneumophila* create a niche permissive for intracellular replication is held by mutant bacteria that reside in conventional phagosomes. Intracellular growth mutants of *L. pneumophila* that are unable to prevent endocytic maturation of their phagosomes have been identified by several laboratories.[10,13-15,21-24] Genetic analysis of these mutants has led to the identification of 24 different genes (Fig. 1), which are clustered on two unlinked regions of the bacterial chromosome.[25,26] These genes are called *dot*, for defect in organelle trafficking, and *icm*, for intracellular multiplication. *L. pneumophila dot/icm* mutants reside in phagosomes that can not evade fusion with lysosomes, indicating that these genes play an important role in controlling biogenesis of vacuole that supports intracellular replication.[22,27,28]

Significant regions of sequence similarity are found between many of the *dot/icm* encoded products and structural components of bacterial type IV secretion machines.[25,26] These specialized protein secretion systems are found in a variety of gram negative bacteria.[29] Two distinct type IV subgroups have been reported. Classification is based on sequence similarities and organization of the genes encoding the type IV transporter. The VirB proteins encoded on the *Agrobacterium tumefaciens* Ti plasmid are the prototypical subunits for the type IVA transporters. Type IVB transporters are defined by the Tra/Trb proteins encoded on self-transmissible plasmids such as R64 and ColIb-P9.

To demonstrate that the *dot/icm*-encoded apparatus is functionally similar to other type IV secretion systems, it was shown that *L. pneumophila* can mobilize plasmids containing the IncQ origin of transfer into other gram negative bacteria by a process that requires *dot/icm* gene function.[25,26] Although these data show that DNA transfer can be mediated by the *dot/icm* encoded apparatus, it is important to note that transfer requires the covalent attachment of a protein molecule at the 5' terminus of the DNA strand.[30] Most likely, this terminal protein is the primary substrate recognized by the transfer system, suggesting that the *dot/icm*-encoded apparatus can also mediate protein transfer. These data support a model in which the *dot/icm* genes encode a transfer system that can inject molecules from *L. pneumophila* into a variety of recipient cells (Fig. 2).

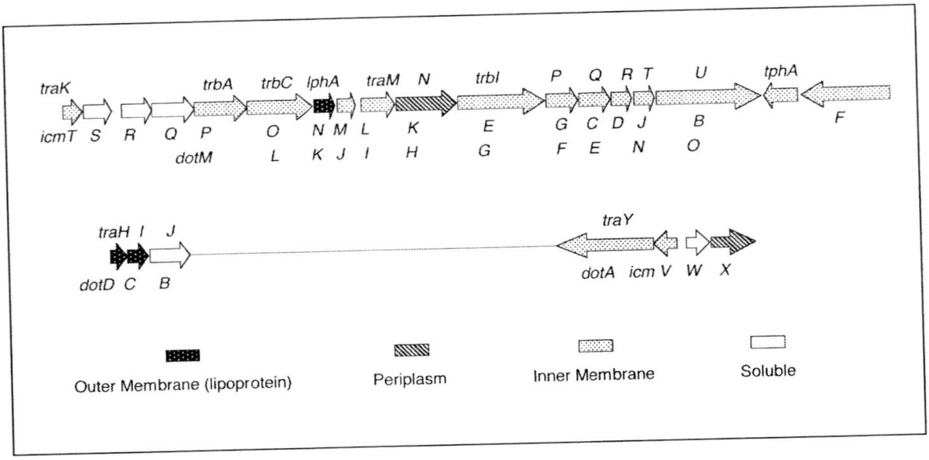

Figure 1. The *L. pneumophila dot/icm* genes encode a type IVB transporter. Genetic organization of the 24 *dot/icm* genes located on two regions of the *L. pneumophila* chromosome is shown schematically. Indicated above the *dot/icm* genes are similar Trb and Tra proteins encoded on the self transmissible IncI plasmids. The shading of each gene indicates the predicted location of the corresponding product. (Adapted from refs. 29, 31, 37)

It was shown recently that at least 15 of the Dot/Icm proteins have an orthologous Tra/Trb protein encoded on IncI1 plasmids (Fig. 1), which places the Dot/Icm apparatus into the type IVB subgroup.[29] It is predicted that orthologous components are playing similar roles in the two secretion systems. It has not been shown that IncI1 encoded products comprise a transporter that can alter endocytic maturation, whereas, the *dot/icm*-encoded apparatus is central to the ability of *L. pneumophila* to regulate phagosome biogenesis. Thus, these orthologous proteins are probably components required for the assembly or function of the core secretion apparatus. It is predicted that when *L. pneumophila* come in contact with eukaryotic host cells, the *dot/icm*-encoded apparatus injects proteins that can affect specific host cellular processes by either mimicking or inhibiting the function of cellular factors required for vesicle trafficking. It is reasonable to assume that the molecules conferring specialized functions to a given secretion apparatus are likely to be unique in each system. For this reason, a number of studies have focused on characterizing Dot/Icm proteins that do not have obvious orthologs in other secretion systems in the hope of learning more about how this secretion apparatus functions, and to identify specific factors injected into host cells by *L. pneumophila* that may directly affect phagosome maturation.

The Dot/Icm Transporter Plays an Essential Role in Biogenesis of a Replicative Organelle

The first *L. pneumophila dot/icm* mutants to be characterized in detail were those defective for a 119 kDa polytopic inner membrane protein encoded by the *dotA* gene. The DotA protein is similar to the ColIb-P9 protein TraY,[31] suggesting that DotA is an essential component of the Dot/Icm transporter. Loss-of-function mutations in one of the core components of the Dot/Icm transporter should eliminate all virulence processes that require this apparatus. Thus, *L. pneumophila dotA* mutants are predicted to be defective in secretion of all putative effector molecules that are injected into host cells by the Dot/Icm transporter.

Phenotypic differences between wild type *L. pneumophila* and isogenic strains containing loss-of-function mutations in the *dotA* gene have been thoroughly analyzed in order to dissect virulence traits that require signaling through the Dot/Icm transporter. From these studies, a

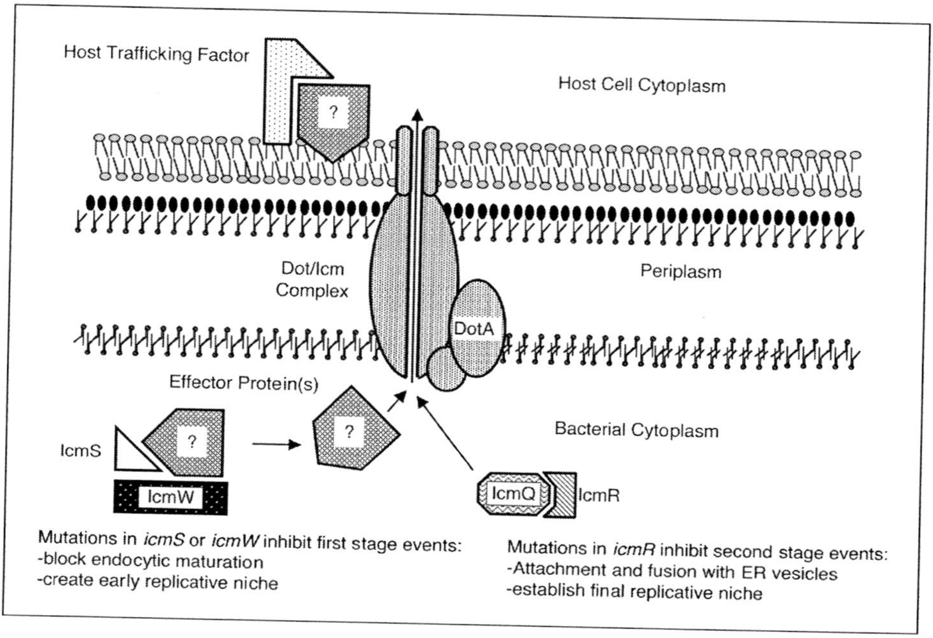

Figure 2. The Dot/Icm transporter controls distinct virulence traits. Most of the *dot/icm* genes are predicted to encode components of the core type IVB translocation apparatus that injects bacterial factors into eukaryotic host cells. Phenotypic analysis of *L. pneumophila* mutants indicates that several *dot/icm* products regulate specific virulence activities. The *icmQ* and *icmR* genes play an important role in the insertion of pores in the host cell membrane upon contact. Genetic evidence indicates that the final stage in biogenesis of a replicative organelle by *L. pneumophila* requires the *icmR* gene. The *icmS* and *icmW* genes are required for initial events that create a replicative niche. Evasion of phagosome lysosome fusion by *L. pneumophila* requires both the *icmW* and *icmS* products. There is evidence to suggest that these genes encode chaperone proteins that pilot secreted proteins to the Dot/Icm apparatus. The secreted proteins have not been identified, however, they are predicted to modulate trafficking of phagosomes containing *L. pneumophila* by affecting the function of eukaryotic host factors. (Adapted from refs. 35, 40)

number of interesting clues regarding the function of the Dot/Icm apparatus have been revealed. First and foremost, it was shown that *dotA* mutants are unable to replicate inside of macrophages.[32] Secondly, electron microscopy revealed that most phagosomes containing *dotA* mutants fuse with lysosomes, unlike compartments containing wild type *L. pneumophila.*[22] These early data indicate that *dotA* function is required for bacterial evasion of lysosome fusion, and that modulation of phagosome trafficking is essential for intracellular growth.

To gain insight into how quickly *L. pneumophila* must alter phagosome trafficking, maturation of vesicles containing either wild type or *dotA* mutant bacteria was examined. Markers enriched on late endosomes, such as the protein LAMP-1, were acquired by a high percentage of phagosomes containing *dotA* mutant bacteria within the first 5 minutes after uptake.[28] In contrast, late endosomal proteins were not observed on most phagosomes containing wild type *L. pneumophila* within this time frame. Furthermore, vacuoles containing wild type *L. pneumophila* remain LAMP-1-negative as the bacteria begin to replicate. Based on these data, it can be concluded that the Dot/Icm transporter is required for a signal to be sent to host cells that alters trafficking of endocytic compartment rapidly. Most importantly, these data show that *L. pneumophila* actually promote bifurcation from the conventional route traveled by endocytic organelles prior to their fusion with late endosomes; thus preventing the phagosome in which these bacteria reside from fusing with lysosomes.

Several lines of evidence suggest that the Dot/Icm transporter plays an important role during the initial stages of infection, but may not be required for multiplication or vacuole maintenance once a replicative organelle has been established. It is possible to phenotypically modulate Dot/Icm function by taking a mutant strain of *L. pneumophila* harboring a defective *dotA* chromosomal allele and transforming this strain with a plasmid that has the functional *dotA* gene strictly regulated by the inducible pTAC promoter.[28] The resulting strain assembles a functional Dot/Icm apparatus when grown in media containing the inducing agent IPTG, whereas, bacteria grown in the absence of IPTG are defective for intracellular growth because they do not produce a functional DotA protein. When these bacteria are first grown extracellularly in broth containing IPTG, they are able to infect and grow within macrophages. As long as the tissue culture medium is supplemented with IPTG these *L. pneumophila* bacteria will remain phenotypically Dot/Icm$^+$ as they multiply intracellularly. However, if the tissue culture medium does not include IPTG, synthesis of the DotA protein will cease upon infection. Interestingly, repressing DotA synthesis after infection does not interfere with the first complete cycle of bacterial replication in macrophages, but bacteria that exit spent host cells are phenotypically Dot/Icm-negative and cannot replicate in neighboring macrophages. If these bacteria are first grown extracellularly in media that lacks IPTG, and then bacteria that are phenotypically Dot/Icm-negative are induced with IPTG immediately after macrophage internalization, they fail to replicate intracellularly. These data suggest that trafficking decisions controlled by the Dot/Icm apparatus must be made prior to the phagosome fusing with late endocytic organelles, since activating this system after *L. pneumophila* have been delivered to a late endocytic compartment will not rescue intracellular growth.

Coinfection studies have provided the most convincing evidence that the Dot/Icm transporter is required to create a replicative organelle, but that bacterial replication within this compartment does not require transporter function.[33] When macrophages are given the opportunity to internalize both wild type *L. pneumophila* and *dotA* mutants, host cells coinfected with both strains are observed. In some of these cells, wild type and mutant bacteria reside in separate phagosomes. When this occurs, wild type *L. pneumophila* will replicate within the phagosomes they have created, whereas, the *dotA* mutants are trapped within endocytic organelles that do not permit multiplication. Coinfection can also occur when wild type and *dotA* mutant bacteria are internalized together, resulting in co-residence in a single phagosome. The fate of the phagosome containing both bacteria is controlled by wild type *L. pneumophila*. Remarkably, these compartments evade endocytic maturation and support growth of both wild type and *dotA* mutant bacteria. When multiplication of only those *L. pneumophila* encoding a functional Dot/Icm transporter is selectively inhibited within co-occupied vacuoles, this can be accomplished by either using a thymine auxotroph or by antibiotic-mediated killing, growth of the *dotA* mutant bacteria within this same compartment does not appear to be affected. These data would suggest that once the Dot/Icm apparatus has effectively established an organelle that permits intracellular multiplication, continuous signaling by this apparatus may not be necessary for expansion and maintenance of the organelle. This would be consistent with the replicative compartment being similar, if not identical, to a preexisting host organelle.

The hypothesis that *L. pneumophila* proteins are injected into macrophages upon contact was substantiated by evidence showing that pores are formed in the plasma membrane of host cells during their encounter with wild type bacteria.[34] Mutations in the *dotA* gene eliminate this pore-forming activity, which demonstrates that the Dot/Icm transporter is essential for this process (Fig. 2). It has been suggested that these pores represent channels in the host cell membrane through which bacterial-derived effector proteins travel.[35-37] There is also evidence to suggest that these pores may play an important role in the late stages of infection and mediate the egress of *L. pneumophila* from exhausted host cells.[38] Interestingly, pore-forming activity is upregulated as *L. pneumophila* enter stationary phase.[39] These data are consistent with both of the functions proposed for the pores since stationary phase bacteria must not only exit the host, but they must also be primed to reinitiate infection. Although a definitive role for these pores remains to be determined, pore formation has become a very useful readout for dissecting Dot/Icm transporter function.

Distinct Virulence Traits Are Regulated by Different Icm Proteins

Like most secretion systems that inject bacterial-derived proteins into eukaryotic cells, the Dot/Icm apparatus has both conserved structural components predicted to be required for an operational transporter and novel factors that presumably provide functions that are unique to the *L. pneumophila* system. Several of these novel *dot/icm*-encoded products are soluble proteins, further distinguishing them from most of the conserved structural components, which have hydrophobic amino acid domains that suggest these proteins are anchored in the bacterial cell membrane (Fig. 1). To understand the role these novel proteins may play in host cell pathogenesis, the proteins IcmQ, IcmR, IcmS, and IcmW have been investigated. Distinct virulence traits that require Dot/Icm transporter function have been delineated by characterizing isogenic mutants missing one of these Icm proteins.

When virulence traits were measured for *icmQ, icmR, icmS* and *icmW* mutants, three phenotypic categories were revealed. The *icmQ* mutant is similar in many respects to a *dotA* mutant.[40] All virulence activities that were shown previously to require DotA protein function were also absent in an *icmQ* mutant. These activities include the ability to form pores in host cell membranes, create a phagosome that rapidly evades fusion with endocytic organelles, and replication in eukaryotic host cells. Thus, the IcmQ protein must be essential for most, if not all, virulence activities required for host cell pathogenesis, suggesting that this protein plays an important role in Dot/Icm transporter function.

Interestingly, loss-of-function mutations in either *icmW* or *icmS* will not completely abolish all virulence activities.[40,41] Pore-forming activity is not affected by mutations in either *icmW* or *icmS*, however, phagosomes containing *icmW* or *icmS* mutants rapidly fuse with endocytic organelles. These data indicate that the IcmW and IcmS proteins are critical determinants of phagosome trafficking. Not surprisingly, the *icmW* and *icmS* mutants are totally defective for replication inside of primary macrophages derived from human and murine hosts,[40,41] and are also unable to replicate inside of protozoan host cells (CRR unpublished data). A macrophage-like cell line called U937, however, will support attenuated growth of *icmW* and *icmS* mutants.[40-42] Limited growth in U937 cells, in combination with the observation that pore formation is not affected, indicates that *icmW* and *icmS* are not core components essential for Dot/Icm transporter function.

The IcmW and IcmS proteins appear to have non-redundant functions. This was shown by constructing a mutant strain of *L. pneumophila* in which both the *icmW* and *icmS* genes were deleted.[40] The *icmW icmS* double mutant is indistinguishable from either single mutant. These data indicate that the IcmW and IcmS proteins are required for a molecular pathway that alters endocytic trafficking of phagosomes containing *L. pneumophila*, which is distinct from that required for pore formation. Yeast two-hybrid analysis has revealed a specific interaction between the *icmW* and *icmS* products,[40] further supporting the hypothesis that these two proteins are involved in a common signaling pathway. It seems likely that the *icmW* and *icmS* products could serve as co-chaperone proteins for substrates secreted into host cells by the Dot/Icm apparatus. Accordingly, the function of secreted proteins that require IcmW and IcmS for injection would be to modulate phagosome biogenesis. It was shown by gel overlay analysis that the IcmS protein can bind directly to a *L. pneumophila* protein with an estimated mass of 130,000 Daltons,[40] making this 130 kDa protein an attractive candidate for an effector protein.

An unexpected phenotype was observed when the *icmR* gene was deleted from the *L. pneumophila* chromosome. Like the *icmW* and *icmS* mutants, *L. pneumophila icmR* mutants are still able to replicate in U937 cells, albeit inefficiently.[40] Pore-forming activity, however, is not detectable in *icmR* mutant bacteria. Interestingly, when *icmR* mutants are internalized by primary murine macrophages, phagosomes in which they reside do not fuse with endocytic organelles as effectively as phagosomes containing other *dot/icm* mutants of *L. pneumophila*. As a result, a small proportion of *icmR* mutants that are internalized by primary macrophages can initiate replication inside of their host vacuole. After 8 hours, most vacuoles containing replicating *icmR* mutants have fused with lysosomes, which ends intracellular proliferation. IcmR also appears to be a chaperone protein. Gel overlay and yeast two hybrid analysis indicate that IcmQ is one protein to which IcmR binds.

From these studies, it appear as if the IcmW and IcmS proteins are essential for early phagosome biogenesis events that create a niche permissive for growth, whereas, the IcmR protein plays a critical role in establishment and maintenance of this replicative vacuole (Fig. 2). Thus, creation of a replicative organelle by *L. pneumophila* is a multi-stage process that requires distinct activities transduced by the Dot/Icm transport apparatus.

The Dot/Icm Transporter Does More than Just Inhibit Maturation of Phagosomes Containing *L. pneumophila*

It has been hypothesized that *L. pneumophila* create an inert vacuole that avoids endocytic maturation.[43,44] According to this hypothesis, the primary function of the Dot/Icm transporter would be to secrete molecules into the host cells that disrupt normal endocytic maturation events. To test this hypothesis we have disrupted endocytic maturation of phagosomes in primary murine macrophages by ectopically expressing dominant negative variants of the small GTP binding proteins Rab5 and Rab7 (CRR and JCK, unpublished data). During endocytic maturation, Rab5 function is required for fusion of plasma membrane-derived organelles with early endosomes[45-47] and Rab7 function is required for the fusion of early endosomes with late endosomal organelles.[45,48,49] The dominant negative Rab5S34N protein and the dominant negative Rab7T22N protein remain locked in an inactive GDP-bound state, which disrupts the function of their wild type counterparts.[47,48,50-52] Expression of either Rab5S34N or Rab7T22N in primary murine macrophages will enhance the efficiency in which wild type *L. pneumophila* are able to form a replicative organelle (CRR and JCK, unpublished data). Close to 100 percent of all wild type *L. pneumophila* internalized by macrophages expressing the Rab7T22N protein will ultimately create an organelle that supports intracellular growth; whereas, only 70-75 percent of the *L. pneumophila* internalized by macrophages transfected with a control plasmid expressing GFP will be successful in establishing a vacuole that supports growth. These data provide proof-of-concept that the dominant interfering Rab proteins suppress the ability of macrophages to traffic phagosomes containing *L. pneumophila* to a compartment that restricts growth. Interestingly, *L. pneumophila dot/icm* mutants shown previously to replicate in vacuoles created by wild type *L. pneumophila*, are unable to replicate intracellularly when fed singularly to macrophages that are producing these dominant interfering Rab proteins. These findings are important because they show that simply blocking maturation of a plasma membrane-derived vacuole is insufficient to create an environment that supports *L. pneumophila* replication, meaning that the Dot/Icm transporter must be doing more for *L. pneumophila* than simply paralyzing the cells ability to destroy these bacteria after internalization.

Subversion of ER Vesicle Trafficking by *L. pneumophila* Creates a Stable Organelle that Supports Intracellular Growth

In the initial studies defining the cell biology of bacterial infection, electron micrographs taken by Horwitz and colleagues showed that phagosomes containing *L. pneumophila* are morphologically distinct from the endocytic vacuoles in which avirulent organisms reside (Fig. 3).[6] During the first hour after infection, phagosomes containing wild type *L. pneumophila* are found surrounded by smooth cytoplasmic vesicles and mitochondria. As the *L. pneumophila*-containing compartment matures, ribosomes begin to associate with the vacuole and the smooth cytoplasmic vesicles become less abundant. By eight hours, the compartment containing replicating *L. pneumophila* is completely surrounded by ribosomes.

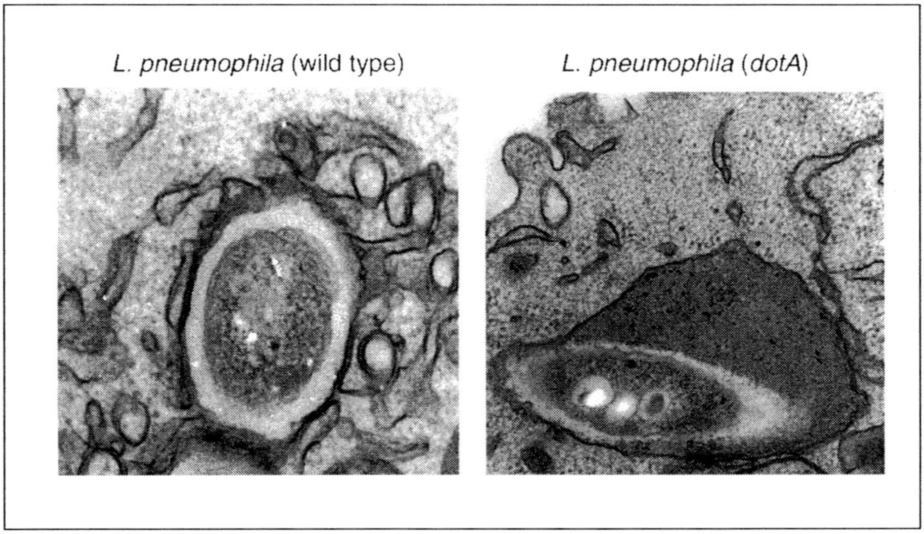

Figure 3. Wild type *L. pneumophila* creates a replicative vacuole by subverting host vesicle trafficking. Morphological differences between phagosomes containing wild type *L. pneumophila* and *dotA* mutants can be observed by electron microscopy. Vacuoles containing wild type bacteria evade fusion with lysosomes and recruit smooth vesicles derived from host ER. Phagosomes harboring *dotA* mutants acquire lysosomal content markers and do not have ER-derived vesicles.

In a recent study, these early morphological observations have been extended by Tilney and colleagues (Lewis J Tilney and CRR, unpublished data). These studies show that the smooth cytoplasmic vesicles reported earlier by Horwitz are actually vesicles derived from host endoplasmic reticulum (ER). ER vesicles can be seen attached to phagosomes containing *L. pneumophila* as early as 5 minutes after uptake. In fact, when vacuoles containing *L. pneumophila* are isolated from disrupted host cells, ER vesicles remain attached, indicating that this is a high affinity interaction. The plasma membrane that surrounds *L. pneumophila* immediately after uptake is thicker than the ER membrane. This difference in membrane thickness is likely due to the presence of cholesterol and sphingolipids, which are abundant in the plasma membrane. Remarkably, the membrane surrounding *L. pneumophila* becomes the same thickness as the ER membrane upon attachment of the ER vesicles, which indicates that lipid exchange occurs between these opposed vesicles. The thinning of the membrane surrounding *L. pneumophila* is rapid, occurring within 15-30 minutes after uptake. Most importantly, phagosomes containing *L. pneumophila dotA* mutants do not have attached ER vesicles and the surrounding membrane remains thick after uptake. These data indicate that the Dot/Icm transporter is required for the physical interactions that occur between vacuoles harboring wild type *L. pneumophila* and host ER.

Analysis of other *dot/icm* mutants reveals that phagosomes containing *icmW* or *icmS* mutants have attached ER vesicles, whereas, *icmR* mutants reside in phagosomes that are not encased by ER. These data are consistent with the hypothesis that biogenesis of a replicative organelle is a multi-stage process (Fig. 4). In the first stage, phagosomes containing *L. pneumophila* evade immediate endocytic maturation. This stage requires the *icmW* and *icmS* gene products, as these mutants reside in phagosomes that acquire LAMP-1 rapidly by fusing with endocytic organelles. Next, *L. pneumophila* promote the attachment of ER-derived vesicles to the surface of their phagosomes. The IcmR protein appears to play an important role in this process. *L. pneumophila icmR* mutants have the ability to form phagosomes that evade fusion with lysos-

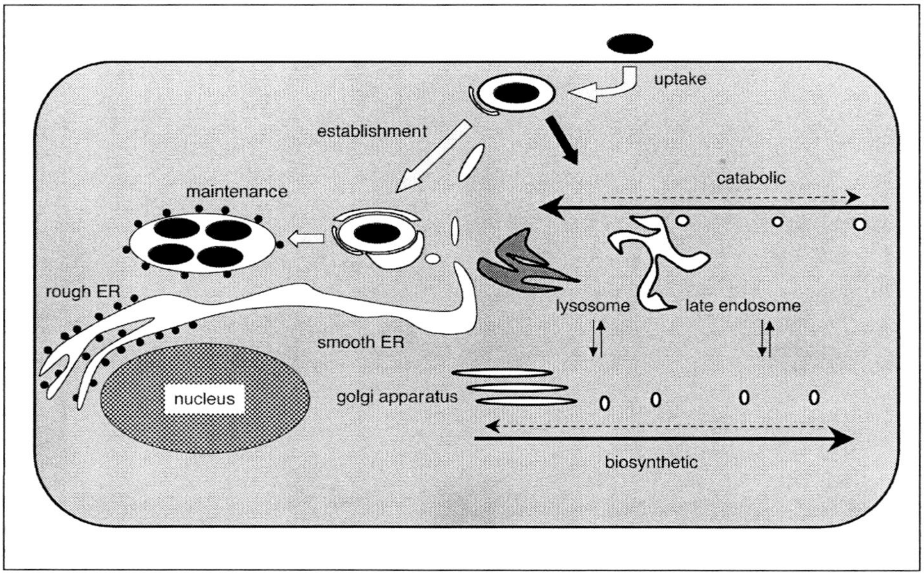

Figure 4. The Dot/Icm transporter orchestrates a multi-stage process that converts endocytic vacuoles containing *L. pneumophila* into replicative vacuoles that are similar to ER. Upon uptake into eukaryotic host cells, *L. pneumophila* evades immediate endocytic maturation. This first stage event enables bacteria to avoid catabolic degradation by the host and gives *L. pneumophila* time to remodel a plasma membrane-derived vacuole into a niche capable of sustaining intracellular growth. This involves the attachment and fusion of vesicles derived from the ER with phagosomes containing *L. pneumophila*. Upon completion of this second stage, *L. pneumophila* reside an a nutrient-rich organelle that is isolated from the endocytic pathway.

omes, but can not recruit ER vesicles. Data obtained in primary macrophages indicates that *icmR* mutants can form a vacuole that supports early replication but that eventual fusion with lysosomes severely limits intracellular proliferation. These data suggest that this second stage event, in which ER vesicles fuse with *L. pneumophila* containing vacuoles, is important for establishing a stable replicative niche. Once this second stage has been completed, *L. pneumophila* appear to be residing in a vacuole that is similar, if not identical to ER, as evidenced by the presence of ribosomes on the cytoplasmic surface of the replicative organelle. Thus, the Dot/Icm apparatus is a weapon used by *L. pneumophila* to beguile the host cell into converting an endocytic vacuole in which the bacteria reside initially into a cellular organelle that is nutrient rich and non-destructive.

Concluding Remarks

Although the picture of how *L. pneumophila* create a replicative niche is becoming clear, the molecular mechanisms used by *L. pneumophila* to parasitize eukaryotic hosts still remain unsolved. Identification of bacterial factors that are delivered into host cells by the Dot/Icm transporter and understanding their biochemical activities will provide important clues on the host cellular machinery that is subverted during the course of infection. The fact that these factors have not been identified in the numerous genetic screens conducted previously using macrophage-like cell lines suggests that a single secreted factor may not be critical for the eventual outcome of infection. This indicates that functional redundancy probably exists among these effector molecules. In addition, a subset of these effector molecules will most likely be proteins that mimic endogenous eukaryotic factors. This means that in certain host cells, *L. pneumophila* may not require a given effector if the cellular equivalent can be subverted effectively. We predict that some of these effectors will have been captured from eukaryotic host cells by horizontal gene transfer. For this reason, a great deal may also be learned about how *L. pneumophila* has evolved as a pathogen once the substrates for the Dot/Icm secretion apparatus have been revealed.

References

1. Horwitz MA, Silverstein SC. Legionnaires' disease bacterium (Legionella pneumophila) multiplies intracellularly in human monocytes. J Clin Invest 1980; 66:441-450.
2. McDade JE, Shepard CC, Fraser DW et al. Legionnaires' disease: Isolation of a bacterium and demonstration of its role in other respiratory diseases. N Engl J Med 1977; 297:1197-1203.
3. Brenner DJ, Steigerwalt AG, McDade JE. Classification of the Legionnaires' disease bacterium: Legionella pneumophila, genus novum, species nova, of the family Legionellaceae, family nova. Ann Intern Med 1979; 90(4):656-658.
4. Glavin FL, Winn WC, Jr., Craighead JE. Ultrastructure of lung in Legionnaires' disease. Observations of three biopsies done during the Vermont epidemic. Ann Intern Med 1979; 90(4):555-559.
5. Horwitz MA. The Legionnaires' disease bacterium (Legionella pneumophila) inhibits phagosome lysosome fusion in human monocytes. J Exp Med 1983; 158:2108-2126.
6. Horwitz MA. Formation of a novel phagosome by the Legionnaires' disease bacterium (Legionella pneumophila) in human monocytes. J Exp Med 1983; 158:1319-1331.
7. Fields BS. The molecular ecology of Legionellae. Trends Microbiol 1996; 4(7):286-290.
8. Horwitz MA. Characterization of avirulent mutant Legionella pneumophila that survive but do not multiply within human monocytes. J Exp Med 1987; 166(5):1310-1328.
9. Liles MR, Edelstein PH, Cianciotto NP. The prepilin peptidase is required for protein secretion by and the virulence of the intracellular pathogen Legionella pneumophila. Mol Microbiol 1999; 31(3):959-970.
10. Edelstein PH, Edelstein MA, Higa F et al. Discovery of virulence genes of Legionella pneumophila by using signature tagged mutagenesis in a guinea pig pneumonia model. Proc Natl Acad Sci USA 1999; 96(14):8190-8195.
11. Feeley JC, Gibson RJ, Gorman GW et al. Charcoal-yeast extract agar: Primary isolation medium for Legionella pneumophila. J Clin Microbiol 1979; 10:437-441.
12. Marra A, Shuman HA. Genetics of Legionella pneumophila virulence. Ann Rev Genet 1992; 26:51-69.
13. Swanson MS, Isberg RI. Identification of Legionella pneumophila mutants that have aberrant intracellular fates. Infect Immun 1996; 64(7):2585-2594.
14. Vogel JP, Roy C, Isberg RR. Use of salt to isolate Legionella pneumophila mutants unable to replicate in macrophages. Ann NY Acad Sci 1996; 797:271-272.
15. Sadosky AB, Wiater LA, Shuman HA. Identification of Legionella pneumophila genes required for growth within and killing of human macrophages. Infect Immun 1993; 61(12):5361-5373.
16. Pope CD, Dhand L, Cianciotto NP. Random mutagenesis of Legionella pneumophila with mini-Tn10. FEMS Microbiol Lett 1994; 124(1):107-111.
17. Gao LY, Harb OS, Kwaik YA. Identification of macrophage-specific infectivity loci (mil) of Legionella pneumophila that are not required for infectivity of protozoa. Infect Immun 1998; 66(3):883-892.
18. Pope CD, O'Connell W, Cianciotto NP. Legionella pneumophila mutants that are defective for iron acquisition and assimilation and intracellular infection. Infect Immun 1996; 64(2):629-636.
19. Mintz CS, Chen J, Shuman H. Isolation and characterization of auxotrophic mutants of Legionella pneumophila that fail to multiply in human monocytes. Infect Immun 1988; 56:1449-1455.
20. Harb OS, Abu Kwaik Y. Essential role for the Legionella pneumophila rep helicase homologue in intracellular infection of mammalian cells. Infect Immun 2000; 68(12):6970-6978.
21. Gao LY, Harb OS, Abu Kwaik Y. Utilization of similar mechanisms by Legionella pneumophila to parasitize two evolutionarily distant host cells, mammalian macrophages and protozoa. Infect Immun 1997; 65(11):4738-4746.
22. Berger KH, Merriam JJ, Isberg RR. Altered intracellular targeting properties associated with mutations in the Legionella pneumophila dotA gene. Mol Microbiol 1994; 14:809-822.
23. Andrews HL, Vogel JP, Isberg RR. Identification of linked Legionella pneumophila genes essential for intracellular growth and evasion of the endocytic pathway. Infect Immun 1998; 66(3):950-958.
24. Segal G, Shuman HA. Characterization of a new region required for macrophage killing by Legionella pneumophila. Infect Immun 1997; 65(12):5057-5066.
25. Vogel JP, Andrews HL, Wong SK et al. Conjugative transfer by the virulence system of Legionella pneumophila. Science 1998; 279(5352):873-876.

26. Segal G, Purcell M, Shuman HA. Host cell killing and bacterial conjugation require overlapping sets of genes within a 22-kb region of the Legionella pneumophila genome. Proc Natl Acad Sci USA 1998; 95(4):1669-1674.
27. Wiater LA, Dunn K, Maxfield FR et al. Early events in phagosome establishment are required for intracellular survival of Legionella pneumophila. Infect Immun 1998; 66(9):4450-4460.
28. Roy CR, Berger K, Isberg RR. Legionella pneumophila DotA protein is required for early phagosome trafficking decisions that occur within minutes of bacterial uptake. Mol Microbiol 1998; 28(3):663-674.
29. Christie PJ, Vogel JP. Bacterial type IV secretion: conjugation systems adapted to deliver effector molecules to host cells. Trends Microbiol 2000; 8(8):354-360.
30. Young C, Nester EW. Association of the virD2 protein with the 5' end of T strands in Agrobacterium tumefaciens. J Bacteriol 1988; 170(8):3367-3374.
31. Wilkins BM, Thomas AT. DNA-independent transport of plasmid primase protein between bacteria by the I1 conjugation system. Mol Microbiol 2000; 38(3):650-657.
32. Berger KH, Isberg RR. Two distinct defects in intracellular growth complemented by a single genetic locus in Legionella pneumophila. Mol Microbiol 1993; 7:7-19.
33. Coers J, Monahan C, Roy CR. Modulation of phagosome biogenesis by Legionella pneumophila creates an organelle permissive for intracellular growth. Nature Cell Biology 1999; 1(7):451-453.
34. Kirby JE, Vogel JP, Andrews HL et al. Evidence for pore-forming ability by Legionella pneumophila. Mol Microbiol 1998; 27(2):323-336.
35. Roy CR. Trafficking of the Legionella pneumophila phagosome. ASM News 1999; 65(6):416-421.
36. Kirby JE, Isberg RR. Legionnaires' disease: The pore macrophage and the legion of terror within. Trends Microbiol 1998; 6(7):256-258.
37. Segal G, Shuman HA. How is the intracellular fate of the Legionella pneumophila phagosome determined? Trends Microbiol 1998; 6(7):253-255.
38. Alli OA, Gao LY, Pedersen LL et al. Temporal pore formation-mediated egress from macrophages and alveolar epithelial cells by Legionella pneumophila. Infect Immun 2000; 68(11):6431-6440.
39. Byrne B, Swanson MS. Expression of Legionella pneumophila virulence traits in response to growth conditions. Infect Immun 1998; 66(7):3029-3034.
40. Coers J, Kagan JC, Matthews M et al. Identification of Icm protein complexes that play distinct roles in the biogenesis of an organelle permissive for Legionella pneumophila intracellular growth. Mol Microbiol 2000; 38(4):719-736.
41. Zuckman DM, Hung JB, Roy CR. Pore-forming activity is not sufficient for Legionella pneumophila phagosome trafficking and intracellular growth. Mol Microbiol 1999; 32(5):990-1001.
42. Segal G, Shuman HA. Legionella pneumophila utilizes the same genes to multiply within Acanthamoeba castellanii and human macrophages. Infect Immun 1999; 67(5):2117-2124.
43. Clemens DL, Horwitz MA. Characterization of the Mycobacterium tuberculosis phagosome and evidence that phagosomal maturation is inhibited. J Exp Med 1995; 181:257-270.
44. Clemens DL, Horwitz MA. Hypoexpression of MHC molecules on Legionella pneumophila phagosomes and phagolysosomes. Infect Immun 1993; 61:2803-2812.
45. Chavrier P, Parton RG, Hauri HP et al. Localization of low molecular weight GTP binding proteins to exocytic and endocytic compartments. Cell 1990; 62(2):317-329.
46. Gorvel JP, Chavrier P, Zerial M et al. rab5 controls early endosome fusion in vitro. Cell 1991; 64(5):915-925.
47. Bucci C, Parton RG, Mather IH et al. The small GTPase rab5 functions as a regulatory factor in the early endocytic pathway. Cell 1992; 70(5):715-728.
48. Meresse S, Gorvel JP, Chavrier P. The rab7 GTPase resides on a vesicular compartment connected to lysosomes. J Cell Sci 1995; 108(Pt 11)):3349-3358.
49. Feng Y, Press B, Wandinger-Ness A. Rab 7: An important regulator of late endocytic membrane traffic. J Cell Biol 1995; 131(6 Pt 1):1435-1452.
50. Li G, Stahl PD. Structure-function relationship of the small GTPase rab5. J Biol Chem 1993; 268(32):24475-24480.
51. Press B, Feng Y, Hoflack B et al. Mutant Rab7 causes the accumulation of cathepsin D and cation-independent mannose 6-phosphate receptor in an early endocytic compartment. J Cell Biol 1998; 140(5):1075-1089.
52. Bucci C, Thomsen P, Nicoziani P et al. Rab7: A key to lysosome biogenesis. Mol Biol Cell 2000; 11(2):467-480.

CHAPTER 10

Phagosome Biogenesis in Relation to Intracellular Survival Mechanisms of Mycobacteria

Lutz Thilo and Chantal de Chastellier

Introduction

The genus *Mycobacterium* includes more than 70 different intracellular bacterial parasites. These range from the obligate intracellular pathogen, *M. leprae*, the etiologic agent of leprosy, and the facultative intracellular parasites *M. tuberculosis, M. bovis* and *M. africanum,* which can cause progressive lung disease in humans or animals, to environmental species such as *M. gordonae, M. fortuitum, M. terrae* and *M. smegmatis*, which are seldom pathogenic for healthy patients. Between these two extremes lies a large number of opportunistic pathogens such as *M. avium, M. intracellulare, M. scrofulaceum, M. simiae, M. kansasii* and *M. marinum,* to cite a few, which can cause, under particular conditions, disseminated systemic disease in humans (for reviews see refs. 1,2). Mycobacteria of the *M. tuberculosis* complex, i.e., *M. tuberculosis, M. bovis* (the agent of bovine tuberculosis) and *M. africanum* (often isolated from patients in Africa), can cause tuberculosis in humans. *M. tuberculosis* itself is a major threat to humans, as it is responsible for more deaths than any other single pathogen, causing 10 million new cases and 3 million deaths per year. [3,4] The relative inefficiency of available vaccines and the emergence of significant levels of multi-drug-resistant *M. tuberculosis* isolates result in severe complications for treatment and control of the disease. The emergence of AIDS has further increased the incidence of tuberculosis, and a growing body of reports have shown that non-tuberculous mycobacteria are able to cause disease in humans (for a review see ref. 5) In this context, *M. avium* has emerged as an important pathogen which is responsible for disseminated, often lethal infections in 25-50% of the patients with AIDS. The development of new strategies for prevention and treatment of tuberculosis, and of non-tuberculous mycobacterial diseases, requires a better understanding of the pathogenesis of infection.

Mycobacteria are phagocytosed by scavenger cells of the host's immune system. Then, instead of being destroyed, virulent mycobacteria survive and multiply inside the phagosomes (Fig. 1). The intention of this article has been to review and discuss possible mechanisms for how mycobacteria can mold the phagosome into a special vacuole that presents a favorable rather than hostile environment for intracellular growth. For this purpose the normal events of phagosome development must be considered in terms of when and how they may present a suitable target for mycobacterial intervention.

Intracellular Survival

When nonpathogenic bacteria are phagocytosed by macrophages, they are killed and degraded inside phagosomes, especially after these have fused with lysosomes that are the digestive organelles of the cell. In the case of phagocytosed virulent mycobacteria, no fusion of

Intracellular Pathogens in Membrane Interactions and Vacuole Biogenesis, edited by Jean-Pierre Gorvel. ©2004 Eurekah.com and Kluwer Academic / Plenum Publishers.

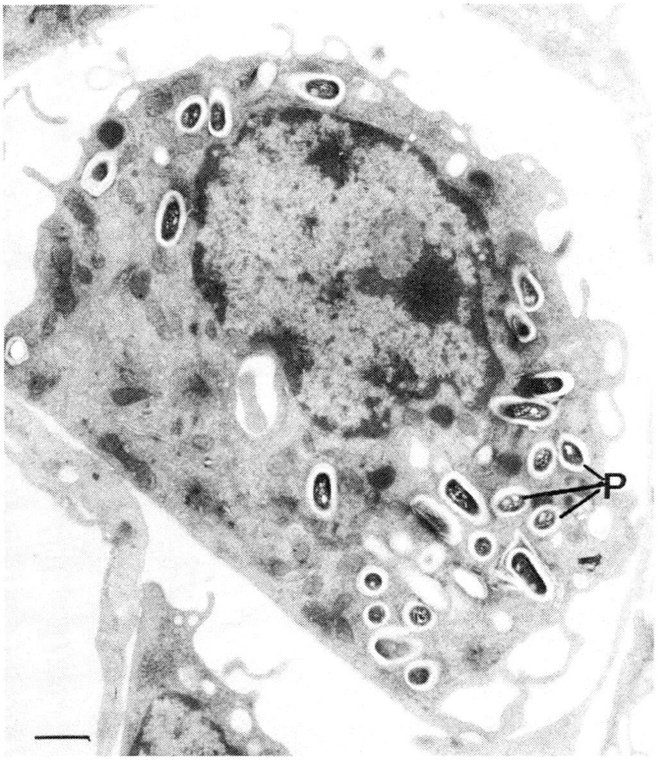

Figure 1. EM micrograph showing a thin section of a bone marrow-derived mouse macrophage at 15 days after infection with *M. avium*. Note that bacteria are morphologically intact and that the majority of the phagosomes (P) in which they are enclosed contain a single bacterium. Bar indicates 1 μm.

phagosomes with lysosomes is observed.[6,7] This inhibition of phagosome–lysosome fusion has been considered to be a key survival strategy because it enables mycobacteria to avoid exposure to the harsh hydrolytic environment inside phagolysosomes. It remains uncertain, however, whether in cases where mycobacteria end up being in phagolysosomes, they are killed or whether they only stop to multiply but survive. An early study has addressed this question and shown that when *M.tuberculosis* is opsonized prior to phagocytic uptake by mouse peritoneal macrophages, not only do about two thirds of all phagosomes fuse with lysosomes, but mycobacteria also remain intact and seem able to multiply in phagolysosomes.[8] In contrast, in the case of *M.avium* in mouse bone marrow-derived macrophages, phagosome-lysosome fusion, as induced by macrophage activation, causes a gradual loss of bacterial viability.[9] The possibility remains that in this case the activated state as such, rather than the resulting phagosome-lysosome fusion leads to killing of mycobacteria. When *M.avium* is made to reside in a common vacuole with lysosomal characteristics together with *Coxiella burnettii* by co-infection of macrophages, the mycobacteria are not killed although it is not known whether bacterial growth is inhibited under these conditions.[10] Extracellular mycobacteria tend to clump together, especially *M.tuberculosis*. As a result, phagocytosis often leads to the uptake of clumps of bacteria, resulting in more than one mycobacterium being inside a single phagosome. Such phagosomes are seen to fuse with lysosomes, as shown in Figure 2.[11,12] Because phagocytic uptake of small bacterial clumps can be expected to represent a common way of infection, delivery to phagolysosomes may be less restrictive to mycobacterial survival than assumed so far.

Attempts to gain an understanding of how mycobacteria prevent phagosome-lysosome fusion have focused on the relationship between phagosomes in general and the organelles of the endocytic pathway. These consist of the plasma membrane, early endosomes, endocytic carrier vesicles, recycling endosomes, late endosomes and lysosomes (for a review see ref. 13 and see Chapter 1 in this volume). Because phagosomes are observed to interact successively with organelles at advancing stages of the endocytic pathway,[14] they are considered to form an integral part of endocytic processing and to be subject to the same general endocytic mechanisms.

As it has turned out, mycobacteria do not interfere with phagosome-lysosome fusion directly, but rather inhibit a preceding step of endocytic processing, namely that of phagosome maturation, without which fusion with lysosomes cannot take place.[11,12,15-18]

The question regarding the mechanism of survival of intracellular mycobacteria, therefore, has now shifted from how mycobacteria prevent phagosome-lysosome fusion to how they prevent phagosome maturation. Several aspects of early-endosome maturation, being generally applicable also to phagosome maturation, have been considered by recent research as possible targets for mycobacterial intervention. Previous review articles on the mycobacterial phagosome have dealt with this topic with additional or different emphasis. [19-21]

Early Intervention by Mycobacteria Is Required to Prevent Phagosome Maturation

The stage of endocytic processing most relevant for interference by mycobacteria is that of early endosomes. Early endosomes are not permanent organelles, but are continuously formed de novo and consumed by a process of maturation[22-23] that changes their compositional and functional characteristics (see below). The process of maturation does not continue to the point where endosomes become lysosomes. Instead, matured endosomes fuse with pre-existing lysosomes, both in vivo[23-26] and in vitro.[27] As this must now be assumed to apply also to phagosomes, frequent statements in the literature, that phagosomes mature into phagolysosomes, are misleading. As is the case for mature endosomes, phagosomes can be considered as fusing with pre-existing lysosomes to give rise to phagolysosomes.

Maturation of early endosomes is a rapid process, occurring within about 3 minutes.[23] Similarly, phagosomes mature rapidly, within 5-15 minutes, after which they start fusing with lysosomes.[15,28,29] Under certain circumstances, phagosome maturation can be made to occur much more slowly, requiring an extended period of more than several hours. This is the case when latex beads with a hydrophobic surface are used as artificial phagocytic particles.[15, 30] By slowing phagosome maturation down from occurring within minutes to ocurring over hours, it has been possible to characterize successive events and compositional changes during this process.[30,31]

Due to the normally short time it takes for maturation of phagosomes, any mechanism by which mycobacteria affect this process must be available at very early stages, either during phagosome formation or within minutes thereafter. To the extent that mycobacteria might initially only slow down the maturation process, subsequent mechanisms for maintaining the phagosome in an immature state would have to be considered.

Choice of Macrophage Surface Receptors for Phagocytic Uptake

The earliest opportunity for mycobacteria to take influence on their intracellular fate could conceivably be the choice of surface receptor type for binding and subsequent phagocytic uptake. Different types of surface receptors can mediate mycobacterial entry, such as different complement receptors, Fc receptors, or the mannose receptor. Factors that play a role in this choice are the mycobacterial species and strain, whether or not the bacterial surface is opsonised with complement or with antibodies, and the state of activation of the phagocytic cell. In general, however, no preferred mode of entry has been identified, and the subsequent intracellular fate

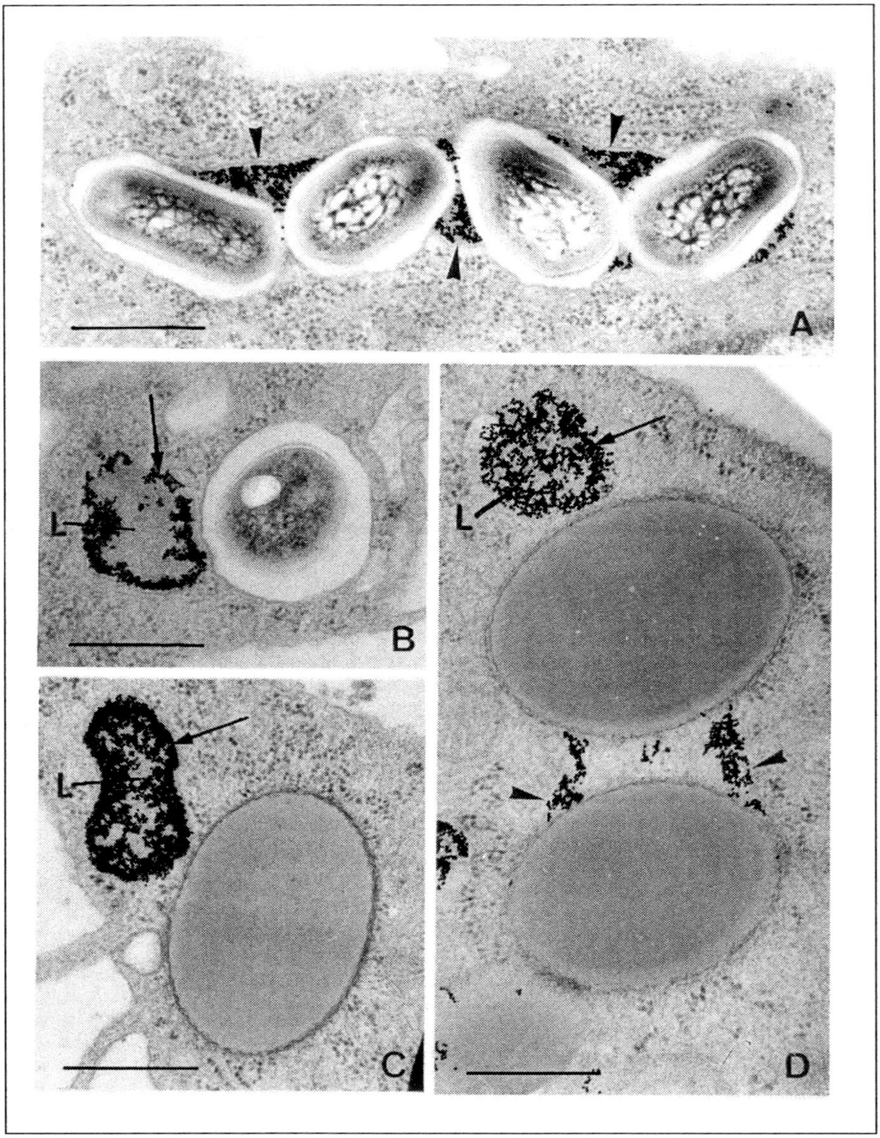

Figure 2. Phagosomes that contain more than one *M. avium* mature and fuse with lysosomes. Prior to infection, lysosomes were loaded with BSA-gold by a 30 minute pulse of endocytic uptake, followed by a 2 hour chase to lysosomes. Macrophages were then given *M. avium* (in A, B) or hydrophobic latex beads of 1 μm in diameter (in C, D). Fusion with lysosomes is inferred from the presence of gold particles (arrow) in the phagosome at 3 hours after phagocytic uptake. Phagosomes containing several bacteria (in A) or beads (in D), as opposed to those containing a single particle (in B, C), fuse with lysosomes (L). Note that the phagosome membrane is not closely apposed to the electron translucent surface layer of the bacterial wall or the bead surface where it spans the region between two adjacent particles (arrowheads in A, D), in contrast to the situation where the phagosome contains a single mycobacterium (in B) or bead (in C) or a Mycobacterium dividing inside a phagosome as in Figure 3. Bar indicates 0.5 μm. Figure 2D reprinted with permission, as in de Chastellier and Thilo.[12]

does not seem to depend directly on the uptake mechanism[32] (for a review see ref. 33). An exception has been mentioned above for antibody-coated *M.tuberculosis*, where phagosome-lysosome fusion is not prevented, but neither are mycobacteria killed as a result.[8] The involvement of particular receptors can have an indirect effect on the mycobacterial fate by stimulating the immune system, which could lead, for example, to the destruction of infected host cells (for a review see ref. 34).

Inhibition of Phagosome Acidification by Mycobacteria

An early event during maturation of early endosomes that could be the target of mycobacterial intervention for the purpose of preventing or slowing down the process of phagosome maturation, is the luminal acidification of these organelles. The acidification of endocytic organelles is the result of the H^+-pumping activity of the vacuolar ATPase in the membrane of these organelles (for a review see ref. 35), although alternative mechanisms of acidification may also contribute.[36]

Because maturation of early endosomes is such a rapid process (see above), it has been difficult to assure that when pH-indicators are applied by endocytic uptake, they remain confined exclusively to early endosomes. By making pH measurements on individual endosomes, using fluorescein-labeled dextran as pH probe, a reliable value for luminal pH of early endosomes has been obtained: Early endosomes become acidified to pH=6.3 within the first 3 minutes.[37] A reduction in this degree of acidification can inhibit dissociation of ligands from receptors and thereby interfere with sorting and recycling of surface-derived receptors (for reviews see refs. 35, 38). Two studies have used the drug bafilomycin A1 to block the H^+-ATPase. Most relevant for the present considerations, is the observation that a block of acidification of early endosomes in Baby Hamster Kidney cells prevents the appearance of endocytic vesicles (which in our opinion represent matured endosomes, see Thilo et al[23]) that can fuse with lysosomes for the delivery of endocytic contents.[39] In contrast, the absence of acidification in hepatoma cells also prevents delivery of endocytic contents to lysosomes, but does not affect sorting of transferrin receptors for recycling from early endosomes,[40] a process that serves as a classical measure for endosome maturation.

Phagosomes that contain virulent mycobacteria are not acidic,[41] or less acidic.[9,15,42-44] This inhibition of phagosome acidification is the result of a limited abundance of the H^+-ATPase in the phagosome membrane.[36,42] The question arises whether this limitation of acidification of the phagosome is related to its inability to mature, rather than serving another aspect of intraphagosomal survival of mycobacteria like, for example, keeping the pH at a level that is non optimal for many lysosomal enzymes (for a review see ref. 45).

It is difficult to compare early events of acidification for early endosomes and phagosomes because of the rapid acidification of early endosomes (within 3 minutes), and the relatively long time it takes for the synchronized formation of a sufficient number of phagosomes for observation. The soonest measurements have been reported for the pH in individual yeast-containing phagosomes in peritoneal macrophages.[46] During the first 2 minutes, phagosomes become slightly alkaline, followed by mild acidification between 2 and 4 minutes. For *M.avium*-containing phagosomes at about 30 minutes after formation, a pH of about 6.3 has been reported.[42] Based on the presented data for later times for the non-mycobacteria-containing phagosomes that acidify normally (up to 90 minutes, see Fig.1 in Sturgill-Koszycki et al[42]), a reasonable (non-linear) interpolation can be made to earlier time points and this suggests that at 5-10 minutes they have a pH not less than that of early endosomes (pH=6.3), and thus do not leave scope for reduced acidification in the case of *M.avium*-containing phagosomes at this early time. Anyhow, because the pH of mycobacteria-containing phagosomes during the first few hours remains the same as that of early endosomes, which mature normally, acidification that is limited to this level only cannot be expected to contribute to an inhibition of phagosome maturation. It has been reported, however, that treatment with ammonia, which can be expected to neutralize acidic endocytic organelles, prevents phagosome-lysosome fusion.[47] Because endosomes only undergo further acidification once they mature, and especially upon subsequent fusion with lysosomes, limited acidification of mycobacteria-containing phagosomes must be considered to reflect their continuing immature status and the resulting inability to fuse with lysosomes, rather than being the cause of this situation.

Upon prolonged intracellular survival of mycobacteria, the luminal pH of phagosomes rises gradually from its initial value of about 6.3 to above 6.6 several days after infection (half-time about 1 day).[15] This additional increase in intraphagosomal pH, therefore, sets in too late to be involved in any early mechanism of interference with phagosome maturation and, therefore, possibly serves another purpose. To the extent that this process reflects an increasing exclusion of the H^+-ATPase from the phagosome membrane, it would be compatible kinetically with a slow change in membrane composition of *M.avium*-containing phagosomes which occurs at a similar rate (see below).

Continued Presence of Rab5 on Mycobacteria-Containing Phagosomes

The Rab proteins are a family of small GTP-binding proteins that are involved at various stages of the endo- and exocytic pathways (for a review see ref. 48). In particular Rab5, which is present on the cytoplasmic side of the plasma membrane, of coated vesicles and of early endosomes, plays a role in regulating fusion among early endocytic vesicles.[49-53] Enhanced activity of Rab5 stimulates endocytosis and fusion of early endosomes and results in enlarged early endosomes,[54,55] while the absence of Rab5 activity causes inhibition of early endosome fusion and leads to the appearance of small tubular and vesicular endocytic structures at the cell periphery.[54,56] Three isoforms of Rab5 have been distinguished with no difference observed regarding their role during endocytosis.[57] The three isoforms of Rab can be phosphorylated differentially by different kinases and this may play a role in their differential regulation.[58] As for early endosomes, Rab5 also plays a role during fusion events of phagosomes.[59] Another small GTP-binding protein, Rab7, is associated with late endosomes where it is involved in the regulation of fusion with lysosomes.[60] During prolonged phagosome maturation, phagosomes lose Rab5 while acquiring Rab7.[30]

Rab5 can be considered as an attractive candidate for manipulation by pathogenic mycobacteria with the aim of preventing phagosome maturation because of its presence on early endocytic structures and its established role in regulation of specific fusion among early endosomes, as well as phagosomes. Therefore, the presence of Rab5 on mycobacteria-containing phagosomes has been investigated. *M.bovis* BCG-containing phagosomes, in a mouse macrophage cell line, retain Rab5 with an increased abundance over a period of about 7 days and do not acquire Rab7.[18] No evidence for the causal nature of this observation is available. According to the general characterization of the mycobacteria-containing phagosome as an immature phagosome (see above), the presence of Rab5 and the absence of Rab7 does not seem to cause the arrest of phagosome maturation, but merely seems to reflect the immature state. Because *M.bovis* BCG is not a mouse pathogen, a subsequent investigation using *M.tuberculosis* infection of cells of a human cell line (HeLa cells, transfected to overexpress Rab5c with the aim of making it accessible to detection by immunoelectron microscopy), may seem more relevant for the possible role of Rab5 in intracellular survival of mycobacteria. In agreement with the first study,[18] the results for *M.tuberculosis*-containing phagosomes also show the continued retention of Rab5 on the phagosome membrane as correlating with their immature status.[61] In contrast to the first study, it has now been shown that *M.tuberculosis*-containing phagosomes, while retaining Rab5, also acquire Rab7 without maturing and fusing with lysosomes, even when a constitutively active mutant of Rab7 is introduced.[62] This study mentions the possibility that the lack of maturation, in spite of the presence of Rab7, could be due to an inhibition of this GTPase by mycobacteria.[62]

As mentioned above, the continued presence of Rab5 on mycobacteria-containing phagosomes seems unlikely to be the cause of the observed block in maturation. The presence of activated Rab5 activity can be expected to speed up maturation rather than block it.

Overexpression of Rab5a indeed causes an accelerated maturation, and the resulting fusion with lysosomes, of phagosomes that contain a hemolysin-defective mutant of the endoparasite, *Listeria monocytogenes*.[63] Whereas these parasites normally delay phagosome maturation in order to escape from the phagosome before it fuses with lysosomes, this escape is blocked in the mutant. When expression of Rab5 is blocked, phagosome maturation is delayed and the lack of fusion with lysosomes enables survival of these mutant parasites.[63] Similarly, the involvement of Rab5 during fusion between latex bead-containing phagosomes and endocytic organelles has recently been demonstrated and, as for early endosomes, an active mutant form of Rab5 leads to uncontrolled fusion with endosomes which results in the formation of enlarged phagosomes.[64]

Instead of preventing phagosome maturation by the retention of Rab5 on the phagosome membrane, mycobacteria actually may have to restrict Rab5 activity to the point where phagosomes do not mature, but not so much as to cause complete cessation of fusion with early endosomes. Continued fusion with early endosomes will remain necessary for the maintenance of the phagosome membrane and to extend it during intraphagosomal growth and division of mycobacteria (see Figs. 1 and 3). Investigating the possible modulation of Rab5 activity by mycobacteria may be a promising future approach to gain understanding of the cause for inhibition of phagosome maturation. In this regard, the differential activation of Rab5 via phosphorylation by specific kinases as mentioned above,[58] could be of interest. Because Rab5 is active in its GTP-bound state, GTPase activating proteins (GAP) can down regulate Rab activity. Tuberin is a protein that has GAP activity specifically for Rab5 and when it down regulates Rab5's activity by stimulating it to convert to its inactive GDP-bound state, the rate of endocytic uptake is reduced.[65]

Phagosome Morphology

According to our own observations for *M.avium*-infected bone marrow-derived macrophages, an early requirement for the prevention of phagosome maturation is that the phagosome membrane is closely apposed to the mycobacterial surface all round. [12] Whenever a phagosome contains more than one mycobacterium, or bacterial clumps, it matures rapidly to a stage where it fuses with lysosomes (Figs. 2A vs 2B, and see Fig. 5a in de Chastellier, et al[15]). A similar observation has been reported by other authors. [11] A close apposition is not given where the phagosome membrane spans two adjacent mycobacteria in the same phagosome (Fig. 2A). This is in stark contrast to the case of a single dividing mycobacterium inside a phagosome in which case a close apposition is maintained down into the site of septum formation (Fig. 3).

We have simulated a delay of phagosome maturation for up to about 3 hours by using latex beads (1 μm) that maintain a closely apposed phagosome membrane (Fig. 4A) as a result of their hydrophobic surface.[12] In contrast, beads with a hydrophilic surface do not maintain the phagosome membrane in close all round apposition (Fig. 4B) and phagosomes mature rapidly to start fusing with lysosomes.[12] Above, we have referred to previous studies with hydrophobic bead-containing phagosomes for which the delayed maturation over a period of about 12 hours has enabled the characterization of gradual changes in phagosome composition and function.[30,31]

Based on these observations, we have proposed a mechanism for how a closely apposed phagosome membrane can lead to a delay of phagosome maturation (Fig. 5).[12] A characteristic of early endosomes is that they fuse among each other and intermingle contents and membrane, and this process also involves immature phagosomes. Once matured, neither endosomes nor phagosomes fuse with immature organelles (discussed in Thilo et al[23] and de Chastellier and Thilo[12]). Homotypic fusion between early endosomes/phagosomes must involve factors that mediate and regulate these specific fusion events. At present, Rab5 is factor to be considered for this role (see above). For fusion with early endosomes to cease upon maturation, such fusion-mediating factors must become inactive or lost. We propose that recycling is involved in gradually reducing the number of fusion-mediating factors from the maturing early endosome/phagosome until it is no longer able to fuse with early endosomes. The gradual loss of

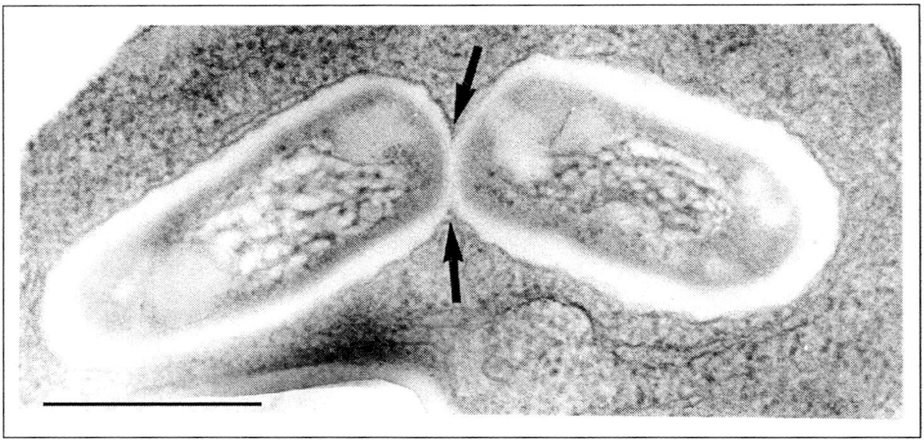

Figure 3. Intraphagosomal division of *M. avium* at day 7 postinfection. The phagosome membrane is closely apposed to the electron translucent surface layer of the bacterial wall, including at the site of septum formation (arrows). This membrane morphology contrasts strongly with the nonapposed membrane where it spans adjacent bacteria, as shown in Figure 2A. Bar indicates 0.5 μm. Reprinted with permission, as in de Chastellier and Thilo.[12]

Rab5 from the maturing phagosome[30] could serve as a case in point. However, as long as fusion with early endosomes continues, intermingling of membrane can be expected to keep adding new fusion-mediating factors as derived from early endosomes. For a net loss to occur, it will be necessary that the loss by recycling is faster than the addition by intermingling. Any condition that slows down the recycling rate can be expected to prolong the presence of fusion-mediating factors and thus slow down the maturation process. Early endosomes are tubular-vesicular structures[66] and the tubular domains are the site for sorting and recycling of cell surface-derived receptors.[67] We propose that an all round closely apposing phagosome membrane prevents the formation of tubular domains and thereby reduces the efficiency of recycling of fusion-mediating factors.

Maturation of early endosomes into endocytic carrier vesicles depends on an intact cytoplasmic coat protein, COPI, as observed via temperature-sensitive mutants of the coat component epsilon-COP[68-70] (for a review see ref. 71). This coat is involved in the budding of vesicles as part of the maturation process. The role of these coatamer proteins has been examined regarding their requirement for maturation of phagosomes.[72] A temperature-sensitive defect in the epsilon-COP subunit leads to a relatively higher abundance of transferrin receptor on Fc-opsonized red blood cell-containing phagosomes, while the relative abundance of Fc receptors on the phagosome membrane is the same at permissive and restricted temperature. However, at both temperatures, and for both receptor types, the clearance from the maturing phagosome occurs with the same half-time of about 20 minutes. Concurrently, in both cases, phagosomes acquire the lysosomal membrane protein, LAMP-1, indicating fusion with lysosomes. This rather slow maturation, therefore, seems not to be dependent on functional COPI coat proteins. Quantitatively the same result is observed for phagosomes in a macrophage cell line when COPI assembly is inhibited by the drug, brefeldin A.[72] So far it seems that COPI is nonessential for phagosome maturation and therefore disqualifying COPI as a likely target for mycobacterial intervention.

Another coat protein, the Tryptophan-Aspartate rich Coat Protein (TACO), which normally is only transiently associated with the cytoplasmic side of phagosomes, has been shown to remain associated with the phagosome when these contain live mycobacteria, and in which

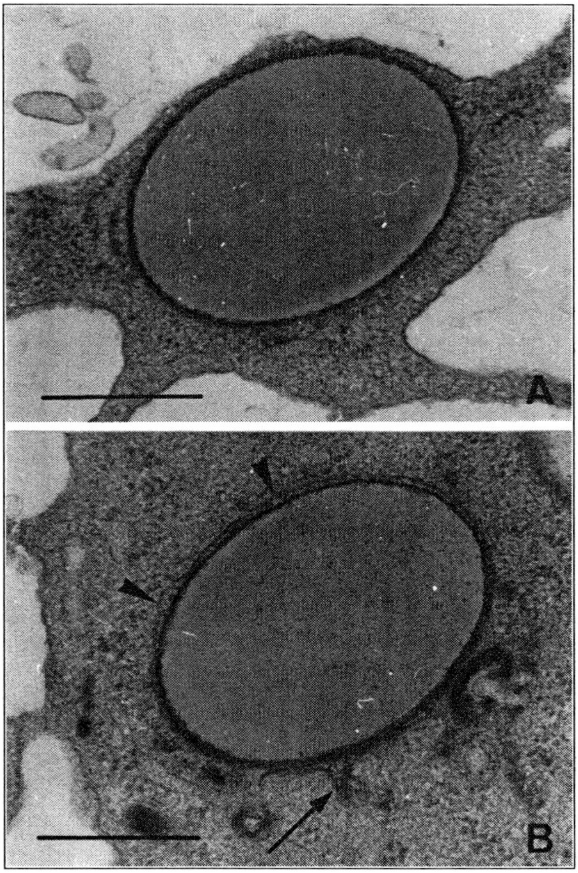

Figure 4. Phagosome containing a latex bead with a hydrophobic (A), or hydrophilic (B) surface at 2 hours after phagocytic uptake. The hydrophobic bead surface maintains the phagosome in all round close apposition (A), whereas the phagosome that contains the bead with the hydrophilic surface (B) displays a loose membrane that allows the formation of tubular membrane domains (arrow). Bar indicates 0.5 μm. Reprinted with permission, as in de Chastellier and Thilo.[12]

case phagosomes do not fuse with lysosomes.[73] In macrophages (Kupffer cells) that do not express TACO, *M.bovis* BCG cannot inhibit phagosome maturation and is destroyed in lysosomes. This strongly suggests that TACO may be an important target for mycobacteria to interfere with phagosome maturation. The role played by TACO is discussed by Jean Pieters in Chapter 12 of this volume.

Composition of the Phagosome Membrane

Due to the close apposition between the mycobacterial surface and the phagosome membrane, it seems reasonable to expect that mycobacteria exert their influence on phagosome maturation by specific interaction with phagosome membrane constituents. It is, therefore, of interest to find the membrane molecules involved. Several studies have analyzed the phagosome membrane composition in terms of the presence or absence of membrane markers typical for endocytic membranes (for a review see ref. 21). Proteins that are typically associated with

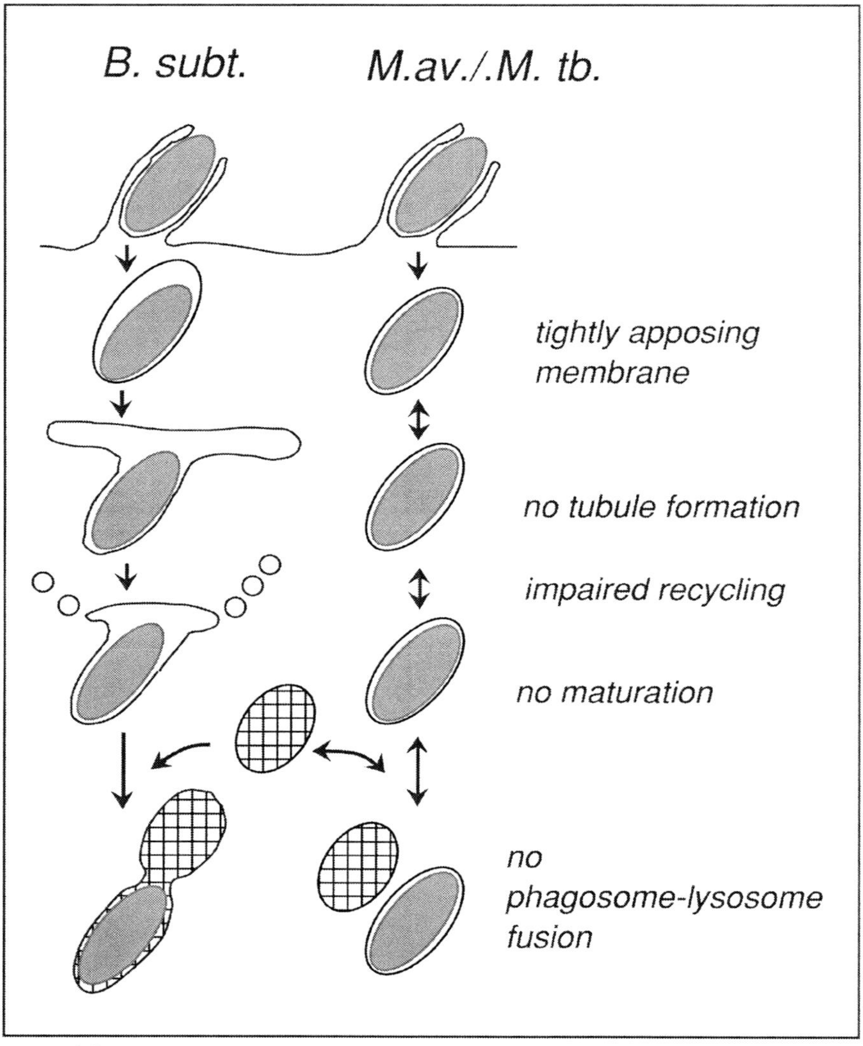

Figure 5. Proposed mechanism for how a phagocytic particle can delay phagosome maturation by maintaining the phagosome membrane closely apposed all round. Virulent, but not avirulent or killed *M. avium*, maintain the phagosome membrane in close apposition indefinitely (see Fig. 3). For kinetic considerations, see appendix to Figure 7 in de Chastellier and Thilo.[12] Reprinted with permission, as in de Chastellier and Thilo.[12]

early, but not late, endosomes (e.g., transferrin receptor and Rab5), are present on the *Mycobacterium*-containing phagosome. Up to now, no reason for this has been discovered and the presence of these proteins merely seems to reflect the status of an immature phagosome. A notable difference is the limited presence of the H^+-ATPase in comparison to early endosomes, but in view of phagosome acidification being the same as that of early endosomes, the reason for this is not known (see above).

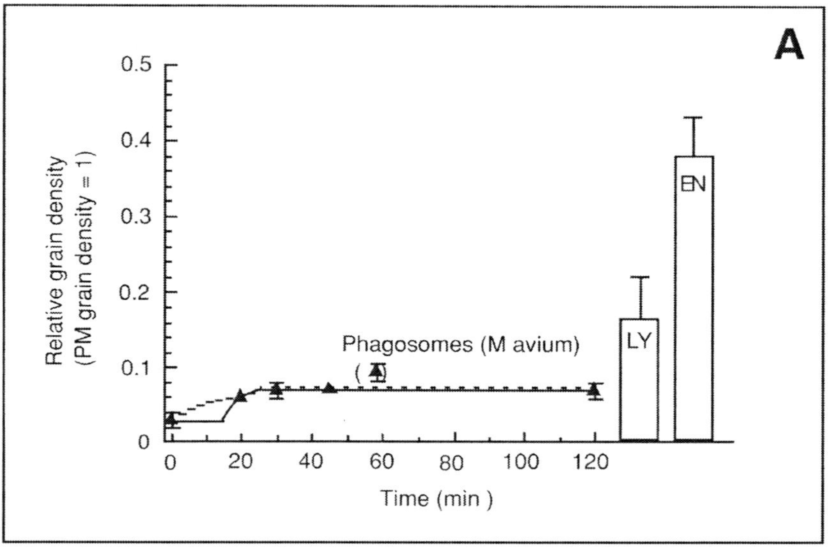

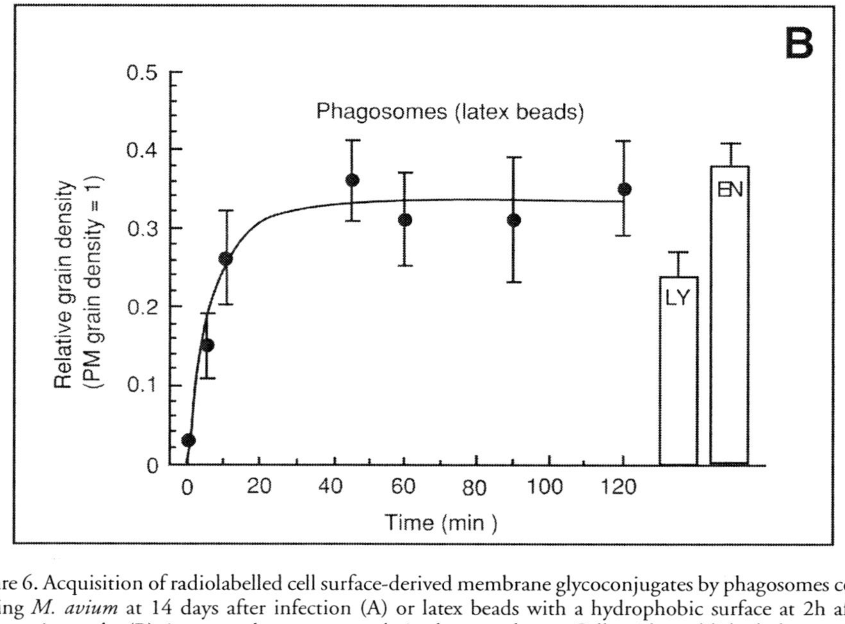

Figure 6. Acquisition of radiolabelled cell surface-derived membrane glycoconjugates by phagosomes containing *M. avium* at 14 days after infection (A) or latex beads with a hydrophobic surface at 2h after phagocytic uptake (B), in mouse bone marrow-derived macrophages. Cells with established phagosomes have been labeled on their surface, followed by membrane traffic for the indicated times to allow redistribution of label to endocytic membranes. The indicated steady-state concentration (grain density) of label on early endosomes (EN), lysosomes (LY) and phagosomes is a measure for their membrane composition in terms of cell surface glycoconjugates. The phagosome membrane containing *M. avium* (A) is about 3-fold depleted in comparison to membrane of early endosomes and of immature phagosomes with hydrophobic latex beads (B). Reprinted with permission, as in de Chastellier et al.[15] Recent observations[74a] show that there is no delay in the acquisition of membrane marker, i.e., the dashed line in (A) represents the correct time course.

A two–dimensional electrophoretic comparison between latex bead- and *M. avium*-containing phagosome membranes has not shown major differences.[74b] However, our own studies have shown that with regard to cell surface-derived glycoconjugates, *M. avium*-containing phagosomes at day 15 after infection are depleted about 3-fold in comparison to early endosomes with which they continue to fuse and exchange contents (Fig. 6A).[15] This depletion is not seen for phagosomes that contain latex beads with a hydrophobic surface and, therefore, remain immature for at least 3 hours, fusing with early endosomes, but not with lysosomes (Fig. 6B).[15] We have found recently that this depletion becomes established only after the first few days following infection (half-time about 2.5 days), too late to play a role in preventing phagosome maturation during the first 5-15 minutes after phagocytic uptake.[74a] This difference in membrane composition, between *M. avium*-containing phagosomes and early endosomes with which they continue to fuse, must be actively established and maintained by the action of live mycobacteria. Dead *M. avium* (unpublished results), or hydrophobic latex beads, also display a closely apposed phagosome membrane, but cannot establish a difference in membrane composition and cannot maintain the phagosome in an immature state beyond several hours.

Depolymerization of Actin Filaments by M. avium *in Macrophages*

Microtubules, actin filaments, and associated proteins, are elements of the cytoskeleton and are involved in coordinated interactions between the organelles of the endocytic pathway.[13,75,76] Actin filaments or actin-binding proteins can also be seen in association with phagosomes.[77-79] Drugs that disrupt the actin filament network, such as cytochalasin D, have been extensively used to show that actin filaments play a role in endocytic uptake (for a review see ref. 80), and that they facilitate transfer of endocytic tracers to late compartments of the endocytic pathway.[25,81] Several pathogens have developed strategies to model the actin cytoskeleton to their own advantage for invading the cells in which they multiply (for reviews see refs. 82,83).

A recent study[79] shows that infection of macrophages by virulent *M. avium,* leads to a marked disorganization of the actin filament network in comparison to uninfected macrophages or macrophages that contain non-pathogenic bacteria (*Bacillus subtilis, Mycobacterium smegmatis*), or hydrophobic latex beads. The most salient features are the formation of large patches of actin at day 1 after infection and, starting from day 2, the progressive disappearance of the small filaments concurrent with the appearance of large numbers of tiny punctuated structures. The mycobacterial molecules involved in actin depolymerization are unknown at present. The fact that actin is disrupted only starting from day one after infection suggests that bacteria must first either reorganize their surface constituents or synthesize new molecules and perhaps also modify the composition of the phagosome membrane (see above). The molecular mechanism by which mycobacteria affect the actin filament network are not known. Several molecules may be considered, such as the actin motor myosin-I that codistributes with organelles of the endocytic pathway,[85] actin-binding proteins such as ezrin and moesin that are important components of the actin assembly/disassembly machinery,[77] or RhoD that induces rearrangements of the actin filament network and plays an important role in endosome movement.[86]

Because actin depolymerization becomes established only from one day after infection onwards, this process cannot be considered as a mechanism to prevent early steps of phagosome maturation, but may play a role during later stages of infection. Because they behave like early endosomes, nonmaturing phagosomes continue to fuse and intermingle contents and membrane with early endosomes. An intact filament network appears to be required for this process. During early stages after infection of macrophages with *M. avium,* phagosomes acquire newly endocytosed tracer (horseradish peroxidase, HRP) through intermingling immediately,[84] as is typical for early endosomes[87-89] However, when the tracer is added to cells at later time points, at 1 or up to 15 days following infection, phagosomes in spite of remaining nonmatured (fusing with early endosomes, but not with lysosomes) do not acquire the endocytic tracer until about 10 to 20 minutes after it has been added to the cells.[15] In addition, interactions between early endosomes and *M. avium*-containing phagosomes appear increasingly restricted at these later times as bacteria multiply.[15] These effects can be simulated by using drugs that either inhibit actin polymerization (cytochalasin D), or that promote actin polymerization (jasplakinolide) in cases where actin has been depolymerized as a result of mycobacterial action.[84] In contrast to these observations, *M. avium*-induced depolymerization of actin filaments does not cause a delay in intermingling of cell surface-derived membrane marker between early endosomes and *M. avium*-containing phagosomes (results to be published). We propose that by disrupting the actin filament network, *M. avium* affects the mode of interaction between phagosomes and early endosomes in an unknown way that impairs efficient transfer of endocytic contents without preventing intermingling of those surface-derived membrane constituents that are specifically retained in the phagosome membrane at late stages after infection (see above). The importance of this phenomenon for *M. avium* survival awaits further investigation.

Summary

Endocytic processing of phagosomes forms an integral part of events along the endocytic pathway and operates with the same molecular machinery. Accordingly, newly formed phagosomes interact with early endosomes by fusion/fission events while they rapidly mature into a state where they are no longer able to fuse with early endosomes, but are now able to fuse with pre-existing lysosomes instead. Intra-phagosomal mycobacteria inhibit phagosome maturation and, as a result, phagosomes do not fuse with lysosomes. Because early endosomes and phagosomes mature within minutes, mycobacteria must have inhibitory mechanisms in place at very early stages after phagocytic uptake. Concepts for mechanism(s) underlying this inhibition are only beginning to emerge, based on the characterization of the mycobacterium-containing phagosome in terms of molecular components and general composition, functional comparison with early endosomes, and phagosome morphology. (i) Receptors involved during formation of phagosomes do not seem to have a direct influence on subsequent phagosome maturation. (ii) The early phagosome membrane displays a low abundance of the proton-pump ATPase , but this is unlikely to play a role because acidification remains similar as for maturing early endosomes, and it has not been shown whether inactivation of this pump affects maturation. (iii) The retention of the coat protein, TACO, on the cytoplasmic side of the phagosome membrane can play a role in preventing maturation because in host cells where TACO is absent mycobacteria cannot prevent fusion with lysosomes. (iv) The early endosome-specific GTPase, Rab5, is retained on the phagosome membrane, but no evidence is available that the presence of Rab5 causes the inhibition of phagosome maturation rather than merely reflecting the immature state of the phagosome. Because Rab5 does play a role in phagosome maturation, the regulation of Rab5 activity by known cellular factors could be a target for mycobacterial intervention. (v) Mycobacteria must keep the phagosome membrane closely apposed to their surface all round. Where this is not the case, as when clumps of bacteria reside together in one phagosome, which causes the membrane to span the region between adjacent mycobacteria, phagosomes are rapidly seen to start fusing with lysosomes. It is proposed that a close all round apposition prevents formation of tubular membrane domains and this slows down membrane recycling as part of the maturation process. This mechanism also results in slowing down maturation of phagosomes that contain latex beads with a hydrophobic surface which results in a closely apposed phagosome membrane. (vi) The protein composition of the mycobacteria-containing phagosome membrane does not appear substantially different from that of latex-bead containing phagosomes. However, regarding cell surface-derived glycoconjugates, mycobacteria-containing phagosomes are depleted about 3-fold as compared to early endosomes and latex bead-containing phagosomes. This depletion is established only several days after infection and cannot play a role in early prevention of phagosome maturation. (vii) In mycobacteria-infected macrophages, the actin filament network becomes depolymerized after about one day. This is too late to affect early phagosome maturation.

Acknowledgments

We would like to acknowledge financial support from the French Institut National de la Santé et de la Recherche Médicale, the Ministère de la Recherche et de la Technologie and the Fondation pour la Recherche Médicale to CC and the South African Foundation for Research and Development to LT for the experimental work cited from our laboratories. We thank Corinne Béziers-Guigue for help in preparing the manuscript.

References

1. Wolinsky E. Nontuberculous mycobacteria and associated diseases. Am Rev Respir Dis 1979; 119:107-159.
2. Collins FM. Mycobacterial disease, immunosuppression, and acquired immunodeficiency syndrome. Clin Microbiol Rev 1989; 2:360-377.
3. Bloom BR, Murray JL. Tuberculosis: Commentary on a reemergent killer. Science 1992; 257:1055-1064.
4. Heifets LB, Cangelosi GA. Drug susceptibility testing of Mycobacterium tuberculosis: A neglected problem at the turn of the century. Int J Tuberc Lung Dis 1999; 7:564-581.
5. Falkinham JO III. Epidemiology of infection by nontuberculous mycobacteria. Clin Microbiol Rev 1996; 9:177-215.
6. Armstrong JA, D'Arcy Hart P. Response of cultured macrophages to Mycobacterium tuberculosis with observations on fusion of lysosomes with phagosomes. J Exp Med 1971; 134:713-740.
7. Fréhel C, de Chastellier C, Lang T et al. Evidence for inhibition of fusion of lysosomal and prelysosomal compartments with phagosomes in macrophages infected with pathogenic Mycobacterium avium. Infect Immun 1986; 52:252-262.
8. Armstrong JA, D'Arcy Hart P. Phagosome-lysosome interactions in cultured macrophages infected with virulent tubercle bacilli. Reversal of the usual nonfusion pattern and observations on bacterial survival. J Exp Med 1975; 142:1-16.
9. Schaible UE, Sturgill-Koszycki S, Schlesinger PH et al. Cytokine activation leads to acidification and increases maturation of Mycobacterium avium-containing phagosomes in murine macrophages. J Immunol 1998; 160:1290-1296.
10. de Chastellier C, Thibon M, Rabinovitch M. Construction of chimeric phagosomes that shelter Mycobacterium avium and Coxiella burnettii (phase II) in doubly infected mouse macrophages: An ultrastructural study. Eur J Cell Biol 1999; 78:580-592.
11. Clemens DL, Horwitz MA. Characterization of the Mycobacterium tuberculosis phagosome and evidence that phagosomal maturation is inhibited. J Exp Med 1995; 181:257-270.
12. de Chastellier C, Thilo L. Phagosome maturation and fusion with lysosomes in relation to surface property and size of the phagocytic particle. Eur J Cell Biol 1997; 74:49-62.
13. Mellman I. Endocytosis and molecular sorting. Annu Rev Cell Dev Biol 1996; 12:575-625.
14. Pitt A, Mayorga LS, Schwartz AL et al. Transport of phagosomal components to an endosomal compartment. J Biol Chem 1992; 267:126-132.
15. de Chastellier C, Lang T, Thilo L. Phagocytic processing of the macrophage endoparasite, Mycobacterium avium, in comparison to phagosomes which contain Bacillus subtilis or latex beads. Eur J Cell Biol 1995; 68:167-182.
16. Sturgill-Koszycki S, Schaible NE, Russell DG. Mycobacterium-containing phagosomes are accessible to early endosomes and reflect a transitional state in normal phagosome biogenesis. EMBO J 1996; 15:6960-6968.
17. Clemens DL, Horwitz MA. The Mycobacterium tuberculosis phagosome interacts with early endosomes and is accessible to exogenously administered transferrin. J Exp Med 1996; 184:1349-1355.
18. Via LE, Deretic D, Ulmer RJ et al. Arrest of mycobacterial phagosome maturation is caused by a block in vesicle fusion between stages controlled by rab 5 and rab7. J Biol Chem 1997; 272:13326-13331.
19. Clemens DL. Characterization of the Mycobacterium tuberculosis phagosome. Trends Microbiol 1996; 4:113-118.
20. Russell DG, Sturgill-Koszycki S, Vanheyningen T et al. Why intracellular parasitism need not be a degrading experience for Mycobacterium. Phil Trans R Soc Lond B 1997; 352:1303-1310.
21. Deretic V, Fratti RA. Mycobacterium tuberculosis phagosome. Mol Microbiol 1999; 31:1603-1609.
22. Dunn KW, Maxfield FR. Delivery of ligands from sorting endosomes to late endosomes occurs by maturation of sorting endosomes. J Cell Biol 1992; 117:301-310.

23. Thilo L, Stroud E, Haylett T. Maturation of early endosomes and vesicular traffic to lysosomes in relation to membrane recycling. J Cell Sci 1995; 108:1791-1803.
24. Racoosin EL, Swanson JA. Macropinosome maturation and fusion with tubular lysosomes in macrophages. J Cell Biol 1993; 121:1011-1020.
25. van Deurs B, Holm PK, Kayser L et al. Delivery to lysosomes in the human carcinoma cell line HEp-2 involves an actin filament-facilitated fusion between mature endosomes and pre-existing lysosomes. Eur J Cell Biol 1995; 66:309-323.
26. Futter CE, Pearse A, Hewlett LJ et al. Multivesicular endosomes containing internalized EGF-EGF receptor complexes mature and then fuse directly with lysosomes. J Cell Biol 1996; 132:1011-1023.
27. Mullock BM, Perez JH, Kuwana T et al. Lysosomes can fuse with a late endosome compartment in a cell-free system from rat liver. J Cell Biol 1994; 126:1173-1182.
28. Pitt A, Mayorga LS, Stahl PD et al. Alterations in the protein composition of maturing phagosomes. J Clin Invest 1992; 90:1978-1983.
29. Oh YK, Swanson JA. Different fates of phagocytosed particles after delivery into macrophage lysosomes. J Cell Biol 1996; 132: 585-593.
30. Desjardins M. Biogenesis of phagolysosomes: the 'kiss and run' hypothesis. Trends Cell Biol 1995; 5:183-186.
31. Desjardins M, Huber LA, Parton RG et al. Biogenesis of phagolysosomes proceeds through a sequential series of interactions with the endocytic apparatus. J Cell Biol 1994; 124: 677-688.
32. Zimmerli S, Edwards S, Ernst JD. Selective receptor blockade during phagocytosis does not alter the survival and growth of Mycobacterium tuberculosis in human macrophages. Am J Respir Cell Mol Biol 1996; 15:760-770.
33. Ernst JD. Macrophage receptors for Mycobacterium tuberculosis. Infect Immun 1998; 66:1277-1281.
34. Aderem A, Underhill DM. Mechanisms of phagocytosis in macrophages. Annu Rev Immunol 1999; 17:593-623.
35. Mellman I, Fuchs R, Helenius A. Acidification of the endocytic and exocytic pathways. Annu Rev Biochem 1986; 55:663-700.
36. Hackam DJ, Rotstein OD, Zhang W et al. Regulation of phagosomal acidification. J Biol Chem 1997; 272:29810-29820.
37. Yamashiro DJ, Maxfield FR. Kinetics of endosome acidification in mutant and wild-type Chinese hamster ovary cells. J Cell Biol 1987; 105:2713-2721.
38. Yamashiro DJ, Maxfield FR. Regulation of endocytic processes by pH. Trends Pharmacol Sci 1988; 9:190-193.
39. Claque MJ, Urbe S, Aniento F et al. Vacuolar ATPase activity is required for endosomal carrier vesicle formation. J Biol Chem 1994; 269:21-24.
40. van Weert AWM, Dunn KW, Geuze HJ et al. Transport from late endosomes to lysosomes, but not sorting of integral membrane proteins in endosomes, depends on the vacuolar proton pump. J Cell Biol 1995; 130: 821-834.
41. Crowle AJ, Dahl R, Ross E et al. Evidence that vesicles containing living, virulent Mycobacterium tuberculosis or Mycobacterium avium in cultured human macrophages are not acidic. Infect Immun 1991; 59:1823-31.
42. Sturgill-Koszycki S, Schlesinger PH, Chakraborty P et al. Lack of acidification in Mycobacterium phagosomes produced by exclusion of the vesicular proton-ATPase. Science 1994; 263:678-681.
43. Oh YK, Straubinger RM. Intracellular fate of Mycobacterium avium: Use of dual-label spectrofluorometry to investigate the influence of bacterial viability and opsonization on phagosomal pH and phagosome-lysosome interaction. Infect Immun 1996; 64:319-325.
44. Hackam DJ, Rotstein OD, Zhang W et al. Host resistance to intracellular infection: Mutation of natural resistance-associated macrophage protein 1 (Nramp1) impairs phagosomal acidification. J Exp Med 1998; 188:351-364.
45. Holtzman E. Lysosomes. In: Siekevitz P, ed. Cellular Organelles. New York: Plenum Press, 1989.
46. Geisow MJ, D'Arcy Hart P, Young MR. Temporal changes of lysosome and phagosome pH during phagolysosome formation in macrophages: studies by fluorescence spectroscopy. J Cell Biol 1981; 89:645-652.
47. Gordon AH, Hart PD, Young MR. Ammonia inhibits phagosome-lysosome fusion in macrophages. Nature 1980; 286:79-80.
48. Zerial M, Stenmark H. Rab GTPases in vesicular transport. Curr Opin Cell Biol 1993; 5:613-620.
49. Mills IG, Jones AT, Clague MJ. Regulation of endosome fusion. Mol Membr Biol 1999; 16:73-79.
50. Mohrmann K, van der Sluijs P. Regulation of membrane transport through the endocytic pathway by rab GTPases. Mol Membr Biol 1999; 16:81-87.
51. Somsel Rodman J, Wandinger-Ness A. Rab GTPases coordinate endocytosis. J Cell Sci 2000; 113:183-192.

52. Armstrong J. How do Rab proteins function in membrane traffic? Int J Biochem Cell Biol 2000; 32:303-307.
53. Gonzales Jr L, Scheller RH. Regulation of membrane trafficking: structural insights from a Rab/ Effector complex. Cell 1999; 96:755-758.
54. Bucci C, Parton RG, Mather IH et al. The small GTPase rab5 functions as a regulatory factor in the early endocytic pathway. Cell 1992; 70:715-728.
55. Stenmark H, Parton RG, Steele-Mortimer O et al. Inhibition of rab5 ATPase activity stimulates membrane fusion in endocytosis. EMBO J 1994; 13:1287-1296.
56. Li G, Barbieri MA, Colombo MI et al. Structural features of the GTP-binding defective rab5 mutants required for their inhibitory activity on endosome fusion. J Biol Chem 1994; 269:14631-14635.
57. Bucci C, Lutcke A, Steele-Mortimer O et al. Cooperative regulation of endocytosis by three Rab5 isoforms. FEBS Lett 1995; 366:65-71.
58. Chiariello M, Bruni CB, Bucci C. The small GTPases Rab5a, Rab5b and Rab5c are differentially phosphorylated in vitro. FEBS Lett 1999; 453:20-24.
59. Roberts RL, Barbieri MA, Ullrich J et al. Dynamics of rab5 activation in endocytosis and phagocytosis. J Leukoc Biol 2000; 68:627-632.
60. Bottger G, Nagelkerken B, van der Sluijs P. Rab4 and Rab7 define distinct nonoverlapping endosomal compartments. J Biol Chem 1996; 271:29191-29197.
61. Clemens DL, Lee BY, Horwitz MA. Deviant expression of Rab5 on phagosomes containing the intracellular pathogens Mycobacterium tuberculosis and Legionella pneumophila is associated with altered phagosomal fate. Infect Immun 2000; 68:2671-2684.
62. Clemens DL, Lee BY, Horwitz MA. Mycobacterium tuberculosis and Legionella pneumophila phagosomes exhibit arrested maturation despite acquisition of Rab7. Infect Immun 2000; 68:5154-5166.
63. Alvarez-Dominguez C, Stahl PD. Increased expression of Rab5a correlates directly with accelerated maturation of Listeria monocytogenes phagosomes. J Biol Chem 1999; 274:11459-11462.
64. Duclos S, Diez R, Garin J et al. Rab5 regulates the kiss and run fusion between phagosomes and endosomes and the acquisition of phagosome leishmanicidal properties in RAW 264.7 macrophages. J Cell Sci 2000; 113:3531-3541.
65. Xiao GH, Shoarinejad F, Jin F et al. The tuberous sclerosis 2 gene product, tuberin, functions as a Rab5 GTPase activating protein (GAP) in modulating endocytosis. J Biol Chem 1997; 272:6097-6100.
66. Griffiths G, Back R, Marsh M. A quantitative analysis of the endocytic pathway in baby hamster kidney cells. J Cell Biol 1989; 109:2703-2720.
67. Geuze HJ, Slot JW, Schwartz AL. Membranes of sorting organelles display lateral heterogeneity in receptor distribution. J Cell Biol 1987; 104:1715-1723.
68. Whitney JA, Gomez M, Sheff D et al. Cytoplasmic coat proteins involved in endosome function. Cell 1995; 83:703-713.
69. Daro E, Sheff D, Gomez M et al. Inhibition of endosome function in CHO cells bearing a temperature-sensitive defect in the coatomer (COPI) component epsilon-COP. J Cell Biol 1997; 139:1747-1759.
70. Gu F, Aniento F, Parton RG et al. Functional dissection of COP-I subunits in the biogenesis of multivesicular endosomes. J Cell Biol 1997; 139:1183-1189.
71. Gu F, Gruenberg J. Biogenesis of transport intermediates in the endocytic pathway. FEBS Lett 1999; 452:61-66.
72. Botelho RJ, Hackam DJ, Schreiber AD et al. Role of COPI in phagosome maturation. J Biol Chem 2000; 275:15717-15727.
73. Ferrari G, Langen H, Naito M et al. A coat protein on phagosomes involved in the intracellular survival of mycobacteria. Cell 1999; 97:435-447.
74a. de Chastellier C, Thilo L. Pathogenic Mycobacterium avium remodels the phagosome membrane in macrophages within days after infection. Eur J Cell Biol 2002; 81:17-25.
74b. Sturgill-Koszycki S, Haddix PL, Russell DG. The interaction between Mycobacterium and the macrophage analyzed by two dimensional polyacrylamide gel electrophoresis. Electrophoresis 1997; 18:2558-2565.
75. Cole NB, Lippincott-Schwartz J. Organization of organelles and membrane traffic by microtubules. Curr Opin Cell Biol 1995; 7:55-64.
76. Gruenberg J, Maxfield FR. Membrane transport in the endocytic pathway. Curr Opin Cell Biol 1995; 7:552-563.
77. Defacque H, Egeberg M, Habermann A et al. Involvement of ezrin/moesin in de novo actin assembly on phagosomal membranes. EMBO J 2000; 19:199-212.

78. Desjardins M, Celis JE, van Meer G et al. Molecular characterization of phagosomes. J Biol Chem 1994; 269:32194-84.
79. Guérin I, de Chastellier C. Pathogenic mycobacteria disrupt the macrophage actin filament network. Infect Immun 2000; 68:2655-2662.
80. Gottlieb TA, Ivanov IE, Adesnik M et al. Actin microfilaments play a critical role in endocytosis at the apical but not the basolateral surface of polarized epithelial cells. J Cell Biol 1993; 120:695-710.
81. Durrbach A, Louvard D, Coudrier E. Actin filaments facilitate two steps of endocytosis. J Cell Sci 1996; 109:457-465.
82. Higley S, Way M. Actin and cell pathogenesis. Curr Opin Cell Biol 1997; 9:62-69.
83. Dramsi S, Cossart P. Intracellular pathogens and the actin cytoskeleton. Annu Rev Cell Dev Biol 1998; 14:137-166.
84. Guérin I, de Chastellier C. Disruption of the actin filament network affects delivery of endocytic contents to phagosomes with early endosome characteristics: the case of phagosomes with pathogenic mycobacteria. Eur J Cell Biol 2000; 79:735-749.
85. Durrbach A, Collins K, Matsudaira P et al. Brush border myosin-I truncated in the motor domain impairs the distribution and the function of endocytic compartments in a hepatoma cell line. Proc Nat Acad Sci (USA) 1996; 93:7053-7058.
86. Murphy C, Saffrich R, Grummt M et al. Endosome dynamics regulated bya Rho protein. Nature 1996; 384:427-432.
87. de Chastellier C, Lang T, Ryter A et al. Exchange kinetics and composition of endocytic membranes in terms of plasma membrane constituents: a morphometric study in macrophages. Eur J Cell Biol 1987; 44:112-123.
88. Steinman RM, Brodie SE, Cohn Z.A. Membrane flow during endocytosis. A stereological analysis. J Cell Biol 1976; 68:665-687.
89. Ward DM, Ajioka R, Kaplan J. Cohort movement of different ligands and receptors in the intracellular endocytic pathway of alveolar macrophages. J Biol Chem 1989; 264:8164-8170.

Chapter 11

Molecular Mechanisms Regulating Membrane Traffic in Macrophages:

Lessons from the Intracellular Pathogen Mycobacterium spp.

Jean Pieters

Material that is engulfed by macrophages through the process of phagocytosis is usually delivered to late endosomal and lysosomal organelles in order to be degraded. The molecular requirements involved in the transfer of material through the phagosomal, endosomal and lysosomal pathways are beginning to be unraveled. Recently, several model systems investigating the interaction of pathogenic microbes with macrophages have contributed significantly to our knowledge on these processes. This Chapter will highlight recent insights into the biology of macrophage membrane trafficking processes obtained by analyzing its interaction with pathogenic mycobacteria. Mycobacteria have co-evolved with their hosts for millions of years and therefore have been able to develop survival mechanisms by hijacking normal cellular processes. Understanding these processes will not only help to understand mycobacterial pathogenicity, but also appear to delineate certain concepts previously unappreciated.

The Endocytic Pathway

Uptake of material by macrophages can be achieved by three mechanisms: receptor mediated endocytosis, fluid phase endocytosis as well as phagocytosis. Receptor mediated uptake and fluid phase endocytosis are quite similar; the main difference being that receptor mediated endocytosis is highly selective, whereas fluid phase uptake is not. In addition, after the initial uptake phase, both routes concentrate the cargo (or receptor-ligand complexes) into early endosomes.

Receptor mediated uptake processes are now relatively well understood. These are regulated by the coordinated assembly and disassembly of different coat proteins which include clathrin, a trimeric scaffold protein that can self-assemble in cage-like lattices.[18,49,57] Clathrin coat formation is thought to be initiated by the assembly of "adaptor proteins" at the plasma membrane,[42] followed by clathrin binding to assembled adaptor protein complexes. Fission of the vesicle from the plasma membrane might be regulated by the GTPase dynamin and results in the generation of a clathrin coated, ~50 nm vesicle.[49,52] Clathrin subsequently dissociates from these vesicles by a process termed uncoating, and these uncoated vesicles can then fuse with other endosomal organelles to deliver their cargo to late stages of the endosomal pathway, sequentially reaching early endosomes, late endosomes and lysosomes where cargo can be degraded by lysosomal hydrolyses (see Fig. 1).

Whereas clathrin is important for receptor-mediated internalization, the molecular mechanisms involved in non-receptor mediated internalization, or fluid phase pinocytosis, are less

Intracellular Pathogens in Membrane Interactions and Vacuole Biogenesis, edited by Jean-Pierre Gorvel.

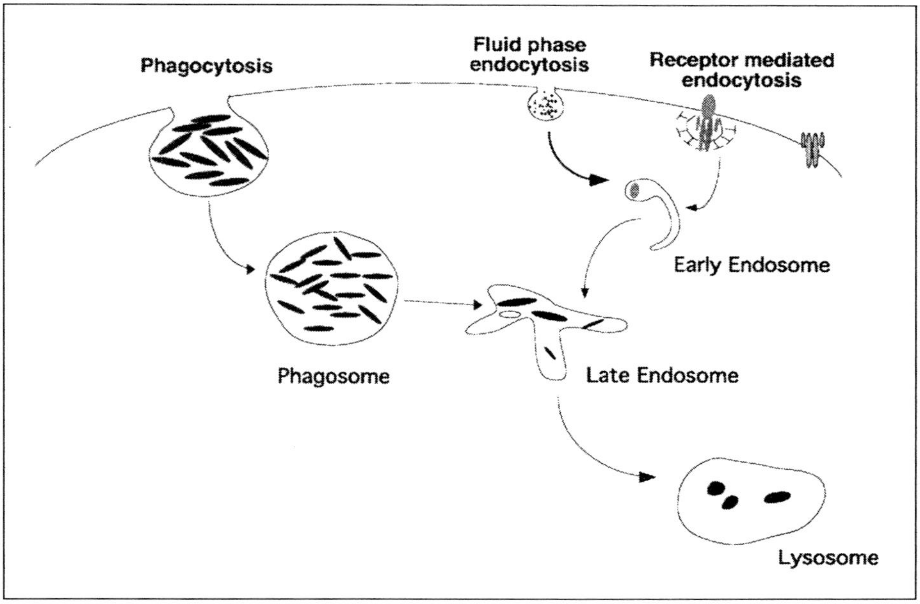

Figure 1. Endosomal and phagosomal pathways.
Virtually all cells can internalize fluids as well as soluble molecules via fluid phase endocytosis, whereas a more specific way to internalize molecules occurs through receptor mediated endocytosis. After internalization, material is transported along the endosomal/lysosomal pathway to lysosomes where it can be degraded. Particulate material such as viable and killed bacteria are taken up by phagocytosis, a process that can occur in macrophages, dendritic cells and neutrophils. Following formation of the phagosome, cargo is delivered to lysosomes for destruction.

well defined. Emerging evidence suggests that such uptake occurs in a clathrin-independent manner.[56] It is however likely that all membrane fission and fusion processes are somehow governed by similar principles that are illustrated by the role for clathrin in delivering material to the endocytic pathway: a regulated assembly into a coat structure that helps to deform the membrane in such a way that it can pinch off, followed by release of this coat structure from the membrane and assembly of a fusiogenic complex.[35,62]

The Phagosomal Pathway

Phagocytosis is a specialized form of endocytosis in which relatively large particles, including intact microorganisms, are internalized into large (>250 nm) vesicles (see also Fig. 1). In contrast to the other forms of endocytosis mentioned above, the capacity to phagocytose is restricted to certain cell types only such as macrophages, neutrophiles and dendritic cells. Phagocytosis is usually mediated by cell surface receptors, that bind directly to determinants on the particles to be ingested. These receptors include Fc receptors, scavenger receptor, mannose receptors and complement receptors.[10,19,22,23] Different receptors are being used for different microorganisms and particles to become phagocytosed. Interestingly, the precise mode of uptake may contribute to the fate of the internalized microorganisms: entry via Fc or mannose receptors causes signal transduction events leading to the activation of the phagocyte, thereby causing a rapid killing of the internalized microbe.[25,27] In contrast, when the same microbes are internalized e.g., via complement receptors, the phagocytes are not activated and the microorganisms remain viable inside the macrophage.[6,8,28]

After binding, the particle is internalized by invagination of the plasma membrane. How phagosome formation is achieved remains largely unknown, although a number of signal transduction events have been implicated.[1,28]

Importantly, all endocytic pathways converge at the late endosome, from where the phagocytosed material is being shuttled into lysosomes for degradation (Fig. 1).[43] While the later stages of endocytosis and phagocytosis may be regulated by the same mechanisms, the early stages are likely to be quite distinct from one another.

The above mentioned routes of how material is transferred from the extracellular environment into degradative lysosome seems rather simple. However, as is clear from the model depicted in Figure 1, this trafficking prompts a multitude of questions, ranging from how large particles are internalized at the cell surface, to what the molecular mechanisms involved in the fission and fusion processes along the phagosomal/lysosomal pathway precisely are.

A great deal of insight has come from reconstituting some of the events in in vitro systems, that are highly amenable to manipulation allowing the addition as well as depletion of different cellular compounds. From several Chapters in this book (Chapters 1 and 2) one can appreciate the power of such reconstitution systems. On the other hand, a complementary approach has relied on the analysis of the interaction of pathogenic mycobacteria with host macrophages as a model system for the study of phagocytosis. The remainder of this Chapter will describe how these studies have contributed to our understanding of some of the mechanisms involved in membrane trafficking in macrophages.

The Model System: Interaction of Pathogenic Mycobacteria with Macrophages

Pathogenic mycobacteria include some of the most notorious pathogens known today such as *Mycobacterium tuberculosis* and *Mycobacterium leprae*. Although the interest in mycobacteria as infectious organisms has declined over the past decades, as these microbes became relatively well amenable to antibiotic treatment, *M. tuberculosis,* continues to cause disease in ~8 million people each year resulting in someone to die every 10 seconds (see http://www.stoptb.org/tuberculosis/#facts.html). Most frightening, development of resistance to various anti-mycobacterial drugs is currently contributing to a dramatic increase in the incidence of mycobacterial diseases such as tuberculosis and leprae.[7,38] In addition, some of the less virulent mycobacteria *(M. avium, M. intracellulare, M. bovis* BCG) can become a major health problem for immunocompromised individuals such as HIV infected persons.[33]

The key to the persistence and therefore pathogenicity of mycobacteria is believed to be caused by their capacity to avoid delivery to lysosomes. Indeed, while normally bacteria that are internalized by macrophages through phagocytosis are rapidly destroyed in lysosomes, mycobacteria remain within so-called mycobacterial phagosomes. These mycobacterial phagosomes fail to fuse with lysosomes, and it has been known for a long time that this capacity to avoid lysosomal degradation is dependent on processes governed by viable mycobacteria, as killed mycobacteria are rapidly transferred to lysosomes,[3,32] suggesting that mycobacterial products manage to interfere with the normal trafficking routes after phagocytosis. In addition, by remaining within macrophage phagosomes these bacteria can avoid the generation of antigenic peptides in endosomal organelles,[58,44] thereby severely compromising a normal immuneresponse.[41]

This capacity to withstand lysosomal delivery is central to the persistence of pathogenic mycobacteria, allowing them to survive for prolonged times within macrophages. The survival strategies developed by pathogenic mycobacteria are likely to be unique among pathogenic bacteria. First, mycobacteria have a very restricted host range, which is especially true for *Mycobacterium tuberculosis,* by far the most virulent of the different mycobacteria. As a result, *M. tuberculosis* has most probably evolved survival strategies that are shaped by its long standing habitat, the human macrophage. Second, although the mycobacterial phagosome might simply represent a (slightly modified) plasma membrane,[32] no other microbes have been observed to reside within a similar vacuole.

Mycobacteria have been around for very long periods of time and therefore have had ample time to adjust to their host cells. Although it is becoming more and more apparent that in general, pathogenic microbes have evolved to make use of normal cellular pathways, which are being hijacked and turned into their own advantage, pathogenic mycobacteria seem to be extremely successful in implementing such strategies. In principle, a pathogen can interfere with any of the mechanisms that are normally involved in their own clearance, which includes the process of uptake as well as the mechanisms that govern the transit of the pathogen to lysosomal organelles.

Point of Entry: Crucial Role for Plasma Membrane Cholesterol

Uptake of mycobacteria, as for most microbes in fact, can be achieved by multiple receptor molecules. Widely used receptors for mycobacteria are the complement receptors, for both opsonized and non-opsonized entry.[10,48,50] However, also mannose receptors that bind glycosylated structures on the bacterial surface,[47] Fc receptors that can internalize IgG-opsonized bacteria[2] and scavenger receptors[21,65] have been implicated in mycobacterial uptake. Interestingly, whereas in case of Fc receptor ligation, the small GTPases Cdc42 and Rac are activated which are known to induce JNK and p38 MAP kinase activity that contribute to inflammatory responses such as macrophage activation,[15,37] triggering complement receptors leads to the activation of Rho but not Rac and Cdc42, and therefore bactericidal mechanisms are not upregulated.[9,11,16] Thus, mycobacterial uptake via complement receptors ensures that the mycobacteria enter a macrophage under relatively hospitable conditions.

Quite serendipitously, it was established that entry into macrophages in a complement receptor dependent manner also requires the presence of the plasma membrane steroid cholesterol. Cholesterol is thought to be a structural component of cellular membranes, and strongly accumulates at the site of mycobacterial uptake (see Fig. 2). When macrophage plasma membranes were however depleted of cholesterol, it appeared that virtually all cellular functions were unaffected, but that these depleted macrophages were rendered unable to internalize mycobacteria[26] (see also video sequences at www.bii.ch/pieters/movies.html). This inhibition of uptake is specific, as other microorganisms can still enter cholesterol depleted macrophages.

The precise reason for the inability of cholesterol depleted macrophages to phagocytose mycobacteria remains unknown. It is possible that the extremely glycolipid rich mycobacterial cell wall contains components that interact directly with cholesterol. In accordance with this, mycobacteria specifically bind to cholesterol, indicating the presence of a high affinity cholesterol binding site at the mycobacterial cell surface.[26] The identification of the mycobacterial cell wall constituents responsible for cholesterol binding might lead to strategies to interfere with cholesterol-mediated mycobacterial entry. Alternatively, cholesterol provides an essential 'platform' for mediating the signaling events that accompany phagocytosis of mycobacteria.[27] It is indeed conceivable that signals have to be transduced locally to mediate the quite dramatic deformation of the plasma membrane at the site where the microbes enter. Cholesterol may play an essential role in transmitting or coordinating the signals necessary for mycobacterial uptake.[10,53] The selectivity of the cholesterol requirement for mycobacteria is also an indication of the unique manner in which mycobacteria have evolved to interact with their mammalian host cells.

Survival Inside the Macrophage: How to Avoid Lysosomal Delivery

As the persistence of mycobacteria inside the host organisms is due to their capacity to inhibit their delivery to lysosomes, the key to understanding mycobacterial pathology lies in delineating the mechanisms of inhibition of phagosome-lysosome fusion processes.

It has been known for a long time that this inhibition of lysosomal delivery was dependent on processes excerpted by the mycobacteria themselves.[2,3] In the early seventies, it was shown using electron microscopy, that while killed mycobacteria were readily found in lysosomes, living mycobacteria are present in vacuoles that are morphologically distinct from lysosomes,

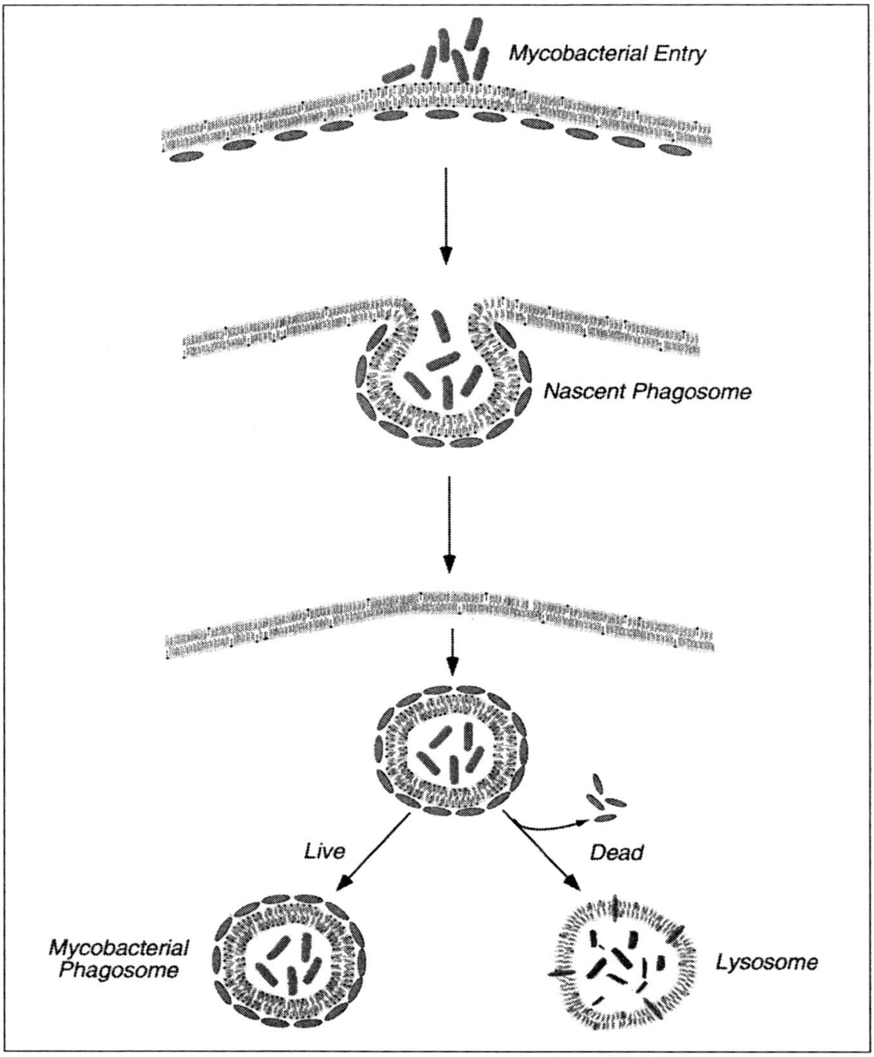

Figure 2. Working model for TACO action.
Mycobacteria (*purple*) enter the macrophage via cholesterol-enriched domains present in the plasma membrane. During phagocytosis, TACO (*yellow*), which also binds via cholesterol, is recruited around the nascent phagosome. Living mycobacteria have gained the capacity to retain TACO at the phagosomal membrane thereby preventing fusion of the phagosome with lysosomes. Killed mycobacteria cannot retain TACO at the phagosomal membrane and are therefore transported to lysosomes. Thus, entry of mycobacteria via cholesterol-enriched domains in the plasma membrane ensures their subsequent delivery to TACO coated phagosomes, thereby allowing these bacteria to survive within macrophages Modified from ref. 45.

which were called mycobacterial phagosomes.[3] These experiments have been confirmed and extended using more modern techniques, and it is firmly established that living mycobacteria, once phagocytosed by macrophages reside within organelles that contain markers of the plasma membrane as well as of early, but not of late endosomal compartments.[4,12-14,31,32,63]

In addition, most markers of late endosomes/lysosomes are missing from the mycobacterial phagosome. Mycobacterial phagosomes also retain the early endosomal marker rab 5 and do not acquire rab 7 (see Chapter 1).[60] Rab7 is a member of the small GTPases that localizes to late endosomes and is believed to be involved in the regulation of membrane traffic between early and late endosomes.[46,64]

The characteristic limited acidification of mycobacterial phagosomes is an additional indication that mycobacterial phagosomes lack interaction with late endosomal/lysosomal organelles.[17,40,51] Material that is transported along the endosomal/lysosomal pathway usually encounters an increased acidified milieu due to the enrichment of V-ATPases in late endosomes/lysosomes.[36] This lack of acidification may be related to a reduced concentration of vacuolar proton-ATPase in the phagosomes,[55] although the activity of this pump can be detected in the phagosomal membrane and also contributes to the limited acidification observed.[29,30] It should be noted, however, that the macrophages plasma membrane also contains V-ATPases which might become co internalized during mycobacterial entry.[54]

Thus, the mycobacterial phagosome seems to contain virtually all markers of the plasma membrane and some of early endosomes. One feature that definitely distinguishes it from an early endosomal organelle is in fact its size. The latter may be a direct result from the way it is being generated; rather than having clathrin-based structures coordinating the invagination of a defined plasma membrane area, the mycobacterial population that is being internalized defines the size of the resulting vacuole.

Role of the Coat Protein TACO for Mycobacterial Survival

Understanding how pathogenic mycobacteria manage to interfere with the normal phagocytic transport routes has in fact proven to also shed some light on the molecular mechanisms involved in phagocytosis. How can we get a clue towards the mechanisms of how living mycobacteria inhibit the normal trafficking processes? A biochemical approach to this problem has given some more insight. By analyzing the different macrophage proteins present in phagosomes containing living mycobacteria, a molecule termed TACO was identified that was absent from phagosomes harboring killed bacilli. TACO was not present in any of the endosomal/lysosomal organelles purified from uninfected cells.[24] Further analysis showed that TACO is actively retained at the mycobacterial phagosome and thereby prevents maturation into or fusion with lysosomes, allowing the mycobacteria to survive within the phagosome (see Fig. 2).

A role for TACO in mycobacterial survival in vivo is suggested by the finding that the major clearance site for mycobacterial infections, the liver, lacks TACO. Usually, infectious organisms are cleared in the liver by the Kupffer cells.[39,61] These liver macrophages lack the expression of TACO and accordingly are able to completely eradicate the mycobacteria in phagolysosomes.[24] Thus, it seems that whereas at the single cell level the mycobacteria have prevailed in that they efficiently circumvent the macrophage defense mechanism, at the level of the whole organism evolutionary pressure may have ensured the availability of an efficient clearance site, namely the liver macrophages.

What is the mechanism involved in TACO mediated inhibition of lysosomal delivery? As presently the normal cellular function of TACO remains unknown we can only speculate. Being a coat protein of the plasma membrane as well as phagosomes (as indicated by its location observed by immunofluorescence microscopy, TACO might play a role in defining the plasma membrane and therefore also the phagosomal membrane. Once material has been phagocytosed, TACO would have to be released from the membrane in order to allow the fission as well as fusion machinery to be recruited onto the membrane. It is precisely at this latter step that mycobacteria have managed to interfere.[24] To avoid lysosomal delivery, mycobacteria actively retain TACO on the phagosomal membrane, and possibly merely try to mimic the plasma membrane.[32]

A major question remains why mammalian cells have retained a molecule actively used by pathogenic mycobacteria for their own advantage. The most likely explanation is that TACO performs an important, as yet unknown function in the uninfected cells. It is interesting to note that TACO shows homology to an actin-binding protein identified in *Dictyostelium discoideum*, termed coronin.[20] When knocked out in *D. discoideum*, the resulting organisms showed a reduction in cytokinesis, phagocytosis and cell motility.[20,34] Although TACO does not interact with the actin cytoskeleton (our unpublished results), and in mammalian cells does not play a role in phagocytosis per se,[24] the homology of TACO with coronin might indicate a function in cell locomotion. Indeed, the processes involved in cell locomotion and phagocytosis are quite similar,[5,59] and TACO may be a crucial component of the cytoskeleton of highly motile cells. As such, it may function both in the invagination of large pieces of plasma membrane as well as in the formation of protrusions of the plasma membrane that are typically generated during cell locomotion.[59]

Conclusion

The past few years have seen a re-appreciation of interdisciplinary research, especially in the field of cell biology. The analysis of the mechanisms of pathogenicity in a variety of microbes has led to the realization that very often, all these microbes do is to try to interfere or hijack the normal cellular processes. Not only do these investigations contribute to a better understanding of microbial pathogenicity, but in addition, these studies teach us a great deal about normal cellular functions. Most importantly it may uncover previously unappreciated mechanisms, in a similar manner as certain drugs have done in the past. The here described role of the widely known component cholesterol to provide a site of entry for pathogenic mycobacteria to enter a highly shielded vacuole within the macrophage is a prime example of the usefulness of pathogens-host interaction as model systems to study cell biology.

Acknowledgements

I thank my collegues Giorgio Ferrari and John Gatfield for continued support and input over the past few years, and Wolfgang Philipp for critical evaluation of this manuscript. The Basel Institute for Immunology was founded and supported by Fa. Hoffmann-La Roche Ltd., Basel, Switzerland. Work in my laboratory is funded by the Swiss National Science Foundation (SNF), the World Health Organization (WHO) and the Maeyenfisch Stiftung.

References

1. Araki N, Johnson MT, Swanson JA. A role for phosphoinositide 3-kinase in the completion of macropinocytosis and phagocytosis by macrophages. J Cell Biol 1996; 135:1249-1260.
2. Armstrong JA, Hart PD. Phagosome-lysosome interactions in cultured macrophages infected with virulent tubercle bacilli. Reversal of the usual nonfusion pattern and observations on bacterial survival. J Exp Med 1975; 142:1-16.
3. Armstrong JA, Hart PDA. Response of cultured macrophages to Mycobacterium tuberculosis, with observations on fusion of lysosomes with phagosomes. J Exp Med 1971; 134:713-740.
4. Barker LP, George KM, Falkow S et al. Differential trafficking of live and dead Mycobacterium marinum organisms in macrophages. Infect Immun 1997; 65:1497-1504.
5. Berlin RD, Oliver JM, Walter RJ. Surface functions during Mitosis I: Phagocytosis, pinocytosis and mobility of surface-bound Con A. Cell 1978; 15:327-341.
6. Berton G, Laudanna C, Sorio C et al. Generation of signals activating neutrophil functions by leukocyte integrins: LFA-1 and gp150/95, but not CR3, are able to stimulate the respiratory burst of human neutrophils. J Cell Biol 1992; 116:1007-1017.
7. Bloom BR. Tuberculosis. Back to a frightening future. Nature 1992; 358:538-539.
8. Blystone SD, Brown EJ. Integrin receptors of phagocytes. In: Gordon S, ed. Phagocytosis and Pathogens: The Host. Stamford: JAI Press Inc., 1999:103-147.
9. Bokoch G. Regulation of the phagocyte respiratory burst by small GTP binding proteins. Trends Cell Biol 1995; 5:109-113.
10. Brown DA, London E. Structure and function of sphingolipid- and cholesterol-rich membrane rafts. J Biol Chem 2000; 275:17221-17224.
11. Brown EJ. Complement receptors and phagocytosis. Curr Opin Immunol 1991; 3:76-82.
12. Caron E, Hall A. Identification of two distinct mechanisms of phagocytosis controlled by different Rho GTPases. Science 1998; 282:1717-1721.

13. Clemens DL, Horwitz MA. Hypoexpression of major histocompatibility complex molecules on Legionella pneumophila phagosomes and phagolysosomes. Infect Immun 1993; 61:2803-2812.
14. Clemens DL, Horwitz MA. Characterization of the Mycobacterium tuberculosis phagosome and evidence that phagosomal maturation is inhibited. J Exp Med 1995; 181:257-270.
15. Clemens DL, Horwitz MA. The Mycobacterium tuberculosis phagosome interacts with early endosomes and is accessible to exogenously administered transferrin. J Exp Med 1996; 184:1349-1355.
16. Coso OA, Chiariello M, Yu JC et al. The small GTP-binding proteins Rac1 and Cdc42 regulate the activity of the JNK/SAPK signaling pathway. Cell 1995; 81:1137-1146.
17. Cox D, Chang P, Zhang Q et al. Requirements for both Rac1 and Cdc42 in membrane ruffling and phagocytosis in leukocytes. J Exp Med 1997; 186:1487-1494.
18. Crowle AJ, Dahl R, Ross E et al. Evidence that vesicles containing living, virulent Mycobacterium tuberculosis or Mycobacterium avium in cultured human macrophages are not acidic. Infect Immun 1991; 59:1823-1831.
19. Crowther RA, Pearse BM. Assembly and packing of clathrin into coats. J Cell Biol 1981; 91:790-797.
20. Daeron M. Fc receptor biology. Annu Rev Immunol 1997; 15:203-234.
21. de Hostos EL, Bradtke B, Lottspeich F et al. Coronin, an actin binding protein of Dictyostelium discoideum localized to cell surface projections, has sequence similarities to G protein beta subunits. Embo J 1991; 10:4097-4104.
22. de Hostos EL, Rehfuess C, Bradtke B et al. Dictyostelium mutants lacking the cytoskeletal protein coronin are defective in cytokinesis and cell motility. J Cell Biol 1993; 120:163-173.
23. Ernst JD. Macrophage receptors for Mycobacterium tuberculosis. Infect Immun 1998; 66:1277-1281.
24. Ezekowitz RA, Williams DJ, Koziel H et al. Uptake of Pneumocystis carinii mediated by the macrophage mannose receptor. Nature 1991; 35:155-158.
25. Fanger NA, Voigtlaender D, Liu C et al. Characterization of expression, cytokine regulation, and effector function of the high affinity IgG receptor Fc gamma RI (CD64) expressed on human blood dendritic cells, J Immunol 1997; 158:3090-3098.
26. Ferrari G, Naito M, Langen H et al. A coat protein on phagosomes involved in the intracellular survival of mycobacteria. Cell 1999; 97:435-447.
27. Fraser IP, Ezekowitz RAB. Mannose receptor and phagocytosis. In: Gordon S, ed. Phagocytosis: The Host. Stamford: JAI Press, Inc, 1999:87-101.
28. Gatfield J, Pieters J. Essential role for cholesterol in entry of mycobacteria in macrophages. Science 2000; 288:1647-1650.
29. Greenberg S. Signal transduction of phagocytosis. Trends Cell Biol 1995; 5:93-99.
30. Greenberg S. Fc receptor mediated phagocytosis. In: Gordon S, ed. Phagocytosis: The Host. Stamford: JAI Press, Inc., 1999:149-191.
31. Gudewicz PW, Beezhold DH, Van Alten P et al. Lack of stimulation of post-phagocytic metabolic activities of polymorphonuclear leukocytes by fibronectin opsonized particles. J Reticuloendothel Soc 1982; 32:143-54.
32. Hackam DJ, Rotstein OD, Zhang W et al. Host resistance to intracellular infection: Mutation of natural resistance-associated macrophage protein 1 (Nramp1) impairs phagosomal acidification. J Exp Med 1998; 188:351-364.
33. Hackam DJ, Rotstein OD, Zhang WJ et al. Regulation of phagosomal acidification. Differential targeting of Na+/H+ exchangers, Na+/K+-ATPases, and vacuolar-type H+-atpases. J Biol Chem 1997; 272:29810-29820.
34. Hasan Z, Pieters J. Subcellular fractionation by organelle electrophoresis: Separation of phagosomes containing heat-killed yeast particles. Electrophoresis 1998; 19:1179-1184.
35. Hasan Z, Schlax C, Kuhn L et al. Isolation and characterization of the mycobacterial phagosome: Segregation from the endosomal/lysosomal pathway. Mol Microbiol 1997; 24:545-553.
36. Kaufmann SH. Is the development of a new tuberculosis vaccine possible? [In Process Citation]. Nat Med 2000; 6:955-960.
37. Maniak M, Rauchenberger R, Albrecht R et al. Coronin involved in phagocytosis: Dynamics of particle-induced relocalization visualized by a green fluorescent protein Tag. Cell 1995; 83:915-924.
38. McNew JA, Parlati F, Fukuda R et al. Compartmental specificity of cellular membrane fusion encoded in SNARE proteins [see comments]. Nature 2000; 407:153-159.
39. Mellman I, Fuchs R, Helenius A. Acidification of endocytic and exocytic pathways. Ann Rev Biochem 1985; 55:663-700.
40. Minden A, Lin A, Claret FX et al. Selective activation of the JNK signaling cascade and c-Jun transcriptional activity by the small GTPases Rac and Cdc42Hs. Cell 1995; 81:1147-1157.
41. Murray CJ, Salomon JA. Modeling the impact of global tuberculosis control strategies. Proc Natl Acad Sci USA 1998; 95:13881-13886.

42. North RJ. T cell dependence of macrophage activation and mobilization during infection with Mycobacterium tuberculosis. Infect Immun 1974; 10:66-71.
43. Oh YK, Alpuchearanda C, Berthiaume E et al. Rapid and complete fusion of macrophage lysosomes with phagosomes containing Salmonella typhimurium. Infect Immun 1996; 64:3877-3883.
44. Pancholi P, Mirza A, Bhardwaj N et al. Sequestration from immune CD4+ T cells of mycobacteria growing in human macrophages. Science 1993; 260:984-986.
45. Pearse BM, Robinson MS. Clathrin, adaptors, and sorting. Annu Rev Cell Biol 1990; 6:151-171.
46. Pieters J. Processing and presentation of phagocytosed antigens to the immune system. In: Gordon S, ed. Phagocytosis: The Host. Stamford: JAI Press Inc., 1999:79-406.
47. Pieters J. MHC Class II restricted antigen processing and presentation. Adv Immunol 2000; 75:159-208.
48. Pieters J. Evasion of host cell defense mechanisms by pathogenic bacteria. Curr Opin Immunol 2001; 13:37-44.
49. Robinson MS, Watts C, Zerial M. Membrane dynamics in endocytosis. Cell 1996; 84:13-21.
50. Schlesinger LS. Macrophage phagocytosis of virulent but not attenuated strains of Mycobacterium tuberculosis is mediated by mannose receptors in addition to complement receptors. J Immunol 1993; 150:2920-2930.
51. Schlesinger LS. Entry of Mycobacterium tuberculosis into mononuclear phagocytes. Curr Top Microbiol Immunol 1996; 215:71-96.
52. Schmid SL. Clathrin-coated vesicle formation and protein sorting: an integrated process. Annu Rev Biochem 1997; 66:511-548.
53. Schorey JS, Carroll MC, Brown EJ. A macrophage invasion mechanism of pathogenic mycobacteria. Science 1997; 277:1091-1093.
54. Segal AW, Abo A. The biochemical basis of the NADPH oxidase of phagocytes. Trends Biochem Sci 1993; 18:43-47.
55. Sever S, Muhlberg AB, Schmid SL. Impairment of dynamin's GAP domain stimulates receptor-mediated endocytosis [see comments]. Nature 1999; 398:481-486.
56. Simons K, Ikonen E. Functional rafts in cell membranes. Nature 1997; 387:569-572.
57. Stevens TH, Forgac M. Structure, function and regulation of the vacuolar (H+)-ATPase. Annu Rev Cell Dev Biol 1997; 13:779-808.
58. Sturgill-Koszycki S, Schlesinger PH, Chakraborty P et al. Lack of acidification in Mycobacterium phagosomes produced by exclusion of the vesicular proton-ATPase. Science 1994; 263:678-681.
59. Synnes M, Prydz K, Lovdal T et al. Fluid phase endocytosis and galactosyl receptor-mediated endocytosis employ different early endosomes. Biochim Biophys Acta 1999; 1421:317-328.
60. ter Haar E, Musacchio A, Harrison SC et al. Atomic structure of clathrin: A beta propeller terminal domain joins an alpha zigzag linker. Cell 1998; 95:563-573.
61. Tulp A, Verwoerd D, Dobberstein B et al. Isolation and characterization of the intracellular MHC class II compartment. Nature 1994; 369:120-126.
62. Valerius NH, Stendahl O, Hartwig JH et al. Distribution of actin-binding protein and myosin in polymorphonuclear leukocytes during locomotion and phagocytosis. Cell 1981; 24:195-202.
63. Via LE, Deretic D, Ulmer RJ et al. Arrest of mycobacterial phagosome maturation is caused by a block in vesicle fusion between stages controlled by rab5 and rab7. J Biol Chem 1997; 272:13326-13331.
64. Wardle EN. Kupffer cells and their function. Liver 1987; 7:63-75.
65. Weber T, Zemelman BV, McNew JA et al. SNAREpins: Minimal machinery for membrane fusion. Cell 1998; 92:759-772.
66. Xu S, Cooper A, Sturgill-Koszycki S et al. Intracellular trafficking in Mycobacterium tuberculosis and Mycobacterium avium-infected macrophages. J Immunol 1994; 153:2568-2578.
67. Zerial M, Stenmark H. Rab GTPases in vesicular transport. Curr Opin Cell Biol 1993; 5:613-620.
68. Zimmerli S, Edwards S, Ernst JD. Selective receptor blockade during phagocytosis does not alter the survival and growth of Mycobacterium tuberculosis in human macrophages. Am J Respir Cell Mol Biol 1996; 15:760-770.

CHAPTER 12

Chlamydia

Isabelle Jutras, Agathe Subtil, Benjamin Wyplosz and Alice Dautry-Varsat

Introduction

Chlamydia are very successful pathogens of humans and animals.[1] As obligate intracellular parasites, they have evolved to establish a unique niche within their host cells. Like other intracellular parasites, they must fulfill several essential functions: survive in the extracellular milieu and have the ability to enter a host cell. They have to avoid cellular defense mechanisms, establish in an intracellular environment favorable for their multiplication and finally exit the host cell to start a new infectious cycle. *Chlamydia* are obligate intracellular eubacteria: their developmental cycle takes place entirely within an intracellular parasitophorous vacuole called an inclusion, which is distinct from other identified parasitophorous vacuoles. Here, we discuss the cell biology of *Chlamydia* infection, i.e., its entry in the host cell, the characteristics of the inclusion, and some effects of the presence of *Chlamydia* on their host cells.

Chlamydia are very distant phylogenetically from other eubacteria.[2] The genus *Chlamydia* consists of three main species: *C. trachomatis, C. pneumoniae* and *C. psittaci*. The *C. trachomatis* species responsible for human diseases are clustered into different serotypes. The serotypes A, B, C cause the serious eye infection trachoma which is the leading cause of preventable blindness responsible for about 6 millions current cases worldwide.[3] The serotypes L1, L2 and L3 are the cause of the rare systemic disease lymphogranuloma venereum (LGV). The *C. trachomatis* serotypes D-K are the most common sexually transmitted bacterial species, and the sequelae of the infections are responsible for pelvic inflammatory disease, ectopic pregnancy and infertility. The murine strain MoPn of *C. trachomatis* offers an animal model of the sexually transmitted infection.

The more recently discovered *C. pneumoniae* causes human respiratory tract infections.[4] Epidemiological studies show a high seroprevalence, increasing with age, of over 50% among the general adult population in many countries of the world. *C. pneumoniae* is responsible for about 10 % of pneumonia and 5 % of bronchitis in adults but most of the infections are asymptomatic. Since 1988, *C. pneumoniae* infection has been associated with atherosclerosis by serological and histopathological studies. The bacterium has also been recovered from atherosclerotic lesions of coronary and carotid arteries.[5] Numerous studies have been and are performed (including clinical trials with antibiotics) to determine if there is a link between *C. pneumoniae* infections and increased risk or aggravation of atherosclerosis.

C. psittaci can infect a wide variety of birds and non human mammals.[6] It causes sporadic zoonotic oubreaks of pneumonia in humans and is responsible for miscarriage in sheep, cattle and goats worldwide. The GPIC strain infects guinea pig and represents an experimental model of genital tract infection very close to the human disease.

The defining characteristic of *Chlamydia* spp. is their unique obligate intracellular developmental cycle (Fig. 1). The extracellular form, the small (0.3 μm) elementary body (EB) is infectious and is thought to be metabolically inert.[7] Its primary function is to survive in the extracellular environment and to invade a susceptible host cell. These EBs enter epithelial host cells where they differentiate into large (about 1 μm) reticulate bodies (RB). RBs are metabolically

Intracellular Pathogens in Membrane Interactions and Vacuole Biogenesis,
edited by Jean-Pierre Gorvel. ©2004 Eurekah.com and Kluwer Academic / Plenum Publishers.

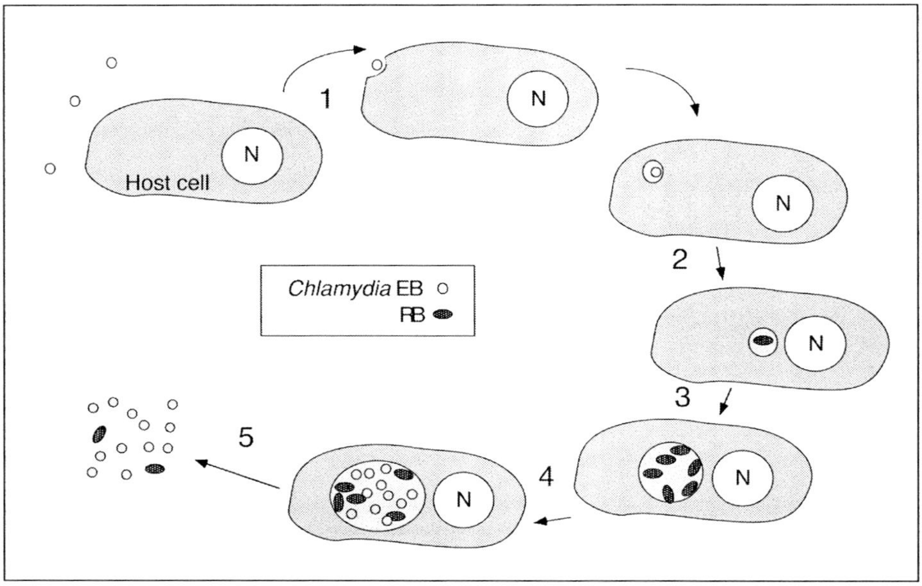

Figure 1. The developmental cycle of chlamydiae. The infectious form, the elementary body (EB), enters by a mechanism probably related to actin-dependent phagocytosis in epithelial cells (1) and is included in a membrane-bound compartment -called an inclusion- in which the whole cycle takes place within 24 to 72 hours, depending on the species. Within the first hours of infection, the inclusion moves towards the microtubule organizing center near the nucleus and the EB differentiates into the replicative form, the reticulate body (RB) (2). Concomitantly to RB multiplication, the volume of the inclusion increases (steps 3 to 5). The composition of its membrane is largely undetermined but it appears to include various lipids imported from the host cell as well as bacterial proteins secreted by a type III mechanism. RBs differentiate back to EBs in an asychronous manner (4) and at the end of the cycle, bacteria are released into the extracellular medium by an as yet unknown mechanism (5).

active and non-infectious and represent the replicating form of the bacteria. Within a couple of hours following internalization, EBs begin differentiating into RBs: the electron-dense nucleoid unique to *Chlamydia* and characteristic of the EB form disappears and the chromatin becomes dispersed. About 8-12 hours later, the inclusion contains mainly RBs, which start dividing by binary fission after a few hours. At some point, the inclusions contain an almost pure population of RBs that can thereafter be seen redifferentiating into condensed EBs. The developmental cycle is somewhat asynchronous and EBs start accumulating in the inclusion while some RBs are still dividing. Depending upon the species and strain, the developmental cycle is complete by 36 hours to 72 hours and EBs are released from the infected cell.

The lack of genetic tools to manipulate *Chlamydia* has hampered efforts to understand molecular aspects of the infection by these bacteria. Sequencing *Chlamydia* genomes has brought an important step forward and much knowledge about these bacteria has been and will be obtained from comparative genomic analyses.[8] Although much information was obtained in the pre-genomic period, completion of 5 genomes, two from *C. trachomatis* and 3 from *C. pneumoniae* has opened new avenues of research, in particularly in proteomics. It will lead to the identification of the whole chlamydial proteins, allow comparison between the different species and open the door to evolutionary biology and understanding of the bacterial diversity. For instance, a family of proteins with unknown functions unique to *Chlamydia* and localized to the inclusion was found (Inc family) and a type III secretion apparatus similar to that of other Gram-negative bacteria such as *Yersinia* or *Salmonella* was discovered.

Chlamydia Entry

Only the infectious form of *Chlamydia*, i.e., the EB, is capable of surviving in the extracellular environment and of entering new host cells. These two properties rely on the nature and structure of the EB cell wall. First, it contains several cysteine-rich outer membrane proteins which are cross-linked by disulfide bonds. The disulfide bond network, which is less developed in the RB form, imparts rigidity to the infectious form and makes it more resistant to the extracellular environment.[7] Second, EBs are able to bind a variety of eukaryotic cells. As we will discuss later, one proposed mechanism for this property is the presence of a heparan-like compound on the EB surface, which may allow the use of ubiquitous receptors for attachment and invasion. However, differences between species and biovars in their abilities to attach and to enter cells have been described.[9-11] In fact, extant entry mechanisms probably appeared relatively late in chlamydial phylogeny, when ancestors of the different kinds of present-day chlamydiae diverged from each other and evolved different relationships with new hosts and new host cells.[7] Consequently, different species and biovars may bind and enter host cells by different mechanisms.

Binding to Host Cells

At 37°C, both attachment and internalization of EBs occur simultaneously and can be measured using radiolabeled or fluorescently labeled bacteria. The rate and extent of association with host cells are proportional to the multiplicity of infection over a very wide range, suggesting that host cell receptor(s) are abundant. Depending on the species and on the multiplicity of infection, EBs bind to host cells in a linear fashion for 1 to 3 hours at 37°C, after which the rate of association declines, suggesting that the binding sites are saturable.[12, 13] In order to determine the mechanism responsible for *Chlamydia* binding, several of its properties have been described.

Temperature Dependence

Use of different species or strains and of different host cells have lead to somewhat different results. For example, adherence of the MN/Cal 10 strain of *C. psittaci* to L cells at 4°C has been reported to be either as efficient as at 37°C[14] or 1/20 as efficient[15]. The GPIC strain of *C. psittaci* binds to HeLa cells half as efficiently at 4°C as at 37°C.[10] Significant binding at 4°C has also been reported for different *C. trachomatis* strains.[7] In conclusion, although the association of EBs to the host cell may be significantly reduced at 4°C compared to 37°C it is measurable for all strains and allows to dissociate the binding step from the internalization step, since the latter does not occur at 4°C.

Enhancement by Centrifugation

For all species, centrifugation of the inoculum onto the host cell monolayer with a centrifuge force from 900 g to 1700 g increases the percentage of infected cells, although the benefit differs between species and strains. It is minimal with the LVG biovar of *C. trachomatis* and with the 6BC strain of *C. psittaci*[16] and maximal for the infection by *C. pneumoniae* which is very inefficient without centrifugation.[17]

Alteration with Physical and Chemical Agents

The effect of a variety of treatments of EBs (heating, protease or detergent treatment, oxidation with periodate) or of host cells (protease treatment) on the success of the infection has been studied, with different results according to the species and the host cell used.[7] This again probably reflects the variety of binding mechanisms used by different *Chlamydia* species, as well as differences in experimental protocols. In most cases, even moderate heating was found to inhibit *Chlamydia* binding. Treatment of EBs with trypsin and other proteases had different effects according to the species and it is not clear whether a protein component on *Chlamydia* surface is necessary for its binding. A number of potential chlamydial ligands have been proposed to date, including the major outer membrane protein (MOMP), an abundant outer membrane protein (omp2), a heat-labile chlamydial cytadhesin (CCA), and heat-shock proteins.[18-22] Regarding the cellular partner, several studies agree on the implication of trypsin-sensitive proteins (possibly glycoproteins) on the host cell in the mechanism of *Chlamydia* binding.[23,24] Finally, binding of *C. trachomatis* LVG strain as well as trachoma biovars, of *C. psittaci* GPIC strain and of *C. pneumoniae* on epithelial cells were found to be at least partially inhibited by heparin, suggesting a common adhesion mechanism to all species.[10,25,26]

Adhesion via a Heparan Sulfate-Like Compound

The hypothesis that *Chlamydia* bind to host cells by a glycosaminoglycan (GAG) intermediate came from the observation that: (i) binding of heparan sulfate and of heparin neutralized *C. trachomatis* infectivity, (ii) treatment of EBs with heparitinase, which is a GAG lyase specific for heparan sulfate, and with heparan sulfate receptor analogs such as platelet factor 4 and fibronectin all inhibited *C. trachomatis* infection, (iii) infection could be restored after heparitinase treatment by adding purified heparan sulfate.[26] There is now a considerable amount of experimental evidence to suggest that heparan-sulfate is involved in the chlamydial attachment-infectivity process.[26-28] What remains currently controversial is whether heparan sulfate is produced by chlamydiae or by the host cell or both. It was recently shown that infection of three GAG-deficient Chinese hamster ovary (CHO) cell lines by *C. trachomatis* LGV was strongly impaired, suggesting that a host cell heparan sulfate is involved. In the few infected cells, no sulfate-like molecule could be found, indicating that chlamydiae are not able to synthesize these molecules de novo.[11] This is in contradiction with what was reported in earlier studies. (i) Radiolabeled sulfate was incorparated in GAGs in a GAG-deficient CHO cell line infected with *C. trachomatis* LGV.[26] However this cell line still synthesizes about 5% of GAGs compared to the parental cell line, which may account for the synthesis observed upon infection and imputed to *Chlamydia.*[11] (ii) Heparan-sulfate was detected by immunofluorescence using a heparan-sulfate specific monoclonal antibody on inclusions developing in 3 different GAG-deficient CHO cell lines,[29] the same cell lines used in the study quoted above.[11] (iii) Heparin or heparan sulfate coated microspheres bound to and were endocytosed by eukaryotic cells, and this entry was competitively inhibited by chlamydial organisms.[28] The discrepancies between these studies cannot be explained yet.

The Internalization Step

Kinetics of Entry

The kinetics of *Chlamydia* entry into cells are difficult to measure precisely, because *Chlamydia* cannot be easily detached from the host cell surface. Treatment of the cells by trypsin or other proteases has been used to discriminate between cell surface-attached and internalized bacteria, but its effect is usually incomplete.[13] However, several studies agree that for *C. psittaci* and *C. trachomatis* strains, most attached bacteria are internalized within 30 to 60 minutes at 37°C.[15,30] This rate may be slower for *C. pneumoniae*, although it has not been precisely studied.

Chloramphenicol, which shuts down protein synthesis in prokaryotes, had no effect on *C. psittaci* MN/Cal 10 ingestion, showing that internalization does not require protein synthesis from the bacteria.[31] In fact, EBs are metabolically inactive suggesting that they act as passive passengers during a host cell-driven internalization process. However, recent results show that chlamydiae possess a functional type III secretion apparatus.[32-34] In several other pathogens this kind of secretion system allows for the translocation, from an extracellular location, of bacterial proteins across the bacterial cell envelope into a host cell. In *Chlamydia*, it has been shown that, like in *Salmonella*, the type III secretion system is active during the intracellular phase of its cycle.[33] The possibility remains that it also allows for the secretion of presynthesized proteins from the cell-attached EB into the host cell. This hypothesis is supported by the observation of type III apparatus-like appendages at the surface of EBs as well as at the surface of RBs.[35-37] It implies that following attachment, EBs get activated by an unknown signal which triggers type III secretion, an ATP-dependent process.

Mechanism of Entry

Conflicting results were obtained regarding the mechanism of entry into host cells.[7] Cytochalasin D, an inhibitor of actin polymerization, drastically inhibits the entry of all species, indicating an actin-dependent phagocytic way of entry. However, entry by clathrin-coated pits was also suggested by electron microscopy observations.[14,38] Part of the confusion may be due to the possibility that *Chlamydia* use different mechanisms of entry in different cells, as discussed earlier, as well as to technical difficulties. Electron microscopy is a fastidious technique requiring good immunolabeling to identify ultrastructures unambiguously, and extensive statistical analysis. It is also not suitable for the study of dynamic processes such as endocytosis. The recent identification of several of the molecules involved in the formation of clathrin-coated pits and vesicles and the design of specific inhibitors of this process has allowed to test its implication in the entry of *C. psittaci* GPIC strain and of *C. trachomatis* LGV/L2 in HeLa cells. When clathrin-dependent endocytosis was specifically inhibited by overexpression of a dominant negative form of Eps15 or of dynamin I, two proteins essential for the clathrin-dependent pathway, the entry of the two *Chlamydia* strains was unaltered, demonstrating that they use a pathway independent of clathrin-coated pits and vesicles to infect HeLa cells.[39] We have recently shown that *C. pneumoniae* is also internalized by a clathrin-independent pathway, which does not require functional dynamin, in Hep2 cells (B.Wyplosz, manuscript in preparation). Altogether it is now clear that chlamydiae are able to invade cells by a nonclathrin-dependent mechanism. However, for technical reasons, these studies have been conducted in nonpolarized cultured cells. Questions on the kinetics and mechanisms of entry of chlamydiae are very difficult to address in polarized epithelial cells, which are the initial target of chlamydial invasion. Entry of *C. trachomatis* via the apical surface of primary cultures of polarized human endometrial gland cells has been reported.[38] This last study suggested that coated pits were preferentially used as an entry pathway in polarized cells. This observation brings us back to the idea that chlamydiae may have adopted different strategies for cell invasion, depending on the host cell to which they attach. Irrespective of the entry mechanism, early chlamydial protein synthesis would allow to redefine the compartment in which the bacteria have been internalized and make it suitable for the rest of the infectious cycle. In agreement with this model, it was shown that Fc-mediated endocytosis of *C. trachomatis* in HeLa cells gave rise to successful infection.[40] However, chlamydiae are not able to develop in all cell types, as it has been shown that in dendritic cells *C. trachomatis* MoPn and *C. psittaci* GPIC were internalized through macropinocytosis and were subsequently degraded in lysosomes.[41]

Maturation of the Chlamydial Inclusion

Early Intracellular Localization Following Entry

Markers for the endocytic pathway have generally not been detected in the chlamydial inclusion although some conflicting results can be found in the literature. Indeed, the transferrin receptor, an early endosome marker, and the cation-independent mannose-6-phosphate receptor, a late endosome marker, have been reported to be in close association with the vacuole.[42] However closer scrutiny by electron microscopy revealed that the transferrin receptor was located in vesicles surrounding the inclusion rather than within the inclusion membrane.[43] In addition, late endosomes markers such as LAMP1 have not been found to localize to the chlamydial inclusion in several studies using epithelial cells.[44,45] In contrast, chlamydiae colocalize with LAMP1 shortly after their internalization by dentritic cells.[45] Likewise, the pH of the chlamydial endocytic vesicles remains above 6 whereas vesicles containing heat-killed chlamydiae rapidly acidify to 5.3.[46] Thus, if internalized chlamydiae encounter the endocytic pathway, separation from early endocytic compartments must occur rapidly following entry.

Shortly following the entry of *Chlamydia* in cells, *Chlamydia*-containing vacuoles have been reported to localize to a peri-nuclear location, in close apposition to the Golgi apparatus. Importantly, this location and avoidance of fusion with lysosomes necessitates early bacterial protein synthesis.[40] Migration of *C. trachomatis*-containing vesicles has been suggested to be microtubule-dependent as inhibition of dynein, a microtubule-dependent motor, affects infectivity and dynein locates with peri-nuclear *Chlamydia* aggregates.[47]

Within 8 to 12 hours following their internalization, EBs have differentiated into RBs, a process also dependent on bacterial protein synthesis. Hallmarks of this reorganization include increase in volume of the bacteria and loss of infectivity.[7] This differentiation is also accompanied by chlamydial chromatin decondensation and the genome sequence of *Chlamydia* spp. has revealed the presence of genes encoding putative helicases, which could be involved in this process.[48]

Development of the Chlamydial Inclusion and Effects on the Host Cell

Chlamydiae multiply strictly within their membrane-bound inclusion. As the RBs divide and proliferate, the inclusion expands such that it occupies most of the cytoplasm of the infected cell. The characteristics of the inclusion, including its membrane and lumenal components as well as their origin, remain largely undetermined. The development of the inclusion does not appear to be dependent on de novo eukaryotic protein synthesis as the growth of the inclusion proceeds normally in cells treated with cycloheximide. However, growth of the inclusion implies an important source of lipids.

Alteration in Lipid Trafficking

Recently, the use of fluorescently labeled ceramide, which normally undergoes conversion to sphingomyelin in early Golgi compartments, allowed to visualize that sphingomyelin was incorporated in the membrane of the inclusion and in the cell wall of the RBs.[49] *Chlamydia* containing vacuoles appeared capable of intercepting the normal trafficking of sphingomyelin to the plasma membrane within 2 hours following infection.[50] These results suggested a mechanism by which host-derived sphingomyelin could supply the inclusion membrane and support its development. This hypothesis is supported by recent data showing that a mutant CHO cell line deficient in sphingolipid synthesis could not support the proliferation of chlamydiae.[51] Several eukaryotic glycerophospholipids, including phosphatidylinositol and phosphaditylcholine, were subsequently found to be trafficked to the chlamydial inclusion.[52] The localization of some model proteins which, after synthesis, undergo trafficking through the Golgi apparatus i.e., vesicular stomatitis virus G-protein, transferrin receptor and human histocompatibility class I antigen, has been assessed in *Chlamydia*-infected cells but none of these has been found to label the inclusion.[53] Notably, known coat proteins such as clathrin were not found to be associated with the inclusion.[50] Thus, host proteins which are trafficked to the inclusion have yet to be identified.

Modification of the Inclusion Membrane by Chlamydial Proteins

While host proteins have not been localized to the inclusion, many chlamydial proteins associated with the inclusion membrane have now been identified and hence these have been termed Inc proteins. Although these proteins show little sequence homology, similarities in their hydropathy profiles indicate a structural homology, which has served to define the Inc family of proteins. Inc proteins comprise a large bilobed hydrophobic region of 40-70 amino acids. The *C. psittaci* IncA, B and C were the first chlamydial proteins to be localized to the inclusion membrane[54,55] and homologues of these have been identified in the genomes of other chlamydial species. Using antisera raised against the membrane fraction of *C. trachomatis*-infected cells, four additional inclusion membrane proteins (IncD-G) have been identified.[56] The search for ORFs encoding proteins with a bilobed hydrophobic pattern in the genomic sequences of *C. trachomatis* and *C. pneumoniae* have revealed respectively, over 40 and at least 60 candidate Inc proteins.[34,57]

Incorporation of chlamydial Inc proteins in the inclusion membrane has been postulated to involve the type III secretion system. Using a heterologous secretion system, *C. pneumoniæ* IncA-C as well as three other proteins exhibiting the characteristic hydrophobic domain, have been shown to undergo type III dependent secretion in *Shigella flexneri.*[34] These results strongly support a model whereby Inc proteins are inserted in the inclusion membrane through a type III secretion system. One of the components of the putative apparatus is the chlamydial protein CopN, the equivalent of *Yersinia* YopN, which is secreted by the type III pathway, exposed on the surface of *Yersinia* and presumed to function as a Ca^{2+}-sensitive regulatory plug, controlling the secretion channel.[58] CopN has been shown to be present in EBs as well as in RBs and to label the inclusion membrane in *Chlamydia*-infected cells.[33] In addition, CopN appears capable of undergoing type III-dependent secretion in a Ca^{2+}-sensitive manner when expressed in *Yersinia.*[33] These results are consistent with the assembly process of the apparatus occuring on RBs prior to their conversion to EBs. An important issue still awaiting to be elucidated is whether type III secretion occurs from EBs upon infection and functions throughout the infection process.

Recently a chlamydial protein lacking the characteristic hydrophobic profile of the Inc proteins was localized to the bacterial inclusion and named Cap1.[59] Thus another chlamydial protein, which is not structurally related to Inc proteins, appears to be associated with the inclusion membrane. In addition, chlamydial proteins not associated with the inclusion membrane are possibly translocated in the cytosol of infected cells,[60] although none have yet been identified.

Fusion of Chlamydial Inclusions

The various chlamydial species can be distinguished by the fusogenicity of the inclusion in which they replicate. While multiple infections with most strains of *C. trachomatis* typically give rise to the fusion of the *Chlamydia* containing-vacuoles to form a single inclusion, *C. trachomatis* serovar L2 or *C. psittaci* strains form multiple inclusions that do not fuse.[61,62] Fusion of *C. trachomatis* inclusions was shown to be temperature sensitive and dependent on bacterial protein synthesis.[61] A role of the chlamydial inclusion protein IncA in homotypic vacuole fusion or septation has been proposed based on the observation that microinjection of antibodies against IncA modified the structure of a single inclusion to a multilobed inclusion in *C. trachomatis* infected cells.[63] Involvement of IncA in vacuole fusion is further supported by the finding that IncA seemed absent from the inclusion membrane of non-fusing variants isolated from different serovars of *C. trachomatis.*[64] In addition, sequencing of the IncA gene from one of these variants revealed mutations in the sequence susceptible of altering the hydrophobicity profile of the protein.[64] In contrast, such a role in inclusion fusion could not be assigned to *C. psittaci* IncA,[63] indicating that homologous proteins between the *Chlamydia* spp. may have diverse functions.

Protein Phosphorylation

Increased host protein phosphorylation has been consistently observed in cells infected with *Chlamydia*, notably a group of proteins with molecular weights around 70 kDa and proteins at 100 kDa and 140 kDa.[65,66] These changes have been reported to occur throughout the infection cycle, at early time points following the entry of *Chlamydia* and as late as 18 hours after infection. However, since identical phosphorylation profiles have been observed following the entry of heparin–coated beads,[28] it appears that these modifications to host proteins are not specifically elicited by chlamydial entry in cells. In addition, immunofluorescence studies have revealed phosphorylated proteins associated with the inclusion membrane,[65] but it is unclear whether these are host or bacterial proteins. At least one chlamydial protein located in the inclusion membrane, IncA, has been shown to be phosphorylated by the host cell.[67]

Host Cell Apoptosis

Dual pro- and anti-apoptotic effects of chlamydial infection on host cells have been reported. Indeed, at early infection stages, *C. trachomatis* has been shown to protect its host cells against apoptosis induced by a wide range of proapoptotic stimuli, including TNF-α, granzyme B/perforin and Fas antibody.[68] This antiapoptotic effect of *Chlamydia* involves inhibition of both caspase-3 activity and cytochrome c release from mitochondria, and was dependent on chlamydial protein synthesis. In contrast, at late stages of infection, starting 24 hours after the onset of infection, *C. psittaci* induces apoptosis by a mechanism independent of known caspases but dependent on chlamydial protein synthesis.[69] *Chlamydia*-induced apoptosis was also demonstrated in vivo, in the murine genital tract.[70] It has been suggested that antiapoptotic activity may protect infected cells from the host inflammatory response at the site of infection, thus allowing the intracellular development of *Chlamydia* to pursue for several hours, while subsequent proapoptotic activity could rather contribute to the inflammatory response and to spreading of the infection.[69] The mechanism used by *Chlamydia* to modulate apoptosis of the host cell remains unknown, although it can be speculated that this process involves secreted bacterial effectors.

Late Events in Chlamydia *Development and Exit from Host Cell*

Multiplication of RBs proceeds until cell lysis occurs although by 15 to 25 hours after infection, increasing numbers of RBs differentiate back to EBs. Newly formed EBs are typically located in the lumen of the inclusion while RBs continue to grow at the periphery.[7] This differentiation step takes place asynchronously in the RB population of an inclusion and cell lysis results in the release of a mixture of RBs and EBs. Although the signal which triggers differentiation remains unknown, these observations suggest it may originate from the RBs themselves rather than from the growth conditions within the inclusion.[7] Transition from the replicating RB form to the inert EB is accompanied by chromatin condensation, a process necessitating the expression of chlamydial histone proteins.[71] Finally, the signal responsible for cell lysis and release of chlamydiae has not been determined and can be hypothesized to involve host cell apoptosis induced by *Chlamydia*.

Conclusion

The developmental cycle of chlamydiae has been known for a long time, however, many important and basic cell biology questions are still open : 1) How does internalization take place and how does it depend on the *Chlamydia* species and the host cells? 2) How does the inclusion avoid fusion with endocytic compartments? 3) What signal transduction pathways are activated at the different steps of the development cycle? 4) We know that the inclusion receives lipids, probably coming from the Golgi. It is very surprising that cellular proteins trafficked to the inclusion have not been also uncovered. What is the pathway they follow? Which host proteins are involved? 5) Finally, we don't know how the bacteria get out of their hosts. Despite the difficulties due to the lack of genetic tools to analyze *Chlamydia* infection, many fundamental questions concerning these pathogen-host interactions can be addressed using cell biology approaches, specially now that a wealth of information is coming from the sequencing of *Chlamydia* genomes.

References

1. Stephens RS. Chlamydia. Intracellular biology, pathogenesis, and immunity. In: Stephens RS, ed. Washington: American Society for Microbiology, 1999:1-321.
2. Everett KDE, Bush RM, Andersen AA. Emended description of the order Chlamydiales, proposal of Parachlamydiaceae fam.nov. and Simkaniaceae fam. nov., each containing one monotypic genus, revised taxonomy of the family Chlamydiaceae, including a new genus and five new species, and standards for the identification of organisms. Int J Syst Bacteriol 1999; 49:415-440.
3. Stamm WE. Chlamydia trachomatis infections: Progress and problems. J Infect Dis 1999; 179:S380-383.
4. Kuo C-C, Jackson L, Campbell L et al. Chlamydia pneumoniae (TWAR). Clin Microbiol Rev 1995; 8:451-461.
5. Maass M, Bartels C, Engel P et al. Endovascular presence of viable Chlamydia pneumoniae is a common phenomenon in coronary artery disease. J Am Coll Cardiol 1998; 31:827-832.
6. Gregory DW, Schaffner W. Psittacosis. Semin Respir Infect 1997; 12:7-11.
7. Moulder JW. Interaction of chlamydiae and host cells in vitro. Microbiol Rev 1991; 55:143-190.

8. Rockey DD, Lenart J, Stephens RS. Genome sequencing and our understanding of chlamydiae. Infect Immun 2000; 68:5473-5479.
9. Davis CH, Wyrick PB. Differences in the association of Chlamydia trachomatis serovar E and serovar L12 with epithelial cells in vitro may reflect biological differences in vivo. Infect Immun 1997; 65:2914-2924.
10. Gutiérrez-Martin CB, Ojcius DM, Hsia R-c et al. Heparin-mediated inhibition of Chlamydia psittaci adherence to HeLa cells. Microb Pathog 1997; 22:47-57.
11. Taraktchoglou M, Pacey AA, Turnbull JE et al. Infectivity of Chlamydia trachomatis serovar LGV but not E is dependent on host cell heparan sulfate. Infect Immun 2001; 69:968-976.
12. Byrne GI. Kinetics of phagocytosis of Chlamydia psittaci by mouse fibroblasts (L cells): Separation of the attachment and ingestion stages. Infect Immun 1978; 19:607-612.
13. Söderlund G, Kihlström E. Attachment and internalization of a Chlamydia trachomatis lymphogranuloma venereum strain by McCoy cells: kinetics of infectivity and effect of lectins and carbohydrates. Infect Immun 1983; 42:930-935.
14. Hodinka RL, Davis CH, Choong J et al. Ultrastructural study of endocytosis of Chlamydia trachomatis by McCoy cells. Infect Immun 1988; 56:1456-1463.
15. Friis RR. Interaction of L cells and Chlamydia psittaci: Entry of the parasite and host responses to its development. J Bacteriol 1972; 110:706-721.
16. Lee CK. Interaction between a trachoma strain of Chlamydia trachomatis and mouse fibroblasts (McCoy cells) in the absence of centrifugation. Infect Immun 1981; 31:584-591.
17. Kuo C-C, Grayston JT. Factors affecting viability and growth in HeLa 229 cells of Chlamydia sp. strain TWAR. J Clin Microbiol 1988; 26:812-815.
18. Joseph TD, Bose SK. A heat-labile protein of Chlamydia trachomatis binds to HeLa cells and inhibits the adherence of chlamydiae. Proc Natl Acad Sci USA 1991; 88:4054-4058.
19. Raulston JE, Davis CH, Schmiel DH et al. Molecular characterization and outer membrane association of a Chlamydia trachomatis protein related to the hsp70 family of proteins. J Biol Chem 1993; 268:23139-23147.
20. Raulston JE. Chlamydial envelope components and pathogen-host cell interactions. Mol Microbiol 1995; 15:607-616.
21. Swanson AF, Kuo C-C. Binding of the glycan of the major outer membrane protein of Chlamydia trachomatis to HeLa cells. Infect Immun 1994; 62:24-28.
22. Ting L-M, Hsia R-C, Haidaris CG et al. Interaction of outer envelope proteins of Chlamydia psittaci GPIC with the HeLa cell surface. Infect Immun 1995; 63:3600-3608.
23. Byrne GI. Requirements for ingestion of Chlamydia psittaci by mouse fibroblasts (L cells). Infect Immun 1976; 14:645-651.
24. Hatch TP, Vance Jr DW, Al-Hossainy E. Attachment of Chlamydia psittaci to formaldehyde-fixed and unfixed L cells. J Gen Microbiol 1981; 125:273-283.
25. Kuo C-C, Grayston JT. Interaction of Chlamydia trachomatis organisms and HeLa 229 cells. Infect Immun 1976; 13:1103-1109.
26. Zhang JP, Stephens RS. Mechanism of Chlamydia trachomatis attachment to eukaryotic host cells. Cell 1992; 69:861-869.
27. Su H, Raymond L, Rockey DD et al. A recombinant Chlamydia trachomatis major outer membrane protein binds to heparan sulfate receptors on epithelial cells. Proc Natl Acad Sci USA 1996; 93:11143-11148.
28. Stephens RS, Fawaz FS, Kennedy KA et al. Eukaryotic cell uptake of heparin-coated microspheres: a model of host cell invasion by Chlamydia trachomatis. Infect Immun 2000; 68:1080-1085.
29. Rasmussen-Lathrop SJ, Koshiyama K, Phillips N et al. Chlamydia-dependent biosynthesis of a heparan sulphate-like compound in eukaryotic cells. Cell Microbiol 2000; 2:137-144.
30. Ward ME, Murray A. Control mechanisms governing the infectivity of Chlamydia trachomatis for HeLa cells: mechanisms of endocytosis. J Gen Microbiol 1984; 130:1765-1780.
31. Tribby, II, Friis RR, Moulder JW. Effect of chloramphenicol, rifampicin, and nalidixic acid on Chlamydia psittaci growing in L cells. J Infect Dis 1973; 127:155-163.
32. Bavoil P, Hsia R-C. Type III secretion in Chlamydia: A case of déjà vu? Mol Microbiol 1998; 28:860-862.
33. Fields KA, Hackstadt T. Evidence for the secretion of Chlamydia trachomatis CopN by a type III secretion mechanism. Mol Microbiol 2000; 38:1048-1060.
34. Subtil A, Parsot C, Dautry-Varsat A. Secretion of predicted Inc proteins of Chlamydia pneumoniæ by a heterologous type III machinery. Mol Microbiol 2001; 39:792-800.
35. Chang JJ, Leonard KR, Zhang YX. Structural studies of the surface projections of Chlamydia trachomatis by electron microscopy. J Med Microbiol 1997; 46:1013-1018.

36. Matsumoto A. Surface projections of Chlamydia psittaci elementary bodies as revealed by freeze-deep-etching. J Bacteriol 1982; 151:1040-1042.
37. Matsumoto A. Electron microscopic observations of surface projections on Chlamydia psittaci reticulate bodies. J Bacteriol 1982; 150:358-364.
38. Wyrick PB, Choong J, Davis CH et al. Entry of genital Chlamydia trachomatis into polarized human epithelial cells. Infect Immun 1989; 57:2378-2389.
39. Boleti H, Benmerah A, Ojcius D et al. Chlamydia infection of epithelial cells expressing dynamin and Eps15 mutants: clathrin-independent entry into cells and dynamin-dependent productive growth. J Cell Sci 1999; 112:1487-1496.
40. Scidmore MA, Rockey DD, Fischer ER et al. Vesicular interactions of the Chlamydia trachomatis inclusion are determined by chlamydial early protein synthesis rather than route of entry. Infect Immun 1996; 64:5366-5372.
41. Ojcius D, Bravo de Alba Y, Kanellopoulos J et al. Internalization of Chlamydia by dendritic cells and stimulation of Chlamydia-specific T cells. J Immunol 1998; 160:1297-1303.
42. Van Ooij C, Apodaca G, Engel J. Characterization of the Chlamydia trachomatis vacuole and its interaction with the host endocytic pathway in HeLa cells. Infect Immun 1997; 65:758-766.
43. Taraska T, Ward DM, Ajioka RS et al. The late chlamydial inclusion membrane is not derived from the endocytic pathway and is relatively deficient in host proteins. Infect Immun 1996; 64:3713-3727.
44. Heinzen RA, Scidmore MA, Rockey DD et al. Differential interaction with endocytic and exocytic pathways distinguish parasitophorous vacuoles of Coxiella burnetii and Chlamydia trachomatis. Infect Immun 1996; 64:796-809.
45. Ojcius DM, Bravo Y, Kanellopoulos JM et al. Internalization of Chlamydia-derived by dendritic cells and stimulation of Chlamydia-specific T cells. J Immunol 1998; 160:1297-1303.
46. Schramm N, Bagnell CR, Wyrick PB. Vesicles containing Chlamydia trachomatis serovar L2 remain above pH 6 within HEC-1B cells. Infect Immun 1996; 64:1208-1214.
47. Clausen J, Christiansen G, Holst H et al. Chlamydia trachomatis utilizes the host cell microtubule network during early events of infection. Mol Microbiol 1997; 25:441-449.
48. Kalman S, Mitchell W, Marathe R et al. Comparative genomes of Chlamydia pneumoniae and C. trachomatis. Nature Genet 1999; 21:385-389.
49. Hackstadt T, Scidmore MA, Rockey DD. Lipid metabolism in Chlamydia trachomatis-infected cells: directed trafficking of Golgi-derived sphingolipids to the chlamydial inclusion. Proc Natl Acad Sci USA 1995; 92:4877-4881.
50. Hackstadt T, Rockey DD, Heinzen RA et al. Chlamydia trachomatis interrupts an exocytic pathway to acquire endogenously synthesized sphingomyelin in transit from the Golgi apparatus to the plasma membrane. EMBO J 1996; 15:964-977.
51. Van Ooij C, Kalman L, Van Ijzendoorn S et al. Host cell-derived sphingolipids are required for the intracellular growth of Chlamydia trachomatis. Cell Microbiol 2000; 2:627-637.
52. Wylie JL, Hatch GM, McClarty G. Host cell phospholipids are trafficked to and then modified by Chlamydia trachomatis. J Bacteriol 1997; 179:7233-7242.
53. Scidmore MA, Fischer ER, Hackstadt T. Sphingolipids and glycoproteins are differentially trafficked to the Chlamydia trachomatis inclusion. J Cell Biol 1996; 134:363-374.
54. Rockey DD, Heinzen RA, Hackstadt T. Cloning and characterization of a Chlamydia psittaci gene coding for a protein localized in the inclusion membrane of infected cells. Mol Microbiol 1995; 15:617-626.
55. Bannantine J, Stamm W, Suchland R et al. Chlamydia trachomatis IncA is localized to the inclusion membrane and is recognized by antisera from infected humans and primates. Infect Immun 1998; 66:6017-6021.
56. Scidmore-Carlson MA, Shaw EI, Dooley CA et al. Identification and characterization of a Chlamydia trachomatis early operon encoding four novel inclusion membrane proteins. Mol Microbiol 1999; 33:753-765.
57. Bannantine JP, Griffiths RS, Viratyosin W et al. A secondary structure motif predictive of protein localization to the chlamydial inclusion membrane. Cell Microbiol 2000; 2:35-47.
58. Hueck C. Type III protein secretion systems in bacterial pathogens of animals and plants. Microbiol Mol Biol Rev 1998; 62:379-433.
59. Fling SP, Sutherland RA, Steele LN et al. CD8(+) T cells recognize an inclusion membrane-associated protein from the vacuolar pathogen Chlamydia trachomatis. Proc Natl Acad Sci USA 2001; 98:1160-1165.
60. Zhong GM, Liu L, Fan T et al. Degradation of transcription factor RFX5 during the inhibition of both constitutive and interferon gamma-inducible major histocompatibility complex class I expression in Chlamydia-infected cells. J Exp Med 2000; 191:1525-1534.

61. Van Ooij C, Homola E, Kincaid E et al. Fusion of Chlamydia trachomatis-containing inclusions is inhibited at low temperatures and requires bacterial protein synthesis. Infect Immun 1998; 66:5364-5371.
62. Rockey DD, Fischer ER, Hackstadt T. Temporal analysis of the developing Chlamydia psittaci inclusion by use of fluorescence and electron microscopy. Infect Immun 1996; 64:4269-4278.
63. Hackstadt T, Scidmore-Carlson MA, Shaw EI et al. The Chlamydia trachomatis IncA protein is required for homotypic vesicle fusion. Cell Microbiol 1999; 1:119-130.
64. Suchland RJ, Rockey DD, Bannantine JP et al. Isolates of Chlamydia trachomatis that occupy nonfusogenic inclusions lack IncA, a protein localized to the inclusion membrane. Infect Immun 2000; 68:360-367.
65. Fawaz FS, van Ooij C, Homola E et al. Infection with Chlamydia trachomatis alters the tyrosine phosphorylation and/or localization of several host cell proteins including cortactin. Infect Immun 1997; 65:5301-5308.
66. Birkelund S, Johnsen H, Christiansen G. Chlamydia trachomatis serovar L2 induces protein tyrosine phosphorylation during uptake by HeLa cells. Infect Immun 1994; 62:4900-4908.
67. Rockey DD, Grosenbach D, Hruby DE et al. Chlamydia psittaci IncA is phosphorylated by the host cell and is exposed on the cytoplasmic face of the developing inclusion. Mol Microbiol 1997; 24:217-228.
68. Fan T, Lu H, Hu H et al. Inhibition of apoptosis in Chlamydia-infected cells: blockade of mitochondrial cytochrome c release and caspase activation. J Exp Med 1998; 187:487-496.
69. Ojcius DM, Souque P, Perfettini J-L et al. Apoptosis of epithelial cells and macrophages due to infection with the obligate intracellular pathogen Chlamydia psittaci. J Immunol 1998; 161:4220-4226.
70. Perfettini J-L, Darville T, Gachelin G et al. Effect of Chlamydia trachomatis infection and subsequent TNFα secretion on apoptosis in the murine genital tract. Infect Immun 2000; 68:2237-2244.
71. Barry CE, Hayes SF, Hackstadt T. Nucleoid condensation in Escherichia coli that express a chlamydial histone homolog. Science 1992; 256:377-379.

CHAPTER 13

Host Cell Signaling Induced by the Pathogenic *Neisseria* Species

Andreas Popp, Oliver Billker and Thomas F. Meyer

Summary

The two pathogenic *Neisseria* species, *N. gonorrhoeae* and *N. meningitidis*, attach to and penetrate the mucosa of the human urogenital tract and nasopharynx, respectively. The molecular events involved in the interaction of these bacteria with the mucosal barrier and with cells of the immune system have been subject to intense studies over the past decade. Several factors involved in the molecular cross-talk between bacteria and host cells have been unravelled and this has substantially expanded our knowledge of the strategies of these two highly adapted pathogens to colonise the human body. For example, attachment via the neisserial type IV pili is emerging as a complex series of events possibly involving several adhesion steps and intense exchange of signals between bacteria and host cells. Moreover, receptors for important adhesins and invasins, such as the Opa proteins, were identified. Signaling pathways initiated either by activated receptors or the transfer of the neisserial PorB protein, were characterised and found to control important steps such as adhesion, invasion, host cell apoptosis and cytokine release. This Chapter will summarise the recent developments and report on our current understanding of these molecular mechanisms of the neisserial infection.

Infection of Mucosal Surfaces by Pathogenic *Neisseria* Species

The genus *Neisseria* comprises several commensal species and the two human-specific pathogens *N. gonorrhoeae* (gonococci) and *N. meningitidis* (meningococci), that all colonise mucosal surfaces. Many virulence factors that allow distinct bacterium host cell interactions are common to both *N. gonorrhoeae* and *N. meningitidis*. A major difference between these two pathogens is a polysaccharide capsule unique to meningococci, enabling this pathogen to spread via the aerosol route and to replicate in the blood stream, while gonococci are transmitted by intimate contact between individuals and usually are more quickly eradicated once they have reached the vascular system. As a consequence, the two pathogens colonise different habitats within the human body and give rise to diseases with different symptoms, and severity.

N. meningitidis colonises successfully the human oro- and nasopharynx of about 3-30% of the human population, in most cases without causing any symptoms.[1] Asymptomatic carriage is so common, that meningococci may be considered part of the normal flora of the human pharynx. However, in rare cases, but especially in infants under 5 years of age, adolescents and young adults, meningococci can disseminate in the body via the vascular system, cross the blood brain barrier and give rise to life-threatening diseases, such as septicaemia and meningitis. Fulminate meningococcal meningitis and septicaemia can often kill a previously healthy individual within only a few hours after the onset of symptoms, which typically include high fever, headache and a stiff neck.[2,3]

Intracellular Pathogens in Membrane Interactions and Vacuole Biogenesis, edited by Jean-Pierre Gorvel.

N. gonorrhoeae is the causative agent of the sexually transmitted disease gonorrhea, one of the most ancient of historically described medical illnesses.[4] Gonococcal infection is usually limited to the male urethra and the female cervix. However, other mucosal tissues inside and outside the urogenital tract can be colonised, including the conjunctiva of the eye, the pharynx and the rectum. In most cases, infection is accompanied by a massive inflammatory response with infiltration of polymorphonuclear neutrophilic granulocytes (PMNs) into the infected tissue leading to rupture of epithelia, exfoliation of infected cells and the typical symptoms of uncomplicated gonorrhea, a purulent urethral or cervical discharge.[4] Asymptomatic carriage and complications occur frequently in women. Ascending of gonococci into the upper genital tract and infection of the fallopian tubes may result in pelvic inflammatory disease (PID), a common cause of infertility in women and one of the pressing reasons why gonorrhea has to be controlled, particularly in third world countries.[5] Disseminated gonococcal infection is rare and mostly develops from symptom-less local infections and can affect the skin, joints, brain and heart.[6]

The high degree of adaptation to the human body is responsible for the fact, that still no appropriate animal model for neisserial infections exists. Our current knowledge of the molecular pathology of neisserial infections is, therefore, mainly based on in vitro organ cultures, microscopy of exudate specimens and in vitro cell culture models.

Probably the most remarkable feature of the two pathogenic *Neisseria* species is the high frequency with which their surface structures can undergo variation. The on/off switching and antigenic variation of prominent virulence factors is mediated by a variety of genetic mechanisms and results in a continuously heterogeneous population of bacteria.[7] This strategy probably serves the important function of shielding the pathogens from the attacks of the humoral arm of the immune system. On the other hand, variation results in the expression of different combinations of virulence factors with distinct functional characteristics on the bacterial surface that may be required for an optimal adaptation to the diverse niches and microenvironments within the human body.

No prominent exotoxins are produced by the two neisserial pathogens. Furthermore, no classical type III or type IV secretion systems that in many other pathogens are key components for the modulation of host cell signaling by direct injection of bacterial proteins into the host cell cytosol[8,9] have been identified in *N. gonorrhoeae* and *N. meningitidis*. However, several neisserial virulence factors mediate the intimate interaction with a variety of host cell receptors enabling the pathogens to effect adherence, invasion, transcytosis and phagocytosis in multiple cell types in vitro. Moreover, at least one protein, the neisserial PorB, can translocate into host cells and substantially modify signaling there.

This Chapter will focus on our current knowledge of host cell receptors and signaling pathways engaged by *N. gonorrhoeae* and *N. meningitidis*, their influence on cellular functions and their possible contribution to neisserial virulence.

Establishing Attachment via Type IV Pili

The first critical step during neisserial colonisation is the adherence to the apical pole of epithelial cells. It is generally assumed that this interaction is mediated by the neisserial type IV pili.[10,11] Type IV pili are fibrous, hair-like protrusions that extend several µm in length from the bacterial surface and are found in many gram-negative bacteria. The neisserial type IV pili are involved in many different functions related to virulence including adherence, the natural competence for DNA uptake,[12] and a form of flagella-independent movement termed twitching motility.[13]

The neisserial pili are composed of a major structural subunit, the pilin or PilE protein, which is assembled into a helical pilus fiber.[14] Several other proteins which are localised in the neisserial inner and outer membrane or are associated with the pilus fibre play important roles in pilus assembly and functions.[15-17]

In human organ culture models piliated *N. gonorrhoeae* and *N. meningitidis* attach selectively to the microvilli of non-ciliated columnar epithelial cells. [10] The early localised adherence proceeds within a few hours to the formation of dense neisserial microcolonies each comprising about 10-100 bacteria (Fig. 1).[18-20] This process often coincides with an elongation of microvilli which extend towards the microcolonies.[18,20,21] At later time-points of infection microcolonies have disbanded, bacteria somehow have spread over the epithelial surface and are in close contact with the host cell membrane.[10,18,20,22] The dispersal of microcolonies and tight adherence is accompanied by a marked reduction of microvilli on infected cells and by a loss in neisserial piliation.[19,20,23] Recently, several studies expanded our view of pilus-mediated adhesion which emerges as a series of events that require intimate cross-talk between bacteria and host cells.[18-20,23-25] However, we are only just beginning to discover the functions of the diverse factors involved in this complex process.

The initial contact with the epithelium is most likely established by the PilC protein, which has been identified as a tip-located adhesin but can additionally be found in the outer membrane.[26,27] A crucial role for PilC in adherence is indicated by the fact, that pre-treatment of human epithelial and endothelial cells with purified PilC competitively blocks pilus-mediated neisserial binding and suggests that *N. gonorrhoeae* and *N. meningitidis* possess identical pilus receptors.[15,28] The membrane cofactor protein (MCP; CD46) was identified as a pilus receptor[29] but whether this involves direct interactions with PilC or other pilus components remains to be shown. CD46 is a transmembrane glycoprotein and plays an important role in controlling the complement system. It prevents the deposition of the central complement components C3b and C4b on the cell surface and serves as a cofactor for their cleavage by the protease factor I.[30]

Several findings demonstrate that the initial pilus binding causes host cell responses such as Ca^{2+} fluxes in epithelial cells, that may be important to stabilise pilus binding.[31,32] However, a clear function for the observed signaling events in the early or subsequent steps of pilus binding remains to be shown.

Another host cell response is the formation of so-called pilus-induced cortical plaques beneath the bacterial microcolonies.[23,24] These structures are characterised by the accumulation of filamentous actin and actin-associated proteins, several membrane glycoproteins (for example CD44v3, EGF-R, ICAM-1), and tyrosine phosphorylated host cell proteins at the sites of microcolony formation.[24] The signals that lead to cortical plaque formation are unknown as are their consequences for pilus mediated attachment. It can be speculated, that the accumulation of receptors beneath the microcolonies somehow contributes to the disbanding of microcolonies and the tight adherence of individual bacteria to the host cell membrane.

Recent experiments using bacteria with mutations in the pilT locus suggest that retraction of the neisserial pili is critically involved in pilus mediated adherence. PilT is localised at the inner bacterial membrane and is considered to be part of the basal machinery that is responsible for pilus assembly and disassembly.[16,17] PilT mutants have a defect in natural competence for DNA uptake and twitching motility.[16] By using lazer tweezers Merz and colleagues have clearly demonstrated that neisserial pili can be retracted and pull gonococci towards antibody coated beads or bacterial colonies and that pilus retraction essentially depends on the PilT protein. [25] PilT mutants have more pili than normal wild-type strains and adhere in great numbers to epithelial cells. However they have a partial defect in cortical plaque formation and are unable to spread from the microcolony, fail to establish intimate contact and do not show reduction in piliation.[20,24] Together these data provide striking evidence for an essential role of pilus retraction in several steps of pilus mediated attachment, like the dispersing of microcolonies and tight adherence.

The sequence of events leading to intimate attachment requires differential regulation of bacterial genes. In meningococci, upregulation of PilC1 is required for efficient initial adherence and occurs after cell contact, even in the absence of pili.[33] Expression declines to basal levels at the stage of intimate adhesion through the action of the transcriptional regulator,

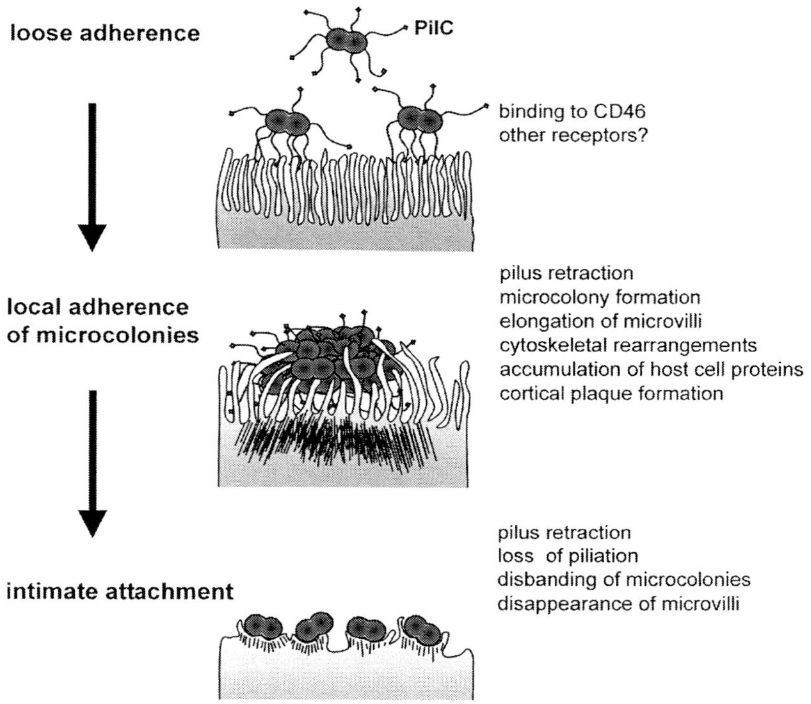

Figure 1. Schematical model of the different steps in pilus mediated attachment of pathogenic *Neisseria* species to epithelial cells.
The pilus tip-located adhesin PilC most likely establishes the first contact with the epithelium. CD46 was suggested as a pilus receptor but it is unclear if it interacts with PilC or other pilus components. Initial interactions with microvilli progresses to the formation of compact neisserial microcolonies. This process is often accompanied with an extension of microvilli towards the microcolony. Pilus induced cortical plaques beneath the microcolonies contain accumulation of actin and actin associated proteins and distinct host cell receptor proteins, like the EGF-receptor, CD44v3 and ICAM-1. In organ cultures and in vitro cell culture models, adherence proceeds within several hours to a tight adherence of individual bacteria that are found in intimate contact with the host cell membrane. The disbanding of microcolonies coinicides with a loss in bacterial piliation and a marked reduction of host cell microvilli. Retraction of the type IV pili could contribute to several steps in the adhesion process, like microcolony formation, induction of cortical plaques, dispersing of microcolonies and tight adherence.

CrgA, that interacts with the pilC1 promoter. CrgA is required for meningococcal intimate attachment to epithelial cells. It accounts for the repression of pilC1 expression to basal levels during intimate adhesion and for the downregulation of capsulation during prolonged attachment that most likely allows better direct interaction between bacterial outer membrane proteins and the host cell surface.[34] However, the intimate attachment following pilus binding does not require Opa or Opc proteins. This strongly indicates the existence of other, as yet uncharacterized adhesion factors which might well be upregulated specifically after pilus contact.

The Role of Opa Proteins in the Interaction with Host Cells

The opacity associated (Opa) outer membrane proteins are the second group of major neisserial adhesins and are involved in a variety of bacteria-host cell interactions. Opa proteins can mediate the attachment to and invasion into epithelial and endothelial cells and trigger the opsonin-independent phagocytosis by PMNs in vitro.[35] There is a huge repertoire of Opa variants in neisserial populations.[36] Up to 11 different Opa variants can be found in one gonococcal strain,[37,38] whereas 3-4 different variants can be expressed by a meningococcal strain.[39] Each Opa protein is encoded by a separate gene. Expression from each *opa* allele is phase variable due to a slipped strand mispairing mechanism that alters the reading frame by modifying the number of pentanucleotide (CTCTT) coding repeats in the N-terminal leader sequence.[37,39] As a consequence the Opa proteins are either fully synthesised or the translation stops early after initiation. This mechanism continuously generates a heterogeneous population of *Neisseria* expressing either none, one or multiple different Opa proteins as long as bacteria replicate.

Two classes of receptors have been identified for the neisserial Opa proteins. The majority of Opa variants interacts with members of the carcinoembryonic antigen related cellular adhesion molecules (CEACAM).[40-42] A minor number of Opa variants mediates the interaction with heparansulfate proteoglycans (HSPG)[43,44] HSPGs serve as receptors for many microbial pathogens.[45] They consist of a protein core with at least one covalently attached heparansulfate (HS) side-chain. HS are composed of the repetitive subunits glucuronic-acid and N-acetylglucosamine and are sulfated at multiple sites. HSPGs are predominantly located in the extracellular matrix and on the cell surface. The two major families of cell surface HSPGs are the GPI-linked glypicans and the transmembrane proteins of the syndecan family. Moreover, a few other transmembrane proteins with a HS side chain exist, for example CD44v3, a splice variant of the hyaluronate receptor.[46,47] Opa-HSPG interaction involves the HS side chains and is mainly charge based. In the meningococci the HSPG binding function is mediated by the Opc protein which is structurally and antigenically different from Opa proteins.[48] Only a small minority of the Opa proteins of *N. gonorrhoeae* interact with sufficient affinity with HSPGs to mediate an uptake into epithelial cells and monocytes.[38,49] HSPG-mediated interactions with monocytes are poorly characterised. However, several different mechanisms have been described for the Opa-HSPG mediated internalisation into epithelial cells. Invasion into Chang conjunctiva cells depends on signal transduction by lipid second messengers.[50] Binding of *N. gonorrhoeae* expressing HSPG-specific Opa proteins leads to an increased production of diacylglycerol (DAG) from phosphatidylcholine (PC) by the PC-specific phospholipase C (PC-PLC). DAG subsequently activates the acidic sphingomyelinase (ASM) which produces ceramide from sphingomyelin (SM). Blocking PC-PLC or ASM by pharmacological inhibitors causes the inhibition of neisserial invasion, indicating that these two enzymes play critical roles in the uptake. Invasion is further inhibited by preincubation of Chang cells with cytochalasin D indicating that it depends on a functional actin cytoskeleton (Fig. 2).[51] Both DAG and ceramide have many molecular targets[52,53] and the signal cascade that leads to a reorganisation of the actin cytoskeleton therefore requires further analysis. However, the signaling pathway is not involved in the integrin dependent uptake pathway used by an *Escherichia coli* strain expressing the *Yersinia* invasin, as it is in the same cell line not affected by a loss of ASM activity.[50] Recent data suggest that the neisserial porin PorB is an additional player involved in this kind of invasion, since mutation of surface exposed loops strongly decreases HSPG dependent invasion into Chang cells.[54]

In many other epithelial cell lines (e.g., HeLa, CHO, Hep-2) uptake via the naturally expressed HSPG receptors is rather inefficient. However, it can be enhanced substantially by an additional interaction with vitronectin and fibronectin, two soluble glycoproteins found in the extracellular matrix and the blood stream. [55, 56] Vitronectin and fibronectin probably act as bridging molecules between the gonococci, HSPGs and their natural integrin receptors, e.g., the $\alpha_v\beta_5$ and $\alpha_5\beta_1$ integrins, leading to receptor accumulation beneath the bacteria and internalisation via a mechanism that depends on the activity of the protein kinase C (PKC).[57,58]

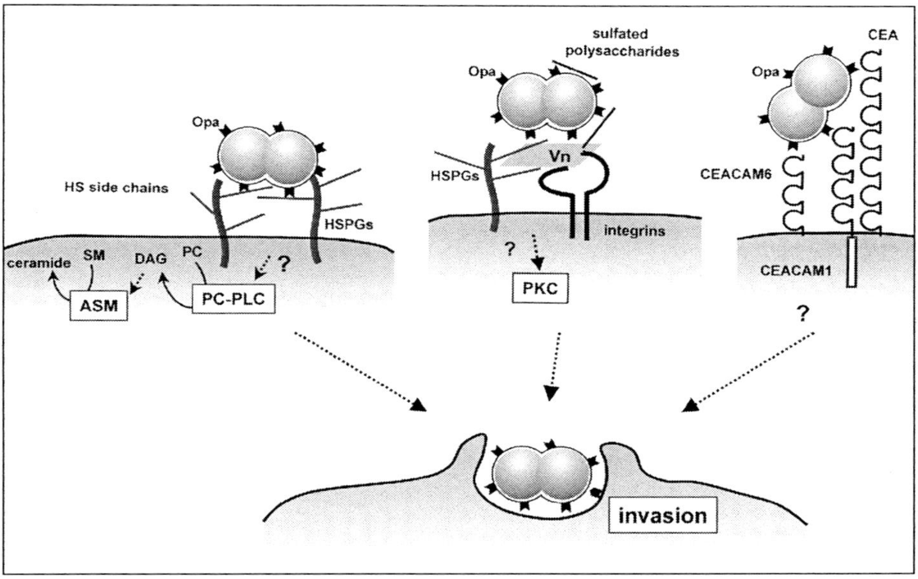

Figure 2. Mechanism of Opa protein interaction with host cell receptors on epithelia. Opa interactions with HSPG and CEACAM receptors can mediate the attachment but also the efficient invasion into various epithelial cell lines in vitro. In Chang conjunctiva cells Opa-HSPG interactions lead to a rapid activation of PC-PLC that produces DAG from PC. DAG was shown to activate ASM which then produces ceramide from SM and can mediate the invasion of *N. gonorrhoeae* via a further unknown signaling pathway that involves rearrangements of the actin cytoskeleton. In HeLa, CHO or Hep-2 cells ingestion of gonococci depends on a concerted interaction of neisserial Opa proteins with the HSPGs, the proteins vitronectin (Vn) or fibronectin (not shown) and integrin receptors and is dependent on the activity of PKC. Interaction with CEACAM1, CEA and CEACAM6 can mediate internalisation into various transfected epithelial cell lines and transcytosis across a polarized epithelial monolayer. The molecular mechanisms regulating this process have not been characterised until now.

The interaction of Opa proteins with vitronectin and fibronectin is probably indirect and occurs via soluble, sulfated polysaccharides which can bind simultaneously to Opa proteins as well as vitronectin and fibronectin.[59,60]

Syndecans are cell surface HSPGs with a transmembrane and a cytoplasmic domain. They are multifunctional proteins that are involved in cell-cell and cell-matrix interactions, and can modulate the ligand dependent activation of primary signaling receptors at the cell surface.[46,47] Syndecan-1 is the major syndecan in epithelial cells, whereas syndecan-4 shows a wider tissue distribution. Overexpression in HeLa cells demonstrated that both syndecans can participate in the internalisation of Opa-expressing gonococci into epithelial cells in vitro.[61] Transfection of mutant syndecans and the use of pharmacological inhibitors revealed a crucial role of the cytoplasmic tails for gonococcal uptake and implied the importance of PKC, phosphatidylinositol-4,5-bisphosphate (PIP_2) and an unknown link to the actin cytoskeleton for syndecan-4 mediated invasion.[61] The role of other HSPGs, e.g., of syndecan-2, which predominates in fibroblasts, or syndecan-3, which abounds in neuronal cells, or the glypicans in promoting neisserial invasion is so far unknown.

Four members of the CEACAM family can function as receptors for the great majority of neisserial Opa variants. These are the GPI-linked CEA (CD66e) and CEACAM6 (NCA, CD66c) as well as the transmembrane proteins CEACAM1 (BGP, CD66a) and CEACAM3 (CGM1, CD66d).[35,62] Different Opa variants bind to distinct subsets of CEACAM receptors with

CEACAM1 and CEA being recognised by most Opa proteins.[63,64] The Opa-CEACAM interaction is a pure protein-protein interaction and the Opa binding site could be identified in the GFCC′-β-sheet of the conserved N terminal domain of the CEACAM receptors.[65-67] CEACAM receptors belong to the immunoglobulin superfamily and are differentially expressed within the human body, including important target tissues of neisserial infection, like the urogenital tract, rectum, cells of the pharyngeal region and of the immune system.[68,69,70] The function of CEACAM receptors is largely unknown. Their frequent dysregulation in malignant cells and most experimental data suggest an important role as regulatory and sensory molecules in cell-cell interactions.[71,72]

Both the transmembrane and the GPI-linked CEACAM receptors can mediate the internalisation of Opa-expressing *Neisseria* into transfected epithelial cells.[63,64,73] Moreover, in the T84 colon carcinoma cell line, which has been used as a model for an epithelial barrier, Opa binding to CEACAM1, CEA and CEACAM6 on the apical pole of the polarized cells promotes transcytosis accross the monolayer via a transcellular route.[74] The underlying mechanisms of epithelial cell invasion through CEACAM receptors are still unknown. However, these observations may have important implications in neisserial virulence since in mucosal epithelia CEACAM receptors are typically expressed on the apical pole where they could serve as important invasion receptors for gonococci and meningococci. Moreover, CEACAM expression can be upregulated by neisserial LPS and proinflammatory cytokines that are present at high local concentrations during neisserial infection.[75,76] This may have important implications for the course of neisserial infections.

It is intriguing to note that Opa proteins also target the pathogenic *Neisseria* species to human monocytes and neutrophils, resulting in phagocytosis and presumably in intracellular killing in vitro. Whether in vivo such opsonin-independent interactions with phagocytes are beneficial or detrimental to the internalised bacteria or the colonising population as a whole is subject to speculation. PMNs express high levels of CEACAM1, CEACAM3 and CEACAM6 which in non-activated cells are mostly present in intracellular granules.[77,78] Upon PMN stimulation, CEACAM receptors become readily exposed on the cell surface and serve as receptors for gonococci expressing the appropriate Opa variants.[40,41,42] Opa-CEACAM mediated interaction of gonococci with the in vitro-differentiated human myelomonocytic JOSK-M cell-line expressing CEACAM1 and CEACAM6 leads to tyrosine phosphorylation of host cell proteins, activation of the Src-family kinases Hck and Fgr, the Rho family GTPase Rac-1, p21-activated kinase (PAK), Jun N-terminal kinas (JNK) and ASM (Fig. 3).[79,80] Activation of the Src-family kinases, Rac, and ASM is required for non-opsonised phagocytosis in this model. Upregulation of Hck and Fgr is accompanied by a downregulation of the activity of the SH2 domain containing tyrosine phosphatase SHP-1.[81] These observations are entirely consistent with the proposed stimulatory function of CEACAM receptors in neutrophil activation[82-84] and with the observed association of Src-family kinases and SH2-domain containing phosphatases with the cytoplasmic domain of CEACAM1.[85,86] These findings demonstrate that neisserial binding to CEACAM-receptors activates specific signaling pathways that result in the activation of phagocytes and the internalisation of bacteria. This notion is supported by a recent study in which CEACAM3 was expressed in a chicken B-cell line and found to mediate particularly efficient phagocytosis of Opa+ gonococci by signaling through an immunoreceptor tyrosine-based activation motif (ITAM) in its cytoplasmic domain, the ITAM-binding tyrosine kinase Syk, and phospholipase C-γ.[87] These data show that nonopsonic phagocytosis via CEACAM3 is highly reminiscent of phagocytosis of IgG-opsonized particles by the low affinity Fc-receptor Fcγ-RIIa, which is also ITAM-dependent.[90] Interestingly, CEACAM1 contains putative immunoreceptor tyrosine-based inhibitory motifs (ITIM) in it′s cytoplasmic domain. Signaling through ITAM and ITIM motifs has opposite effects in the regulation of important leukocytes functions, such as antigen receptor signaling in lymphocytes and Fcγ-receptor mediated phagocytosis by myeloid cells.[88,89] Whether in neutrophils the putative ITIM in the cytoplasmic tail of CEACAM1 could counteract signaling by CEACAM3 (or other phagocytic

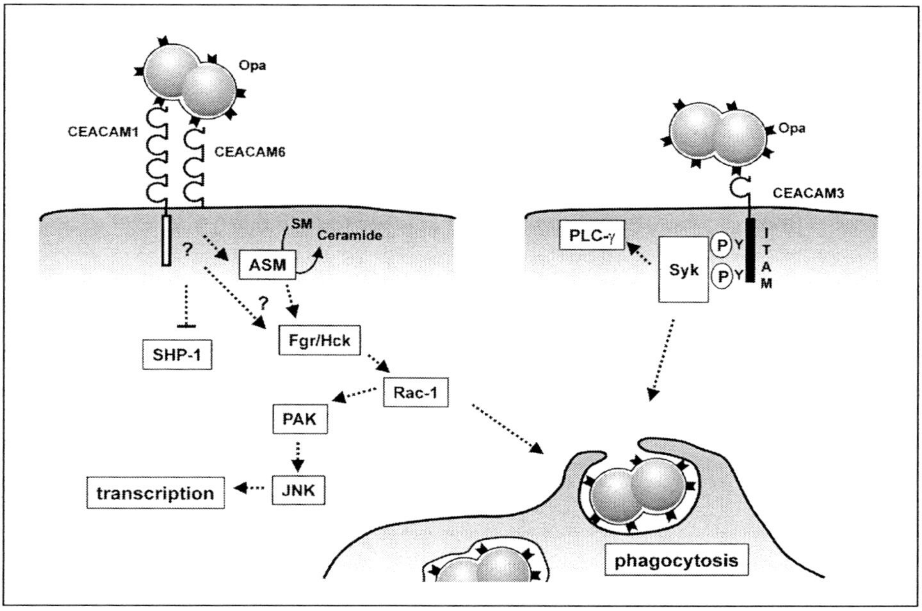

Figure 3. Opa mediated interactions with CEACAM receptors on phagocytes.
Opa-CEACAM interaction in differentiated JOSK-M cells leads to an early activation of the ASM and the Src family kinases Fgr and Hck. This results in the activation of Rac-1, a small GTP binding protein which can regulate cytoskeletal rearrangements. Rac-1 might also activate p21 activated kinase (PAK) and JNK and thereby modulate transcription. This signaling pathway is accompanied by the downregulation of the activity of the phosphatase SHP-1 and is independent of the kinase Syk. Gonococcal binding to CEACAM3-transfected DT40-B cells results in a rapid phagocytosis of adherent gonococci. Ingestion depends on tyrosine residues localised in an ITAM motif in the receptors cytoplasmic tail and a signaling pathway that involves the kinase Syk which is known to bind phosphorylated tyrosine residues in ITAM motifs and the phospholipase C-γ.

receptors), is not known and will be a fascinating area of future research that will not only contribute to our comprehension of host-pathogen interactions but will also improve our limited understanding of CEACAM receptor functions.

PorB Influence on Host Cell Functions

Porins are the most abundant proteins in the outer membrane of *Neisseria*. Meningococci can produce up to two porins simultaneously: Expression of PorA (class I protein) is subject to phase variation, while PorB is essential and constitutively expressed. Gonococci only express PorB. Depending on the strain, PorB belongs to one of two distinct and mutually exclusive families designated as class 2 and class 3 in meningococci and $PorB_{IA}$ and $PorB_{IB}$ in gonococci. Like porins of other gram-negative bacteria, the neisserial porins form trimeric β-pleated barrels in the bacterial outer membrane. [90] Apart from the normal porin functions the PorB proteins of pathogenic *Neisseria* species have some unusual properties that suggest an important role in virulence. PorB was shown to contribute to serum resistence of *N. gonorrhoeae* by interfering with components of the classical and alternative pathway of complement activation.[91,92] Moreover, PorB has been shown to influence such diverse processes as degranulation, oxidative burst and phagosomal maturation in phagocytes, invasion of epithelial cells and apoptosis in various cell types in vitro.[54,93-96] The key to the understanding of at least some of the diverse properties of PorB lies in its unusual ability to translocate from the bacterial outer membrane into artificial membranes and into the plasma membrane of human cells.[97,98] Translocation requires intimate contact between the neisserial outer membrane and the cytoplasmic membrane and therefore depends on prior attachment via pili or Opa proteins. However, translocation can also be observed with purified PorB indicating that other virulence factors are necessary for establishing a tight contact but not for the translocation process.[98] The translocation of PorB occurs vectorially, in a way that the loops which are surface-exposed in bacteria face the host cell cytoplasm after translocation. Once inserted into the host cell membrane host cell cytosolic nucleoside trisphosphates (ATP or GTP) regulate voltage gating and ion selectivity of the porin channel.[99] Both the gonococcal and the meningococcal porin were shown to target the mitochondria of the host cell upon infection and to interfere with the regulation of apoptosis at the mitochondrial signaling checkpoint.[100,101] Following the translocation of gonococcal PorB to the mitochondria, pro-apoptogenic factors are released into the cytosol that activate cystein proteases, like the caspases, thereby initiating the apoptotic program.[95,101] Interestingly, PorB shows striking similarities with the mitochondrial porin VDAC (voltage dependent anion channel), which is part of a protein complex that regulates the release of pro-apoptogenic factors. In marked contrast to these findings are those of Massari and colleagues who also observed translocation of meningococcal PorB to the mitochondria, but report an anti-apoptotic effect of PorB after treatment with staurosporine.[100] The circumstances that lead to these opposite effects of PorB clearly need further analysis, especially when considering that triggering or preventing apoptosis could be beneficial or detrimental depending on the particular microenvironment colonised by the *Neisseria* during the course of infection.

Stress Response Signaling in Epithelial and Endothelial Cells

Epithelial cells not only are the first barrier colonised by pathogenic *Neisseria* species. Recent in vitro studies suggest that they may also form an important integral component of the innate immune response. In various epithelial cell lines gonococci activate the early response transcription factor nuclear factor kappa B (NF-κB) and the basic domain/leucine zipper transcription factor activator protein 1 (AP-1). [102] This results in the rapid upregulation of proinflammatory cytokine mRNAs and the release of granulocyte-macrophage colony-stimulating factor (GM-CSF), tumour necrosis factor-α (TNFα), interleukin-8 (IL-8), IL-6, monocyte chemoattractant protein 1 (MCP-1), transforming growth factor-β (TGFβ), and IL-1β.[102] The signal transduction pathway leading to cytokine induction through AP-1 was partially elucidated. In infected epithelial cells the AP-1 transcriptional complex becomes activated through phosphorylation by the stress response kinase c-Jun N-terminal kinase (JNK).[103] Upstream of JNK a mitogen-activated protein kinase cascade is regulated indirectly by p21-activated kinase (PAK). Consistent with this, the process is sensitive to inhibition by Toxin B of *Clostridium difficile*[103] which inhibits specifically the small GTPases of the Rho family that activate PAK.

TNFα, IL-1, IL-6 have been implicated in the defence of bacterial infections and promote bactericidal activity of leukocytes. On the other hand, TNFα is also a key mediator of tissue damage in gonococcal infections[75,76] that may promote bacterial dissemination by rupture and exfoliation of epithelial cells. TNFα is also a major mediator of septic shock during meningococcal disease. The neisserial components that induce cytokine production in epithelial cells are not entirely clear. In monocytes, in endothelial cells expressing toll-like receptor 2 (TLR-2) and TLR-4 and in primary endometrial epithelial cells cytokine induction by gonococci is contact-independent and can be mediated by LPS.[104,105] In contrast, in epithelial cell lines that are non-responsive to LPS (presumably because they lack CD14, TLR-2 and TLR-4) cytokine induction requires gonococcal adherence via pili, HSPG-binding Opa proteins or CEACAM-binding Opa.[102,105a]

These data suggest that in vivo both LPS-dependent and independent mechanisms may contribute to cytokine induction and that epithelial cells may participate in the first line of host defense. Whether the observed early upregulation of a variety of proinflammatory cytokines in vitro would be beneficial or detrimental for the human host is not clear. Interestingly, expression of the Opa-receptor CEACAM1 is itself under the control of NF-κB. In both epithelial and endothelial cells infection can therefore upregulate CEACAM1 expression, resulting in increased adherence of bacteria to their target cells.[102,105a] Receptor upregulation could be a mechanism by which local mucosal inflammation induced by other bacterial or viral pathogens could predispose a person for neisserial colonisation or invasive disease. These observations also raise the intriguing possibility that CEACAM1-mediated host cell invasion could be a mechanisms of immune evasion that could be particularly efficient in the presence of a local proinflammatory cytokine profile. Furthermore, CEACAM1 is capable of undergoing homotypic and heterotypic interactions with CEACAM family members expressed by human neutrophils and CEACAM receptors are implicated in the adherence of PMNs to cytokine stimulated endothelial cells.[82,106,107] The upregulation of this and other adhesion molecules on infected tissue could thus increase the infiltration of neutrophils which may exacerbate tissue damage and disease but may also contribute to pathogen control.

Conclusions

The pathogenic *Neisseria* species are paradigmatic for an evolutionary highly adapted pathogen host relationship. Their key strategy of continuously changing the functional characteristics and combination of prominent virulence factors by phase and antigenic variation enables these two pathogens to escape the immune system, to use different routes of infection, and to generate diverse subpopulations of bacteria that are specialised to the different environmental niches within the human body. At present, it is still difficult to counteract this strategy used also by a variety of other bacteria, viruses and parasites. The analysis of the molecular events controlling critical steps in neisserial infection gave us fundamental insights into the gonococcal and meningococcal disease processes and general principles of molecular host-pathogen interactions. Research over the past decade provided us with the knowledge that host cell functions are indeed as important as bacterial virulence factors in order to successfully establish colonisation. Moreover, an intense cross-talk with sending and sensing of signals between bacteria and host cells seems to control the critical steps of infection. The discovery of relatively consistent molecular interactions, such as the dependence of pilus mediated attachment on the PilC protein and of the Opa binding site on CEACAM receptors indicates that we are probably on the right way to discover molecular targets that promise successful medical intervention to combat or prevent infections caused by the pathogenic *Neisseria* species or pathogens utilising similar strategies.

References

1. Caugant, DA, Hoiby EA, Magnus P et al. Asymptomatic carriage of Neisseria meningitidis in a randomly sampled population. J Clin Microbiol 1994; 32:323-330.
2. Rosenstein, NE, Perkins BA, Stephens DS et al. Meningococcal disease. N Engl J Med 2001; 344:1378-1388.
3. van Deuren M, Brandtzaeg P, van der Meer JW. Update on meningococcal disease with emphasis on pathogenesis and clinical management. Clin Microbiol Rev 2000; 13:144-166
4. Handsfield HH. Neisseria gonorrhoeae. In: Mandell GL, Douglas RG, Bennet JE, eds. Principles and Practice of Infectious Diseases. New York: Churchill Livingstone, 1990:1613-1631.
5. Munday PE. Clinical aspects of pelvic inflammatory disease. Hum Reprod 1997; 12:121-126.
6. Kerle KK, Mascola JR, Miller TA. Disseminated gonococcal infection. Am Fam Physician 1992; 45:209-214.
7. Meyer TF, Pohlner J, van Putten JPM Biology of the pathogenic Neisseriae. Curr Top Microbiol Immunol 1994; 192:283-317.
8. Christie PJ, Vogel JP. Bacterial type IV secretion: Conjugation systems adapted to deliver effector molecules to host cells. Trends Microbiol 2000; 8:354-360.
9. Hueck CJ. Type III protein secretion systems in bacterial pathogens of animals and plants. Microbiol Mol Biol Rev 1998; 62:379-433.
10. McGee ZA, Stephens DS, Hoffman LH et al. Mechanisms of mucosal invasion by pathogenic Neisseria. Rev Infect Dis 1983; 5 Suppl 4:S708-S714.

11. Swanson J. Studies on gonococcus infection. IV. Pili: Their role in attachment of gonococci to tissue culture cells. J Exp Med 1973; 137:571-589.
12. Fussenegger M, Rudel T, Barten R et al. Transformation competence and type-4 pilus biogenesis in Neisseria gonorrhoeae—A review. Gene 1997; 192:125-134.
13. Wall D, Kaiser D. Type IV pili and cell motility. Mol Microbiol 1999; 32:1-10.
14. Parge HE, Forest KT, Hickey MJ et al. Structure of the fibre-forming protein pilin at 2.6 A resolution. Nature 1995; 378:32-38.
15. Scheuerpflug I, Rudel T, Ryll R et al. Roles of PilC and PilE proteins in pilus-mediated adherence of Neisseria gonorrhoeae and Neisseria meningitidis to human erythrocytes and endothelial and epithelial cells. Infect Immun 1999; 67:834-843.
16. Wolfgang M, Lauer P, Park HS et al. PilT mutations lead to simultaneous defects in competence for natural transformation and twitching motility in piliated Neisseria gonorrhoeae. Mol Microbiol 1998; 29:321-330.
17. Wolfgang M, van Putten JPM, Hayes SF Components and dynamics of fiber formation define a ubiquitous biogenesis pathway for bacterial pili. EMBO J 2000; 19:6408-6418.
18. Griffiss JM, Lammel CJ, Wang J et al. Neisseria gonorrhoeae coordinately uses Pili and Opa to activate HEC-1-B cell microvilli, which causes engulfment of the gonococci. Infect Immun 1999; 67:3469-3480.
19. Pujol C, Eugene E, de Saint ML et al. Interaction of Neisseria meningitidis with a polarized monolayer of epithelial cells. Infect Immun 1997; 65:4836-4842.
20. Pujol C, Eugene E, Marceau M et al. The meningococcal PilT protein is required for induction of intimate attachment to epithelial cells following pilus-mediated adhesion. Proc Natl Acad Sci USA 1999; 96:4017-4022.
21. Stephens DS, Hoffman LH, McGee ZA. Interaction of Neisseria meningitidis with human nasopharyngeal mucosa: attachment and entry into columnar epithelial cells. J Infect Dis 1983; 148:369-376.
22. Shaw, JH, Falkow S. Model for invasion of human tissue culture cells by Neisseria gonorrhoeae. Infect Immun 1988; 56:1625-1632.
23. Merz, AJ, So M. Attachment of piliated, Opa- and Opc-gonococci and meningococci to epithelial cells elicits cortical actin rearrangements and clustering of tyrosine-phosphorylated proteins. Infect Immun 1997; 65:4341-4349.
24. Merz AJ, Enns CA, So M. Type IV pili of pathogenic Neisseriae elicit cortical plaque formation in epithelial cells. Mol Microbiol 1999; 32:1316-1332.
25. Merz AJ, So M, Sheetz MP. Pilus retraction powers bacterial twitching motility. Nature 2000; 407:98-102.
26. Rahman M, Kallstrom H, Normark S et al. PilC of pathogenic Neisseria is associated with the bacterial cell surface. Mol Microbiol 1997; 25:1-25.
27. Rudel T, Scheurerpflug I, Meyer TF. Neisseria PilC protein identified as type-4 pilus tip-located adhesin. Nature 1995; 373:357-359.
28. Ryll RR, Rudel T Scheuerpflug I et al. PilC of Neisseria meningitidis is involved in class II pilus formation and restores pilus assembly, natural transformation competence and adherence to epithelial cells in PilC-deficient gonococci. Mol Microbiol 1997; 23:879-892.
29. Kallstrom H, Liszewski MK, Atkinson JP. Membrane cofactor protein (MCP or CD46) is a cellular pilus receptor for pathogenic Neisseria. Mol Microbiol 1997; 25:639-647.
30. Liszewski MK, Farries TC, Lublin DM et al. Control of the complement system. Adv Immunol 1996; 61:201-283
31. Ayala BP, Vasquez B, Clary S et al. The pilus-induced Ca2+ flux triggers lysosome exocytosis and increases the amount of Lamp1 accessible to Neisseria IgA1 protease. Cell Microbiol 2001; 3:265-275.
32. Kallstrom H, Islam MS, Berggren PO et al. Cell signaling by the type IV pili of pathogenic Neisseria. J.Biol.Chem. 1998; 273:21777-21782.
33. Taha MK, Morand PC, Pereira Y et al. Pilus-mediated adhesion of Neisseria meningitidis: the essential role of cell contact-dependent transcriptional upregulation of the PilC1 protein. Mol Microbiol 1998; 28:1153-1163.
34. Deghmane AE, Petit S, Topilko A et al. Intimate adhesion of Neisseria meningitidis to human epithelial cells is under the control of the crgA gene, a novel LysR-type transcriptional regulator. EMBO J 2000; 19:1068-1078.
35. Dehio C, Gray-Owen SD, Meyer TF. The role of neisserial Opa proteins in interactions with host cells. Trends Microbiol 1998; 6:489-495.
36. Malorny B, Morelli G, Kusecek B et al. Sequence diversity, predicted two-dimensional protein structure, and epitope mapping of neisserial Opa proteins. J Bacteriol 1998; 180:1323-1330.

37. Stern A, Brown M, Nickel P et al. Opacity genes in Neisseria gonorrhoeae: Control of phase and antigenic variation. Cell 1986; 47:61-71.
38. Kupsch EM, Knepper B, Kuroki T et al. Variable opacity (Opa) outer membrane proteins account for the cell tropisms displayed by Neisseria gonorrhoeae for human leukocytes and epithelial cells. EMBO J 1993; 12:641-650.
39. Stern A, Meyer TF. Common mechanism controlling phase and antigenic variation in pathogenic neisseriae. Mol Microbiol 1987; 1:5-12.
40. Chen T, Gotschlich EC. CGM1a antigen of neutrophils, a receptor of gonococcal opacity proteins. Proc Natl Acad Sci USA 1996; 93:14851-14856.
41. Gray-Owen SD, Dehio C, Haude A et al. CD66 carcinoembryonic antigens mediate interactions between Opa-expressing Neisseria gonorrhoeae and human polymorphonuclear phagocytes. EMBO J 1997; 16:3435-3445.
42. Virji M, Makepeace K, Ferguson DJ et al. Carcinoembryonic antigens (CD66) on epithelial cells and neutrophils are receptors for Opa proteins of pathogenic neisseriae. Mol Microbiol 1996; 22:941-950.
43. Chen T, Belland RJ, Wilson J et al. Adherence of pilus-Opa+ gonococci to epithelial cells in vitro involves heparan sulfate. J Exp Med 1995; 182:511-517.
44. van Putten JPM, Paul SM. Binding of syndecan-like cell surface proteoglycan receptors is required for Neisseria gonorrhoeae entry into human mucosal cells. EMBO J 1995; 14:2144-2154.
45. Rostand KS, Esko JD. Microbial adherence to and invasion through proteoglycans. Infect Immun 1997; 65:1-8.
46. Carey DJ. Syndecans: Multifunctional cell-surface co-receptors. Biochem J 1997; 327:1-16.
47. Zimmermann P, David G. The syndecans, tuners of transmembrane signaling. FASEB J 1999; 13:S91-S100.
48. de Vries FP, van der Ende A, van Putten JPM et al. Invasion of primary nasopharyngeal epithelial cells by Neisseria meningitidis is controlled by phase variation of multiple surface antigens. Infect Immun 1996; 64:2998-3006.
49. Knepper B, Heuer I, Meyer TF et al. Differential response of human monocytes to Neisseria gonorrhoeae variants expressing pili and opacity proteins. Infect Immun 1997; 65:4122-4129.
50. Grassme H, Gulbins E, Brenner B et al. Acidic sphingomyelinase mediates entry of N. gonorrhoeae into nonphagocytic cells. Cell 1997; 91:605-615.
51. Grassme H, Ireland RM, van Putten JPM. Gonococcal opacity protein promotes bacterial entry-associated rearrangements of the epithelial cell actin cytoskeleton. Infect Immun 1996; 64:1621-1630.
52. Lennartz MR. Phospholipases and phagocytosis: The role of phospholipid-derived second messengers in phagocytosis. Int J BiochemCell Biol 1999; 31:415-430.
53. Levade T, Jaffrezou JP. Signaling sphingomyelinases: Which, where, how and why? Biochim Biophys Acta 1999; 1438:1-17.
54. Bauer FJ, Rudel T, Stein M et al. Mutagenesis of the Neisseria gonorrhoeae porin reduces invasion in epithelial cells and enhances phagocyte responsiveness. Mol Microbiol 1999; 31:903-913.
55. Duensing TD van Putten JPM. Vitronectin mediates internalization of Neisseria gonorrhoeae by Chinese hamster ovary cells. Infect Immun 1997 65:964-970.
56. Gomez-Duarte OG, Dehio M, Guzman CA et al. Binding of vitronectin to opa-expressing Neisseria gonorrhoeae mediates invasion of HeLa cells. Infect Immun 1997; 65:3857-3866.
57. Dehio M, Gomez-Duarte OG, Dehio C et al. Vitronectin-dependent invasion of epithelial cells by Neisseria gonorrhoeae involves alpha(v) integrin receptors. FEBS Lett 1998; 424:84-88.
58. van Putten, JPM, Duensing TD, Cole RL. Entry of OpaA+ gonococci into HEp-2 cells requires concerted action of glycosaminoglycans, fibronectin and integrin receptors. Mol Microbiol 1998; 29:369-379.
59. Duensing TC, Wing JS, van Putten JPM. Sulfated polysaccharide-directed recruitment of mammalian host proteins: a novel strategy in microbial pathogenesis. Infect Immun 1999; 67:4463-4468.
60. van Putten, JPM, Hayes SF, Duensing TD. Natural proteoglycan receptor analogs determine the dynamics of Opa adhesin-mediated gonococcal infection of Chang epithelial cells. Infect Immun 1997; 65:5028-5034.
61. Freissler E, Meyer auf der Heyde, A David G et al. Syndecan-1 and syndecan-4 can mediate the invasion of OpaHSPG-expressing Neisseria gonorrhoeae into epithelial cells. Cell Microbiology 2000; 2:69-82.
62. Beauchemin N, Draber P, Dveksler G et al. Redefined nomenclature for members of the carcinoembryonic antigen family. Exp Cell Research 1999; 252:243-249.
63. Bos MP, Grunert F, Belland RJ. Differential recognition of members of the carcinoembryonic antigen family by Opa variants of Neisseria gonorrhoeae. Infect Immun 1997; 65:2353-2361.

64. Gray-Owen SD, Lorenzen DR, Haude A. Differential Opa specificities for CD66 receptors influence tissue interactions and cellular response to Neisseria gonorrhoeae. Mol Microbiol 1997; 26:971-980.
65. Bos MP, Hogan D, Belland RJ. Homologue scanning mutagenesis reveals CD66 receptor residues required for neisserial Opa protein binding. J Exp Med 1999; 190:331-340.
66. Popp A, Dehio C, Grunert F et al. Molecular analysis of neisserial Opa protein interactions with the CEA family of receptors: Identification of determinants contributing to the differential specificities of binding. Cell Microbiol 1999; 1:169-181.
67. Virji M, Evans D, Hadfield A et al. Critical determinants of host receptor targeting by Neisseria meningitidis and Neisseria gonorrhoeae: Identification of Opa adhesiotopes on the N-domain of CD66 molecules. Mol Microbiol 1999; 34:538-551.
68. Bamberger AM, Riethdorf L, Nollau P. Dysregulated expression of CD66a (BGP, C-CAM), an adhesion molecule of the CEA family, in endometrial cancer. Am J Pathol 1998; 152:1401-1406.
69. Ogihara S, Kato H. Endocrine cell distribution and expression of tissue-associated antigens in human female paraurethral duct: Possible clue to the origin of urethral diverticular cancer. Int J Urol 2000; 7:10-15.
70. Prall F, Nollau P, Neumaier M et al. CD66a (BGP), an adhesion molecule of the carcinoembryonic antigen family, is expressed in epithelium, endothelium, and myeloid cells in a wide range of normal human tissues. J Histochem Cytochem. 1996; 44:35-41.
71. Hammarstrom S, Olsen A, Teglund S et al. The nature and expression of the human CEA family. In: Stanners CP, ed. Cell Adhesion and Communication Mediated by the CEA Family. Amsterdam: Haarwood Academic Publishers, 1998:10-36.
72. Obrink B. CEA adhesion molecules: Multifunctional proteins with signal-regulatory properties. Curr Opin Cell Biol 1997; 9:616-626.
73. Muenzner P, Dehio C, Fujiwara T et al. Carcinoembryonic antigen family receptor specificity of Neisseria meningitidis Opa variants influences adherence to and invasion of proinflammatory cytokine-activated endothelial cells. Infect Immun 2000; 68:3601-3607.
74. Wang J, Gray-Owen SD, Knorre A et al. Opa binding to cellular CD66 receptors mediates the transcellular traversal of Neisseria gonorrhoeae across polarized T84 epithelial cell monolayers. Mol Microbiol 1998; 30:657-671.
75. McGee ZA, Clemens CM, Jensen RL et al. Local induction of tumor necrosis factor as a molecular mechanism of mucosal damage by gonococci. Microb Pathog 1992; 12:333-341.
76. McGee ZA, Jensen RL, Clemens CM et al. Gonococcal infection of human fallopian tube mucosa in organ culture: relationship of mucosal tissue TNF-alpha concentration to sloughing of ciliated cells. Sex Transm Dis 1999; 26:160-165.
77. Ducker, TP, Skubitz KM. Subcellular localization of CD66, CD67, and NCA in human neutrophils. J Leukoc Biol. 1992; 52:11-16.
78. Kuroki M, Yamanaka T, Matsuo Y. Immunochemical analysis of carcinoembryonic antigen (CEA)-related antigens differentially localized in intracellular granules of human neutrophils. Immunol Invest 1995; 24:829-843.
79. Hauck CR, Meyer TF, Lang F et al. CD66-mediated phagocytosis of Opa52 Neisseria gonorrhoeae requires a Src-like tyrosine kinase- and Rac1-dependent signaling pathway. EMBO J 1998; 17:443-454.
80. Hauck CR, Grassme H, Bock J et al. Acid sphingomyelinase is involved in CEACAM receptor-mediated phagocytosis of Neisseria gonorrhoeae. FEBS Lett 2000; 478:260-266.
81. Hauck CR, Gulbins E, Lang F et al. Tyrosine phosphatase SHP-1 is involved in CD66-mediated phagocytosis of Opa52-expressing Neisseria gonorrhoeae. Infect Immun 1999; 67:5490-5494.
82. Kuijpers, TW, Hoogerwerf M, van der Laan LJ et al. CD66 nonspecific cross-reacting antigens are involved in neutrophil adherence to cytokine-activated endothelial cells. 1992; J Cell Biol 118:457-466.
83. Kuijpers TW, van der Schoot CE, Hoogerwerf M et al. Cross-linking of the carcinoembryonic antigen-like glycoproteins CD66 and CD67 induces neutrophil aggregation. J Immunol 1993; 151:4934-4940.
84. Skubitz KM, Campbell KD, Skubitz SP. 1996. CD66a, CD66b, CD66c, and CD66d each independently stimulate neutrophils. J Leukoc Biol 1996; 60:106-117.
85. Brummer J, Neumaier M, Gopfert C et al. Association of pp60c-src with biliary glycoprotein (CD66a), an adhesion molecule of the carcinoembryonic antigen family downregulated in colorectal carcinomas. Oncogene 1995; 11:1649-1655.
86. Skubitz KM, Campbell KD, Ahmed K et al. CD66 family members are associated with tyrosine kinase activity in human neutrophils. J Immunol 1995; 155:5382-5390.

87. Chen T, Bolland TS, Chen I, et al The CGM1a (CEACAM3/CD66d) mediated phagocytic pathway of Neisseria gonorrhoeae expressing opacity (Opa) proteins is also the pathway to cell death. J Biol Chem 2001; 276:17413-17419
88. Daeron M. Fc receptor biology. Annu Rev Immunol 1997; 15:203-34.
89. Ravetch JV, Lanier LL. Immune inhibitory receptors. Science 2000; 290:84-89.
90. van der Ley P, Heckels JE, Virji M et al. Topology of outer membrane porins in pathogenic Neisseria spp. Infect Immun 1991; 59:2963-2971.
91. Ram S, McQuillen DP, Gulati S et al. Binding of complement factor H to loop 5 of porin protein 1A: a molecular mechanism of serum resistance of nonsialylated Neisseria gonorrhoeae. J Exp Med 1998; 188:671-680.
92. Ram S, Cullinane M, Blom AM et al. Binding of C4b-binding Protein to Porin. A molecular mechanism of serum resistance of Neisseria gonorrhoeae. J Exp Med 2001; 193:281-296.
93. Lorenzen DR, Gunther D, Pandit J et al. Neisseria gonorrhoeae porin modifies the oxidative burst of human professional phagocytes. Infect.Immun. 2000; 68:6215-6222.
94. Mosleh I, Huber LA, Steinlein P et al. Neisseria gonorrhoeae porin modulates phagosome maturation. J Biol Chem 1998; 273:35332-35338.
95. Muller A, Gunther D, Dux F et al. Neisserial porin (PorB) causes rapid calcium influx in target cells and induces apoptosis by the activation of cysteine proteases. EMBO J 1999; 18:339-352.
96. van Putten JPM, Duensing TD, Carlson J. Gonococcal invasion of epithelial cells driven by P.IA, a bacterial ion channel with GTP binding properties. J.Exp.Med. 1998; 188:941-952.
97. Lynch EC, Blake MS, Gotschlich EC et al. Studie on Porins: Spontaneously transferred from whole cells and from proteins of Neisseria gonorrhoeae and Neisseria meningitidis. Biophysic J 1984; 45:104-107.
98. Weel JF, van Putten JPM. Fate of the major outer membrane protein P.IA in early and late events of gonococcal infection of epithelial cells. Res Microbiol 1991; 142:985-993.
99. Rudel T, Schmid A, Benz R et al. Modulation of Neisseria porin (PorB) by cytosolic ATP/GTP of target cells: parallels between pathogen accommodation and mitochondrial endosymbiosis. Cell 1996; 85:391-402.
100. Massari P, HoY Wetzler LM. Neisseria meningitidis porin PorB interacts with mitochondria and protects cells from apoptosis. Proc Natl Acad Sci USA 2000; 97:9070-9075.
101. Muller A, Gunther D, Brinkmann V et al. Targeting of the pro-apoptotic VDAC-like porin (PorB) of Neisseria gonorrhoeae to mitochondria of infected cells. EMBO J 2000; 19:5332-5343.
102. Naumann M, Wessler S, Bartsch C et al. Neisseria gonorrhoeae epithelial cell interaction leads to the activation of the transcription factors nuclear factor kappaB and activator protein 1 and the induction of inflammatory cytokines. J Exp Med. 1997; 186:247-258.
103. Naumann M, Rudel T, Wieland B et al. Coordinate activation of activator protein 1 and inflammatory cytokines in response to Neisseria gonorrhoeae epithelial cell contact involves stress response kinases. J Exp Med 1998; 188:1277-1286.
104. Christodoulides M, Everson JS, Liu BL et al. Interaction of primary human endometrial cells with Neisseria gonorrhoeae expressing green fluorescent protein. Mol Microbiol 2000; 35:32-43.
105. Muenzner P, Naumann M, Meyer TF et al. Pathogenic Neisseria trigger expression of their CEACAM1 (CD66a) receptor on primary endothelial cells by activating the immediate early respnse transcription factor NF-kB. J Biol Chem 2001; 276:24331-24340.
105a. Muenzer P, Billker O, Meyer TF et al. Nuclear factor NF-κB directs carcinoembryonic antigen-related cellular adhesion molecule 1 receptor expression in Neisseria gonorrhoeae-infected epithelial cells. J Biol Chem 2002; 277:7468-7446.
106. Benchimol S, Fuks, Jothy S et al. Carcinoembryonic antigen, a human tumor marker, functions as an intercellular adhesion molecule. Cell 1989; 57:327-334.
107. Oikawa S, Inuzuka C, Kuroki M et al. A specific heterotypic cell adhesion activity between members of carcinoembryonic antigen family, W272 and NCA, is mediated by N-domains. J Biol Chem 1991; 266:7995-8001.

CHAPTER 14

Against Gram-Negative Bacteria:
The Lipopolysaccharide Case

Ignacio Moriyón

Introduction

It is estimated that bacteria and eucaryotes have coexisted for about 1400 millions years on Earth. The immune system is one of the results of this coexistence, and its cornerstone the ability to distinguish self from non-self, an ability shared by both the innate and adaptive immune systems. Adaptive immunity can detect subtle variations in just one or few components of macromolecules, and this provides a powerful mechanism for identification of foreign or altered macromolecules. On the other hand, the far less sophisticated innate immunity relies on the recognition of overall molecular patterns (MPs) absent from self and repeatedly used by microbes, and the molecules that take part in such recognition and activate the innate immune system have been termed pattern recognition receptors (PRR).[1] Recent evidence shows that mechanisms of recognition of foreign MPs have been highly conserved throughout evolution[2] and this stresses that, as they act before the adaptive system becomes fully active, such mechanisms constitute an indispensable first line of defense against microbial invaders. Moreover, evolution has tuned fine mechanisms of interplay by which the innate system bolsters adaptive immunity.

Pathogenic prokaryotes belong to the *Bacteria* (or *Eubacteria*) domain (or *Imperium*). Although the ecophysiological diversity of prokaryotes is astounding and largest among living beings, the structural variations they display are comparatively few and are concentrated on the architecture of the cell envelopes. Cell envelopes of *Bacteria* belong to two main types: those based on a single membrane, and those based on two membranes enclosing a periplasmic space. With few exceptions, thick peptidoglycan layers surround the single-membrane based envelopes, and thinner peptidoglycan layers are enclosed within the periplasmic space in the two-membrane design. Additional external layers can be present in both types. In contrast with bacteria carrying single membranes with a thick outer peptidoglycan layer, those displaying the two-membrane envelopes are almost always negative by the Gram staining method, and are thus called gram-negative bacteria. Because of functional constraints, several molecular details of the gram-negative cell envelope are highly conserved and carry peculiar MPs. This review deals with the MPs displayed by what is perhaps the most characteristic macromolecule of the gram-negative envelope, the glycolipid known as lipopolysaccharide (LPS) or endotoxin. The innate immune system responds to minute amounts of LPS releasing proinflammatory and inflammatory cell mediators that act as the messengers of an early warning system, and endotoxic shock is not but the pathophysiological condition derived from the massive release of such mediators caused by an overload of the system.

LPS Structure, Role in the Outer Membrane (OM) and MP

Taxonomically, most gram-negative bacteria belong to the *Proteobacteria*, also called purple gram-negative bacteria because it is thought that they share a common photosynthetic ancestor.

Intracellular Pathogens in Membrane Interactions and Vacuole Biogenesis, edited by Jean-Pierre Gorvel. ©2004 Eurekah.com and Kluwer Academic / Plenum Publishers.

However, there are also gram-negative bacteria, including pathogens like some spirochetes, outside this main group. *Proteobacteria* are divided into five main sections, and the γ-*Proteobacteria* include *Escherichia coli* and *Salmonella*, the gram-negatives whose OM is often taken as the archetype.

Molecular Structure

The molecular dimensions estimated for the LPS of *Salmonella* and the proposed structure of *E. coli* LPS K-12 lipid A and inner core and are presented in Figure 1. The lipid A is a glucosamine disaccharide phosphorylated at C1 and C4' and carrying four ester- and amide-linked 3-hydroxy-myristoil groups, two of which (those corresponding to the non-reducing sugar) bear lauric and myristic acyl chains in acyloxyacyl linkages.[3] Although this very hydrophobic structure is needed to provide a firm anchorage to the large hydrophilic moiety, in *E. coli* and *Salmonella* additional stabilization by divalent cations interacting with the inner section of the hydrophilic moiety is necessary for OM stability (see below). Based on biological and structural criteria, this hydrophilic moiety is divided in two sections: the O-polysaccharide and the core oligosaccharide which links the former to the lipid A and in which an inner and an outer part can be distinguished.[4] The inner core (Fig. 1) is made up of distinctive sugars (3-deoxy-D-*manno*-oct-2-ulosonic acid [Kdo] and D-*glycero*-D-*manno*-heptose) that are either acidic or substituted in part with additional phosphates, thus creating a dense negatively charged area at physiological pH. Molecular modeling and X-ray diffraction studies show that the inner core forms a compact entity with an unusually high partial density.[5] Dense negative charge and tight packing make the inner core a unique structure. There is firm evidence that negatively charged groups in core and lipid A are equilibrated by divalent cations such as Mg^{2+} to yield the natural salt forms of LPS. The outer core is also made of a few sugars (mostly neutral) and shows little variability[4] but the O-polysaccharide, made of monosaccharide or oligosaccharide repeating units, shows enormous intra and inter species diversity. O-polysaccharides are not of a defined length and the LPS molecules of a given bacterial strain can have from a few to several hundred O sugars. Although not in nature, mutants lacking the O-polysaccharide and elements of the core (but not Kdo I and II [Fig. 1]) are viable under laboratory conditions. Their LPS is referred to as rough (R) LPS in opposition to the smooth (S) LPSs carrying the complete core and O-polysaccharides.

Supramolecular Structure

Like other amphiphiles, monomeric LPS molecules form aggregates above a given concentration (the critical aggregate concentration). To understand the LPS properties derived from this fact is important because both soluble effectors and PRRs encounter different proportions of LPS aggregates and monomeric units depending on the conditions. For an authoritative and thorough discussion of the biophysical aspects of LPS, the reader is directed to a recent review by Seydel et al.[6]

The LPS supramolecular structure depends on the molecular conformation and overall structure of the individual LPS molecules as well as on the LPS/water ratio, the divalent cation content and the temperature. The properties of the aggregate depend on both the lipid A and the hydrophilic O-polysaccharide as the hydrophilic/hydrophobic balance would determine the critical aggregate concentration (higher for the LPSs with longer and more hydrophilic O-polysaccharides), its size and other features. Using *S. typhimurium* (*S. enterica* serotype *typhimurium*) LPSs, it has been shown that the gel to liquid-crystalline $\beta \leftrightarrow \alpha$ T_c of LPS (the transition temperature at which the acyl chains of the lipid A "melt") depends not only of the lipid structure but also on the O-polysaccharide being lowest (about 30°C) for LPS containing only Kdo I and II and highest for the LPS bearing the longest O-polysaccharides.[7] This is an important supramolecular property since it is well known that the biological properties of amphipatic aggregates (such as biological membranes) depend on their fluidity. Molecular conformation and aggregate properties have been carefully examined for the lipid A. Under

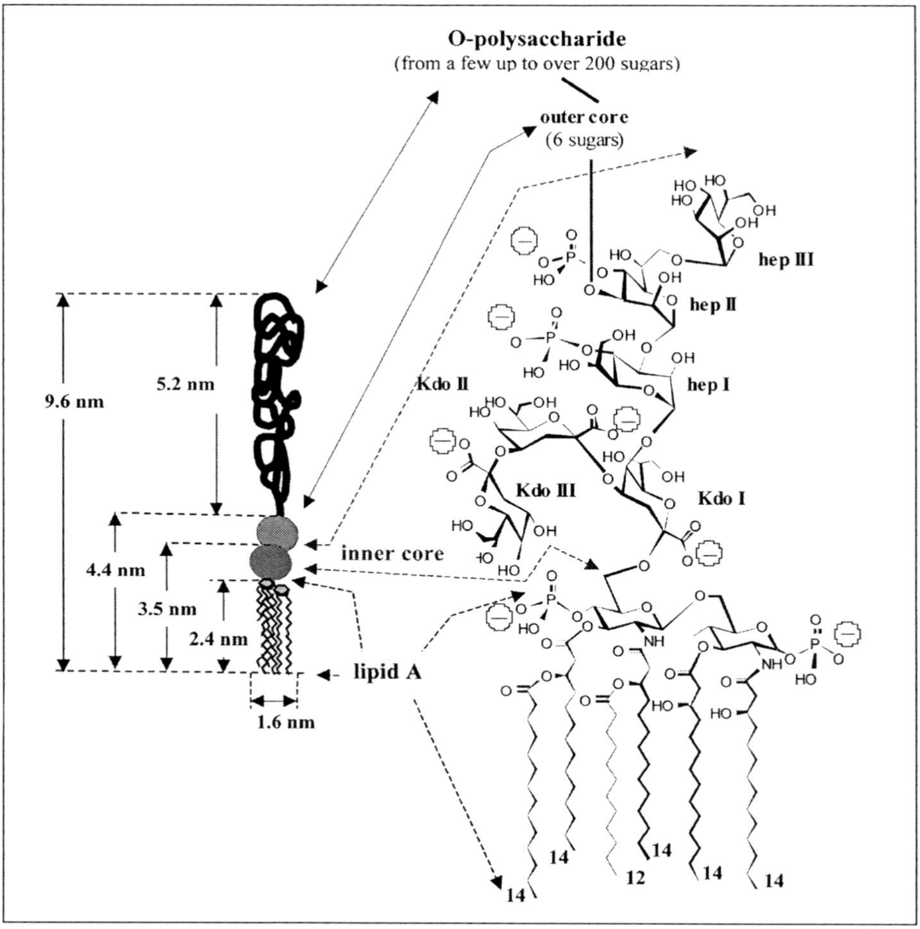

Figure 1. The structure of LPS. A schematic model showing the main sections and dimensions calculated for the LPS of *Salmonella* is shown on the left (adapted from refs. 5 and 10). The degree of coiling of the O-polysaccharide under physiological conditions is not known and actual dimensions may be larger when fully hydrated.[6] The proposed structure of the *E. coli* K-12 lipid A[3] and inner core[4] is shown on the right. Inner core variations on *E. coli* are limited to the presence of glucosamine or acetylglucosamine linked to the third D-*glycero*-D-*manno*-heptose (hep III), the presence of P-ethanolamine linked to the phosphate of hep I, or to the substitution of the third molecule of 3-deoxy-D-*manno*-oct-2-ulosonic acid (Kdo III) by either galactose of rhamnose.[4] Rough (R) LPS mutants lack the O-polysaccharide or the O-polysaccharide plus sugars of the core up to Kdo II.

physiological or near physiological conditions (i.e., high water content, appropriate divalent cation concentration and temperature above T_c) bisphosphorylated and hexaacylated lipids A (such as that of Fig. 1) show a truncated cone molecular conformation and form hexagonal (H_{II}) aggregates, whereas lipids A adopting a cylindrical conformation (such as those of *Rhodobacter capsulatus* or synthetic tetraacylated lipid A) form lamellar (L) aggregates.[8] Interestingly, the molecular conformation of lipid A correlates with bioactivity since hexagonal (H_{II}) aggregates are endotoxic and potent cytokine inducers whereas lamellar (L) aggregates are inactive.[8,9] The molecular conformation also depends on the number, nature and precise location of negative charges in the lipid A-core of LPS.[9] Obviously, aggregational and conformational effects modulate the interaction with the membranes of animals cells at both LP receptor(s) and lipid matrix levels.[6]

LPS and Outer Membrane Function

Because of the presence of LPS in the outer leaflet, the OM of *E. coli* is structurally asymmetric. LPS is strongly anchored by its lipid A section and the hydrophilic moiety extends outwards. The densely acylated moiety of the lipid A is highly ordered[10] and has a rigidifying effect which is complemented by the bridging of the phosphoryl groups and acidic sugars in the lipid A-inner core by divalent cations and polyamines.[11-13] Strong interactions with the outer membrane proteins (Omps) and intercatenary interactions among the sugars in neighboring O-polysaccharide chains also exist.[14] It is postulated that all these LPS-LPS and LPS-Omp interactions exclude phospholipids from the outer leaflet, and this creates a profound difference with the inner leaflet. For example, there is a highly asymmetric electrical potential barrier in phospholipid-LPS OM models[15] and the fluidity of the outer leaflet of the OM is considerably less than that of the inner phospholipid leaflet.[6,14] These properties create a barrier at the outer leaflet of the OM which retards diffusion of hydrophobic noxious substances, thus bolstering the action of efflux pumps,[16] and makes the selective channels of porins the normal entry pathway for hydrophilic solutes.[11,14,17] This barrier action and selective permeability is the key function of the OM and one without which these bacteria only survive under artificial conditions or not at all. The structural constraints that selective permeability impose on LPS structure are obvious and specifically relate to the inner core and lipid A sections.

LPS and the Innate Immune System

The molecular structure of LPS creates a unique MP: that of an amphiphile with a very hydrophobic section adjacent to a densely charged and compact oligosaccharide. Although other constituents of the gram-negative envelope (phospholipids, ornithine or serine containing lipids, and lipoproteins) are amphipatic and some carry negative charge, none of them shows such a strongly marked MP. Thus, it is not surprising that the LPS MP is both recognized by PRRs and the target of many innate immunity effector molecules. It is obvious that the core and lipid A acidic sugars and phosphates are essential parts of this MP but the role of the acyl chains is not merely that of a hydrophobic carrier: the fact that the lipid A precursors and analogues lacking the acyloxyacyl substitutions have considerably less endotoxicity or even behave as antagonists[18] strongly suggests that the acyloxyacyl configuration is a major determinant of the LPS MP.

Since the above-describe MP is maintained and extended on the surface of the OM, many of such effectors (see below) act directly on the bacteria by breaking LPS-LPS interaction, thus bringing about the disruption of the permeability barrier and, in many cases, a significant release of LPS and variable amounts of tightly bound Omps and other OM components.[13] Indeed, it was established long ago that the "endotoxin protein" (i.e., the proteins co-extracting with LPS and which modulate its activity and recognition by the immune system) are mostly Omps.[19] It is thus important to consider that what is presented to the innate immune system in vivo are in all likelihood LPS and Omps mixed aggregates which can also include the host antimicrobial peptides (see below) that cause their release.

Soluble Innate Immunity Effectors that Recognize the LPS MP

Soluble pattern recognition molecules that target on LPS include at least some complement (C') components, collectins (such as the mannose binding proteins and the lung surfactants proteins)[20] and cationic antimicrobial peptides and proteins.

Collectins and Complement

It has been known for a long time that purified LPSs have an anticomplementary activity (i.e., C' availability in a given serum disappears when incubated with LPS), and this is explained by the fact that both the alternative and the classical pathways of C' activation can be triggered directly by LPS and by gram-negative cell surfaces. C' components have been identified in all vertebrate classes, in protochordates and in some lower invertebrates indicating the primordial nature and importance of this system.[21] The interaction of C' factors with LPS has been recently reviewed by Tesh,[22] and I will only summarize here the most significant facts with regard to recognition of LPS by the innate immunity side of the C' system.

LPSs with O-chains rich in mannose, or mannose derived aminosugars, trigger C' activation by the classical pathway,[23-25] as these O-chains form complexes with the mannose binding protein and associated serine protease which can cleave C2 and C4 (Fig. 2) and perhaps interact directly with C3. Conceptually, this activation belongs to the lectin-type of PRRs and it depends on the properties of the carbohydrate recognition properties of the mannose binding protein[20] and not on the LPS MP described above. The lung surfactant proteins (SP) are also known to bind to LPS and to induce bacterial killing[26,27] but the mechanisms involved have not been studied in detail. SP-A but not SP-B has been reported to promote LPS uptake and deacylation (see below) by alveolar macrophages.[28]

Normal serum has a bactericidal activity on many gram-negative bacteria that depends on the activation of C' by the alternative pathway. In this pathway, control of the spontaneously produced metastable C3b by factors H and I to produce inactive C3bi is thwarted when C3b is bound to an activator surface and stabilized by properdine, thereby leading to amplification of C3b production and to the terminal C' pathway (Fig. 2). Many microbial polysaccharides act as efficient activators and, on the other hand, host components containing sialic acid but also the sialic acid-containing LPSs of the mucosal pathogens *Haemophilus* and *Neisseria*[29] enhance H activity (Fig. 2). Although the processes involved are not fully understood, activation of the alternative pathway by LPS is related to the O-polysaccharide. Studies with liposomes carrying *Salmonella* LPS and bound C3b have shown that the regulatory activity of factor H decreases when tested on LPS of greater polysaccharide size.[30] Moreover, C3b deposition, but not its breakdown, depends on the fine structure of the particular *Salmonella* LPS O-polysaccharide[31] and the lipid A-core sections do not seem to be required for C' activation via the alternative pathway.[32] On the other hand, enterobacterial R-LPS (devoid of the O-polysaccharide) triggers the classical pathway by binding directly to C1q and this interaction depends in part on the anionic MP of the LPS core (Fig. 2).[22] C1q is a highly basic protein (pI 9.2) and reagents that block its histidine residues (cationic) prevent binding to R-LPS. Moreover, hydrolytic removal of Kdo from heptose-less *E. coli* R-LPS abrogates binding. With regard to the role of the O-chain in this kind of activation, it has been shown that C1q binds poorly to S-LPS and this strongly suggests a steric hindrance in the access to inner targets, an O-polysaccharide effect clearly shown with even smaller molecules.[33]

Antimicrobial Peptides and Proteins

Antimicrobial peptides represent an ancient mechanism of defense and/or destruction of microbes, and are present in plants, vertebrates, invertebrates and in at least some protists which feed on bacteria.[2,34-36] They are found within intracellular compartments, on the mucosae and in secretions and body fluids.[13,36] So far, about 500 of such peptides have been identified (http://bbcm1.univ.trieste.it/~tossi/pag1.htm) and most show a net positive charge of up to +7 due to an excess of arginine, lysine and histidine, and also contain about 50% of hydrophobic amino acids.[37] Many of the antimicrobial peptides produced by mammals have other complementary roles in immunity (induction of cytokines and cell receptors and chemotactic and opsonizing activities), but membrane destabilization is one of their main activities. The ability to interact with membranes is explained by their overall structure, although not all have the same potency.[13] Mammalian defensins, that are perhaps the most investigated ones, are protease-resistant peptides of about 3,5 kDa stabilized by internal disulfide bonds and with triple stranded β-sheets connected by a loop with a β-hairpin hydrophobic finger and protruding cationic chains (Fig. 3). This creates an amphipatic pattern which targets defensins to nega-

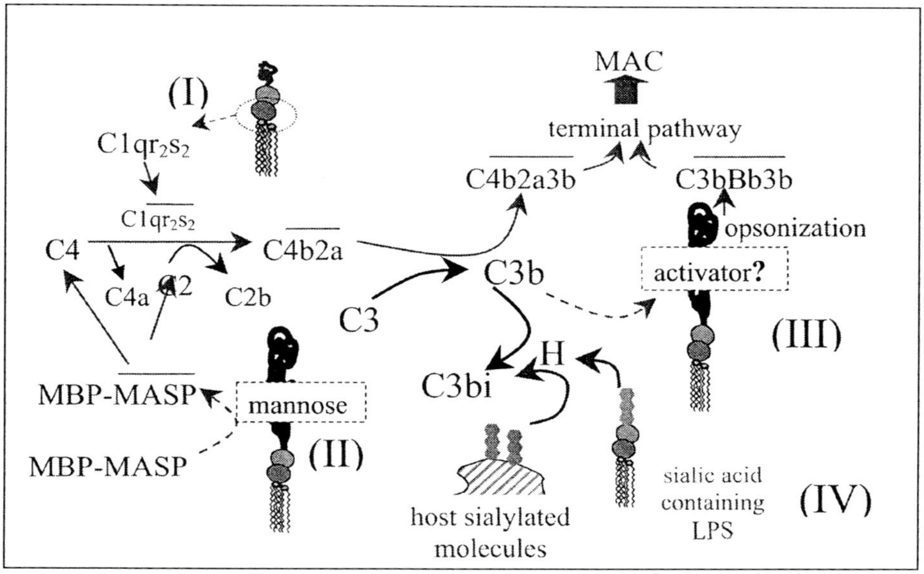

Figure 2. LPS and the C' system. (I) the absence of the O-chain in R-LPS allows interaction of the Kdo region with C1qr2s2 and activation by the classical pathway; (II), in some S-LPSs, the O-polysaccharide contains mannose or closely related sugars at a density high enough to enable binding of MBP and its associated serine proteinase (MASP) which can cleave C4 and C2 and activate the classical pathway; (III), in other LPSs, composition of the O-polysaccharide is such that it efficiently binds spontaneously produced C3b and prevents its conversion to inactive C3bi by H, thus triggering the alternative pathway; (IV), some LPSs have short oligosaccharide sialylated stubs instead of long O-polysaccharides and, like host molecules containing sialic acid, promote C3b conversion to C3bi.

tively charged membranes. Other antimicrobial peptides are less structured but they keep the cationic nature and seem to be able to fold into amphipatic α-helices upon contact with target membranes. Although some cationic peptides have different activities (antifungal, antiparasite or antiviral), in many cases those properties are related to the recognition of the MP created by LPS in the OM of gram-negative bacteria. In fact, since the pioneering work of Modrzakowski and Spitznagel[38] many of these peptides have been shown to bind to the inner section of LPS and to disrupt the OM permeability barrier.[13] Bacteria have also developed antibiotic peptides which recognize the same LPS MP and share with eukaryotic cationic peptides some aspects of the mechanism of action. Polymyxins, a group of cationic lipopeptides (Fig. 3), have been extensively studied and have provided clues about the mechanism of MP recognition and membrane action. Indeed, it has been known for a long time that polymyxins can both compete with lysosomal cationic peptides for binding to LPS[39] and block LPS endotoxicity.[40] Polymyxins act on model membranes by a "self-promoted uptake" mechanism due to its cationic detergent-like nature.[41,42] According to this model, in the intact OM some cationic peptides displace the divalent cations bridging the lipid A-core negatively charged groups (affinities for LPS can be at least 3 order of magnitude higher for the peptides than for divalent cations) and produce a disturbance that favors insertion of their hydrophobic section into the OM. This disturbance somehow creates the conditions for OM passage (hence "self-promoted transport") and, although a general model for the subsequent events cannot be proposed,[43] further penetration could be driven by the anionic groups in phosphatidylglycerol and cardiolipine and by the inward negative gradient of the cytoplasmic membrane.[36]

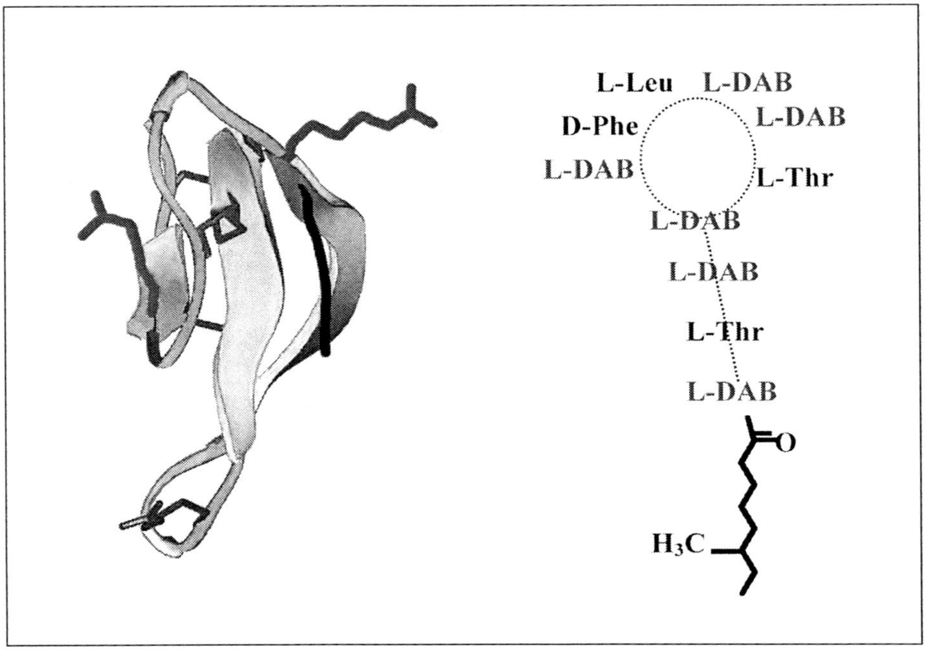

Figure 3. Human neutrophil defensin 1 and polymyxin B. The protein folds as three β-sheets interconnected by β-turns and random structures. About 50% af the amino acid residues are hydrophobic but positively charged lysine rich domains presumably protrude from the structure tail (adapted from ref. 36). Polymyxin B is a circular peptide rich in cationic diaminobutyric acid with a hydrophobic 6-methylotanoic acyl. The characteristics ofboth molecules (polycationic and amphiphilic) allow recognition of the LPS MP (polyanionic and amphiphilic) shown in Figure 1 and, under some circumstances, block LPS recognition by cognate PRRs.

In addition to the above-described peptides, some host proteins contain cationic amino acid stretches that enable them to recognize the anionic sections in the LPS MP, although they also have other and possibly more important activities. The NH_2-terminal section of lactoferrin is able to bind to LPS and to prevent its interaction with PRRs of monocytes and other cells[44] and, upon release by hydrolysis, the corresponding peptide has a potent microbicidal action on gram-negative and gram-positive bacteria.[45,46] Many other proteins, most notably serum lipid carriers,[47] can also interact with LPS once released from bacteria. However, the bactericidal/permeability-increasing protein (BPI) stands out among all because of its abundance and properties.

BPI is a 50-55 kDa lysine-rich glycoprotein found mostly in primary granules of polymorphonuclear leukocytes where it represents about 1% of the total protein content and plays a main role in the oxygen-independent killing of microbes.[48] BPI preferentially acts on gram-negative bacteria and breaks the OM permeability barrier[13,49] and, although only causes release of LPS at a high concentration[50] and shows little ability to deaggregate LPS,[51] it is highly bactericidal. It avidly binds to the inner LPS sections[52] and has anti-endotoxic activity associated with the NH_2-terminal section.[53,54] Although enterobacterial LPSs show various degrees of binding to BPI, this relates in part to shielding by the O-polysaccharide.[55] Physicochemical measurements show that the NH_2-terminal portion interacts and neutralizes the negative groups carried by the lipid A-inner core and displaces Mg^{2+} counterions,[56] and studies with *Rhizobium* LPSs (which carry part of the core negative charge in carboxyl and sulfate groups rather than in phosphates) demonstrate that it is this overall lipid A-core MP, and not the fine structure, what determines binding.[57] The crystal structure of BPI[58] reveals an almost symmetrical boomerang-shaped molecule of 135 by 35 by 35 A in which a central proline-rich linker connects two opposite NH_2- and COOH-terminal domains. The NH_2-terminal section carries the antibacterial activity[52] but the holoprotein also has opsonizing activity linked to the COOH-terminal portion.[59] Interestingly, BPI copurifies with phosphatidylcholine which, like LPS, carries negatively charged phosphate and hydrophobic acyl chains. In BPI, phosphatidylcholine is lodged in two apolar pockets, one in each domain, near of which there are both basic and acidic chains. Although this could account for the binding to zwitterionic molecules and for the recognition of the typical LPS MP,[58] it is not known how this would be achieved because lipid A has larger size than phosphatidylcholine. Nevertheless, the refined structure of BPI should explain why it is about 1000 times more potent than defensin HNP-1.[60] Significantly, BPI belongs to a family of proteins that have evolved to efficiently bind several types of lipids and that includes the cholesterol-transport protein, the phospholipid transport protein and the specific LPS binding protein (LBP [see below]).

Interaction of LPS with Monocyte-Macrophages

LPS has a potent stimulatory action on several types of cells, including phagocytes. Experimental evidence shows that multiple LPS binding mechanisms and receptors may exist in different cell types or even in an single type. Of particular relevance in immunity is the response of macrophages to LPS as these cells not only ingest and destroy bacterial invaders and process their products but also signal their presence and stimulate other macrophages and cells such as T lymphocytes, polymorphonuclear neutrophils (PMNs), epithelial cells and fibroblasts. LPS activates macrophages to produce IL-1β, IL-6, IL-12, IL-18, TNF-α, IFN-γ, NO and other reactive oxygen derivatives. This activation is mediated mostly by both serum LPS-binding proteins and cell surface receptors[2,61] (Fig. 4). Moreover, LPSs can interact with cell membranes or be taken up by pathways that not necessarily lead to activation (Fig. 4). The mechanisms discussed below are those that depend clearly on the LPS MP and/or belong to the innate immune system and have been characterized in some detail.

The LBP-CD14-TLR Pathway of Cell Activation

This pathway has been clearly shown to be involved in cell activation and stands out by its exquisite sensitivity to minute amounts of LPS. The current model (Fig. 4) is that soluble LBP takes up LPS and interacts with glycosylphosphatidylinositol-anchored membrane CD14 (mCD14) of myeloid cells and with soluble CD14 (sCD14). In macrophages, the LPS-mCD14 complex activates a signal-transducing membrane molecule whereas endothelial cells first bind the sCD14-LPS complexes which then act in a similar way. This activation sequence does not exclude other possible effects mediated by LBP (Fig. 4) such as the sequestering of LPS and its neutralization into lipoprotein complexes,[62,63] the direct internalization of the LBP-LPS when LPS aggregates have a large size,[64,65] or the receptor-independent insertion into the lipid mediated by LBP (see below).

LBP is an acute-phase glycoprotein of about 60 kDa that belongs to the same family as BPI.[66] Moreover, sequence analysis show that 102 amino acids are completely conserved in BPI and LBP, and most of the conserved regions correspond to those that conform the two apolar pockets where phosphatidylcholine was found in BPI.[67] There are also 5 conserved lysine residues clustered at the NH_2-terminal domain[67] which in both BPI and LBP is involved in LPS binding.[68,69] However, whereas BPI is bactericidal and has endotoxin-neutralizing ability (see above), LBP binds poorly to live gram-negative bacteria,[70] can deaggregate LPS,[51] interacts with CD14[71,72] and promotes macrophage activation.[66] It is striking that two closely related proteins recognizing the LPS MP have so different capacities and this relates to the structure of the COOH-terminal section which in LBP is responsible for CD14 interaction.[68] It is not known whether this last interaction depends on LPS binding by LBP and, although it is clear that sCD14 is able to form complexes with LPS molecules with a 1:1 or 1:2 stoichiom-

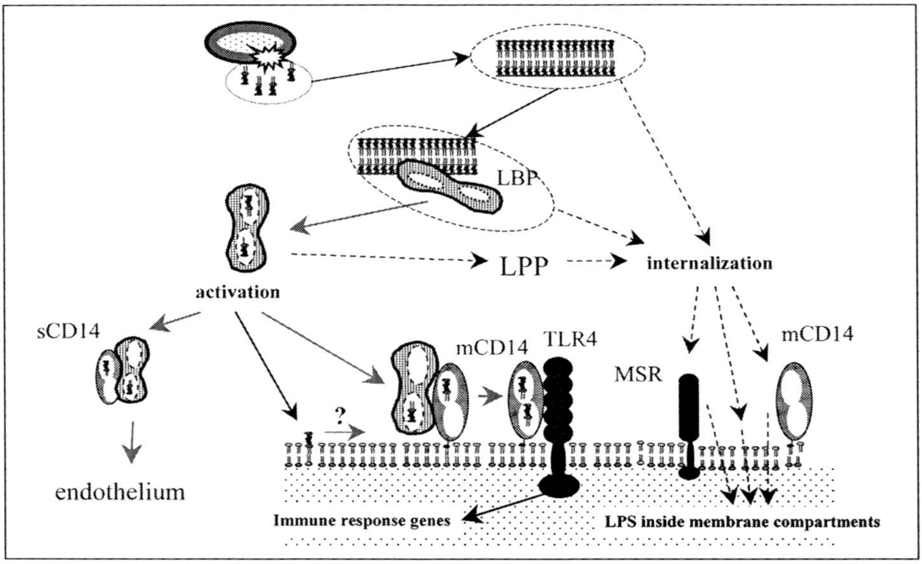

Figure 4. LPS-cell interactions. LPS is released either spontaneously or by the action of polycationic effectors of teh immune system and forms aggregates (often including Omps) which may be directly internalized by receptor-dependent (MSR, CD14 and possibly others) or other mechanisms. However, in cell activation pathways (red arrows), the LPS MP is recognized by the NH_2-domain of LBP and, although large LBP-LPS aggregates may also be internalized, LBP can deaggregate LPS and transport monomers possibly in two binding sites (by analogy with BPI). In macrophage activation, LBP transfers LPS to GPI-anchored mCD14 which once loaded with LPS may interact with the extra-cellular leucine-rich domain of TLR4. In turn, TLR4 triggers a cascade of signals through its Toll/IL-1 domain, thereby activating the immune response genes. Moreover, LBP can transfer LPS to sCD14 which is recognized by endothelial cells. Also possible is the transference to serum lipoprotein (LPP) and the intercalation of LPS into the membranes mediated by LBP.

etry and that the process is catalyzed by LBP,[73-75] the mechanism of catalysis and the molecular basis of the CD14-LPS interactions are also unknown. Although CD14 also plays a role in internalization (see below), it is generally accepted that activation occurs at the cell surface and not in internal compartments (see below). Evidence in favor of this is the fact that LBP-LPS internalization can be dissociated from cell activation by using monoclonal antibodies anti-LBP and anti-CD14. While the first monoclonal does not block cell activation, anti-CD-14 blocks activation but not uptake.[64] Other investigators have also found that activation is not blocked by agents that prevented uptake by macropinocytosis[76] (see below) and, conversely, agents blocking activation do not prevent uptake.[65]

Since it is not an integral membrane protein, mCD14 cannot act as a signal transducer. Moreover, the similarly important role of sCD14 in activating non myeloid cells also shows that at least a third component located in the cell membrane recognizes the mCD14-LPS complex and acts as the signal transducer. This molecule has been identified as a member of the Toll family of cell receptors.[77] First described in *Drosophila* as an element involved in embrionary morphogenesis, Toll has an intracytoplasmatic domain homologous to that of IL-1 receptor and its activation triggers a cascade of signals involving elements that participate in the control of genes encoding insect antimicrobial peptides.[2] This cascade of signals closely resembles that involved in the cytokine-induced NF-κB-dependent transcriptional activation of effector proteins of mammals.[2] It seems that sepsis would lead in *Drosophila* to the activation of proteolytic cascades in the hemolymph that would eventually cleave proteins like Spaetzle (a member of a growth factor and cytokine-like proteins) whose fragments can interact with Toll. Toll like receptors (TLR) are found in plants, insects, and vertebrates and they have a conserved IL-1 cytoplasmatic domain and a leucine-rich repeat extracellular domain that varies among different TLRs. Macrophages and monocytes express at least TLR2, TLR4 and TLR5[78] and the weight of the evidence shows that TLR4 is involved in cell activation by at least *Salmonella* and *E. coli* lipids A or synthetic analogues.[79-81] The hyporesponsiveness to LPS, which is a trait of mouse strains C3H/HeJ and C57BL/10ScCr and of some human individuals, has also been shown to be linked to mutations in TLR4.[82]

It has to be noted that the LBP-CD14-TLR4 pathway could not account for all the situations emerging from infection. As pointed out above, Omps are often included within LPS aggregates and are known to modulate LPS biological activities.[19] At least some bacterial lipoproteins are known to activate cells through TLR receptors other than TLR4[83,84] and Omps and LPS can act sinergically in activating cells.[85,86]

Intercalation of LPS into Membranes

The amphipatic MP of LPS makes it a "detergent-like" molecule able to interact directly with lipid membranes. As discussed above, LPS free monomers and aggregates exist depending on the concentration and the properties of the particular LPS molecules. Thus, direct intercalation of LPS monomers into lipid membranes would be an expected event below the critical aggregate concentration. Above this concentration, aggregates could be split by cationic peptides and proteins and deaggregated into small units by lipid carrier molecules which could then favor membrane intercalation. It is suggestive that LBP (see below) has been shown to transport and promote the intercalation of LPS and lipid A (and also other amphiphiles) into lipid membranes in a CD14-independent manner, and models for macrophage activation at high LPS concentration have been proposed on this bases.[6,9,87] Receptor-independent intercalation events could also take place in the internal compartments where killing by BPI and cationic peptides release important amounts of LPS. However, nothing is known about this seemingly important aspect of LPS biology.

Internalization and Detoxification

LPS aggregates coming from extracellularly lysed bacteria or in the form of OM blebs can be taken up by phagocytes. In addition, LPS aggregates most probably containing Omps and bactericidal peptides are produced directly within the internal compartments where bacteria are degraded. It has been reported that the presence of Omps in the LPS aggregates accelerates their uptake[88] but, although the presence of mixed LPS-Omp aggregates is probably closer to the situation during infection, this observation has received little attention.

The mechanisms of binding, internalization and the intracellular trafficking of LPS are not well understood. Kitchens and Munford[89] have recently reviewed the literature on LPS internalization by phagocytes and stressed that the difficulties in obtaining a clear picture lay at least in part in the relatively great variety of experimental systems and methods used to study LPS internalization as well as in the complexities derived from the coexistence of free and aggregated LPS forms. Receptor-independent mechanisms seem less efficient, and I will summarize here only the evidence concerning the best characterized molecules that have been shown to act as LPS receptors.

Macrophage scavenger receptors (MSR) have been implicated in LPS binding and both their presence in phylogenetically distant groups[90] and molecular features[91] show that they are PRRs. They are trimeric integral membrane glycoproteins implicated in a variety of cellular processes and host defense and of the three classes known, only class A specifically binds bacterial products. Bovine class A MRS, that have been studied in detail, have six domains, and the three closest to the NH_2-terminal end have anchoring and structural roles. The fourth is a a-helical coiled structure that plays a role in the conformational changes leading to ligand dissociation at low pH[92] and, since MSRs are linked to the endocytic pathway, this event would take place in the acidic environment of endosomal compartments. The fifth domain is the ligand-binding section and contains a cluster of lysine residues which may be negatively charged at physiological pH.[93] The COOH-terminal domain shows some variability and does not seem essential for ligand binding. MSR A binds some types of bacterial cells and bacterial polyanionic molecules, including LPS whose core-lipid A MP is a suitable ligand. It has been shown that MSRs can bind *E. coli* lipid A and that, although this leads to lipid A detoxification (see below), this does not activate peritoneal macrophages.[94] Thus, the MSR role with regard to LPS may be restricted to the clearance and subsequent detoxification of plasma LPS aggregates in a way preventing excessive cytokine production and septic shock.[95]

Cells expressing mCD14 can internalize LPS, either as monomers or as aggregates[65,76,89,96,97] but the information on how internalization proceeds is contradictory. Whereas some investigators have found that the dominant route of entry is via non-clathrin coated invaginations with little or no role for macropinocytosis,[89,96] others have reported that mCD14 and LPS are co-internalized by this second mechanism.[76] As judged by experiments carried out in C3H/HeJ mice, TLR4 does not cooperate in the internalization of LPS mediated by CD14.[65] This, and the other evidence discussed above, show that internalization and activation are most likely independent processes. Thus, the CD14 internalization is also likely to represent part of the mechanisms leading to clearance and detoxification.

Internalized LPS has been located in phagosomes, endosomes, the Golgi complex, mitochondria, the nucleus and the cytoplasm, but the meaning of many of these observations is unclear and some may be artifacts caused by an LPS overload (for a critical review, see ref. 89). The intracellular compartments in which LPS is localized after entry by receptor-dependent mechanisms have not been characterized. However, it has been clearly established that PMNs and monocyte-macrophages are able to dephosphorylate lipid A and also have an acyloxyacyl hydrolase that partially deacylates lipid A.[89] These modifications are known to abrogate endotoxicity and, by comparison with lipid A synthetic analogues, Matsuura et al[98] have also suggested that the *in vivo* deacylation may represent a mechanism to enhance the adjuvant activity of lipid A. There is little evidence that animal cells are able to degrade the O-polysaccharides or the core oligosaccharides. Wuorela et al[99] found that *Y. enterocolitica* O:3 LPS showed and altered electrophoretic pattern after 7 days of incubation of killed bacteria within peripheral blood monocytes and interpreted this to mean a partial degradation of the O-polysaccharide. Fox et al[100] found differences in the state of aggregation of *E. coli* LPS after being processed in Kupffer cells and suggested that this was due to changes in the lipid and O-polysaccharide moieties. However, direct chemical evidence for definite changes in the LPS sugar moiety is lacking.

The LPS of Facultative Intracellular Pathogens

Since the macromolecules carrying MPs play essential roles, microbial pathogens either overcome PRRs by masking the surface macromolecules carrying the cognate MPs or by colonizing niches where they escape from being detected or reached, or produce toxins that severely damage the host or interfere in subtle ways with the immune system. Paradoxically, some gram-negative bacteria have evolved to thrive within the very cells that both respond vigorously to LPS and carry antimicrobial peptides and proteins. For this purpose, these intracellular pathogens use several mechanisms. The examples considered here use type III (*Salmonella*)[101] or type IV (*Brucella* and *Legionella*)[102,103] secretion systems to inject proteins into the host cell which interfere with their processing (see Chapters in section IV of this book). In addition, they complement this strategy by altering their LPS MPs.

The Salmonella *Case*

The *Salmonellae* are *Enterobacteriaceae* that cause several diseases in humans and animals. The mechanisms by which they cause disease are complex and only partially known, in part because of the complexity of the regulatory systems that control the coordinate expression of multiple virulence genes. One of these systems, PhoP-PhoQ, plays a key role as it controls genes encoding factors that are involved in resistance to antimicrobial peptides, magnesium transport, type III secretion of proteins and resistance to noxious OM permeants such as bile salts (reviewed in ref. 104). As expected, part of these gene products control fine details in the structure of the LPS. Indeed, *S. typhimurium* was the first pathogenic bacteria in which lipid A modifications were linked to resistance to antimicrobial peptides.[39]

Changes in LPS MP

At low Mg^{2+} concentration (thought to represent that in intracellular compartments) PhoQ becomes activated and catalyzes PhoP phosphorylation. Phosphorylated PhoP binds to promoters of PhoP-repressed and PhoP-activated genes.[105,106] The latter include genes involved in resistance to some antimicrobial peptides and proteins plus the genes of a second two-component regulatory system (PmrA-PmrB) that controls resistance to polymyxin.[107] Although many aspects and links within this network remain to be fully understood, activation of all these genes results in changes in lipid A structure[108,109] and, perhaps concurrently, in OM permeability to bile salts.[110] Figure 5 compares the *S. typhimurium* lipid A structures that can be derived from the MALDI-TOF mass spectra obtained by Guo et al [108] in their analysis of lipid A regulation by PhoP-PhoQ. Arabinosamine (4-amino-4-deoxy-L-arabinopyranose) linked to the phosphate at C4' and variations in the degree of acylation and type of acyl chains are the changes observed. As discussed below, they significantly alter the LPS MP.

Sensitivity to Antimicrobial Peptides and Proteins

Because of the presumed decrease in electronegative charge of lipid A,[39] arabinosamine substantially alters the LPS MP. Not surprisingly, arabinosamine linked to lipid A is clearly implicated in resistance to polymyxin not only in *S. typhimurium*[111] but also in other enteric bacteria.[112-114] That this modified lipid A is synthesized under conditions thought to emulate the environment of macrophage compartments where antimicrobial peptides act is an observation of obvious meaning. Moreover, detailed analysis of the phenotypes resulting from expression of the PhoP-activated *pagP* gene shows that changes in the acylation pattern significantly alter the resistance to some, but not all, antimicrobial peptides.[109] PagP is specifically required for the palmitoylation that leads to the synthesis of the heptaacylated form of lipid A (Fig. 5), and $PagP^-$ mutants unable to synthesize it are comparatively sensitive to some cationic antimicrobial peptides (although no differences in polymyxin sensitivity are observed) in viability and OM permeability tests.[109] Taking into account the mechanism of action of these agents, it seems that increased lipid A acylation alters the hydrophobic interactions in the OM to the extent that it becomes resistant to permeabilization by some of them. How other more subtle modifications in the acylation pattern depicted in Figure 5 affect interaction with polycationic peptides is not known.

Biological Activity

The effects that the changes in *S. typhimurium* LPS MP have on the endotoxic-related properties were studied by Guo et al.[108] These authors tested E-selectin expression by cultured human endothelial cells after exposure to a constitutive $PhoP^C$ mutant that permanently produces heptaacylated and arabinosamine substituted lipid A forms, to a defective $PhoP^-$ mutant unable to do so, and to the parental wild type strain. They found a comparatively reduced or increased stimulation of E-selectin by $PhoP^C$ and $PhoP^-$ bacteria, respectively. They further tested the LPSs purified from the same bacteria on monocytes and found, respectively, a 40% reduction and a 30 to 40% increase in E-selectin and TNF-α expression with respect to the effects of wild type LPS. These results strongly suggest that, complementary to other virulence mechanism, *S. typhimurium* modulates lipid A to promote its survival by lowering cytokine production.[108] However, at which level of the activating pathways this occurs is not known. Because of the

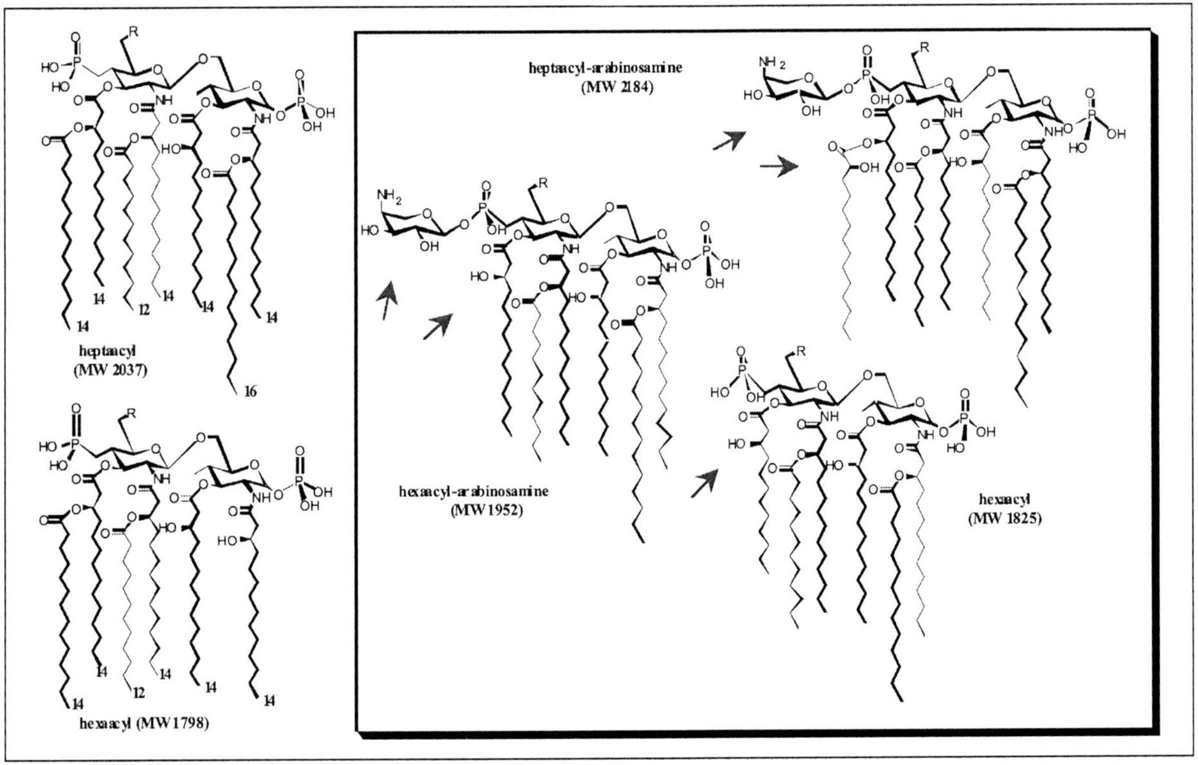

Figure 5. Lipid A structures of *Salmonella typhimurium*. Under most in vitro growing conditions, only the hepta and hexaacyl structures on the left are observed. Activation of the PhoP-PhoQ two-component regulatory system by growth in low Mg^{2+} media leads to the synthesis of the structures presented within the frame and other in minor proportions (heptaacyl lacking arabinosamine [MW 2052] and hexaacyl carrying myristoyl in 3'). These structures show altered LPS MPs (arrows) because of the reduction in lipid A negative charge caused by arabinosamine and/or the change in acyl groups. The heptaacylated form observed under standard conditions is also under the regulation of PhoP since it is not produced by $PhoP^-$ mutants (structures have been drawn based on the MALDI-TOF spectra of refs. 108 and 109).

analogies between the mechanisms of LBP recognition and antimicrobial peptide action, it seems likely that at least the initial steps of the LBP-CD14-TLR pathway are affected.

The Legionella-Brucella *Case*

Legionellae are human pathogens but they are primarily environmental bacteria that have the suggestive capacity to multiply within protozoa that prey on microbes.[115] Like *E. coli* and *Salmonella*, the *legionellae* belong to the γ-*Proteobacteria*. On the other hand, the *brucellae* have no other known niche than those in the hosts in which they cause a severe disease, and belong to the α-*Proteobacteria*, along with soil bacteria, plant symbionts and plant pathogens [116]. The closest known phylogenetic relatives of *Brucella*, the members of the genus *Ochrobactrum*, are in fact environmental bacteria that only occasionally infect compromised patients (reviewed in ref. 117). Comparison of the LPS properties of *E. coli, L. pneumophila* and the *Brucella-Ochrobactrum* cluster reveals features suggestive of convergent adaptation to intracellular pathogenicity.

Unconventional LPS MP

Figure 6 shows the structure of the lipid A of *L. pneumophila*[118] and of the *B. abortus-O. intermedium* cluster.[119] Their similarities (and the differences with the lipids A of *E. coli* [Fig. 1] and *Salmonella* [Fig. 5]) are evident: the diaminoglucose (rather than glucosamine) backbone, the fact that all acyl substitutions on the disaccharide are in amide linkages (rather than in both amide and ester linkages), and the long (C18 to C22) or very long (C28) acyl chains in acyloxyacyl linkages. These very long fatty acids have been proposed to span the OM and to be anchored on the inner surface by the hydroxyl group close to the end of the acyl chain,[120] perhaps adding in this way additional stability to the OM. With regard to the sugar moiety, the complete structure of the LPS core is only known for *L. pneumophila* serogroup 1 (Philadelphia 1 strain)[118] but a partial structure of the LPS core of *Ochrobactrum* (the core of the R strain *O. intermedium* 3301^T) is available[121] and the qualitative composition of the *B. abortus* LPS core can be inferred from several analyses.[119,122] These data are presented in Figure 7 where it can be seen that, although Kdo remains as a characteristic core sugar, there is no heptose or phosphate. In the LPS-core of *O. intermedium*, but not in that of *B. abortus*, there is galacturonic acid, which would carry a negative charge under physiological conditions. Accordingly, the overall negative charge of these LPSs is predicted to be very low, and this has been experimentally confirmed for *B. abortus*.[119] Also worth to note is that the outer core oligosaccharide of *L. pneumophila* LPS has an overall hydrophobic character due to the *N*- and *O*-acetyl groups and the presence of 6-deoxysugars.[118] Hydrophobicity is also a property of the O-polysaccharide of *L. pneumophila* (a homopolymer of α-(2→ 4) linked 5-acetamidino-7-acetamido-8-*O*-acetyl-3,5,7,9,-tetradeoxy-D-*glycero*-L-*galacto*-nonulosonic acid [legionaminic acid] in serogroup 1) and of *B. abortus* (a homopolymer of α-(1→2) linked 4,6 dideoxy-4-formamido-D-mannose [perosamine]).[118,123] Legionaminic acid lacks free hydroxyl groups and the *N*-formylated perosamine has just one. These features make these S-LPSs phenol-soluble in opposition to most S-LPSs that partition into water when extracted by the hot water-phenol protocol. However, the *B. abortus* O-polysaccharide has a dual character since it is also readily soluble in water where it shows a marked tendency to self-aggregate at low ionic strength.[124] The O-polysaccharide of *O. anthropi* LMG 3331^T is a linear repeating hydrophilic disaccharide of N-acetyl-glucosamine and rhamnose.[125] It is obvious that these three LPSs have a more markedly altered MP than that of the LPS of *S. typhimurium* grown in low Mg^{2+} media. This is particularly true for the two pathogens due to the peculiarities of their O-chains.

Interaction with Collectins and Complement

To the best of my knowledge, the anticomplementary activity of these LPSs has been studied only in *B. abortus* and it has been found to be comparatively reduced (Table 1) owing to a failure to activate the alternative C' pathway (pathway III, Fig. 3).[126,127] The O-polysaccharide

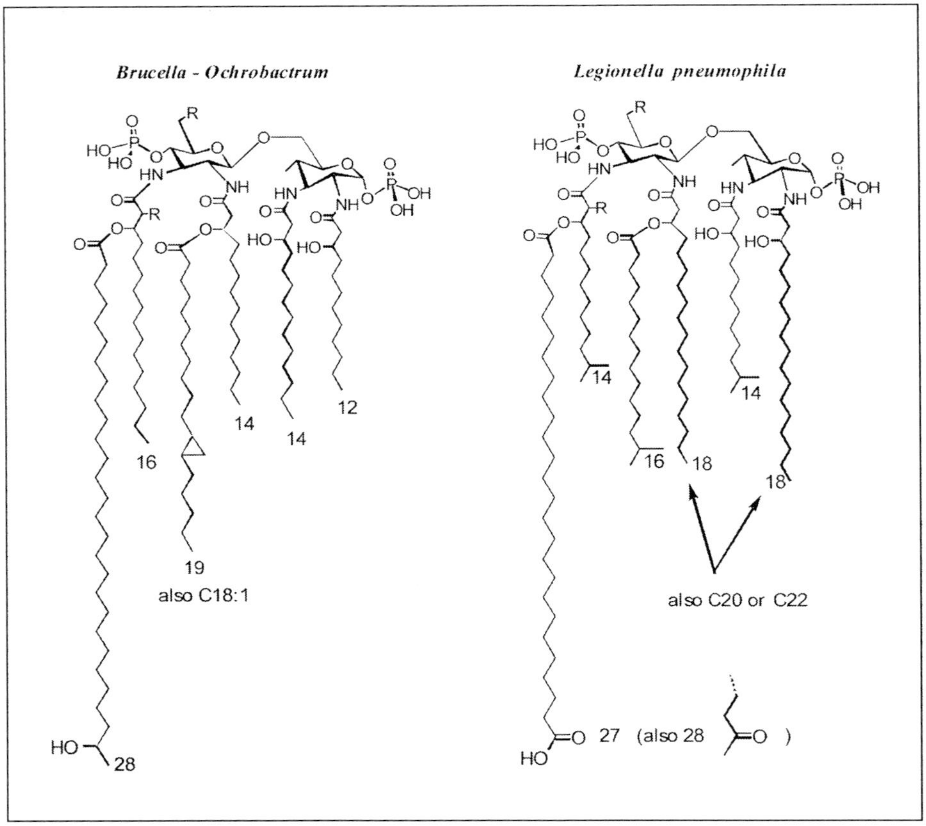

Figure 6. Proposed structures of the lipids A of the *Brucella-Ochrobactrum* cluster and of *Legionella pneumophila*. The structures do not take into account heterogeneity due to different degrees of acylation or additional microheterogeneity due to variations in the type of acyl substitutions (based on the data of refs. 118 and 119).

of *B. abortus* has been shown to play a role in C' resistance[128-130] and, although an interpretation in molecular terms is difficult, it is tempting to speculate that this relates to the absence in *N*-formyl-perosamine of free amino groups susceptible of nucleophilic attack by activated C3b. In fact, pure *N*-formyl-perosamine homopolymers show reduced or no binding of C3b when C' is activated by specific IgG.[131] This suggests a mechanism to avert not only lysis by MAC but also a C3b opsonization (Fig. 3) that would lead to entry into professional phagocytes by routes other than those favorable to *B. abortus* intracellular multiplication (see Chapter IV. 1.). Although *B. abortus* is an intracellular parasite, enhanced resistance to C' activation could help these bacteria to survive during bacteriemic states before they reach target cells. There is no evidence of interaction between the O-chain of either *L. pneumophila* or *B. abortus* and collectins, although *B. abortus* has been shown to bind to human B cells by a lectin-type mechanism.[132]

Sensitivity to Antimicrobial Peptides and Proteins

To the absence of anionic groups in the LPS outer core of *L. pneumophila* and *B. abortus* corresponds a marked resistance to polymyxin[133-135] and, at least for *B. abortus*, to a variety of polycationic peptides including defensin NP-2, lactoferricin B, the active peptide from macrophage cationic protein 18[133] and crude lysosomal extracts of PMNs.[133,136] This resistance is

***Table 1. Some biological activities of the LPS of* Brucella, L. pneumophila *and* E. coli**

	μg Causing the Indicated Effect			
Effect	***Brucella* LPS**	***L. pneumophila* LPS**	***E. coli / S. enterica* LPS**	**Reference**
LD 50 in C3H/HeJ mice [1]	70	n.d.	>500	150
chick embryo lethality	10	n.d.	0.003	150
local Schwartzman reaction	>1000	weak or none	20	140,150
pyrogenicity in rabbits	>300	weak or none	0.15	140,150
Lymulus lysate gelation	0.001	<0.001	0.002	142,150
inactivation by polymyxin B	no	no	yes	126,140
anti-C′	50 (weak)	n.d.	20 (strong)	150
maximal induction of:				
IFN-γ	200	n.d.	50	142,151
IL-1β	≥10	≥10	0.007	142,151
IL-6	n.d.	≥10	0.010	142
TNF-α	≥10	≥10	0.023	142,151

1. Carrageenan-treated mice.

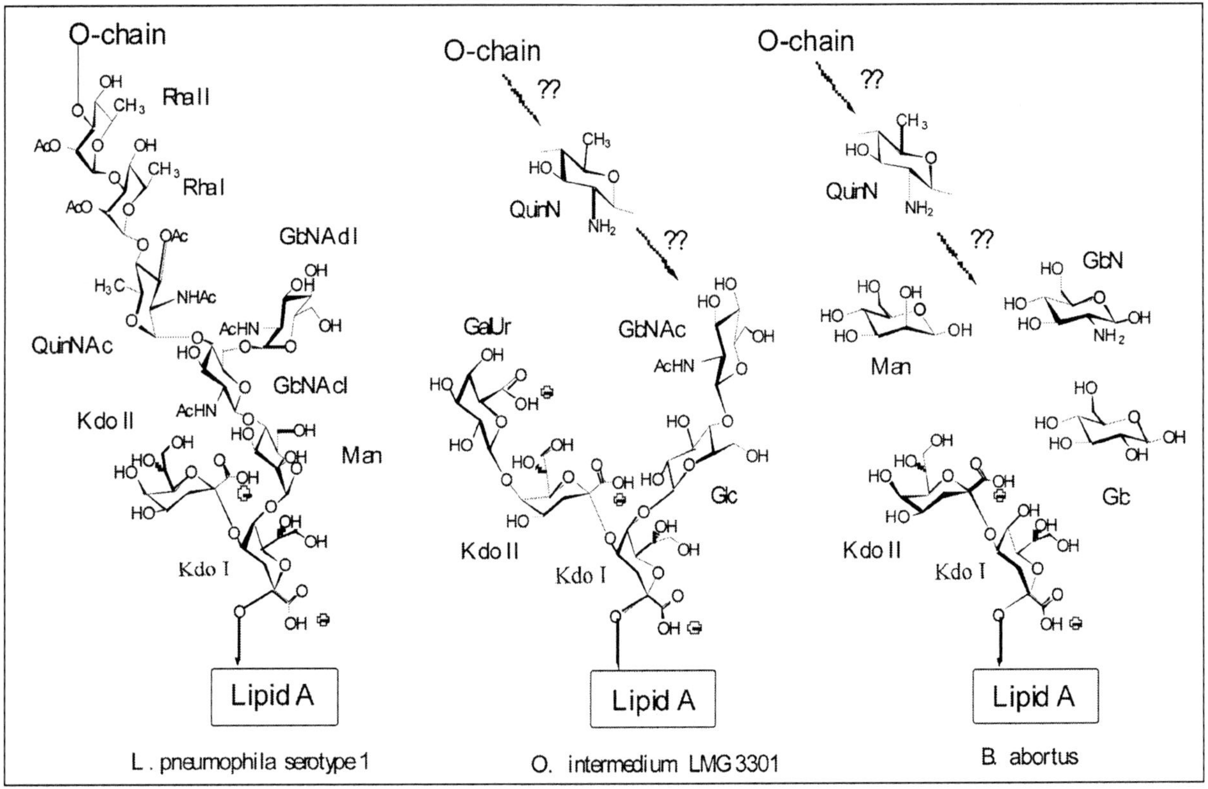

Figure 7. Core oligosaccharides of the LPS of *L. pneumophila, O. intermedium* and *B. abortus*. In marked contrast with *E. coli* (Fig. 1), there is no heptose or phosphate. The core of *L. pneumophila* is hydrophobic because of the *O*-acetyl groups. The LPS cores of both *B. abortus* and *O. intermedium* LPSs carry free amino groups but they differ in the presence of anionic galacturonic acid. The structure presented for *O. intermedium* LMG 3301may be only part of the whole core since sugar analysis of *Ochrobactrum* S-LPS suggest quinovosamine in the core. Positions and stoichiometry of the sugars in the *B. abortus* LPS core have not been determined (data are from references 118 and 119).

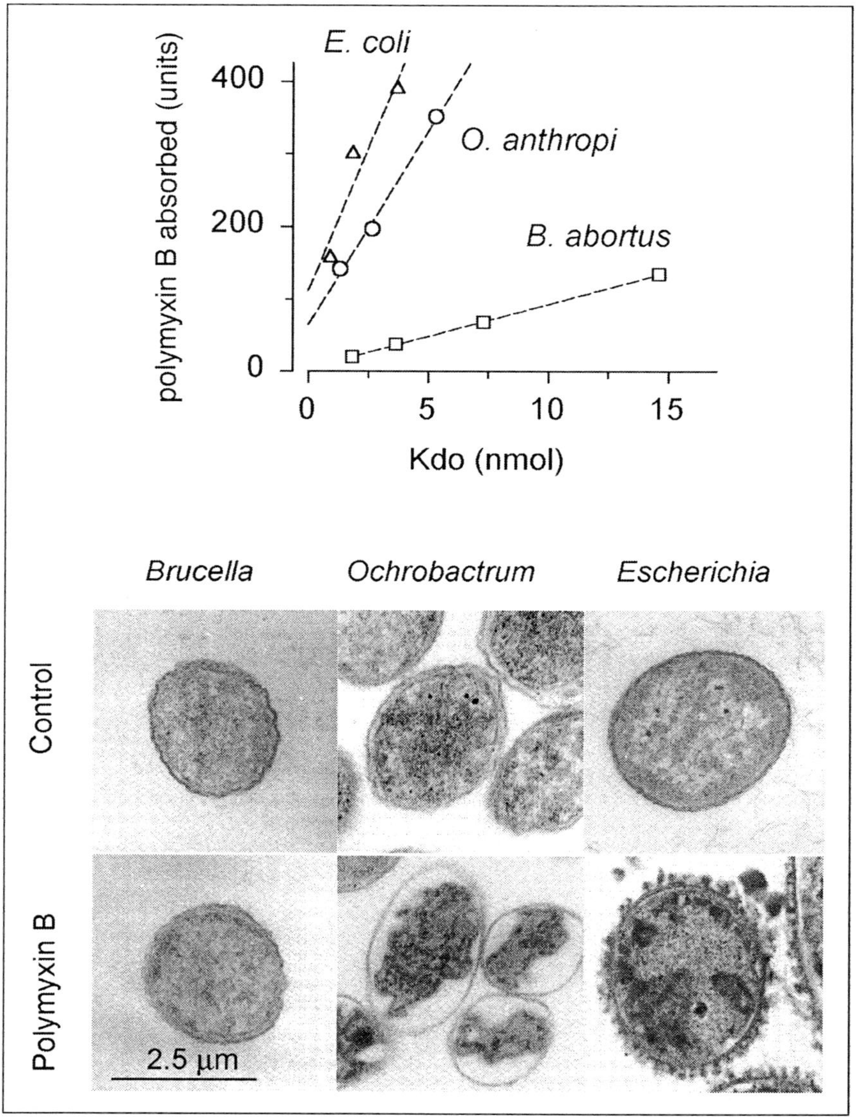

Figure 8. Polymyxin binding by *E. coli, O. anthropi* and *B. abortus* LPSs (upper panel) and its action on the outer membrane of these bacteria (lower panel) (adapted from ref. 119).

due mostly to the core-lipid A LPS section[133,134] since comparisons of R and S *Brucella* strains and the *B. abortus*-O-chain bearing *Yersinia enterocolitica* O:9 shows that the O-chain plays a minor (albeit significant) role.[133] Moreover, the importance of some fine details of the LPS-core structure is revealed when the *B. abortus* and *Ochrobactrum* LPSs are compared.[119] *Ochrobactrum*, although more sensitive to polymyxin than *B. abortus*, has a LPS that binds less antibiotic than the LPS of *E. coli* (Fig. 8), and the reduction in binding in these three bacteria parallels the absence of phosphate in the LPS-core of *Ochrobactrum* and *B. abortus* and of galacturonate in the latter. In keeping with what is observed in *S. typhimurium*, it is also likely that the lipid moiety is involved in polycation resistance. Indirect proof for this is provided by the observation that the OM of *Ochrobactrum* keeps its morphological appearance in the presence of polymyxin despite obvious internal membrane damage (Fig. 8). This shows that the higher hydrophobicity that can be predicted from the chemical structure of the lipid A somehow blocks the detergent-like action of polymyxin without altering the self-promoted way of entry. In all likelihood, resistance to polycationic peptides and proteins mediated by their unusual LPS is a virulence mechanism of *L. pneumophila* and *B. abortus*.

Biological Activity

Lipid A is the structure directly involved in cell activation and, therefore, it is not surprising that the LPSs of *L. pneumophila* and *B. abortus* show altered endotoxicity. In fact, the first hints that these LPSs had a different structure came not from detailed chemical analyses but from the study of their biological activities.[137-140] Altered endotoxicity is manifested in both cases as a marked reduction in the ability to stimulate susceptible host cells (Table 1), and the obvious implication is that these LPSs escape detection by the corresponding PRRs, at least at doses at which the LPS of other gram-negatives would have caused a strong (and sometimes lethal) activation of the immune system. This deficient stimulation has been proposed as a virulence mechanism for both *B. abortus*[141] and *L. pneumophila*[118] and it is obvious that it is particularly useful for an intracellular parasite. Although direct comparisons have not been made, the data suggest that the 30-40% reduction in bioactivity associated with the lipid A modifications of *S. typhimurium* (see above) are much less pronounced than those existing between the LPSs of *L. pneumophila* and *B. abortus* and of classical enteric bacteria which often reach several orders of magnitude (Table 1). This is in keeping with their more markedly altered MP.

The basis for such a reduced stimulation have been partially studied. It has been shown that binding of labeled *E. coli* LPS to CHO cells transfected with human CD14 cDNA is not reduced by competition with *L. pneumophila* LPS or lipid A in the presence of LBP under conditions in which such binding is blocked by the homologous unlabelled molecule.[142] Moreover, incubation of *E. coli* and *L. pneumophila* LPS with sCD14 results in complex formation only in the first case, no matter the presence of LBP.[142] These observations show that at least in comparative terms there is no direct or LBP-mediated interaction (Fig. 4) between CD14 and *L. pneumophila* LPS, and they strongly suggest that this is the main cause of its low endotoxicity.[142] Although similar experiments have not been performed, *B. abortus* LPS is as active in normal mice as in the endotoxin-resistant C3H/HeJ mice (Table 1) that carry a mutated TLR4.[32] This suggests that, as in the *L. pneumophila* LPS case, the PRR represented by the LBP-CD14-TLR4 system is not efficient in detecting *B. abortus* LPS. It remains to be determined whether the endotoxicity of *L. pneumophila* and *B. abortus* LPSs is mediated by this pathway at high LPS concentrations or whether alternative mechanisms are involved. Experiments with knockout mice should help to understand the involvement of the different elements of this pathway in the endotoxicity of these LPSs.

Analysis of the supramolecular structure of *B. abortus* LPS offers some data that could explain the inefficiency of the LBP-CD14-TLR4 system. The LPS aggregates of *B. abortus* have a larger size, a lower overall charge (the zeta potential) and a more restricted acyl-chain fluidity at temperatures above the T_c than the LPSs aggregates of *E. coli*.[119] These features suggest that the action of LBP on such aggregates could be hindered at two levels. First, reduced charge would hamper interaction with the polycationic amino-terminal domain of LBP. Second, the stronger hydrophobic effect that can be predicted from the lipid A structure (compare Figs. 1, 5 and 6) would necessarily increase the energy necessary for deaggregation. A fact consistent with the first hypothesis is the resistance to polycation binding commented above and illustrated in this context by the inefficiency of polymyxin in inactivating these LPSs (Table 1). In support of the second hypothesis, it is known that the acyl chains in the lipid A modulate some aspects of the LBP interaction with the components of the LBP-CD14 system. Partially deacylated enterobacterial-type lipids A (which are not stimulatory or endotoxic) bind to CD14 via LBP in a similar manner as the complete lipid A molecule,[143] and indirect evidence suggests that the lipid A of *Rhodobacter sphaeroides* (which is not endotoxic and carries acyl chains of 10 to 14 carbons) also binds to CD14.[144] On the other hand, the LPSs of *Helicobacter pylori* and *Porphyromonas (Bacteroides) gingivalis* are transferred to CD14 by LBP at a significantly slower rate than that of *E. coli*.[145] The lipids A of these bacteria carry acyl chains of 15 to 18 carbons[3] suggesting that their overall hydrophobicity is higher than that of the lipids A of either *E. coli* or *R. sphaeroides*. Therefore, these data also suggest a link between increased hydrophobicity and deficient recognition by the LBP-CD14-TLR4 system. Indeed, in the light of the above-summarized studies in which a relationship between molecular conformation of lipid A and endotoxicity was established,[8,9] it would be very interesting to determine the molecular conformation of the LPS of *L. pneumophila* and *B. abortus*.

Internalization and Detoxification

These aspects have not been investigated with *L. pneumophila* LPS, but there are two recent studies on the fate and activity of the *B. abortus* LPS after internalization by murine peritoneal macrophages.[146,147] After internalization under conditions favoring the LPS receptor-dependent pathways, *B. abortus* LPS moves slowly from early endosomes to lysosomes, and from here it is recycled to the cell surface where it forms very large patches. These patches are reminiscent of the comparatively large aggregates observed in water suspensions[119] and both the tendency of N-formyl-perosamine polymers to aggregate and the lipid A structure may account for them. Since it is not released from the cell surface, this LPS is in all likelihood firmly anchored by its lipid A moiety (rather than forming typical LPS aggregates). Moreover, as far as it has been determined it remains chemically unscathed (see next paragraph) and it is not cleared for at least 3 months. The same observations are made when attenuated S *B. abortus* is destroyed within macrophages: the bacterial LPS finally appears as large patches in the cell membrane showing that this reflects, in fact, something likely to happen during natural infection.[146] The LPS patches are enriched in MHC class II molecules, and Forestier et al[148] have suggested that this may relate to the impairment of MHC class II presentation of peptides observed by these authors. Significantly, the low toxicity of *B. abortus* LPS for macrophages[148] means that these cells are still able to support the growth of the pathogen while being handicapped in their antigen presentation function. The receptors for *B. abortus* S-LPS have not been identified (as noted above, a lectin-like mechanism seems to promote its binding to B lymphocytes) and it is possible that the same supramolecular features that hypothetically hinder activation of the LBP-CD14 pathway by *B. abortus* S-LPS could favor its internalization as large LPS aggregates are internalized faster.[65] An additional factor that remains to be investigated is the influence that the LPS tightly bound Omps of *Brucella*, and Omp3a and 3b in particular, could have on the internalization, processing and modulation of the biological activities of this LPS.

Although largely speculative, it is worth to comment on some possible implications of the structure of *L. pneumophila* and *B. abortus* LPSs in its processing by professional phagocytes. It is conceivable that these LPSs are more resistant to intracellular degradation than the classical enterobacterial-type LPSs. The predicted stronger hydrophobic effect would make more difficult the action of deacylating enzymes as these LPS molecules would form larger and more strongly bound aggregates. In addition, it may be that the chemical structures of *B. abortus* and *L. pneumophila* lipid A are not as sensitive to the deacylating enzyme of professional macrophages as other more conventional lipids A. Based on structural considerations, Zahringer et al[118] have suggested that the *L. pneumophila* lipid A is resistant to the action of esterases and amidases of amoebae. Since the activity of such esterases depends on the nature of the fatty acid and the amidases require previous removal of acyloxyacyl groups, it seems likely that the unusually long acyl chains would make removal of acyloxyacyl groups inefficient and this, in turn, would prevent amidases from acting. To what extent aggregation and acyl chain length affect degradation of lipid A by the acyloxyacyl hydrolase of professional phagocytes has not been tested but, as commented above, it is remarkable that *B. abortus* LPS is not cleared by mouse macrophages in at least 3 months after administration.[146]

Final Remarks

It is clear that the LPSs of several characteristic intracellular parasites depart in significant aspects from the classical LPS MP and that this is manifested as a failure of the innate immune system to recognize and act upon these molecules. The LPS of *Brucella*, and perhaps that of *Legionella*, stands out as a structure gathering a related and remarkable set of properties: resistance to antimicrobial peptides and proteins, low anticomplementary activity, low endotoxicity and toxicity for the cells where these bacteria grow, and the ability to present antigens in the MHC II context. Many aspects of the interaction of LPSs bearing conventional MPs with cells remain to be clarified and, in these studies, LPSs with altered MPs will represent a tool useful for a more precise definition of how PRR focused on LPS actually work.

Given the critical role of LPS in the OM permeability barrier, one may ask whether this function is also modified in the intracellular parasites and, if so, why. In *S. typhimurium*, there is the suggestive observation that PhoP-PhoQ regulates resistance to bile salts and LPS structure, but the picture is obscured by the fact that efflux pumps active on bile salts have been described in enterobacteria and they might also be regulated. In the case of *B. abortus*, the answer is that, in contrast to *S. typhimurium*, the OM is always readily penetrated by hydrophobic permeants.[149] This permeability can be explained on the sole basis of the LPS properties.[134,149] Comparison with *Ochrobactrum*, which keeps the OM barrier to hydrophobic permeants, suggests that in this cluster of the α-*Proteobacteria* only little structural changes are necessary to become simultaneously resistant to antimicrobial polycations and permeable.[119] Thus, the *brucellae* may have evolved from an α-*Proteobacteria* environmental microbe[116] that, by colonizing those host environments where no noxious hydrophobic permeants exist, would have been easily able to reduce at a minimum acidic core sugars under the selective pressure of antimicrobial peptides. Likewise, the *legionellae* may have evolved towards pathogenicity in mammals by gaining first the ability to multiply within free-living phagocytic eukaryotes taking advantage of an LPS structure that, although favoring this adaptation, was already present in environmental bacteria. In such *Legionella* and *Brucella* ancestors the LPS structure would represent a scaffold on which acquisition of additional virulence factors would lead to pathogenicity. These considerations, although speculative, may help to stress the complex and contingent ways by which intracellular pathogens emerge in nature.

Acknowledgements

Research at the Department of Microbiology of the University of Navarra is supported in part by grants PIUNA 1997, AGL2000-0305-CO2-01 and QLK-CT-2002-01001 ANEPID.

References

1. Janeway CA. The immune system evolved to discriminate infectious nonself from noninfectious self. Immunol Today 1992; 13(1):11-16.
2. Hoffmann JA, Kafatos FC, Janeway CA et al. Phylogenetic perspectives in innate immunity. Science 1999; 284(5418):1313-1318.
3. Zahringer U, Lindner B, Rietschel ET. Chemical structure of lipid A: Recent advances in structural analysis of biologically active molecules. In: Brade H, Opal SM, Vogel SN et al, eds. Endotoxin in Health and Disease. New York: Marcel Dekker, 1999:93-114.
4. Holst O. Chemical structure of the core region of lipopolysaccharides. In: Brade H, Opal SM, Vogel SN et al, eds. Endotoxin in Health and Disease. New York: Marcel Dekker, 1999:115-154.
5. Kastowsky M, Gutberlet T, Bradaczek H. Molecular modelling of the three-dimensional structure and conformational flexibility of bacterial lipopolysaccharide. J Bacteriol 1992; 174(14):4798-4806.
6. Seydel U, Wiese A, Schromm AB et al. A biophysical view on the function and activity of endotoxins. In: Brade H, Opal SM, Vogel SN et al, eds. Endotoxin in Health and Disease. New York: Marcel Dekker, 1999:195-219.

7. Brandenburg K, Seydel U. Investigation into the fluidity of lipopolysaccharide and free lipid A membrane systems by Fourier-transform infrared spectroscopy and differential scanning calorimetry. Eur J Biochem 1990; 191(1):229-236.
8. Brandenburg K, Mayer H, Koch MH et al. Influence of the supramolecular structure of free lipid A on its biological activity. Eur J Biochem 1993; 218(2):555-563.
9. Schromm AB, Brandenburg K, Loppnow H et al. The charge of endotoxin molecules influences their conformation and IL-6-inducing capacity. J Immunol 1998; 161(10):5464-5471.
10. Labischinski H, Barnickel G, Bradaczek H et al. High state of order of isolated bacterial lipopolysaccharide and its possible contribution to the permeation barrier property of the outer membrane. J Bacteriol 1985; 162(1):9-20.
11. Nikaido H, Vaara M. Molecular basis of bacterial outer membrane permeability. Microbiol Rev 1985; 49:1-32.
12. Koski P, Vaara M. Polyamines as constituents of the outer membranes of Escherichia coli and Salmonella typhimurium. J Bacteriol 1991; 173(12):3695-3699.
13. Vaara M. Agents that increase the permeability of the outer membrane. Microb Rev 1992; 56(3):395-411.
14. Lugtenberg B, Van Alphen L. Molecular architecture and functioning of the outer membrane of Escherichia coli and other gram-negative bacteria. Biochim Biophys Acta 1983; 737(1):51-115.
15. Seydel U, Eberstein W, Schroder G et al. Electrostatic potential barrier in asymmetric planar lipopolysaccharide/phospholipid bilayers probed with the valinomycin-K+ complex. Z Naturforsch C 1992; 47(9-10):757-761.
16. Thanassi DG, Cheng LW, Nikaido H. Active efflux of bile salts by Escherichia coli. J Bacteriol 1997; 179(8):2512-2518.
17. Vaara M. Lipopolysaccharide and the permeability of the bacterial outer membrane. In: Brade H, Opal SM, Vogel SN et al, eds. Endotoxin in Health and Disease. New York: Marcel Dekker, 1999:31-38.
18. Rossignol DP, Hawkins LD, Christ WJ et al. Synthetic endotoin antagonists. In: Brade H, Opal SM, Vogel SN et al, eds. Endotoxin in Health and Disease. New York: Marcel Dekker, 1999:699-717.
19. Goldman RC, White D, Leive L. Identification of outer membrane proteins, including known lymphocyte mitogens, as the endotoxin protein of Escherichia coli 0111. J Immunol 1981; 127(4):1290-1294.
20. Weis WI, Taylor ME, Drickamer K. The C-type lectin superfamily in the immune system. Immunol Rev 1998; 163:19-34.
21. Gross PS, Al SW, Clow LA et al. Echinoderm immunity and the evolution of the complement system. Dev Comp Immunol 1999; 23(4-5):429-442.
22. Tesh VL. Complement-mediated lipopolysaccharide release. In: Brade H, Opal SM, Vogel SN et al, eds. Endotoxin in Health and Disease. New York: Marcel Dekker, 1999:77-92.
23. Schweinle JE, Ezekowitz RA, Tenner AJ et al. Human mannose-binding protein activates the alternative complement pathway and enhances serum bactericidal activity on a mannose-rich isolate of Salmonella. J Clin Invest 1989; 84(6):1821-1829.
24. Jiang GZ, Sugiyama T, Kato Y et al. Binding of mannose-binding protein to Klebsiella O3 lipopolysaccharide possessing the mannose homopolysaccharide as the O-specific polysaccharide and its relation to complement activation. Infect Immun 1995; 63(7):2537-2540.
25. Devyatyarova JM, Rees IH, Robertson BD et al. The lipopolysaccharide structures of Salmonella enterica serovar Typhimurium and Neisseria gonorrhoeae determine the attachment of human mannose-binding lectin to intact organisms. Infect Immun 2000; 68(7):3894-3899.
26. Brogden KA, Cutlip RC, Lehmkuhl HD. Complexing of bacterial lipopolysaccharide with lung surfactant. Infect Immun 1986; 52(3):644-649.
27. Brogden KA. Ovine pulmonary surfactant induces killing of Pasteurella haemolytica, Escherichia coli, and Klebsiella pneumoniae by normal serum. Infect Immun 1992; 60(12):5182-5189.
28. Stamme C, Wright JR. Surfactant protein A enhances the binding and deacylation of E. coli LPS by alveolar macrophages. Am J Physiol 1999; 276(3 Pt 1):L540-L547
29. MacLeod Griffiss J, Schneider H. The chemistry and biology of lipooligosaccharides: the endotoxins of bacteria of the respiratory and genital mucosae. In: Brade H, Opal SM, Vogel SN et al, eds. Endotoxin in Health and Disease. New York: Marcel Dekker, 1999:179-194.
30. Kraus D, Medof ME, Mold C. Complementary recognition of alternative pathway activators by decay-accelerating factor and factor H. Infect Immun 1998; 66(2):399-405.
31. Grossman N, Joiner KA, Frank MM et al. C3b binding, but not its breakdown, is affected by the structure of the O-antigen polysaccharide in lipopolysaccharide from Salmonellae. J Immunol 1986; 136(6):2208-2215.

32. Grossman N, Svenson SB, Leive L et al. Salmonella O antigen-specific oligosaccharide-octyl conjugates activate complement via the alternative pathway at different rates depending on the structure of the O antigen. Mol Immunol 1990; 27(9):859-865.
33. Peterson AA, Haug A, McGroarty E. Physical properties of short- and long-O-antigen-containing fractions of lipopolysaccharide from Escherichia coli O111:B4. J Bacteriol 1986; 165:116-122.
34. Leippe M. Antimicrobial and cytolytic polypeptides of amoeboid protozoa—Effector molecules of primitive phagocytes. Dev Comp Immunol 1999; 23(4-5):267-279.
35. Andra J, Leippe M. Pore-forming peptide of Entamoeba histolytica. Significance of positively charged amino acid residues for its mode of action. FEBS Lett 1994; 354(1):97-102.
36. Hancock RE, Chapple DS. Peptide antibiotics. Antimicrob Agents Chemother 1999; 43(6):1317-1323.
37. Hancock RE, Diamond G. The role of cationic antimicrobial peptides in innate host defences. Trends Microbiol 2000; 8(9):402-410.
38. Modrzakowski MC, Spitznagel JK. Bactericidal activity of fractionated granule contents from human polymorphonuclear leukocytes: antagonism of granule cationic proteins by lipopolysaccharide. Infect Immun 1979; 25(2):597-602.
39. Shafer WM, Casey SG, Spitznagel JK. Lipid A and resistance of Salmonella typhimurium to antimicrobial granule proteins of human neutrophil granulocytes. Infect Immun 1984; 43(3):834-838.
40. Rifkind D, Palmer JD. Neutralization of endotoxin toxicity in chick embryos by antibiotics. J Bacteriol 1966; 92(4):815-819.
41. Schroder G, Brandenburg K, Seydel U. Polymyxin B induces transient permeability fluctuations in asymmetric planar lipopolysaccharide/phospholipid bilayers. Biochemistry 1992; 31(3):631-638.
42. Wiese A, Munstermann M, Gutsmann T et al. Molecular mechanisms of polymyxin B-membrane interactions: Direct correlation between surface charge density and self-promoted transport. J Membr Biol 1998; 162(2):127-138.
43. Wu M, Maier E, Benz R et al. Mechanism of interaction of different classes of cationic antimicrobial peptides with planar bilayers and with the cytoplasmic membrane of Escherichia coli. Biochemistry 1999; 38(22):7235-7242.
44. Elass RE, Legrand D, Salmon V et al. Lactoferrin inhibits the endotoxin interaction with CD14 by competition with the lipopolysaccharide-binding protein. Infect Immun 1998; 66(2):486-491.
45. Bellamy W, Takase M, Wakabayashi H et al. Antibacterial spectrum of lactoferricin B, a potent bactericidal peptide derived from the N-terminal region of bovine lactoferrin. J Appl Bacteriol 1992; 73(6):472-479.
46. Yamauchi K, Tomita M, Giehl TJ et al. Antibacterial activity of lactoferrin and a pepsin-derived lactoferrin peptide fragment. Infect Immun 1993; 61(2):719-728.
47. Van Deventer SJ, Pajkrt D. Interactions of lipopolysaccharides and lipoproteins. In: Brade H, Opal SM, Vogel SN et al, eds. Endotoxin in Health and Disease. New York: Marcel Dekker, 1999:379-388.
48. Weiss J, Victor M, Stendhal O et al. Killing of gram-negative bacteria by polymorphonuclear leukocytes: role of an O2-independent bactericidal system. J Clin Invest 1982; 69(4):959-970.
49. Wiese A, Brandenburg K, Carroll SF et al. Mechanisms of action of bactericidal/permeability-increasing protein BPI on reconstituted outer membranes of gram-negative bacteria. Biochemistry 1997; 36(33):10311-10319.
50. Weiss J, Muello K, Victor M et al. The role of lipopolysaccharides in the action of the bactericidal/permeability-increasing neutrophil protein on the bacterial envelope. J Immunol 1984; 132(6):3109-3115.
51. Tobias PS, Soldau K, Iovine NM et al. Lipopolysaccharide (LPS)-binding proteins BPI and LBP form different types of complexes with LPS. J Biol Chem 1997; 272(30):18682-18685.
52. Ooi CE, Weiss J, Elsbach P et al. A 25-kDa NH2-terminal fragment carries all the antibacterial activities of the human neutrophil 60-kDa bactericidal/permeability-increasing protein. J Biol Chem 1987; 262(31):14891-14894.
53. Marra MN, Wilde CG, Griffith JE et al. Bactericidal/permeability-increasing protein has endotoxin-neutralizing activity. J Immunol 1990; 144(2):662-666.
54. Ooi CE, Weiss J, Doerfler ME et al. Endotoxin-neutralizing properties of the 25 kD N-terminal fragment and a newly isolated 30 kD C-terminal fragment of the 55-60 kD bactericidal/permeability-increasing protein of human neutrophils. J Exp Med 1991; 174(3):649-655.
55. Capodici C, Chen S, Sidorczyk Z et al. Effect of lipopolysaccharide (LPS) chain length on interactions of bactericidal/permeability-increasing protein and its bioactive 23-kilodalton NH_2-terminal fragment with isolated LPS and intact Proteus mirabilis and Escherichia coli. Infect Immun 1994; 62(1):259-265.

56. Wiese A, Brandenburg K, Lindner B et al. Mechanisms of action of the bactericidal/permeability-increasing protein BPI on endotoxin and phospholipid monolayers and aggregates. Biochemistry 1997; 36(33):10301-10310.
57. Gazzano-Santoro H, Parent JB, Conlon PJ et al. Characterization of the structural elements in lipid A required for binding of a recombinant fragment of bactericidal/permeability-increasing protein rBPI23. Infect Immun 1995; 63(6):2201-2205.
58. Beamer LJ, Carroll SF, Eisenberg D. Crystal structure of human BPI and two bound phospholipids at 2.4 angstrom resolution. Science 1997; 276(5320):1861-1864.
59. Iovine NM, Elsbach P, Weiss J. An opsonic function of the neutrophil bactericidal/permeability-increasing protein depends on both its N- and C-terminal domains. Proc Natl Acad Sci USA 1997; 94(20):10973-10978.
60. Levy O, Ooi CE, Elsbach P et al. Antibacterial proteins of granulocytes differ in interaction with endotoxin. Comparison of bactericidal/permeability-increasing protein, p15s, and defensins. J Immunol 1995; 154(10):5403-5410.
61. Fenton MJ, Golenbock DT. LPS-binding proteins and receptors. J Leukoc Biol 1998; 64(1):25-32.
62. Wurfel MM, Kunitake ST, Lichenstein H et al. Lipopolysaccharide (LPS)-binding protein is carried on lipoproteins and acts as a cofactor in the neutralization of LPS. J Exp Med 1994; 180(3):1025-1035.
63. de Haas CJ, Poppelier MJ, Van Kessel KP et al. Serum amyloid P component prevents high-density lipoprotein-mediated neutralization of lipopolysaccharide. Infect Immun 2000; 68(9):4954-4960.
64. Gegner JA, Ulevitch RJ, Tobias PS. Lipopolysaccharide (LPS) signal transduction and clearance. Dual roles for LPS binding protein and membrane CD14. J Biol Chem 1995; 270(10):5320-5325.
65. Kitchens RL, Munford RS. CD14-dependent internalization of bacterial lipopolysaccharide (LPS) is strongly influenced by LPS aggregation but not by cellular responses to LPS. J Immunol 1998; 160(4):1920-1928.
66. Schumann RR, Leong SR, Flaggs GW et al. Structure and function of lipopolysaccharide binding protein. Science 1990; 249(4975):1429-1431.
67. Beamer LJ, Carroll SF, Eisenberg D. The BPI/LBP family of proteins: a structural analysis of conserved regions. Protein Sci 1998; 7(4):906-914.
68. Abrahamson SL, Wu HM, Williams RE et al. Biochemical characterization of recombinant fusions of lipopolysaccharide binding protein and bactericidal/permeability-increasing protein. Implications in biological activity. J Biol Chem 1997; 272(4):2149-2155.
69. Schumann RR, Lamping N, Hoess A. Interchangeable endotoxin-binding domains in proteins with opposite lipopolysaccharide-dependent activities. J Immunol 1997; 159(11):5599-5605.
70. Lengacher S, Jongeneel CV, Le Roy D et al. Reactivity of murine and human recombinant LPS-binding protein (LBP) within LPS and gram negative bacteria. J Inflamm 1995; 47(4):165-172.
71. Wright SD, Ramos RA, Tobias PS et al. CD14, a receptor for complexes of lipopolysaccharide (LPS) and LPS binding protein. Science 1990; 249(4975):1431-1433.
72. Mathison JC, Tobias PS, Wolfson E et al. Plasma lipopolysaccharide (LPS)-binding protein. A key component in macrophage recognition of gram-negative LPS. J Immunol 1992; 149(1):200-206.
73. Pugin J, Schurer MC, Leturcq D et al. Lipopolysaccharide activation of human endothelial and epithelial cells is mediated by lipopolysaccharide-binding protein and soluble CD14. Proc Natl Acad Sci USA 1993; 90(7):2744-2748.
74. Hailman E, Lichenstein HS, Wurfel MM et al. Lipopolysaccharide (LPS)-binding protein accelerates the binding of LPS to CD14. J Exp Med 1994; 179(1):269-277.
75. Tobias PS, Soldau K, Gegner JA et al. Lipopolysaccharide binding protein-mediated complexation of lipopolysaccharide with soluble CD14. J Biol Chem 1995; 270(18):10482-10488.
76. Poussin C, Foti M, Carpentier JL et al. CD14-dependent endotoxin internalization via a macropinocytic pathway. J Biol Chem 1998; 273(32):20285-20291.
77. Medzhitov R, Preston HP, Janeway CA. A human homologue of the Drosophila Toll protein signals activation of adaptive immunity. Nature 1997; 388(6640):394-397.
78. Muzio M, Bosisio D, Polentarutti N et al. Differential expression and regulation of toll-like receptors (TLR) in human leukocytes: selective expression of TLR3 in dendritic cells. J Immunol 2000; 164(11):5998-6004.
79. Chow JC, Young DW, Golenbock DT et al. Toll-like receptor-4 mediates lipopolysaccharide-induced signal transduction. J Biol Chem 1999; 274(16):10689-10692.
80. Takeuchi O, Hoshino K, Kawai T et al. Differential roles of TLR2 and TLR4 in recognition of gram-negative and gram-positive bacterial cell wall components. Immunity 1999; 11(4):443-451.

81. Hoshino K, Takeuchi O, Kawai T et al. Cutting edge: Toll-like receptor 4 (TLR4)-deficient mice are hyporesponsive to lipopolysaccharide: evidence for TLR4 as the Lps gene product. J Immunol 1999; 162(7):3749-3752.
82. Poltorak A, He X, Smirnova I et al. Defective LPS signaling in C3H/HeJ and C57BL/10ScCr mice: mutations in Tlr4 gene. Science 1998; 282(5396):2085-2088.
83. Brightbill HD, Libraty DH, Krutzik SR et al. Host defense mechanisms triggered by microbial lipoproteins through toll-like receptors. Science 1999; 285(5428):732-736.
84. Aliprantis AO, Yang RB, Mark MR et al. Cell activation and apoptosis by bacterial lipoproteins through toll-like receptor-2. Science 1999; 285(5428):736-739.
85. Zhang H, Peterson JW, Niesel DW et al. Bacterial lipoprotein and lipopolysaccharide act synergistically to induce lethal shock and proinflammatory cytokine production. J Immunol 1997; 159(10):4868-4878.
86. Arbour NC, Lorenz E, Schutte BC et al. TLR4 mutations are associated with endotoxin hyporesponsiveness in humans. Nat Genet 2000; 25(2):187-191.
87. Schromm AB, Brandenburg K, Rietschel ET et al. Lipopolysaccharide-binding protein mediates CD14-independent intercalation of lipopolysaccharide into phospholipid membranes. FEBS Lett 1996; 399(3):267-271.
88. Korn A, Rajabi Z, Wassum B et al. Enhancement of uptake of lipopolysaccharide in macrophages by the major outer membrane protein OmpA of gram-negative bacteria. Infect Immun 1995; 63(7):2697-2705.
89. Kitchens RL, Munford RS. Internalization of lipopolysaccharide by phagocytes. In: Brade H, Opal SM, Vogel SN et al, eds. Endotoxin in Health and Disease. New York: Marcel Dekker, 1999:521-536.
90. Pearson AM. Scavenger receptors in innate immunity. Curr Opin Immunol 1996; 8(1):20-28.
91. Wada Y, Doi T, Matsumoto A et al. Structure and function of macrophage scavenger receptors. Ann NY Acad Sci 1995; 748:226-238.
92. Doi T, Kurasawa M, Higashino K et al. The histidine interruption of an alpha-helical coiled coil allosterically mediates a pH-dependent ligand dissociation from macrophage scavenger receptors. J Biol Chem 1994; 269(41):25598-25604.
93. Doi T, Higashino K, Kurihara Y et al. Charged collagen structure mediates the recognition of negatively charged macromolecules by macrophage scavenger receptors. J Biol Chem 1993; 268(3):2126-2133.
94. Hampton RY, Golenbock DT, Penman M et al. Recognition and plasma clearance of endotoxin by scavenger receptors. Nature 1991; 352(6333):342-344.
95. Gordon S. Macrophage-restricted molecules: role in differentiation and activation. Immunol Lett 1999; 65(1-2):5-8.
96. Kitchens RL, Wang P, Munford RS. Bacterial lipopolysaccharide can enter monocytes via two CD14-dependent pathways. J Immunol 1998; 161(10):5534-5545.
97. Thieblemont N, Thieringer R, Wright SD. Innate immune recognition of bacterial lipopolysaccharide: Dependence on interactions with membrane lipids and endocytic movement. Immunity 1998; 8(6):771-777.
98. Matsuura M, Shimada S, Kiso M et al. Expression of endotoxic activities by synthetic monosaccharide lipid A analogs with alkyl-branched acyl substituents. Infect Immun 1995; 63(4):1446-1451.
99. Wuorela M, Jalkanen S, Toivanen P et al. Yersinia lipopolysaccharide is modified by human monocytes. Infect Immun 1993; 61(12):5261-5270.
100. Fox ES, Thomas P, Broitman SA. Clearance of gut-derived endotoxins by the liver. Release and modification of 3H, 14C-lipopolysaccharide by isolated rat Kupffer cells. Gastroenterology 1989; 96(2 Pt 1):456-461.
101. Hueck CJ. Type III protein secretion systems in bacterial pathogens of animals and plants. Microbiol Mol Biol Rev 1998; 62(2):379-433.
102. Sieira R, Comerci DJ, Sanchez DO et al. A homologue of an operon required for DNA transfer in Agrobacterium is required in Brucella abortus for virulence and intracellular multiplication. J Bacteriol 2000; 182(17):4849-4855.
103. Vogel JP, Isberg RR. Cell biology of Legionella pneumophila. Curr Opin Microbiol 1999; 2(1):30-34.
104. Gunn JS, Ernst RK, McCoy AJ et al. Constitutive mutations of the Salmonella enterica serovar typhimurium transcriptional virulence regulator phoP. Infect Immun 2000; 68(6):3758-3762.
105. Groisman EA. The ins and outs of virulence gene expression: Mg2+ as a regulatory signal. Bioessays 1998; 20(1):96-101.
106. Ernst RK, Guina T, Miller SI. How intracellular bacteria survive: surface modifications that promote resistance to host innate immune responses. J Infect Dis 1999; 179 Suppl 2:S326-S330.

107. Gunn JS, Miller SI. PhoP-PhoQ activates transcription of pmrAB, encoding a two-component regulatory system involved in Salmonella typhimurium antimicrobial peptide resistance. J Bacteriol 1996; 178(23):6857-6864.
108. Guo L, Lim KB, Gunn JS et al. Regulation of lipid A modifications by Salmonella typhimurium virulence genes phoP-phoQ. Science 1997; 276(5310):250-253.
109. Guo L, Lim KB, Poduje CM et al. Lipid A acylation and bacterial resistance against vertebrate antimicrobial peptides. Cell 1998; 95(2):189-198.
110. van Velkinburgh JC, Gunn JS. PhoP-PhoQ-regulated loci are required for enhanced bile resistance in Salmonella spp. Infect Immun 1999; 67(4):1614-1622.
111. Stinavage P, Martin LE, Spitznagel JK. O antigen and lipid A phosphoryl groups in resistance of Salmonella typhimurium LT-2 to nonoxidative killing in human polymorphonuclear neutrophils. Infect Immun 1989; 57(12):3894-3900.
112. Radziejewska LJ, Bhat UR, Brade H et al. The chemical structure of the lipopolysaccharide of A Rc-type mutant of Proteus mirabilis lacking 4-amino-4-deoxy-L arabinose and its susceptibility towards polymyxin B. Adv Exp Med Biol 1990; 256:121-126.
113. Nummila K, Kilpelainen I, Zahringer U et al. Lipopolysaccharides of polymyxin B-resistant mutants of Escherichia coli are extensively substituted by 2-aminoethyl pyrophosphate and contain aminoarabinose in lipid A. Mol Microbiol 1995; 16(2):271-278.
114. Helander IM, Kato Y, Kilpelainen I et al. Characterization of lipopolysaccharides of polymyxin-resistant and polymyxin-sensitive Klebsiella pneumoniae O3. Eur J Biochem. 1996; 237(1):272-278.
115. Fields BS. The molecular ecology of legionellae. Trends Microbiol 1996; 4(7):286-290.
116. Moreno E. Evolution of Brucella. In: Plommet M, ed. Prevention of Brucellosis in the Mediterranean Countries. Wageningen: Pudoc Scientific Publishers, 1992:198-218.
117. Velasco J, Romero C, López-Goñi I et al. Evaluation of the relatedness of Brucella spp. and Ochrobactrum anthropi and description of Ochrobactrum intermedium spp. nov., a new species with a closer relationship to Brucella spp. Int J Syst Bacteriol 1998; 48 Pt 3:759-768.
118. Zahringer U, Knirel YA, Lindner B et al. The lipopolysaccharide of Legionella pneumophila serogroup 1 (strain Philadelphia 1): chemical structure and biological significance. Prog Clin Biol Res 1995; 392:113-139.
119. Velasco J, Bengoechea JA, Brandenburg K et al. Brucella abortus and its closest phylogenetic relative, Ochrobactrum spp., differ in outer membrane permeability and cationic peptide resistance. Infect Immun 2000; 68(6):3210-3218.
120. Bhat UR, Carlson RW, Busch M et al. Distribution and phylogenetic significance of 27-hydroxy-octacosanoic acid in lipopolysaccharides from bacteria belonging to the alpha-2 subgroup of Proteobacteria. Int J Syst Bacteriol 1991; 41(2):213-217.
121. Velasco J, Moll H, Knirel YA et al. Structural studies on the lipopolysaccharide from a rough strain of Ochrobactrum anthropi containing a 2,3-diamino-2,3-dideoxy-D-glucose disaccharide lipid A backbone. Carbohydr Res 1998; 306(1-2):283-290.
122. Moriyon I, Lopez-Goñi I. Structure and properties of the outer membranes of Brucella abortus and Brucella melitensis. Int Microbiol 1998; 1(1):19-26.
123. Caroff M, Bundle DR, Perry MB et al. Antigenic S-type lipopolysaccharide of Brucella abortus 1119-3. Infect Immun 1984; 46(2):384-388.
124. Aragon V, Díaz R, Moreno E et al. Characterization of Brucella abortus and Brucella melitensis native haptens as outer membrane O-type polysaccharides independent from the smooth lipopolysaccharide. J Bacteriol 1996; 178(4):1070-1079.
125. Velasco J, Moll H, Vinogradov EV et al. Determination of the O-specific polysaccharide structure in the lipopolysaccharide of Ochrobactrum anthropi LMG 3331. Carbohydr Res 1996; 287(1):123-126.
126. Moreno E, Berman DT, Boettcher LA. Biological activities of Brucella abortus lipopolysaccharides. Infect Immun 1981; 31(1):362-370.
127. Hoffmann EM, Houle JJ. Failure of Brucella abortus lipopolysaccharide (LPS) to activate the alternative pathway of complement. Vet Immunol Immunopathol 1983; 5(1):65-76.
128. Corbeil LB, Blau K, Inzana TJ et al. Killing of Brucella abortus by bovine serum. Infect Immun. 1988; 56(12):3251-3261.
129. Allen CA, Adams LG, Ficht TA. Transposon-derived Brucella abortus rough mutants are attenuated and exhibit reduced intracellular survival. Infect Immun 1998; 66(3):1008-1016.
130. Eisenschenk FC, Houle JJ, Hoffmann EM. Mechanism of serum resistance among Brucella abortus isolates. Vet Microbiol 1999; 68(3-4):235-244.
131. Aragon, V. Polisacáridos hapténicos de Brucella 1994; Ph.D. Thesis, University of Navarra, Spain.
132. Lee CM, Mayer EP, Molnar J et al. The mechanism of natural binding of bacteria to human lymphocyte subpopulations. J Clin Lab Immunol 1983; 11(2):87-94.

133. Martinez de Tejada G, Pizarro-Cerdá J, Moreno E et al. The outer membranes of Brucella spp. are resistant to bactericidal cationic peptides. Infect Immun 1995; 63(8):3054-3061.
134. Freer E, Moreno E, Moriyón I et al. Brucella-Salmonella lipopolysaccharide chimeras are less permeable to hydrophobic probes and more sensitive to cationic peptides and EDTA than are their native Brucella spp. counterparts. J Bacteriol 1996; 178(20):5867-5876.
135. Edelstein PH, Finegold SM. Use of a semiselective medium to culture Legionella pneumophila from contaminated lung specimens. J Clin Microbiol 1979; 10(2):141-143.
136. Riley LK, Robertson DC. Brucellacidal activity of human and bovine polymorphonuclear leukocyte granule extracts against smooth and rough strains of Brucella abortus. Infect Immun 1984; 46(1):231-236.
137. Redfearn, M.S. An immunochemical study of antigens of Brucella extracted by the Westphal technique 1960; Ph.D. Thesis, University of Wisconsin-Madison.
138. Leong D, Díaz R, Milner K et al. Some structural and biological properties of Brucella endotoxin. Infect Immun 1970; 1:174-182.
139. Fumarola D. Endotoxic potency of the Legionella pneumophila: Recent data. Boll Ist Sieroter Milan 1979; 58(1):100-103.
140. Wong KH, Moss CW, Hochstein DH et al. "Endotoxicity" of the Legionnaires' disease bacterium. Ann Intern Med 1979; 90(4):624-627.
141. Rasool O, Freer E, Moreno E et al. Effect of Brucella abortus lipopolysaccharide on oxidative metabolism and lysozyme release by human neutrophils. Infect Immun 1992; 60(4):1699-1702.
142. Neumeister B, Faigle M, Sommer M et al. Low endotoxic potential of Legionella pneumophila lipopolysaccharide due to failure of interaction with the monocyte lipopolysaccharide receptor CD14. Infect Immun 1998; 66(9):4151-4157.
143. Kitchens RL, Ulevitch RJ, Munford RS. Lipopolysaccharide (LPS) partial structures inhibit responses to LPS in a human macrophage cell line without inhibiting LPS uptake by a CD14-mediated pathway. J Exp Med 1992; 176(2):485-494.
144. Kirikae T, Schade FU, Kirikae F et al. Diphosphoryl lipid A derived from the lipopolysaccharide (LPS) of Rhodobacter sphaeroides ATCC 17023 is a potent competitive LPS inhibitor in murine macrophage-like J774.1 cells. FEMS Immunol Med Microbiol 1994; 9(3):237-243.
145. Cunningham MD, Seachord C, Ratcliffe K et al. Helicobacter pylori and Porphyromonas gingivalis lipopolysaccharides are poorly transferred to recombinant soluble CD14. Infect Immun 1996; 64(9):3601-3608.
146. Forestier C, Moreno E, Pizarro-Cerdá J et al. Lysosomal accumulation and recycling of lipopolysaccharide to the cell surface of murine macrophages, an in vitro and in vivo study. J Immunol. 1999; 162(11):6784-6791.
147. Forestier C, Moreno E, Meresse S et al. Interaction of Brucella abortus lipopolysaccharide with major histocompatibility complex class II molecules in B lymphocytes. Infect Immun 1999; 67(8):4048-4054.
148. Forestier C, Deleuil F, Lapaque N et al. Brucella abortus lipopolysaccharide in murine peritoneal macrophages acts as a down-regulator of T cell activation. J Immunol 2000; 165:5202-5210.
149. Martinez de Tejada G, Moriyon I. The outer membranes of Brucella spp. are not barriers to hydrophobic permeants. J Bacteriol 1993; 175(16):5273-5275.
150. Cherwonogrodzky JW, Dubray G, Moreno E et al. Antigens of Brucella. In: Nielsen K, Duncan JR, eds. Animal Brucellosis. Boca Raton: CRC Press, 1990:19-64.
151. Goldstein J, Hoffman T, Frasch C et al. Lipopolysaccharide (LPS) from Brucella abortus is less toxic than that from Escherichia coli, suggesting the possible use of B. abortus or LPS from B. abortus as a carrier in vaccines. Infect Immun 1992; 60(4):1385-1389.

Chapter 15

Immune Recognition of the Mycobacterial Cell Wall

Steven A. Porcelli and Gurdyal S. Besra

Introduction

Mycobacteria are extraordinarily successful pathogens with a remarkable ability to persist within animal tissues even in the presence of an intact immune system. Pathogenic mycobacterial species are predominantly intracellular parasites that replicate within cells of their animal hosts, thus sequestering themselves in an environment that is protected from many of the immune mechanisms that normally eliminate bacterial invaders.[1] In the case of *M. tuberculosis*, the major host cell type infected is the macrophage. Following receptor mediated binding to the macrophage, *M. tuberculosis* bacilli are endocytosed and delivered to an early endosomal compartment, which then becomes modified into a stable intracellular vesicle that is typically referred to as the mycobacterial phagosome.[2] In this location, the bacillus is protected from the actions of neutralizing or opsonizing antibodies and complement, and the major challenge that it faces come from cell-mediated mechanisms of immunity that detect signals originating from infected cells.

Given the important role of cell-mediated immunity in generating protective responses against mycobacteria, it is not surprising that the vertebrate immune system has evolved a variety of mechanisms that stimulate macrophages and lymphocytes in response to infection by these organisms. In the case of macrophages and other myeloid lineage cells, initial contact with *M. tuberculosis* leads to cellular activation and the rapid production of cytokines, such as interleukin-1 (IL-1) and tumor necrosis factor α (TNFα) which initiate inflammatory cascades, and also to the synthesis of reactive oxygen and reactive nitrogen intermediates that can mediate bactericidal effects.[1] T lymphocytes recognize specific antigens of *M. tuberculosis* and provide a source of interferon-γ (IFNγ) and other cytokines that are critical to the activation of macrophages. In addition, T lymphocytes have been shown to be capable of directly attacking mycobacteria through their secretion of bactericidal granule components.[3] Yet in spite of all this immunological artillery, mycobacteria are difficult or perhaps impossible for the immune system to completely eliminate. Many factors, still mostly poorly characterized or unknown, are likely to contribute to the virulence and persistence of *Mycobacterium tuberculosis* and other pathogenic mycobacteria. However, a major key to the success of these organisms as pathogens is likely to be their unusual cell wall structure and its many interactions with the immune system.

Overview of the Mycobacterial Cell Wall Structure

In terms of its basic organization, the cell wall of mycobacteria is believed to be essentially a tripartite structure in which the two inner layers are composed of a true lipid bilayer that is encased within a lattice of peptidoglycan. This is then surrounded by a complex outer capsule of polysaccharides and lipids that make up the unique waxy coat of these organisms (Fig. 1).[4,5] In general, the traditional view of mycobacterial cell wall structure has been that its overall

Intracellular Pathogens in Membrane Interactions and Vacuole Biogenesis, edited by Jean-Pierre Gorvel.

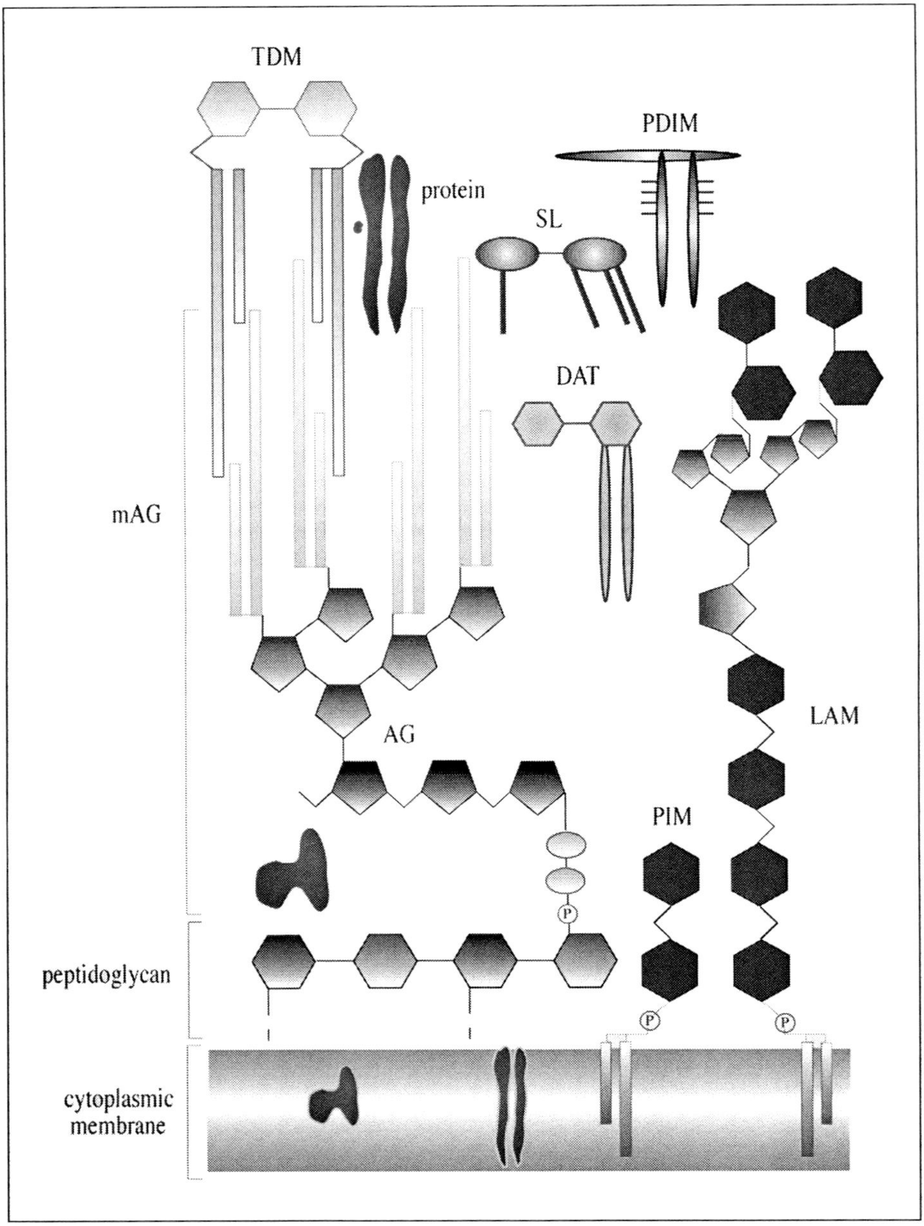

Figure 1. A model of the mycobacterial cell wall and envelope. Abbreviations: TDM, trehalose dimycolate; SL, sulfolipid; PDIM, phthiocerol dimycocerosate; DAT, diacyl trehalose; LAM, lipoarabinomannan; PIM, phosphatidylinositol mannoside; mAG, mycolylarabinogalactan; and AG, arabinogalactan.

arrangement is similar to that of Gram-positive organisms, which have only a single lipid bilayer that acts as a membrane dividing the cytoplasm from the cell wall peptidoglycan and the exterior capsule. However, it is also conceivable that the outer lipid rich layer of the mycobacterial capsule functions as a second outer membrane that organizes the cell wall into a periplasmic space between two membranes as typically seen in Gram-negative bacteria. Whichever view one prefers, it is clear that the cell wall and capsule structure of *M. tuberculosis* is organized in a way which brings a substantial array of immunoreactive substances into contact with cells of the immune system.

The immunologically active components of mycobacterial cell walls include some rather familiar targets that occur as immunogenic components of most bacterial pathogens. These include the protein antigens that provide specific epitopes for recognition by the adaptive immune response, and the basic building blocks of cell wall peptidoglycan that contribute adjuvant activity through activation of microbial pattern recognition receptors of the innate immune system. In addition, a myriad of exotic lipid structures exist in the *M. tuberculosis* cell wall, a number of which are now known to activate both the innate and adaptive arms of the immune response.[4-6] Many of these are likely to be intercalated into the waxy outer layer of the cell wall where they are well positioned to interact directly with membranes of host cells, and emerging evidence indicates that many lipids may also be released or actively secreted by growing mycobacteria.[7-9] The following sections provide an overview of the most extensively characterized components of the mycobacterial cell wall, and a summary of the current understanding of their interactions with the mammalian immune system.

Mycobacterial Peptidoglycans and Muramyl Dipeptide

As in all other eubacteria, mycobacterial peptidoglycans are highly cross-linked polymers of amino sugars and amino acids which include the repeating muramyl dipeptide structure (N-acetyl-muramyl-L-alanyl-D-isoglutamine; MDP) (Fig. 2). This MDP structure is well known as the least common denominator of adjuvant activity in bacterial cell walls, and is likely to be one of the components of killed mycobacteria preparations that imparts to them such potent adjuvant properties (i.e., as in Complete Freund's Adjuvant, a mixture of mineral oil and heat killed mycobacteria).[10,11] However, it is clear that MDP must represent only a component of the adjuvant effect of mycobacterial cells, since other bacteria with abundant amounts of this structure generally do not function as such potent adjuvants.

The major site of action of MDP is believed to be on mononuclear cells such as macrophages and dendritic cells, and this substance in the absence of other cell wall constituents has been shown to modulate surface receptor expression on human mononuclear cells.[12] While the mechanisms leading to cellular activation by MDP are not completely understood, recent evidence points to an important role for the Toll-like receptor family in responses to peptidoglycans. These receptors comprise a family of innate immune signaling receptors that were identified on the basis of their homology to the *Drosophila* Toll protein, a key molecule in the activation of immune defense against fungal pathogens in the fly.[13-15] In studies reported thus far, it has been found that the Toll-like receptor TLR-2, a member of this receptor family that mediates responses of macrophages to multiple structurally diverse ligands, is activated by purified soluble peptidoglycan derived from the Gram-positive organism *Staphylococcus aureus*. Activation of TLR-2 by peptidoglycan leads to nuclear translocation of the transcription factor NF-κB and subsequent upregulation of a variety of genes involved in stimulation of the inflammatory response.[16] Thus, it seems likely that mycobacterial peptidoglycan, and possibly its MDP component, may represent members of the growing list of compounds contained within the mycobacterial cell wall that interact with members of the Toll-like receptor family. The extent to which peptidoglycan and its MDP core sequence are accessible to surface receptors such as TLR-2 is unclear, and it is possible that the capsule of the mycobacterial cell wall may effectively eliminate this interaction during the usual course of infection.

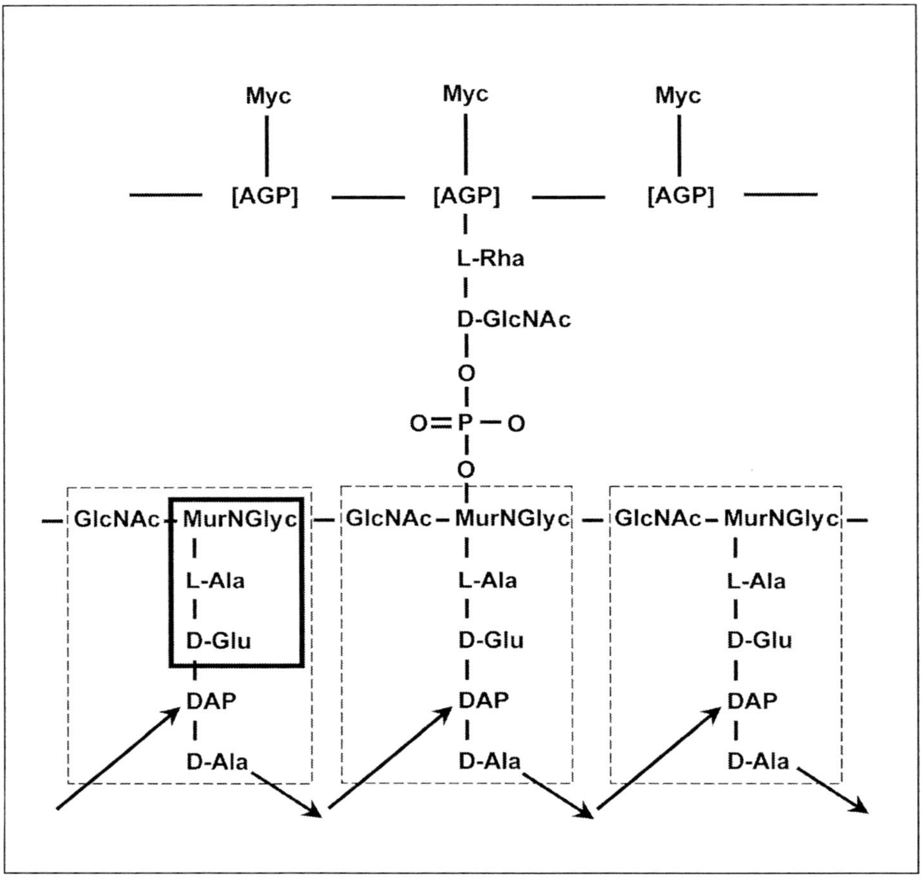

Figure 2. Schematic illustration of the mycolylarabinogalactan-peptidoglycan complex. The dashed lines enclose the repeating disaccharide tetrapeptide motif of one peptidoglycan (PG) chain. This is similar to the basic PG core structure found in other types of bacteria except for two modifications: 1) the presence of an N-glycolyl-muramic acid (MurNGlyc) rather than N-acetylmuramic acid in alternation with N-acetylglucosamine (GlcNAc) in the repeating core polysaccharide, and 2) the presence of interpeptide linkages between two *meso*-diaminopimelate (DAP) residues in addition to the usual D-alanyl (D-Ala) to DAP linkages. In one repeating PG subunit, the minimal adjuvant structure, muramyl dipeptide (MDP), is boxed with solid lines. The arrows represent covalent crosslinks between the tetrapeptides of one chain and those of an adjacent chain. Covalent linkage of the outer capsule of the cell wall is through a phosphodiester bond and a unique disaccharide bridge composed of L-rhamnose (L-Rha) and D-GlcNAc. This is linked to a complex arabinogalactan polymer (AGP), to which mycolic acids (Myc) are covalently linked via ester bonds.

Mycolic Acids and Glycosylated Mycolates

Unlike MDP, the mycolic acids and their glycosylated derivatives are unique structures of mycobacteria and related bacteria of the order *Actinomycetales*. As major constituents of the outer permeability barrier of the mycobacterial cell wall, the amount and specific structural modifications of these lipids have many effects on the ability of mycobacteria to grow both within host cells and in other environments. In addition, there is substantial evidence that

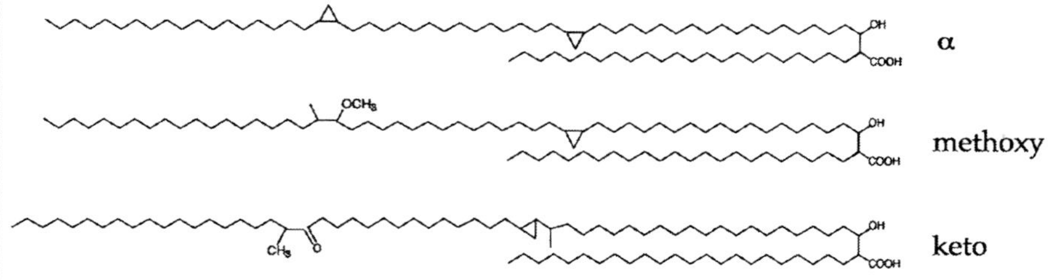

Figure 3. Representative structures of the major mycolic acids of *Mycobacterium tuberculosis*. Methoxymycolates and ketomycolates have either *cis* or *trans* cyclopropane rings, whilst α-mycolates possess only *cis* cyclopropane rings.

mycolic acids, in the free form and as fatty acid esters in mycobacterial glycolipids, are able to activate both innate and adaptive immune responses of mammals.

The unique structural aspects and biosynthesis of mycolic acids have been the focus of many detailed studies.[17] Mycolic acids are high molecular weight α-alkyl, β-hydroxy fatty acids, typically varying in length from approximately C_{60} to C_{90} in *M. tuberculosis* (Fig. 3). The marked abundance of mycolic acids in the mycobacterial cell wall is indicative of the importance of this class of lipids to the organism's growth and pathogenicity. This is underscored by studies on the mechanism of action of the antibiotic isoniazid, still one of the cornerstones of tuberculosis therapy, which indicate that its antituberculous activity is likely to result largely from inhibition of mycolic acid synthesis.[18,19] Several chemical modifications of the basic mycolic acid structure are well characterized, including the presence of oxygen-containing substitutions (e.g., methoxy or keto groups) and cyclopropane rings within the longer meromycolate chain of the branched fatty acid structure. Recent genetic studies have strongly suggested that both of these modifications of mycolic acids are necessary for full expression of virulence of *M. tuberculosis*. Thus, inactivation of the *hma* gene in *M. tuberculosis*, which leads to the inability to synthesize keto- or methoxymycolates, is associated with attenuation of growth in mice.[20] Similarly, mutants of the *pcaA* gene encoding a mycolic acid cyclopropane synthetase show reduced persistence of bacilli in the lungs, spleen and liver, as well as markedly reduced ability to cause mortality in infected mice.[21] While it is possible that the observed phenotypes of these mutants could reflect changes in the fluidity and permeability of the cell wall of *M. tuberculosis*, it remains to be determined whether alterations in the host immune response may also have contributed to the reduction in virulence. In this regard, it is noteworthy that histologic examination of the lungs of mice infected with the *pcaA* mutant of *M. tuberculosis* showed increased lymphocytic infiltration but reduced granulomatous inflammation compared to mice infected with wild type *M. tuberculosis*.[21]

A variety of glycosylated froms of mycolic acids also exist in the mycobacterial cell wall, and abundant evidence has accumulated on the interactions of these with the immune system. The most extensively studied glycolipid in this category is trehalose 6,6'-dimycolate (TDM), which is found as an abundant free glycolipid in the mycobacterial cell wall (Fig. 4). This glycolipid was implicated originally as the component of *M. tuberculosis* responsible for its in vitro growth in serpentine cords, thus leading to its designation as "cord factor". The influence of TDM on the structural organization and physical characteristics of the mycobacterial cell wall remain unclear, and it is still uncertain whether TDM is really the primary factor responsible for the cording morphology in vitro, which has long been believed to correlate with virulence in vivo. However, there is abundant evidence from in vivo and in vitro studies demonstrating that

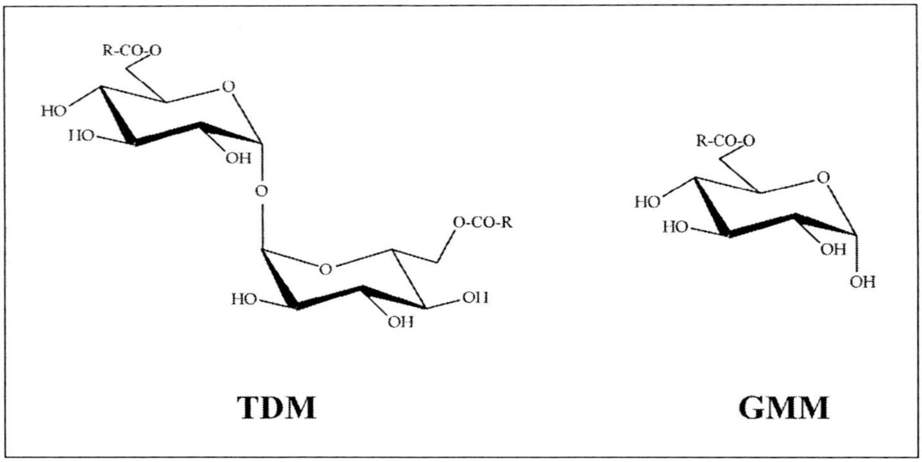

Figure 4. Glucose monomycolate (GMM) and trehalose dimycolate (TDM). The R-group represents the location of the mycolic acid (see Figure 3 for representative structures of *M. tuberculosis* mycolic acids).

TDM and other structurally related glycosylated mycolates have important interactions with the mammalian immune system.

One remarkable property that has been well established for TDM is its ability to induce granuloma formation when injected in the absence of protein antigen into mice or rabbits.[22-25] In a recent study on the effects of systemic administration of purified TDM in rabbits, the effects of a single intravenous injection of a suspension of TDM micelles were documented.[26] Generalized toxicity was observed, as manifested by diarrhea and a delay in normal weight gain in young animals over a three week period of observation. Studies of tissues at necropsy revealed marked thymic atrophy, which was demonstrated to be due to apoptosis in the thymic cortex. The lungs of TDM injected animals showed marked granuloma formation with macrophages and epithelioid cells predominating, along with modest numbers of lymphocytes and occassional multinucleated giant cells. Granuloma formation was also observed in the spleens and livers of these animals. These striking activities of TDM imply that it signals cells of the immune system to initiate the early steps of macrophage activation and clustering that lead to containment of mycobacteria. However, TDM alone appears not to lead to the caseating necrosis that is characteristic of active mycobacterial lesions in humans and other highly susceptible hosts such as rabbits.

While the mechanisms and receptors involved in the activation of granuloma formation by TDM are currently unknown, it is widely accepted that macrophages represent a major target of the immune stimulating activity of TDM. In fact, published data indicate that many effects of TDM in vivo are mediated by macrophages, including its anti-tumor activity and its ability to augment resistance to influenza virus infection in mice.[27,28] It is also clear that macrophages harvested from TDM treated animals are in a primed state, as defined by their ability to produce nitric oxide and develop antitumor activity upon further activation by low doses of lipopolysaccharide.[29] However, it is not yet clear whether activation of macrophages by TDM proceeds by direct interaction with receptors on these cells, as opposed to indirectly as a consequence of activation of other cell types. The general structural similarities between TDM and the lipid A moiety of LPS suggests the possibility that direct activation of macrophages by TDM could occur via Toll-like receptors expressed on these cells, although this point remains to be established.

Interestingly, TDM also appears to stimulate B cells to produce antibodies that react specifically with this glycolipid. This occurs during *M. tuberculosis* infections in humans, as shown by the detection of anti-TDM IgG in the sera of patients with both active and inactive tuberculosis.[30,31] Similar IgG antibodies were also shown to develop in the sera of rabbits or mice immunized with purified TDM administered as a suspension of micelles formed by combining an oil in water emulsion of TDM with detergent. These antibodies showed evidence of fine specificity for the lipid portion of the TDM structure, including strong reactivity for the methoxymycolic acid that makes up a portion of the mycolates esterified to TDM in *M. tuberculosis*. Although antibodies against TDM are of uncertain significance for the protective immune response against the pathogen, these antibodies are of interest as possible markers that could assist in the diagnosis of tuberculosis or other mycobacterial infections. Whether particular B cell subsets, such as B-1 or marginal zone B cells, are primarily responsible for production of antibodies to TDM has not yet been evaluated, nor has the T cell dependence or independence of this antibody response.

Another surprising aspect of the immune response to mycolates and their glycoconjugates is the recent demonstration that these can also be recognized as specific antigens by T lymphocytes. Studies reported thus far indicate that this phenomenon occurs through a novel pathway of antigen presentation that involves the CD1 family of cell surface lipid-binding proteins.[32] Recognition of lipid or glycolipid moieties complexed to CD1 proteins has been shown to be a component of the T cell response to mycobacterial pathogens, and appears to account for a portion of the specific immunological memory that develops following *M. tuberculosis* infection.[33-36] Among the relatively small number of lipid and glycolipid T cell antigens that have been described thus far, both free mycolic acids and the monoglucosylated form of these lipids (glucose-6-monomycolate, GMM) have been identified as specific antigens presented to T cells by CD1b, one of the five related CD1 proteins known to exist in humans.[37,38] The recognition of these antigens by CD1b-restricted T cells has been demonstrated to be highly specific for the polar cap structure of the lipid or glycolipid, and is mediated by specific interactions between clonally variable T cell antigen receptors and complexes formed by the binding of the antigen to a hydrophobic cleft in the CD1b protein.[37,39-42]

Although the specific role played by GMM in the structure and function of the mycobacterial cell wall is currently not known, the finding that T cells isolated from a human subject immune to *M. leprae* respond to GMM as a CD1b-presented antigen suggest that it is produced during growth of mycobacteria in tissues in vivo. Interestingly, in vitro studies have recently shown that GMM is only synthesized at measurable levels when mycobacteria are provided with an exogenous source of free glucose.[43] This requirement for free glucose for the synthesis of GMM is seen for both pathogenic species of mycobacteria such as *M. tuberculosis* and *M. avium*, and also for nonpathogenic saprophytes such as *M. smegmatis* and *M. phlei*. In the case of *M. avium*, it was found that GMM at levels that could be recognized by CD1b-restricted T cells specific for this glycolipid was produced when the bacteria were grown in media supplemented with glucose at concentrations comparable to those normally found in mammalian tissues and body fluids (e.g., 100 mg/dl). Furthermore, glycolipids with chromatographic properties identical to GMM were shown to be present in *M. leprae* bacilli purified from animal tissue, and these were stimulatory for T cells specific for this antigen. Together, these results demonstrated that GMM is likely to be formed by mycobacteria that invade tissues and replicate in vivo where free glucose is abundant, but is not likely to be formed by mycobacteria growing outside of an animal host. Thus, the recognition of GMM by CD1b-restricted T cells in humans has been proposed as a mechanism by which pathogenic mycobacteria that have productively infected tissues can be distinguished from ubiquitous nonpathogenic mycobacteria.

Lipoarabinomannan and Other Mycobacterial Lipoglycans

The high molecular weight lipoglycan known as lipoarabinomannan (LAM) is one of the dominant immunoreactive substances of the mycobacterial outer cell wall. These large and complex molecules, which are ubiquitously found in the envelopes of all mycobacteria species, are heterogeneous but share a tripartite amphipathic structure consisting of a phosphatidyl-*myo*-inositol anchor, a mannan core with a branching arabinan polymer, and the cap motifs that decorate the termini of the branched arabinan (Fig. 5).[44,45] At present, LAMs are subdivided into two categories according to the structure of the cap motifs. These are ManLAM, in which the caps consist of mannose oligosaccharides, and AraLAM (also known as PILAM) in which the mannose caps are absent and the branches of the arabinan may be terminated by phospho-*myo*-inositol residues.[45] ManLAMs have been isolated from *M. tuberculosis, M. leprae* and the *bacille* Calmette-Guerin (BCG) vaccine strain of *M. bovis.*[46,47] In contrast, AraLAMs appear to be characteristic of fast-growing avirulent mycobacterial species.

Many effects of LAMs on the immune response have now been documented, particularly with respect to effects on macrophages and other myeloid lineage cells. Purfied ManLAM has been shown to inhibit antigen processing and IFNγ induced macrophage activation.[48-50] However, ManLAM can also directly activate macrophages leading to their secretion of TNFα, IL-1, chemokines and the production of nitric oxide.[51-53] Interestingly, AraLAM isolated from avirulent mycobacteria has generally been found to be substantially more potent at activating these proinflammatory activities of macrophages, suggesting that ManLAM may be modified to achieve a partial activation of these cells that somehow facilitates the uptake or growth of virulent species such as *M. tuberculosis*. In this regard, an interesting observation that has recently been documented is the ability of ManLAM treatment of macrophages to induce alterations of calcium dependent signaling events related to apoptosis.[54] One outcome of this effect on cellular signaling is an inhibition of macrophage apoptosis that may normally be triggered following *M. tuberculosis* infection of the cell. Although the relevance of this effect in vivo remains to be clarified, it is possible that the enhanced survival of infected cells due to ManLAM stimulation is a strategy that *M. tuberculosis* uses to maintain the protected environment that it occupies once it has successfully parasitized a host macrophage.

Significant progress has been made in understanding the receptors on macrophages and other myeloid lineage cells that interact with LAM. Interactions between ManLAM and the macrophage mannose receptor (MR) have been shown to mediate attachment of *M. tuberculosis* to the cell surface, and this is most likely plays a role in the entry of the bacillus into the cell.[55-58] Interestingly, recent data suggest that the MR may function in certain physiologic settings primarily as a clearance receptor, and as such may help to maintain tissue homeostasis and immune tolerance rather than delivering strongly proinflammatory signalling properties.[59] In fact, engagement of the MR by intact mycobacteria, and presumably by ManLAM, has been demonstrated to lead to bacterial uptake without activation of bactericidal processes such as activation of NADPH oxidase or maturation of phagosomes.[60] Thus, uptake by this route may represent a safe portal of entry for organisms such as *M. tuberculosis*, providing a strategy by which the bacterium can attach to and enter the cell without signaling a strong inflammatory response. Free ManLAM has also been shown to bind to the MR on myeloid dendritic cells in vitro, which leads to the delivery of the lipoglycan to late endosomes and lysosomes.[61] This pathway of ManLAM uptake is believed to be important for the processing and recognition of ManLAM for T cell responses through the CD1b antigen presenting molecule (see below).

Another system of pattern recognition molecules that has been shown to recognize ManLAM is the pulmonary surfactant protein family. In vitro studies show that ManLAM binds specifically to human pulmonary surfactant protein A (hSP-A).[62] This interaction is likely to represent one of the initial events that occurs when *M. tuberculosis* in aerosolized microdroplets is delivered into the alveolar space and initiates primary infection in the lung. The binding of hSP-A to the mycobacterial surface promotes attachment of *M. tuberculosis* to phagocytes,[63] and interactions between hSP-A and the macrophage surface leads to upregulation of the MR which in turn causes increased cellular uptake of *M. tuberculosis.*[64] In addition, mycobacteria that are opsonized by hSP-A also have been found to interact directly with a specific phagocyte surface receptor (SPR210) via the collagen-like domain of hSP-A.[65,66] Binding of ManLAM to hSP-A was shown to be both calcium and carbohydrate dependent, indicating direct involve-

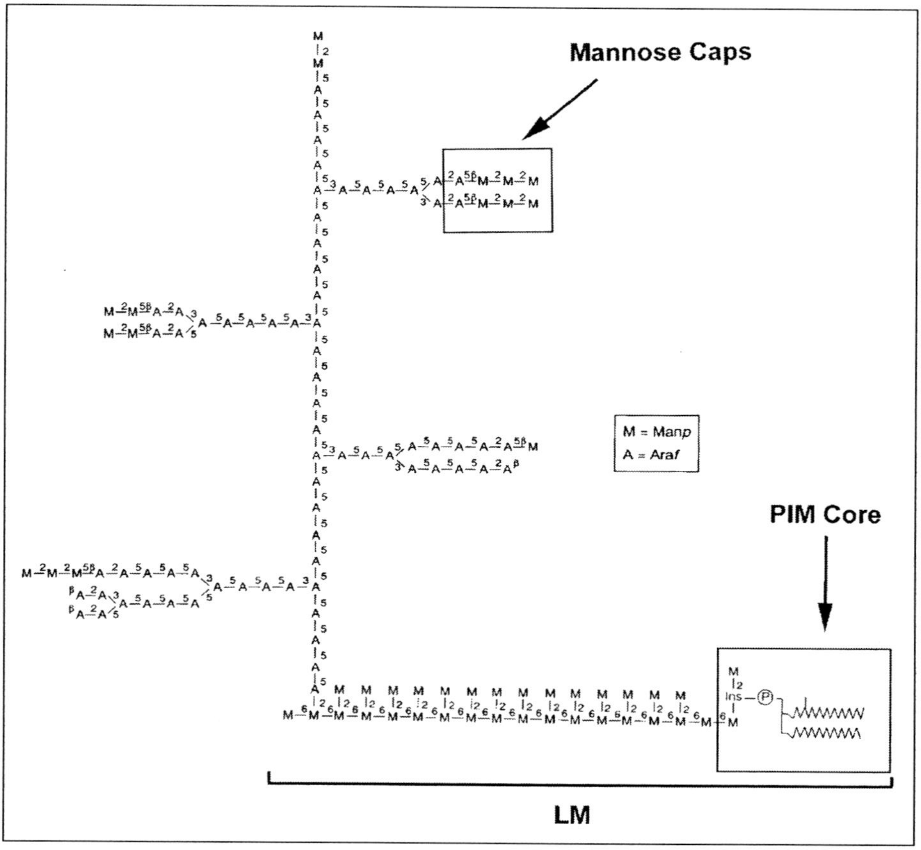

Figure 5. General structure of mannose-capped lipoarabinomannan (LAM) and its related subunit structures. Lipomannan (LM) lacks the large branching arabinan structure of LAM, and phosphatidylinositol mannosides (PIM_2 shown in boxed area) represents the glycolipid anchor of the much larger LAM and LM structures. Arabinose in the furanose form is indicated by A, mannose in the pyranose form by M, D-*myo*-inositol by Ins, and phosphate by P.

ment of the lectin domains of hSP-A. However, binding was also dependent on the presence of the fatty acids of ManLAM, since their removal abrogated the binding to hSP-A.[62] It is not clear at present whether this requirement for the lipid moiety of ManLAM represents a direct interaction of the lipid with hSP-A, or if it is due to the effect of the lipid on clustering of the ligand *via* aggregation or formation of micellar or lamellar structures. Interestingly, *M. tuberculosis* bacilli and ManLAM also bind to another structurally related surfactant protein hSP-D, but in this case the binding appears to inhibit rather than enhance phagocytosis of the bacterium by macrophages.[67] Thus, the net outcome of the initial contact between *M. tuberculosis* and cells in the alveolar lining may depend on the balance between antagonistic effects of different surfactant proteins, including but not necessarily limited to hSP-A and hSP-D.

Signaling through Toll-like receptors expressed on macrophages also appears to be a major pathway through which LAMs and related lipoglycans activate immune mechanisms, although this pathway may be effectively abrogated by the chemical modifications of the ManLAMs found in virulent mycobacteria. Studies on the activation of a mouse macrophage cell line revealed a requirement for TLR-2 in the activation of TNFα secretion by AraLAM.[68] In this experimental system, ManLAM was found to be nonstimulatory, consistent with the idea that its structural modifications are important for preventing its recognition by microbial pattern recognition systems that activate bactericidal mechanisms. Similar results have been reported from studies assessing recognition of LAM by human Toll-like receptors.[69,70] In these experiments, human TLR-2 was found to confer responsiveness to AraLAM isolated from nonpathogenic mycobacteria, but not to ManLAM isolated from *M. tuberculosis* or *M. bovis* BCG. In contrast, TLR-4, which is now established to be a necessary signaling component of the receptor complex that mediates responses to LPS of Gram-negative bacteria, did not confer responsiveness to AraLAM. Interestingly, expression of membrane bound CD14, an essential component of the LPS receptor complex and a necessary cofactor for TLR-4 signaling to LPS, was not required for TLR-2 signaling to AraLAM. However, in these studies there was a requirement for serum to achieve activation by AraLAM through TLR-2. This could indicate that soluble CD14 and its cofactor LPS binding protein (LBP), which are present in serum, can substitute for membrane bound CD14 in the response to AraLAM.[70]

The lipoglycans, including LAM and some of its subunit structures, are also known to be targets of the adaptive immune response to mycobacteria. The ability of LAM to stimulate specific antibody production against its carbohydrate structure is well documented in both animal models and human mycobacterial infections.[71,72] The importance of such antibodies in immunity to mycobacterial pathogens remains unclear, although one study has demonstrated a significant influence of IgM antibodies specific for LAM in modifying the clearance and organ distribution of this lipoglycan in mice.[73] Specific T cell responses against LAM have also been detected using T cell lines derived from the skin lesions of leprosy patients or from the peripheral blood of normal healthy donors.[34,74] Similar to T cell responses against mycolic acids or glycosylated mycolates, T cells specific for LAM recognize this antigen following its presentation by the CD1b molecule. The CD1b-dependent recognition of LAM involves uptake of LAM by the MR on monocyte-derived dendritic cells, which leads to the delivery of LAM to late endocytic compartments where it is presumably processed and becomes compexed to CD1b.[61] In some cases, T cells reactive to LAM also respond to the smaller acylated subunits of this antigen, including lipomannan (LM) and phosphatidyl-*myo*-inositol mannosides (PIMs) containing between two and six mannose residues (Fig. 5).[34,74] These results suggest that uptake and enzymatic processing of LAM may occur in dendritic cells during the course of mycobacterial infection, thus leading to presentation of the lipidated core of LAM by CD1b and the generation of a specific T cell response. Whether such T cell responses have significant effects on the subsequent control of mycobacterial growth in vivo remains to be determined.

Other Potentially Immunoreactive Mycobacterial Cell Wall Lipids

The mycobacterial cell wall lipids discussed above represent the most extensively studied examples in terms of their interactions with the mammalian immune system. However, there exist many more classes and subclasses of mycobacterial lipids that have not yet been subjected to such extensive studies using the methods now available for immunological analysis.[75] At least several of these appear to be particularly interesting to pursue as potentially important modulators of the immune response. For example, among the waxes of the *M. tuberculosis* cell wall is the phthiocerol dimycocerosate (PDIM) family. These consist of the mycocerosic acids, a complex mixture of multi-methyl branched fatty acids, esterified to the hydroxyl groups of phthiocerols and other related long chain alcohols (Fig. 6). Several of the glycosylated variants of lipids in this family have been shown to be among the immunoreactive glycolipids of mycobacteria in as much as they are targets for the antibody response produced during infection. This is particularly true for the phenolic glycolipids of *M. leprae*,[76,77] and also for the related triglycosyl phenol phthiocerol dimycocerosate that has been identified in some isolates of *M. tuberculosis*.[78] In addition, early work on the activities of the *M. leprae* phenolic glycolipids implicated these in the suppression of the cell-mediated immune response to *M. leprae*,[79] al-

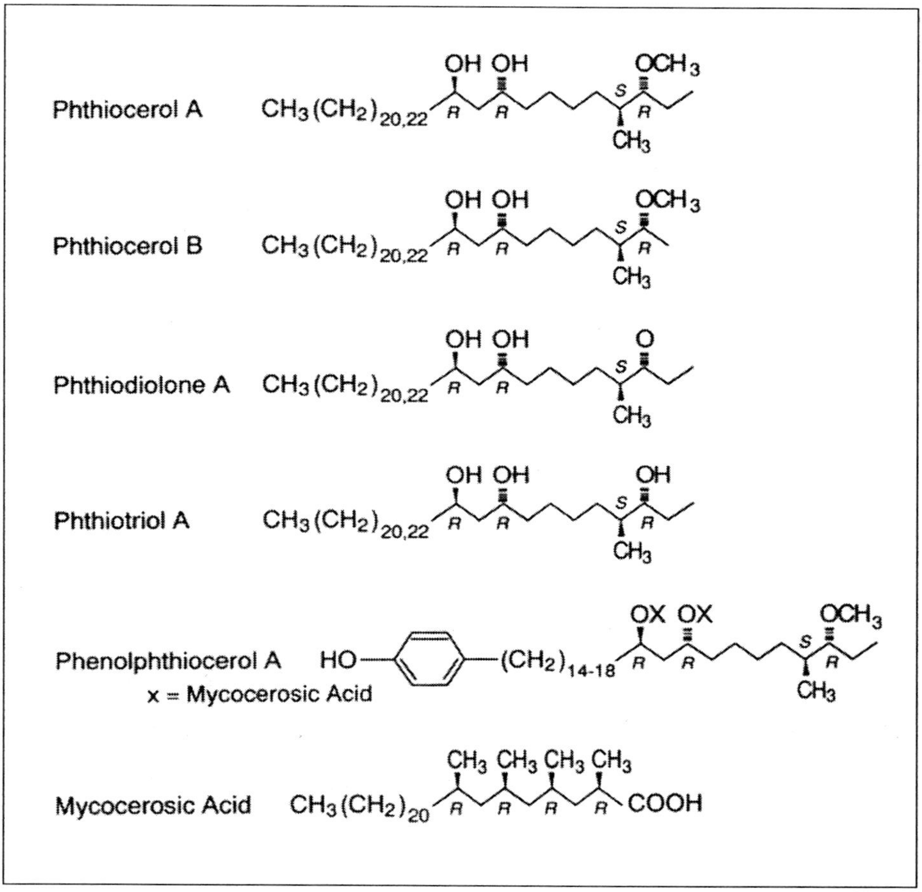

Figure 6. The phthiocerol dimycocerosate (PDIM) family of lipids. Mycocerosic acids (shown at the bottom) are esterified to the hydroxyls on the phthiocerol structures to form PDIM.

though there has been remarkably little follow up on this potentially important observation and the mechanism for the effects of these lipids remains unknown. Very recently, a genetic analysis identified genes involved in the synthesis of PDIM and in its transfer out of the mycobacterial cell as important factors for growth of *M. tuberculosis* in the lungs of infected mice.[9] It is not yet known whether this requirment for PDIM in the in vivo growth of *M. tuberculosis* reflects an effect of the lipid on the host reaction to the organism, as opposed to effects on the permeability or other properties of the mycobacterial cell wall.

A number of acylated trehalose compounds distinct from TDM have also come to light as potentially important immunoreactive substances. For example, the sulfolipids of *M. tuberculosis* represent 2,3,6,6'-tetra-O-acyltrehalose derivatives in which the 2' position is also sulfated (Fig. 7). The acyl groups of these compounds include a mixture of simple and complex fatty acids.[75] A link between these lipids and the virulence of various *M. tuberculosis* isolates was suggested by challenge studies carried out in guinea pigs,[6] and other work has suggested that they may inhibit phagocyte activation in some manner which has yet to be defined.[80]

Recent work has also called attention to isoprenoid lipids as potential targets of the adaptive immune response to mycobacteria. Several isoprenoid lipids have been identified in the myco-

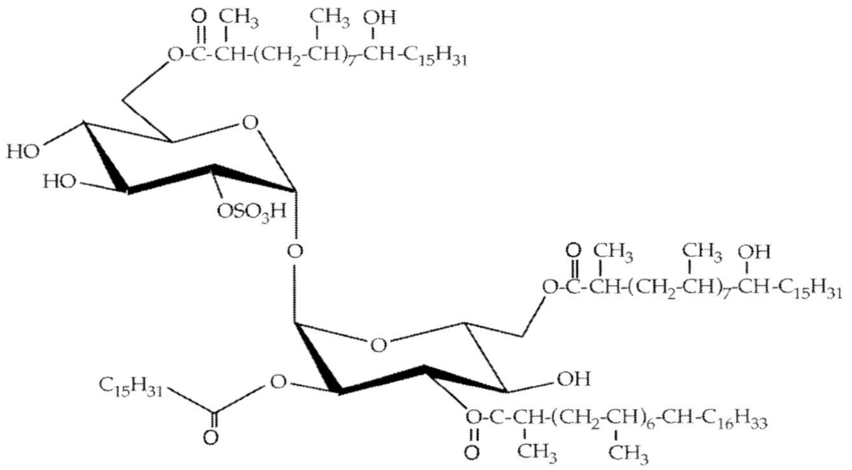

Figure 7. A representative *M. tuberculosis* sulfolipid (SL) structure (2,3,6,6'-Tetra-*O*-acyltrehalose-2'-sulfate).

bacterial cell wall, where they are most likely involved in the synthesis of peptidoglycan and other components of the cell wall.[75,81,82] In a recent study of T cell responses to phospholipids of mycobacteria presented by the human CD1c protein, two novel fully saturated β1-mannosyl-phosphoisoprenoid lipids from *M. tuberculosis* and *M. avium* were isolated and structurally characterized (Fig. 8).[36] Analysis of *M. tuberculosis* infected human subjects indicated that T cells responding to these lipids expand in response to infection, and persist in the circulation as part of the memory T cell population specific for *M. tuberculosis*. The contribution of these CD1c-restricted T cells to the successful control of *M. tuberculosis* by the immune system is not yet known, but their existence provides novel targets for the T cell response that may prove relevant to vaccination or immunotherapy for mycobacterial infection.

Cell Wall Associated Proteins of Mycobacteria

Although lipids and carbohydrates clearly dominate the mycobacterial cell wall structure, there also exist an array of cell wall associated proteins which are likely to have enormous relevance to the host immune response. Since protein antigens provide peptide epitopes that are presented by major histocompatibility complex (MHC) class I and II molecules to antigen specific T cells, they are obviously of premier importance to the adaptive immunity that develops in response to mycobacterial infection. Studies in mice and in immunodeficient humans clearly establish the importance of the T cell response in limiting the growth of pathogenic mycobacteria, and in somehow maintaining the organisms in a dormant state within tissue granulomas.[83-91] In addition, since T cells provide the basis for long term immunological memory, identification of the dominant protein antigens is a central goal in the quest for effective vaccines for the control of tuberculosis and other mycobacterial diseases.

Proteins secreted by *M. tuberculosis* during its growth have thus far received the greatest attention as possible immunodominant antigens that may give rise to protective T cell immunity in experimental animals and in humans.[92] However, a number of reports have also documented the immunogenicity in humans of proteins that are both weakly and strongly associated with the cell walls of *M. tuberculosis* and *M. leprae*.[93,94] Several studies have examined a

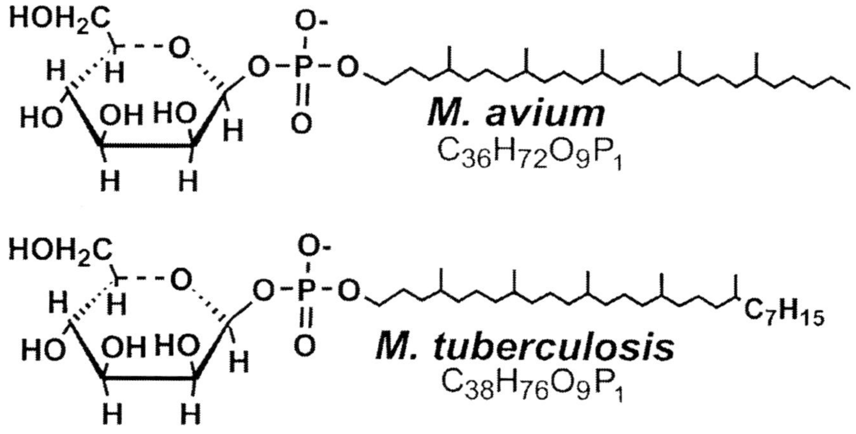

Figure 8. Mannosyl-phosphoisoprenoid lipids from *M. tuberculosis* and *M. avium*. The two β-1-mannosyl polyisoprenoid structures illustrated have been shown to be presented by the CD1c protein to T cells. Their fully saturated hydrocarbon structure is unusual among polyisoprenoid compounds of this class, as is the insertion of one additional methylene group between the phosphate group and the first isoprene unit of the isoprenoid chain.

subset of immunoreactive proteins that are very tightly associated with the cell wall and resist extraction even with strong ionic detergents such as sodium dodecyl sulfate (SDS). These proteins may be physically embedded within the extensively cross-linked cell wall peptidoglycan, or covalently associated with it in some undetermined way. This property has made these proteins difficult to recover and analyze in purified form, particularly since mycobacterial peptidoglycan is highly resistant to muramidase digestion. Studies in mice showed that immunization with a cell wall-protein peptidoglycan complex induced significant levels of protection against challenge with virulent *M. tuberculosis*, and the immunoprotective epitopes present in this material were lost after protease treatment.[95] These findings point to cell wall associated proteins as potential targets of the T cell response that accompanies natural infection with *M. tuberculosis*, and also support their potential to induce protective immunity.

Individual protein species among the tightly cell wall associated proteins of *M. tuberculosis* have also been studied, following their successful isolation from detergent extracted cell wall preparations using chemical degradation of peptidoglycan with trifluoromethanesulfonic acid.[96,97] One such study identified distinct cell wall associated protein species with molecular masses of 71, 60 and 45 kDa, all of which stimulated proliferative responses in vitro of lymphocytes from *M. tuberculosis* immune human subjects.[97] The identities of these proteins and their possible functions in the cell wall remain unknown. Another study using a similar approach isolated a 23 kDa protein that was found by peptide sequencing to have homology to the *E. coli* outer membrane protein OmpF.[96] This protein was shown to be immunogenic in rabbits as it comprised one of the targets for antibodies that developed following immunization of these animals with intact nonviable *M. tuberculosis* bacilli.

Other mycobacterial cell wall associated proteins are less strongly bound to the cell wall, and can be efficiently solubilized by detergent extraction. A few of these extractable cell wall proteins have been definitively identified, and in several cases have been implicated as factors that influence the outcome of the host-pathogen interaction. For example, a 16 kDa heat shock protein, also referred to as an α-crystallin homolog, has been found to be highly expressed in the thickened cell walls of *M. tuberculosis* that has been cultured under anaerobic conditions, and is also induced by stationary phase culture.[98,99] Although the function of this protein is not known, it appears to be important for the growth of *M. tuberculosis* in macrophages, and it has been suggested that it may in some way assist the bacterium during long-term survival in low oxygen tension or under other conditions that lead to dormancy.[100] Interactions of the immune system with this protein are not yet known, but it could potentially provide an attractive target for T cell surveillance of tissues for dormant mycobacteria.

The 19 kDa lipoprotein of *M. tuberculosis* is a second well characterized cell wall associated protein that has recently become a focus of attention in studies of both innate and adaptive immunity. This protein appears to be associated with the cell wall primarily through hydrophobic interactions involving its triacylated lipid-modified aminoterminus, and is thus relatively easily extracted using detergents. It was identified originally as a major antigen of *M. tuberculosis* based on its recognition by antibodies generated in mice immunized with crude mycobacterial extracts,[101] and subsequently was shown to be a target for both T cell and antibody responses during mycobacterial infection in humans.[102-104] Surprisingly, studies in animal challenge models have shown that the expression of the 19 kDa lipoprotein in experimental nonpathogenic live mycobacterial vaccines somehow inhibits the development of protective immunity against *M. tuberculosis*, indicating that this protein may interact with the host immune system in a manner that benefits the pathogen.[105] This finding appears to represent the first proposed example of an immunogenic protein of *M. tuberculosis* that interferes with effective vaccination. Although the mechanism by which this occurs remains obscure, it has obvious important implications for the development of both live attenuated and subunit vaccines against *M. tuberculosis*.

A second important aspect of the response to the 19 kDa lipoprotein antigen is the recent demonstration that it is a major stimulus for the production of IL-12 by human monocytes.[106] This response is triggered through the TLR-2 molecule, the same member of the Toll-like receptor family that has been implicated in the response to mycobacterial LAMs. TLR-2 dependent responses to the 19 kDa antigen and other bacterial lipoproteins are enhanced by CD14 expression and require the acyl chain moiety of the lipoproteins, similar to the findings reported on TLR-2 signaling in response to LAMs. The TLR-2 signaling in response to 19 kDa antigen appears to upregulate genes involved in type 1 inflammatory responses, but it remains to be established whether or not this is beneficial to the host response or a strategy of immune evasion employed by the microorganism.

Another well defined cell wall associated protein that has been the subject of numerous studies is the Ag85 complex of *M. tuberculosis*. The three highly homologous 30-32 kDa proteins (85A, 85B and 85C) that make up this complex are all among the most prominent secreted proteins of *M. tuberculosis*, and make up as much as 30% of the total bacterial protein in culture filtrates. However, a fraction of Ag85 also remains associated with the cell wall through an interaction that has not yet been characterized. These proteins are mycolyltransferases, and they play an essential role in mycobacterial cell wall synthesis by catalyzing the transfer of mycolic acids into trehalose esters and other mycolate-containing cell wall structures.[107,108] In addition, all three Ag85 proteins are known to be fibronectin binding proteins, which may mediate important aspects of bacterial adherence or interfere with leukocyte adhesion.[109]

The Ag85 proteins have also become a focus of immunological studies because they induce delayed hypersensitivity, specific antibodies and protective immune responses in *M. tuberculosis* infected mice and guinea pigs,[110-114] and also elicit responses of cultured peripheral blood mononuclear cells from human subjects with previous *M. tuberculosis* infection.[115,116] The Ag85 proteins are known to be prominent targets for $CD4^+$ T cells that recognize peptides bound to MHC class II molecules, a T cell population that produces abundant amounts of IFNγ and is known to be a component of the protective response to *M. tuberculosis* in mice.[117] Recently, a study using HLA-A*0201 transgenic mice has also provided strong evidence that Ag85 complex proteins may also be dominant targets for the MHC class I restricted responses of $CD8^+$ cytotoxic T cells. This study identified specific peptide epitopes on Ag85B that are

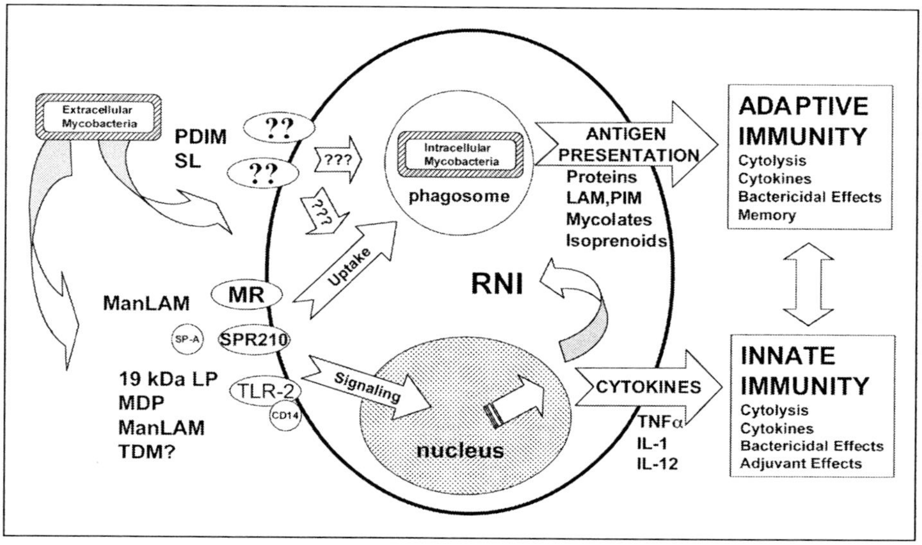

Figure 9. Schematic representation of the multiple interactions of mycobacterial cell wall components with the vertebrate immune system. Several known receptors on macrophages and other myeloid lineage cells (MR, SPR210, TLR-2) and their cofactors (hSP-A, CD14) are illustrated, as well as putative receptors for other compounds that are thought to be immunologically active such as PDIM and SL. Interaction of cell wall components on or shed from extracellular mycobacteria trigger these receptors, leading to uptake of the bacteria and transmembrane signals that activate genes involved in the innate immune response. An important outcome of this initial response is the upregulation of inducible nitric oxide synthase (iNOS or NOS2) and the production of reactive nitrogen intermediates (RNI) that have the ability to directly kill intracellular mycobacteria. Subsequent release and processing of protein and lipid components produced primarily by intracellular mycobacteria lead to antigen presentation via MHC class I and II and CD1 molecules. The resulting T cell responses form the key element of the adaptive immune response to mycobacteria. Substantial interplay between the innate and adaptive responses also exists (symbolized by the two-headed arrow), which may be of crucial importance to determining the net outcome of the infection.

presented by this human HLA molecule, and directly demonstrated a population of Ag85B reactive $CD8^+$ T cells in human subjects with previous *M. tuberculosis* infection using fluorescent peptide/HLA-A*0201 tetramers.[118] These epitopes of Ag85B represent some of the first antigens to be defined for human T cells responding to *M. tuberculosis* antigens presented by the MHC class I pathway, and represent potential subunit components for vaccines against this pathogen.

Concluding Remarks

Mycobacteria have evolved a capsule and cell wall of extraordinary complexity. Certain features of this structure impart a high degree of strength and resilience to the bacilli, and are likely to be adaptations that allow them to survive and persist for long periods in spite of a variety of physical stresses. However, it also seems virtually certain that many of the unique structural components of cell walls of virulent mycobacteria, such as *M. tuberculosis,* have evolved as a result of interactions with the immune systems of their host species. The overview of immunoreactive components of the mycobacterial cell wall provided by this Chapter gives some sense of how great the challenge is to identify and integrate all of the signals that are generated in the immune system during responses to mycobacterial infection (Fig. 9). While certain aspects of the host immune response to mycobacteria undoubtedly promote killing or containment of the bacilli and should thus be beneficial to the host, it appears equally likely that other aspects of the response occur for the specific purpose of subverting host immunity. It remains a major challenge to determine which of the stimuli presented by the *M. tuberculosis* cell wall generate useful immunologic responses that contain or destroy the organism, and which represent decoys or generate potentially deleterious activation of the immune system. Sorting out which signals aid and which interfere with the development of durable protective immunity are key goals in the effort to create a truly safe and efficacious vaccine to *M. tuberculosis* and other mycobacterial pathogens.

Acknowledgements

SAP is supported by grants from the NIH/NIAID, and by a grant from the Irene Diamond Foundation. GSB, who is currently a Lister Institute Jenner Research Fellow, acknowledges support from The Wellcome Trust, The Medical Research Council and the National Institutes of Health. The authors are indebted to Lynn Dover for assistance with the illustrations in this article.

References

1. Kaufmann SHE. Immunity to intracellular bacteria. Ann Rev Immunol 1993; 11:129-163.
2. Clemens DL, Horwitz MA. Characterization of the Mycobacterium tuberculosis phagosome and evidence that phagosomal maturation is inhibited. J Exp Med 1995; 181:257-270.
3. Stenger S, Hanson DA, Teitelbaum R et al. An antimicrobial activity of cytolytic T cells mediated by granulysin. Science 1998; 282:121-125.
4. Brennan PJ, Nikaido H. The envelope of mycobacteria. Annu Rev Biochem 1995; 64:29-63.
5. Daffe M, Draper P. The envelope layers of mycobacteria with reference to their pathogenicity. Adv Microb Physiol 1998; 39:131-203.
6. Goren MB. Immunoreactive substances of mycobacteria. Am Rev Respir Dis 1982; 125(3 Pt 2):50-69.
7. Beatty WL, Rhoades ER, Ullrich HJ et al. Trafficking and release of mycobacterial lipids from infected macrophages. Traffic 2000; 1:235-247.
8. Schaible UE, Hagens K, Fischer K et al. Intersection of group I CD1 molecules and mycobacteria in different intracellular compartments of dendritic cells. J Immunol 2000; 164(9):4843-4852.
9. Cox JS, Chen B, McNeil M et al. Complex lipid determines tissue-specific replication of Mycobacterium tuberculosis in mice. Nature 1999; 402(6757):79-83.
10. Lederer E. Immunostimulation: recent progress in the study of natural and synthetic immunomodulators derived from the bacterial cell wall. Immunology 1980; 80:1194-1211.
11. Adam A, Petit J-F, Lefrancier P et al. Muramyl peptides. Chemical Structure, biological activity and mechanism of action. Mol Cell Biochem 1981; 41:27-47.
12. Heinzelmann M, Polk HC, Jr., Chernobelsky A et al. Endotoxin and muramyl dipeptide modulate surface receptor expression on human mononuclear cells. Immunopharmacology 2000; 48(2):117-128.
13. Medzhitov R, Preston-Hurlburt P, Janeway CA, Jr. A human homologue of the Drosophila Toll protein signals activation of adaptive immunity. Nature 1997; 388(6640):394-397.
14. Medzhitov R, Janeway C, Jr. The toll receptor family and microbial recognition. Trends Microbiol 2000; 8(10):452-456.
15. Aderem A, Ulevitch RJ. Toll-like receptors in the induction of the innate immune response. Nature 2000; 406(6797):782-787.
16. Yoshimura A, Lien E, Ingalls RR et al. Cutting edge: Recognition of Gram-positive bacterial cell wall components by the innate immune system occurs via Toll-like receptor 2. J Immunol 1999; 163(1):1-5.
17. Barry CE, III, Lee RE, Mdluli K et al. Mycolic acids: Structure, biosynthesis and physiological functions. Prog Lipid Res 1998; 37(2-3):143-179.
18. Miesel L, Rozwarski DA, Sacchettini JC et al. Mechanisms for isoniazid action and resistance. Novartis Found Symp 1998; 217:209-220.
19. Slayden RA, Lee RE, Barry CE. Isoniazid affects multiple components of the type II fatty acid synthase system of Mycobacterium tuberculosis. Mol Microbiol 2000; 38(3):514-525.
20. Dubnau E, Chan J, Raynaud C et al. Oxygenated mycolic acids are necessary for virulence of Mycobacterium tuberculosis in mice. Mol Microbiol 2000; 36(3):630-637.
21. Glickman MS, Cox JS, Jacobs WR Jr. A novel mycolic acid cyclopropane synthetase is required for coding, persistence, and virulence of Mycobacterium tuberculosis. Mol Cell 2000; 5(4):717-727.

22. Ozeki Y, Kaneda K, Fujiwara N et al. In vivo induction of apoptosis in the thymus by administration of mycobacterial cord factor (trehalose 6,6'-dimycolate). Infect Immun 1997; 65(5):1793-1799.
23. Kaneda K, Sumi Y, Kurano F et al. Granuloma formation and hemopoiesis induced by C36-48-mycolic acid- containing glycolipids from Nocardia rubra. Infect Immun 1986; 54(3):869-875.
24. Yano I, Tomiyasu I, Kitabatake S et al. Granuloma forming activity of mycolic acid-containing glycolipids in nocardia and related taxa. Acta Leprol 1984; 2(2-4):341-349.
25. Matsunaga I, Oka S, Fujiwara N et al. Relationship between induction of macrophage chemotactic factors and formation of granulomas caused by mycoloyl glycolipids from Rhodococcus ruber (Nocardia rubra). J Biochem (Tokyo) 1996; 120(3):663-670.
26. Hamasaki N, Isowa K, Kamada K et al. In vivo administration of mycobacterial cord factor (Trehalose 6, 6'- dimycolate) can induce lung and liver granulomas and thymic atrophy in rabbits. Infect Immun 2000; 68(6):3704-3709.
27. Lepoivre M, Tenu JP, Lemaire G et al. Antitumor activity and hydrogen peroxide release by macrophages elicited by trehalose diesters. J Immunol 1982; 129(2):860-866.
28. Sazaki K, Yoshida I, Azuma M. Mechanisms of augmented resistance of cyclosporin A-treated mice to influenza virus infection by trehalose-6,6'-dimycolate. Microbiol Immunol 1992; 36(10):1061-1075.
29. Oswald IP, Dozois CM, Petit JF et al. Interleukin-12 synthesis is a required step in trehalose dimycolate-induced activation of mouse peritoneal macrophages. Infect Immun 1997; 65(4):1364-1369.
30. He H, Oka S, Han YK et al. Rapid serodiagnosis of human mycobacteriosis by ELISA using cord factor (trehalose-6,6'-dimycolate) purified from Mycobacterium tuberculosis as antigen. FEMS Microbiol Immunol 1991; 3(4):201-204.
31. Maekura R, Nakagawa M, Nakamura Y et al. Clinical evaluation of rapid serodiagnosis of pulmonary tuberculosis by ELISA with cord factor (trehalose-6,6'-dimycolate) as antigen purified from Mycobacterium tuberculosis. Am Rev Respir Dis 1993; 148(4 Pt 1):997-1001.
32. Porcelli SA, Modlin RL. The CD1 system: antigen-presenting molecules for T cell recognition of lipids and glycolipids. Ann Rev Immunol 1999; 17:297-329.
33. Porcelli S, Morita CT, Brenner MB. CD1b restricts the response of human CD4-8- T lymphocytes to a microbial antigen. Nature 1992; 360:593-597.
34. Sieling PA, Chatterjee D, Porcelli SA et al. CD1-restricted T cell recognition of microbial lipoglycan antigens. Science 1995; 269(5221):227-230.
35. Sieling PA, Jullien D, Dahlem M et al. CD1 expression by dendritic cells in human leprosy lesions: correlation with effective host immunity. J Immunol 1999; 162:1851-1858.
36. Moody DB, Ulrichs T, Muhlecker W et al. CD1c-mediated T-cell recognition of isoprenoid glycolipids in Mycobacterium tuberculosis infection. Nature 2000; 404(6780):884-888.
37. Moody DB, Reinhold BB, Guy MR et al. Structural requirements for glycolipid antigen recognition by CD1b-restricted T cells. Science 1997; 278:283-286.
38. Beckman EM, Porcelli SA, Morita CT et al. Recognition of a lipid antigen by CD1-restricted $\alpha\beta+$ T cells. Nature 1994; 372:691-694.
39. Zeng Z, Castano AR, Segelke BW et al. Crystal structure of mouse CD1: An MHC-like fold with a large hydrophobic binding groove. Science 1997; 277(5324):339-345.
40. Ernst WA, Maher J, Cho S et al. Molecular interaction of CD1b with lipoglycan antigens. Immunity 1998; 8(3):331-340.
41. Porcelli SA, Segelke BW, Sugita M et al. The CD1 family of lipid antigen presenting molecules. Immunol Today 1998; 19(8):362-368.
42. Grant EP, Degano M, Rosat JP et al. Molecular recognition of lipid antigens by T cell receptors. J Exp Med 1999; 189:195-205.
43. Moody DB, Guy MR, Grant E et al. CD1b-mediated T cell recognition of a glycolipid antigen generated from mycobacterial lipid and host carbohydrate during infection. J Exp Med 2000; 192(7):965-976.
44. Chatterjee D, Khoo KH. Mycobacterial lipoarabinomannan: an extraordinary lipoheteroglycan with profound physiological effects. Glycobiology 1998; 8(2):113-120.
45. Vercellone A, Nigou J, Puzo G. Relationships between the structure and the roles of lipoarabinomannans and related glycoconjugates in tuberculosis pathogenesis. Front Biosci 1998; 3:e149-e163.
46. Hunter SW, Fujiwara T, Brennan PJ. Structure and antigenicity of the major specific glycolipid antigen of Mycobacterium leprae. J Biol Chem 1982; 257:15072-15078.
47. Prinzis S, Chatterjee D, Brennan PJ. Structure and antigenicity of lipoarabinomannan from Mycobacterium bovis BCG. J Gen Microbiol 1993; 139:2649-2658.
48. Chan J, Fan XD, Hunter SW et al. Lipoarabinomannan, a possible virulence factor involved in persistence of Mycobacterium tuberculosis within macrophages. Infect Immun 1991; 59:1755-1761.

49. Sibley LD, Adams LB, Krahenbuhl JL. Inhibition of interferon-gamma-mediated activation in mouse macrophages treated with lipoarabinomannan. Clin Exp Immunol 1990; 80:141-148.
50. Sibley LD, Hunter SW, Brennan PJ et al. Mycobacterial lipoarabinomannan inhibits gamma interferon-mediated activation of macrophages. Infect Immun 1988; 56:1232-1236.
51. Zhang Y, Doerfler M, Lee TC et al. Mechanisms of stimulation of interleukin-1 beta and tumor necrosis factor-alpha by Mycobacterium tuberculosis components. J Clin Invest 1993; 91(5):2076-2083.
52. Zhang Y, Rom WN. Regulation of the interleukin-1 beta (IL-1 beta) gene by mycobacterial components and lipopolysaccharide is mediated by two nuclear factor-IL6 motifs. Mol Cell Biol 1993; 13(6):3831-3837.
53. Roach TIA, Barton CH, Chatterjee D et al. Macrophage activation: lipoarabinomannan from avirulent and virulent strains of Mycobacterium tuberculosis differentially induces the early genes c-fos, KC, JE, and tumor necrosis factor-alpha. J Immunol 1993; 150:1886-1896.
54. Rojas M, Garcia LF, Nigou J et al. Mannosylated lipoarabinomannan antagonizes Mycobacterium tuberculosis-induced macrophage apoptosis by altering Ca^{+2}-dependent cell signaling. J Infect Dis 2000; 182(1):240-251.
55. Schlesinger LS. Macrophage phagocytosis of virulent but not attenuated strains of Mycobacterium tuberculosis is mediated by mannose receptors in addition to complement receptors. J Immunol 1993; 150(7):2920-2930.
56. Kang BK, Schlesinger LS. Characterization of mannose receptor-dependent phagocytosis mediated by Mycobacterium tuberculosis lipoarabinomannan. Infect Immun 1998; 66(6):2769-2777.
57. Schlesinger LS, Kaufman TM, Iyer S et al. Differences in mannose receptor-mediated uptake of lipoarabinomannan from virulent and attenuated strains of Mycobacterium tuberculosis by human macrophages. J Immunol 1996; 157(10):4568-4575.
58. Schlesinger LS, Hull SR, Kaufman TM. Binding of the terminal mannosyl units of lipoarabinomannan from a virulent strain of Mycobacterium tuberculosis to human macrophages. J Immunol 1994; 152(8):4070-4079.
59. Linehan SA, Martinez-Pomares L, Gordon S. Macrophage lectins in host defence. Microbes Infect 2000; 2(3):279-288.
60. Astarie-Dequeker C, N'Diaye EN, Le CV et al. The mannose receptor mediates uptake of pathogenic and nonpathogenic mycobacteria and bypasses bactericidal responses in human macrophages. Infect Immun 1999; 67(2):469-477.
61. Prigozy TI, Sieling PA, Clemens D et al. The mannose receptor delivers lipoglycan antigens to endosomes for presentation to T cells by CD1b molecules. Immunity 1997; 6(2):187-197.
62. Sidobre S, Nigou J, Puzo G et al. Lipoglycans are putative ligands for the human pulmonary surfactant protein A attachment to mycobacteria. Critical role of the lipids for lectin-carbohydrate recognition. J Biol Chem 2000; 275(4):2415-2422.
63. Downing JF, Pasula R, Wright JR et al. Surfactant protein A promotes attachment of Mycobacterium tuberculosis to alveolar macrophages during infection with human immunodeficiency virus. Proc Natl Acad Sci USA 1995; 92(11):4848-4852.
64. Gaynor CD, McCormack FX, Voelker DR et al. Pulmonary surfactant protein A mediates enhanced phagocytosis of Mycobacterium tuberculosis by a direct interaction with human macrophages. J Immunol 1995; 155(11):5343-5351.
65. Weikert LF, Edwards K, Chroneos ZC et al. SP-A enhances uptake of bacillus Calmette-Guerin by macrophages through a specific SP-A receptor. Am J Physiol 1997; 272(5 Pt 1):L989-L995.
66. Weikert LF, Lopez JP, Abdolrasulnia R et al. Surfactant protein A enhances mycobacterial killing by rat macrophages through a nitric oxide-dependent pathway. Am J Physiol Lung Cell Mol Physiol 2000; 279(2):L216-L223.
67. Ferguson JS, Voelker DR, McCormack FX et al. Surfactant protein D binds to Mycobacterium tuberculosis bacilli and lipoarabinomannan via carbohydrate-lectin interactions resulting in reduced phagocytosis of the bacteria by macrophages. J Immunol 1999; 163(1):312-321.
68. Underhill DM, Ozinsky A, Smith KD et al. Toll-like receptor-2 mediates mycobacteria-induced proinflammatory signaling in macrophages. Proc Natl Acad Sci USA 1999; 96(25):14459-14463.
69. Means TK, Wang S, Lien E et al. Human toll-like receptors mediate cellular activation by Mycobacterium tuberculosis. J Immunol 1999; 163(7):3920-3927.
70. Means TK, Lien E, Yoshimura A et al. The CD14 ligands lipoarabinomannan and lipopolysaccharide differ in their requirement for Toll-like receptors. J Immunol 1999; 163(12):6748-6755.
71. Sousa AO, Henry S, Maroja FM et al. IgG subclass distribution of antibody responses to protein and polysaccharide mycobacterial antigens in leprosy and tuberculosis patients. Clin Exp Immunol 1998; 111(1):48-55.

72. Beuria MK, Mohanty KK, Katoch K et al. Determination of circulating IgG subclasses against lipoarabinomannan in the leprosy spectrum and reactions. Int J Lepr Other Mycobact Dis 1999; 67(4):422-428.
73. Glatman-Freedman A, Mednick AJ, Lendvai N et al. Clearance and organ distribution of Mycobacterium tuberculosis lipoarabinomannan (LAM) in the presence and absence of LAM-binding immunoglobulin M. Infect Immun 2000; 68(1):335-341.
74. Sieling PA, Ochoa MT, Jullien D et al. Evidence for human CD4+ T cells in the CD1-restricted repertoire: derivation of mycobacteria-reactive T cells from leprosy lesions. J Immunol 2000; 164(9):4790-4796.
75. Besra GS, Chatterjee D. Lipids and Carbohydrates of Mycobacterium tuberculosis. In: Bloom BR, ed. Tuberculosis: Pathogenesis, Protection, and Control. Washington, D.C.: ASM Press, 1994:285-306.
76. Brett SJ, Payne SN, Draper P et al. Analysis of the major antigenic determinants of the characteristic phenolic glycolipid from Mycobacterium leprae. Clin Exp Immunol 1984; 56:89-96.
77. Brennan PJ. The phthiocerol-containing surface lipids of Mycobacterium leprae—A perspective of past and present work. Int J Lepr Other Mycobact Dis 1983; 51(3):387-396.
78. Daffe M, Papa F, Laszlo A et al. Glycolipids of recent clinical isolates of Mycobacterium tuberculosis: Chemical characterization and immunoreactivity. J Gen Microbiol 1989; 135 (Pt 10):2759-2766.
79. Mehra V, Brennan PJ, Rada E et al. Lymphocyte suppression in leprosy induced by unique M. leprae glycolipid. Nature 1984; 308(5955):194-196.
80. Goren MB, Vatter AE, Fiscus J. Polyanionic agents do not inhibit phagosome-lysosome fusion in cultured macrophages. J Leukoc Biol 1987; 41(2):122-129.
81. Besra GS, Sievert T, Lee RE et al. Identification of the apparent carrier in mycolic acid synthesis. Proc Natl Acad Sci U S A 1994; 91(26):12735-12739.
82. Takayama K, Schnoes HK, Semmler EJ. Characterization of the alkali-stable mannophospholipids of Mycobacterium smegmatis. Biochim Biophys Acta 1973; 316(2):212-221.
83. Muller I, Cobbold SP, Waldmann H et al. Impaired resistance to Mycobacterium tuberculosis infection after selective in vivo depletion of L3T4+ and Lyt-2+ T cells. Infect Immun 1987; 55:2037-2041.
84. Flynn JL, Goldstein MM, Triebold KJ et al. Major histocompatibility complex class I-restricted T cells are required for resistance to Mycobacterium tuberculosis infection. Proc Natl Acad Sci USA 1992; 89:12013-12017.
85. Orme IM, Miller ES, Roberts AD et al. T lymphocytes mediating protection and cellular cytolysis during the course of Mycobacterium tuberculosis infection. J Immunol 1992; 148:189-196.
86. Caruso AM, Serbina N, Klein E et al. Mice deficient in CD4 T cells have only transiently diminished levels of IFN-gamma, yet succumb to tuberculosis. J Immunol 1999; 162(9):5407-5416.
87. Sousa AO, Mazzaccaro RJ, Russell RG et al. Relative contributions of distinct MHC class I-dependent cell populations in protection to tuberculosis infection in mice. Proc Natl Acad Sci USA 2000; 97(8):4204-4208.
88. Scanga CA, Mohan VP, Yu K et al. Depletion of CD4(+) T cells causes reactivation of murine persistent tuberculosis despite continued expression of interferon gamma and nitric oxide synthase 2. J Exp Med 2000; 192(3):347-358.
89. Barnes PF, Bloch AB, Davidson PT et al. Tuberculosis in patients with human immunodeficiency virus infection. N Engl J Med 1991; 324:1644-1650.
90. Law KF, Jagirdar J, Weiden MD et al. Tuberculosis in HIV-positive patients: cellular response and immune activation in the lung. Am J Respir Crit Care Med 1996; 153(4 Pt 1):1377-1384.
91. Havlir DV, Barnes PF. Tuberculosis in patients with human immunodeficiency virus infection. N Engl J Med 1999; 340(5):367-373.
92. Andersen P. Effective vaccination of mice against Mycobacterium tuberculosis infection with a soluble mixture of secreted mycobacterial proteins. Infect Immun 1994; 62(6):2536-2544.
93. Barnes PF, Mehra V, Hirschfield GR et al. Characterization of T-cell antigens associated with the cell wall protein-peptidoglycan complex of Mycobacterium tuberculosis. J Immunol 1989; 143:2656-2662.
94. Mehra V, Bloom BR, Torigian VK et al. Characterization of Mycobacterium leprae cell wall-associated proteins using T-lymphocyte clones. J Immunol 1989; 142:2873-2878.
95. Chugh IB, Khuller GK. Immunoprotective behaviour of liposome entrapped cell wall subunit of Mycobacterium tuberculosis against experimental tuberculous infection in mice. Eur Respir J 1993; 6(6):811-815.
96. Hirschfield GR, McNeil M, Brennan PJ. Peptidoglycan-associated polypeptides of Mycobacterium tuberculosis. J Bacteriol 1990; 172(2):1005-1013.

97. Dhiman N, Verma I, Khuller GK. Cellular immune responses to cell wall peptidoglycan associated protein antigens in tuberculosis patients and healthy subjects. Microbiol Immunol 1997; 41(6):495-502.
98. Cunningham AF, Spreadbury CL. Mycobacterial stationary phase induced by low oxygen tension: cell wall thickening and localization of the 16-kilodalton alpha-crystallin homolog. J Bacteriol 1998; 180(4):801-808.
99. Yuan Y, Crane DD, Barry CE, III. Stationary phase-associated protein expression in Mycobacterium tuberculosis: Function of the mycobacterial alpha-crystallin homolog. J Bacteriol 1996; 178(15):4484-4492.
100. Yuan Y, Crane DD, Simpson RM et al. The 16-kDa alpha-crystallin (Acr) protein of Mycobacterium tuberculosis is required for growth in macrophages. Proc Natl Acad Sci USA 1998; 95(16):9578-9583.
101. Engers HD. Results of a World Health Organization-sponsored workshop to characterize antigens recognized by mycobacterium-specific monoclonal antibodies. Infect Immun 1986; 51:718-720.
102. Harris DP, Vordermeier HM, Friscia G et al. Genetically permissive recognition of adjacent epitopes from the 19-kDa antigen of Mycobacterium tuberculosis by human and murine T cells. J Immunol 1993; 150(11):5041-5050.
103. Harris DP, Vordermeier HM, Roman E et al. Murine T cell-stimulatory peptides from the 19-kDa antigen of Mycobacterium tuberculosis. Epitope-restricted homology with the 28-kDa protein of Mycobacterium leprae. J Immunol 1991; 147(8):2706-2712.
104. Faith A, Moreno C, Lathigra R et al. Analysis of human T-cell epitopes in the 19,000 MW antigen of Mycobacterium tuberculosis: Influence of HLA-DR. Immunology 1991; 74(1):1-7.
105. Yeremeev VV, Lyadova IV, Nikonenko BV et al. The 19-kD antigen and protective immunity in a murine model of tuberculosis. Clin Exp Immunol 2000; 120(2):274-279.
106. Brightbill HD, Libraty DH, Krutzik SR et al. Host defense mechanisms triggered by microbial lipoproteins through toll-like receptors. Science 1999; 285(5428):732-736.
107. Belisle JT, Vissa VD, Sievert T et al. Role of the major antigen of Mycobacterium tuberculosis in cell wall biogenesis. Science 1997; 276(5317):1420-1422.
108. Jackson M, Raynaud C, Laneelle MA et al. Inactivation of the antigen 85C gene profoundly affects the mycolate content and alters the permeability of the Mycobacterium tuberculosis cell envelope. Mol Microbiol 1999; 31(5):1573-1587.
109. Armitige LY, Jagannath C, Wanger AR et al. Disruption of the genes encoding antigen 85A and antigen 85B of Mycobacterium tuberculosis H37Rv: effect on growth in culture and in macrophages. Infect Immun 2000; 68(2):767-778.
110. Haslov K, Andersen A, Nagai S et al. Guinea pig cellular immune responses to proteins secreted by Mycobacterium tuberculosis. Infect Immun 1995; 63(3):804-810.
111. Huygen K, Content J, Denis O et al. Immunogenicity and protective efficacy of a tuberculosis DNA vaccine. Nat Med 1996; 2:893-898.
112. Lozes E, Huygen K, Content J et al. Immunogenicity and efficacy of a tuberculosis DNA vaccine encoding the components of the secreted antigen 85 complex. Vaccine 1997; 15(8):830-833.
113. Huygen K, Content J, Denis O et al. Immunogenicity and protective efficacy of a tuberculosis DNA vaccine. Nat Med 1996; 2(8):893-898.
114. Horwitz MA, Lee BW, Dillon BJ et al. Protective immunity against tuberculosis induced by vaccination with major extracellular proteins of Mycobacterium tuberculosis. Proc Natl Acad Sci U S A 1995; 92(5):1530-1534.
115. Havlir DV, Wallis RS, Boom WH et al. Human immune response to Mycobacterium tuberculosis antigens. Infect Immun 1991; 59:665-670.
116. Huygen K, Van Vooren JP, Turneer M et al. Specific lymphoproliferation, gamma interferon production, and serum immunoglobulin G directed against a purified 32 kDa mycobacterial protein antigen (P32) in patients with active tuberculosis. Scand J Immunol 1988; 27:187-194.
117. Thole JE, Janson AA, Cornelisse Y et al. HLA-class II-associated control of antigen recognition by T cells in leprosy: A prominent role for the 30/31-kDa antigens. J Immunol 1999; 162(11):6912-6918.
118. Geluk A, van Meijgaarden KE, Franken KL et al. Identification of major epitopes of Mycobacterium tuberculosis AG85B that are recognized by HLA-A*0201-restricted $CD8^+$ T cells in HLA-transgenic mice and humans. J Immunol 2000; 165(11):6463-6471.

CHAPTER 16

Antigen Presentation by MHC Class II Molecules

Tone F. Gregers, Tommy W. Nordeng and Oddmund Bakke

Introduction

Complex multicellular organisms have evolved defense systems to prevent and clear diseases caused by pathogenic microorganisms. The innate immune system prevents microorganisms to enter the host intracellular space, while the adaptive immune system responds to such an invasion. The adaptive immune system can mount two different responses towards pathogens: The humoral response is based on B cells, which produce antibodies specific for antigenic determinants in immunogens, while the cellular immune response involves the expansion and differentiation of antigen-specific regulatory and effector T cells. Major histocompatibility complex (MHC) molecules perform essential functions in the activation and functioning of the adaptive immune system. While the circulating antibodies will recognize and clear pathogens that are present in the extracellular fluids, MHC molecules are required to monitor intracellular pathogens. The MHC molecules bind peptides intracellularly and display them to T cells. These cells scan the cell surface-exposed MHC to monitor the cell interior for the presence of pathogens.

MHC molecules are highly polymorphic glycoproteins, which can be divided into two major groups: MHC class I and class II molecules. MHC class I mainly associate with peptide fragments derived from nuclear and cytoplasmic proteins and they are only 8-11 amino acids in length. In contrast, MHC class II allows binding of longer peptide fragments (7-25 amino acids). MHC class II has even been shown to bind large polypeptides derived from partially unfolded proteins.[1] Peptide fragments or proteins which bind to MHC class II are mainly derived from proteins degraded in the endosomal compartments. However, MHC class II may also bind endogenous peptides derived from cytoplasmic proteins, but only after they have entered endosomal compartments possibly by translocation or autophagy.[2,3]

During the past fifteen years the knowledge about molecular mechanisms of protein antigen processing and presentation in the context of MHC molecules has been considerably increased. Their polymorphism and high levels of expression have allowed extensive biochemical, structural and genetic studies on pathogen-derived peptides presented to T-lymphocytes. This review points out the recent developments in the MHC class II antigen presentation.

Assembly and Initial Transport of MHC Class II and Invariant Chain

Structure

MHC class II molecules, encoded by polymorphic genes are expressed as noncovalent heterodimers of two transmembrane polypeptides, the a chain (35kD) and the β chain (27kD) (for a review, see ref. 4). The luminal domains of the class II chains have an intrinsic ability to

Intracellular Pathogens in Membrane Interactions and Vacuole Biogenesis, edited by Jean-Pierre Gorvel.

associate,[5,6] but interactions by the transmembrane domains promote the formation of correctly assembled complexes.[7] The extracellular part of the MHC class II molecule forms a groove composed of two α-helices supported by an eight-strand β-pleated sheet. After synthesis and translocation into the endoplasmic reticulum (ER), the α and β chains associate with a third transmembrane glycoprotein, the invariant chain (Ii),[8,9] forming a nonameric $(\alpha\beta Ii)_3$ subunit complex.[10] In humans, there are three pairs of polymorphic class II genes, named human leukocyte antigen (HLA)-DR, HLA-DP and HLA-DQ,[11] and two pairs of non-classical class II genes, HLA-DM[12] and HLA-DO.[13] In the murine system, two major groups of the class II genes exist, I-A and I-E, as well as the non-classical genes encoding H-2M and H-2O.

Ii, a type II transmembrane glycoprotein, exists in different isoforms defined by the primary amino acid sequence (Fig. 1). The major p33 isoform has an N-terminal cytosolic tail of 30 amino acids, a transmembrane domain consisting of amino acid 31-56 and a C-terminal 160 residue luminal domain.[14-16] In humans, another form of Ii exists, the p35, containing an N-terminal cytoplasmic extension of 16 residues, which results from an alternative translation initiation site.[17,18] Two additional isoforms of the p33 and p35 exist due to alternative splicing of the exon 6, giving rise to the p41 and p43 forms, respectively (Fig. 1). This results in a 64 amino acid insertion in the luminal domain encoded by exon 6b.[17-19]

Expression of MHC II and Ii

Coordinate expression of class II and Ii molecules has been reported both in human tissues[20,21] and culture cells.[22] The expression of MHC class II genes is tightly regulated, and in contrast to the ubiquitous expression of MHC class I, class II is mainly restricted to professional antigen presenting cells (APC), including B cells, dendritic cells (DC) and cells of the monocyte-macrophage lineage. In other cells and tissues, the expression of class II molecules can be induced by interferon (INF)-γ.[23] Since the genes encoding Ii and MHC class II are located on different chromosomes, their co-regulation in antigen processing and presentation is essential as illustrated in patients with the bare lymphocyte syndrome (BLS). These patients lack mRNA and proteins of all class II molecules except Ii, although the structural genes are intact.[24,25] In most cases, IFN-γ treatment cannot restore class II expression,[26] and research on BLS led to the discovery of several *trans*-acting factors important for class II expression.[27,28] These factors are mainly DNA binding proteins and ubiquitously expressed.[28,29] The class II transactivator (CIITA) is an important regulator of MHC class II expression, both in cells with constitutive expression and after INF-γ induction.[27] CIITA is a coactivator that lacks DNA binding activity, but has strong transactivation properties.[30,31] It acts as a transcriptional scaffold that interacts with several DNA-binding factors,[32] and INF-γ induced MHC class II expression is mediated via expression of CIITA. Both Ii and HLA-DM (see below) are regulated by IFN-γ and CIITA.[33,34] It is crucial that the CIITA expression is carefully controlled and recently Morris and coworkers[35] have reported that methylation of CpG islands in the CIITA promoter prevents IFN-γ-induced transcription.

Transcription of Ii is controlled by elements common to those controlling class II transcription,[36,37] and an intronic IFN-γ responsive enhancer with homology to the class II promoter has been identified in the murine Ii gene.[38,39]

The fine-tuned regulation of MHC class II expression is controlled by a large number of different stimuli reviewed in ref. 40. Expression of both Ii and class II molecules can be increased in a variety of cell lines by stimulation with cytokines such as tumor necrosis factor (TNF)-α,[41,42] and interleukin (IL)-4.[43] IFN-γ induced transcription of MHC class II has been shown to be inhibited by cAMP and prostaglandins,[44] in addition to IL-1β.[45] Moreover, addition of granulocyte-machrophage colony stimulating factor to macrophage cell lines resulted in elevated Ii expression without class II induction.[46] In APCs Ii is usually produced in excess of class II,[9,47] and overproduction can be further increased by activation of protein kinase C (PKC),[48] resulting in reduced proteolysis of Ii.[49]

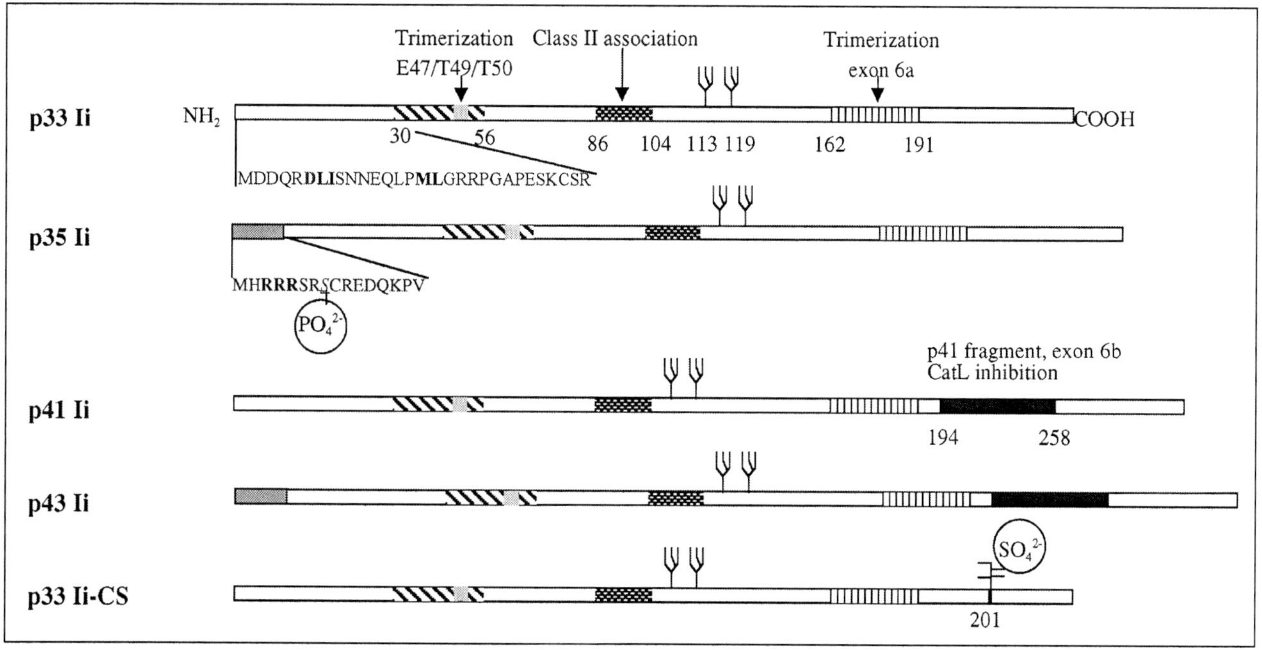

Figure 1. Schematic overview of the Ii isoforms and their modifications.
Ii exists in 4 major different form, Iip33, Iip35, Iip41 and Iip43. Iip33 is the most common and most studied form. Two leucin based sorting motifs (LI and ML, black bold) are resided in the cytoplasmic tail, which are independently sufficient for efficient endosomal transport. The Asp6 (D, red bold) gives a negative charge in this position, which is shown to be important for delayed degradation and enhanced presentation of endocytosed antigen. The Iip35 has a slightly increased molecular weight, due a 16 residue N-terminal extension. A double arginine motif maintains Iip35 in the ER[110] unless the cytoplasmic tail is phosphorylated at Ser8 (*S*, underlined and intalic).[112] Alternative splicing gives an extra exon (6b) in Iip33 and give rise to Iip41. Exon 6b encodes the p41 fragment, which shows homology to cystein protease inhibitors and is shown to interact with and inhibit CatL,[309] thereby regulating protein degradation. The Iip43 isoform is derived by both alternative splicing and alternative initiation of transcription. All four isoforms include a transmembrane domain, a trimerization domain, encoded by exon 6, and the MHC class II interaction site (CLIP). In addition, a hydrophilic spike in the transmembrane domain (E47/T49/T50, green) is also shown to be important for trimerization.[55] All isoforms are N-glycosylated at least in two positions (113 and 119). 2-5 % of the Iip33 population may acquire a chondroitin sulphate moiety at Ser201, which forms a GAG-modified Ii (Ii-CS).

Multimerization of Ii

Ii forms trimers in the ER due to a trimerization domain situated in the lumenal domain (residue 163-183) which corresponds to exon 6 (Fig. 1).[50,51] The three-dimensional structure of the trimerization domain has been determined by nuclear magnetic resonance (NMR) and found to have a cylindrical shape.[52] However, this carboxy-terminal lumenal region is not required for maintaining the nonameric class II-Ii complex since partially degraded Ii still remains trimeric.[53,54] A hydrophilic patch of polar amino acids in the transmembrane region has recently been described as an additional trimerization domain (Fig. 1).[55]

The biological significance of Ii trimerization is uncertain. It has been suggested that multimerization of Ii is required for efficient transport of class II-Ii complexes from the Trans Golgi Network (TGN) to endosomes.[56] Ii trimerization could contribute to the formation of a motif made up of adjacent chains that could be recognized by the protein-sorting machinery. However, cell free studies of monomeric and trimeric Ii constructs by Hofmann et al[57] indicate that interactions between adaptors and sorting signals do not require trimerization of the Ii. Oligomerization is not a prerequisite for Ii-MHC class II association as Ii mutants unable to trimerize still bind to MHC class II, but they are unable to present certain peptides.[58] Thus, trimerization of Ii is certainly important for antigen presentation to T cells. A connection between trimerization and transport properties is further supported by the observation that the formation of Ii-induced enlarged endosomes in transfected cells is dependent on the same region.[51]

Ii Interacts with Class II α and β Chains

The lumenal domain of Ii forms an extended structure, which is important for the interaction between Ii and MHC class II. Ii trimers interact with three newly synthesized MHC class II heterodimers in the ER to form nonameric complexes. Ii interacts with the class II $\alpha\beta$ chains predominantly through a amino acid region corresponding almost exclusively to exon 3 (residues 86-104).[59,60] It is now well established that this Ii region, named class II- associated invariant chain peptide (CLIP), actually occupies the peptide-binding groove of class II molecules.[61] The CLIP region is situated between the two trimerization domains, the lumenal and the transmembrane (Fig. 1). It has therefore been suggested that Ii trimerization based on two independent contact points may impose some structural order onto the CLIP region that can facilitate the assembly of nonameric MHC class II-Ii complexes in the ER.[55] Crystal structure of the MHC class II-CLIP complex has shown that CLIP binds to class II molecules in a manner similar to the binding of an antigenic peptide.[62,63] This promotes correct folding of MHC class II and inhibits premature binding of resident peptides.[60,63-68] CLIP has been subdivided into two functional regions: the C-terminal segment (residues 92-105) occupies the peptide binding groove, whereas the N-terminal segment (residues 81-91) binds outside the groove and seems to be important for the fast off-rate for CLIP.[69,70] However, the exact nature of CLIP binding is clearly different for each member of the polymorphic class II family as the affinities for individual $\alpha\beta$ dimers vary dramatically.[65,66] Thus, the property of CLIP may be to avoid high affinity binding rather than to fit into a polymorphic groove.

Additional regions of Ii besides CLIP have also been reported to maintain the association with class II. Recently, Thayer et al[71] suggested that a region C-terminal to CLIP (residues 103-118) could interact with class II molecules outside the peptide-binding groove under conditions where CLIP has little or no affinity for the class II peptide-binding site. Other interaction sites between Ii and MHC class II have been suggested as well, since Ii was still able to bind the class II molecule even though the class II peptide binding groove was occupied by high affinity peptides and was therefore not free to interact with CLIP.[72] Castellino et al[73] have identified the transmembrane region of Ii as an additional CLIP-independent MHC class II interaction site. In addition, two β-strands in the class II α-chain represent a binding pocket for Ii, and residue 181 within this pocket forms a salt-bridge to the Ii residue number 74.[74] A flexible loop in the MHC class II β-chain may also form electrostatic interactions with the N-terminal portion of Ii (residues 36-57),[74] indicating that several CLIP independent interactions between Ii and MHC class II exits.

Ii Association Facilitates Transport of Class II Molecules from the ER

Correct folding, assembly and post-transcriptional assembly are particularly important for MHC class II and Ii molecules. Changes in these molecules may lead to differences in peptides presented to T cells. Ii has been shown to have a "chaperoning" role in the maturation of class II,[75-78] and the interaction between Ii and class II helps in the transport of class II molecules out of the ER. Mice lacking the Ii gene show aberrant transport of MHC class II molecules, resulting in reduced levels of class II complexes at the surface, and these complexes do not have the typical compact SDS-stable conformation indicative of tight peptide binding (see below). Consequently, mutant cells present protein antigens very poorly and mutant mice are deficient in stimulating CD4+ T cells.[79] However, the interaction between Ii and class II is not an absolute requirement since certain MHC class II haplotypes have been found to be transported out of the ER in the absence of Ii.[76,78,80,81] Class II expressed alone will be localized mainly on the plasma membrane, but is also found to some extent in endocytic compartments.[81-83] Similarly, Ii is transported more efficiently from the ER to the TGN in the presence of class II,[84,85] indicating that exit from the ER is mutually facilitated by class II-Ii assembly.

Cross-linking studies suggest that the class II and Ii molecules are released from the ER as a nonameric complex,[10] but the precise order of assembly is not clear. αβ-dimers either assemble stepwise to pre-existing Ii trimers,[10] or separate α and β chains associate with Ii trimers one at a time to form the nonameric complex.[86,87] However, the contribution of Ii to subunit assembly differs for allelic variants of class II, suggesting that sequential associations of α, β and Ii may be affected by polymorphic differences.[88]

Calnexin, a ubiquitous ER phosphoprotein, associates with unassembled subunits of multimeric complexes[89-93] as well as secretory monomeric glycoproteins.[94] Calnexin association with several membrane glycoproteins depends on interactions involving N-linked glycans. Calnexin associates also with Ii and the α and β chains, and it remains associated with the assembling αβ-Ii complex until the final class II subunit is added to form the nonameric complex.[87,93] Dissociation of calnexin parallels egress of αβ-Ii from the ER, and this suggests that calnexin retains and stabilizes both free subunits and partially assembled class II-Ii complexes until the nonamer is complete. The molecular requirements for association of αβ-Ii complexes with calnexin are uncertain as neither replacement of the transmembrane region of the HLA-DRß subunit with a GPI-anchor or deglycosylation of the complex constituents abolish the association.[95] However, eliminating N-linked carbohydrates on Ii inhibits association with calnexin, and Ii lacking N-linked sugars has enhanced rates of pre-endosomal degradation.[96] When expressed in the absence of Ii, class II molecules are found also to aggregate with the ER resident chaperone BiP.[97] Thus, it is likely that the class II-Ii interaction mediates dissociation of these molecules also from other chaperones known to bind and to retain misfolded or partially folded proteins in the ER.

Post-Transcriptional Modifications of MHC Class II-Ii Chains

Once assembled, the nonameric Ii-class II complex undergoes several post-translational modifications (Fig. 1), such as N- and/or O-linked glycosylation,[98] phosphorylation,[99] acylation[100] and sulfation[101] in the ER and TGN. p33 Ii bears two N-linked and at least two O-linked glycans [102], whereas p41 Ii has two additional N-linked glycans that are tunicamycin resistant.[19] In the absence of Ii, αβ is poorly glycosylated,[103] and Ii alone does not acquire complex type sugars efficiently.[104] The Ii glycosylation pattern was shown not to differ between the class II-negative mutant lymphoblastoid cell line T2 and its class II-positive parental cell line T1,[105] demonstrating that association with class II is not necessarily a major factor influencing glycosylation.

The role of class II glycosylation is not established. One study suggests that N-glycosylation is important for T cell regulation,[106] whereas other studies claim that deglycosylation of the class II molecules does not influence peptide binding or T cell recognition.[107,108]

The p35 and p43 forms of Ii are retained in the ER[85,109] by a calnexin independent mechanism.[95] A double arginine motif in the prolonged N-terminal segment of the cytoplasmic tail has been identified as the ER retention signal (Fig. 1).[110] Interestingly, the tail of p35Ii is associated with 14-3-3 proteins, a family of adaptor proteins that are involved in coordinating signal transduction.[111] Post-translational phosphorylation of Ser 8 in the tail has been demonstrated to be required for the egress of p35Ii out of the ER (Fig. 1).[112] Phosphorylation of Ii is mediated by PKC and may regulate both Ii and class II transport to a peptide loading compartment (PLC).[113,114] These findings suggest that Ii transport may be regulated also directly by signal transduction.

A small portion of the Ii population (2-5%) may acquire a chondroitin sulfate moiety at Ser 201 in p33 to form a sulfate glycosaminoglycan (GAG)-modified Ii (Fig. 1; Ii-CS).[115,116] This Ii-CS has been found to interact with CD44 on T cells and to stimulate their response.[117]

Class II Molecules in the Endocytic Pathway

The Peptide Loading Compartment

MHC class II present antigenic peptides derived from material that has entered the endosomal/phagosomal system in APC (for reviews, see refs. 118 and 119). However the localization of a specific peptide PLC, has been difficult to define. The early studies of Cresswell indicated that MHC class II was transported to the plasma membrane through early endosomes,[120] and that the release of Ii took place in acidic endosomal compartments.[121] Observation by electron microscopy and biochemical studies showed that MHC class II was found throughout the endosomal pathway, including mannose 6-phosphate receptor (M6PR) positive late endosomes[122] and in late endosomes/lysosomes.[122,123] Actually, MHC class II molecules have been found in all maturation stadiums of endosomal compartments. It is now clear that peptides can be loaded efficiently on MHC class II in an Ii independent recycling route, due to the internalization signals resided in the class II β chain.[124,125] Peptide loading in late endocytic compartments is mostly via the Ii-dependent, biosynthetic route (reviewed in ref. 126).

Although it is not possible to determine with great accuracy a typical PLC, it may be possible to describe in morphological terms such a compartment limited to the cell type and the conditions used. Using kinetic parameters of BSA-gold uptake and a unique set of antibodies that recognize HLA-DR with either CLIP or bound peptide, a multimembrane compartment (multi-vesicular compartments), positioned late in the endocytic pathway was found in which HLA-DR with CLIP and HLA-DR with peptide colocalized.[127] These intracellular compartments are most likely where CLIP is exchanged for peptide. Additionally, double labeling showed that these compartments labeled for HLA-DM, which catalyzes this exchange (see below), as well as for markers of late endocytic compartments; the cation independent mannose 6-phosphate receptor (CI-MPR) and CD63. Therefore, one might conclude that antigen loading can occur in a variety of normal and more or less specialized compartments of the endosomal pathway.

Ii-MHC Class II Transport from the TGN To the PLC

The exact transport route of MHC class II-Ii complexes to the endosomal compartments is still unclear. Several reports have suggested that the complex is transported directly from the TGN to PLC.[128-131] However, others have shown an indirect pathway via the plasma membrane.[105,132-135] It might be that both transport routes exist, and that the different routes are regulated depending on cell type and state of activation. Co-internalization of antigens and

MHC class II-Ii complexes in the same endocytic vesicle would ensure that some MHC class II molecules may bind peptides derived from easily degradable antigens. Indeed, antigen processing and peptide binding to MHC class II have been demonstrated in early endosomes.[136-139] On the other hand, direct sorting of MHC class II-Ii complexes to a late endosomal compartment could prevent occupancy of all available class II molecules by peptides from easily degradable antigens. Alternative entry levels of MHC class II-Ii complexes to the endocytic pathway may thus promote the presentation of a broader spectrum of antigenic peptides, regardless of the vulnerability of the endocytosed antigen to protease activity. Dual endosomal routing has also been described for the lysosomal associated membrane protein, LAMP1,[140] lysosomal membrane glycoprotein lgp120 (lgp-A)[141] CI-M6PR.[142]

Sorting Signals in Ii

Several sorting signals have been described for the MHC class II-Ii complex, and these could contribute to the traffic the molecules. Ii alone is efficiently sorted to endosomes if released from the ER[109,143] and the cytoplasmic tail of Ii contains two leucine-based motifs; leucine-isoleucine (LI) in positions 7 and 8 and methionine-leucine (ML) in positions 16 and 17 (Fig. 1). The LI motif is a potent internalization signal although either signal is independently sufficient for endosomal localization of Ii.[133,144] NMR studies on a peptide corresponding to the cytoplasmic tail of Ii showed that the LI motif is located within a regular α-helix.[145] This prediction was supported by biological data showing that mutations of residues neighboring the LI sequence in the helical secondary structure could abolish internalization, whereas mutations of the residues opposite of the LI did not hamper sorting, making this combined motif a putative signal patch.[145] Acidic N-terminal residues are essential for a functional leucine signal.[146,147] It is still unclear why Ii harbors multiple sorting signals, however this is not unique to Ii, because several membrane proteins are known to have multiple, and functionally redundant internalization motifs. It has been suggested for other membrane proteins, that certain signals may be most effective in internalization, while other signals are important for sorting within the endosomal compartment.[148,149] Kang et al[144] have suggested that the LI signal is necessary and sufficient for endocytosis, while either signal is sufficient for lysosomal targeting. Also the transmembrane region of Ii has been implemented in sorting to degradative compartments, but details have not yet been resolved.[128]

Sorting Signals in MHC Class II

Although the major sorting information for endocytic transport of MHC class II-Ii complexes resided within Ii, class II molecules also contain endosomal sorting signals.[81,83,150] Class II molecules can be internalized and recycled,[151] and it has been shown that in the absence of Ii, antigen presentation of certain epitopes is still quite efficient.[152] Pinet et al[153] showed that the cytoplasmic tails of class II heterodimers are essential for internalization and recycling of MHC class II. The class II cytoplasmic tails are also believed to influence on the spectrum of class II-bound peptides.[154,155] Based on the observation that tail-less and full length class II molecules differ in their antigen presenting capacity, others have also reported that the MHC class II cytoplasmic tails are involved in antigen presentation.[153,156-159] In MDCK cells, tail-less MHC class II molecules were detected in multivesicular late endocytic structures and Ii co-expression increased the fraction of endosomal structures containing these molecules.[125] However, as opposed to full-length MHC class II, the tail-less molecules were not detected in early endosomes, in line with previous data showing that both tails are required for MHC II internalization and presentation of peptides by recycled MHC class II molecules.[153] Zhong et al[124] reported that the LL sequence in the C terminus of the mouse I-Ak β-chain was a sorting signal and this has been confirmed for the corresponding FL motif in HLA-DR1.[125] Taken together, these data indicate that information within the MHC class II cytoplasmic tails is important for internalization of class II molecules and for correct transport to the plasma membrane after peptide binding in endosomes.[125]

Internalization of Antigen and Routing to PLC

Invading pathogens can be endocytosed by antigen presenting cells by specific or non-specific mechanisms. The pathway by which the antigen is internalized is dependent on both the antigen itself and the internalizing cell. Large protein aggregates and intact bacteria may be engulfed by macrophages and DC in a process called phagocytosis, while smaller molecules mostly enter the cell via non-specific fluid-phase uptake or more specific receptor-mediated endocytosis.

Fluid-Phase Uptake

At all time, every cell in the body is able to internalize material via fluid-phase uptake, but the efficiency varies among the different cell types. DCs have been shown to have high endocytic capacity known as macropinocytosis during the immature stage. In fact, within an hour, DCs take up a fluid volume close to the cell itself to ensure volume capture and solute concentration in the PLC.[160] Macropinocytosis can be induced in macrophages and epithelial cells by growth factors, and is cytoskeleton-dependent, which differs from micropinocytosis that occurs via clathrin-coated pits.[161-163] On the other hand, B and T cells are very poor in fluid-phase endocytosis. Antigens that enters the cell via fluid phase are first located to early endosomes before they are transported further to late endosomes and eventually to the lysosomes for degradation (reviewed in ref. 164).

Receptor Mediated Endocytosis

Specific uptake involves endocytosis and phagocytosis mediated by antigen specific receptors. Receptor mediated uptake is a well characterized clathrin dependent process which ensures a high intracellular concentration of antigen, which normally circulate extracellularly at relatively low levels.[165] Receptor mediated uptake is also very efficient in targeting the antigen to the PLC, thereby enhancing the antigen presentation by MHC class II molecules.[166,167] The receptor is concentrated in clathrin coated pits after ligand binding and internalized via clathrin coated vesicles due to sorting signals in the cytoplasmic domain of the receptor. The ligand is then delivered to early endosomes, late endosomes and lysosomes, however, the precise routing of the ligand depend on the specific receptor.

The B Cell Receptor

On resting B-cells the B cell receptor (BCR) consists of membrane associated immunoglobulins (mIg, M or D isotype) associated with Igα/Igβ heterodimers bearing in their immunoreceptor tyrosine-based activation motif (ITAM) all the information to induce intracellular signaling and antigen uptake. BCR-ligation induces tyrosine phosphorylation of Igα and recruitment of the tyrosine kinase Syk, and this leads eventually to sorting of the BCR-ligand complex to PLC (for review see ref. 310). The cytoplasmic tails of Igα/Igβ contain sorting information for transport to late or early endosomal compartments, respectively.[166] This raises the possibility that the two tails may direct antigen transport to PLCs occupied by either recycling or newly synthesized MHC class II molecules.[168]

After binding to IgD, it has been shown that the antigen presentation efficiency is increased 100-1000-fold compared to fluid phase uptake in vitro.[169,170] In vivo studies on IgD targeting have, however, varied from initiating B- and T cell responses[171-173] to induction of tolerance or Th2-type T cell responses.[174,175] In contrast, mIgG expressed on memory B-cells is efficiently internalized independently of Igα/Igβ[176,177] due to a tyrosine-based sorting signal in the conserved cytoplasmic tail. B-cell mediated antigen internalization through mIgG can stimulate T cells at concentrations up to 1000 fold lower than those of non-specific B cells.[178,179] Ligand binding to BCR can modulate MHC class II presentation and the immune response in general, in several ways (for reviews, see refs. 180,181 and 182). Studies by the group of Davoust[155,183] have shown that antigen internalized by BCR require newly synthesized MHC class II and Ii for efficient presentation to T cells. BCR ligation was shown to modulate the class II presentation pathway and to induce intracellular accumulation of class II in an Ii-dependent fashion.[183]

The Fc Receptors

Fc receptors (FcR) are a family of molecules that bind to immune complexes through the Fc region of secreted Igs. For each Ig isotype, there is a specific FcR. Apart from these subtypes, two classes of FcR exist: those that initiate cell activation (Fcγ/α/εRI and FcγRIII) and those that generate inhibiting signals (FcγRIIB1 and IIB2). FcγRIIB2 and FcγIII are both involved in internalization of immune complexes, but they relay on different signals. FcγRIIB2 contains a leucine-based sorting signal in the cytoplasmic tail that is recognized for endocytosis.[184] However, separate tyrosine-based signals in the cytoplasmic tail of FcγRIIB2 are also involved in endocytosis and phagocytosis of the receptor-antigen complex.[185] FcγRI and FcγRIII require the association with γ-chains which couple the receptor to their cytoplasmic effectors.[186] The γ-chain contains both an ITAM for cell activation and a tyrosine-based internalization signal which overlaps the ITAM.[186,187] Recent data suggest that different FcγRs are not equivalent in terms of epitope presentation.[188] Certain epitopes from the same antigen were shown to be efficiently presented only after FcγRIII engagement,[188] suggesting that the two receptors were targeted to different PLCs in a process involving signal transduction. Thus, the type of peptides presented from a given antigen appears to depend on which FcγR that mediates the internalization.

FcαR has a particular strong association with the γ-chain,[189] and recently, Shen et al[190] demonstrated that antigen presentation of ovalbumin was enhanced when it was internalized via the FcαR as a complex with IgA. They suggest that the γ-chain ITAM is necessary for transport of ligated FcαR to PLC and for efficient degradation of IgA complexes.

Mannose Receptors

Intracellular sorting of receptors with affinity for mannose is important for antigen presentation. Two different types of receptors contribute to separate processes. The first type, the M6PR is required for transport of proteolytic enzymes from the TGN to endosomes. These enzymes play a key role in the endosomal degradation of antigens and Ii, and are thus required for MHC class II peptide loading (for a review, see ref. 191). M6PR is transported via the plasma membrane and contains two internalization signals in its cytoplasmic tail, which involves phenylalanine and tyrosin.[142,192] A third histidine-leucin signal is important for proper lysosomal enzyme sorting function.[193] The second type of mannose receptors (MRs) is found on the cell surface of macrophages and DCs. These receptors recognize patterns of carbohydrates on the surfaces and cell walls of infectious agents, thereby promoting endocytosis and phagocytosis (for a review, see ref. 194). This has been shown to dramatically increase the efficiency of class II-mediated presentation of such antigens in DCs.[195-198] The surface macrophage MR recycles constitutively between the plasma membrane and the early endosomes due to a recently discovered di-aromatic motif in the cytoplasmic tail.[199] In contrast to the other receptors, which are degraded together with their ligands, MRs release their ligand and allow several rounds of ligand internalization.[200] The DEC-205 multilectin receptor is found on DCs and is a homologue to the macrophage MR. It is localized to coated pits and involved in enhanced uptake.[195] Interestingly, the receptor has recently been found to contain a novel signaling motif in its cytoplasmic domain containing a cluster of acidic residues, which mediates efficient recycling through late endosomes rich in MHC class II and a 100-fold increase in antigen presentation.[201]

Other Receptors Involved in Antigen Uptake and Presentation

Recently it has been clear that several surface molecules important to the innate immune system are also involved in antigen capture and presentation by MHC class II molecules, and may thereby serve as a link between innate and adaptive immunity.

Scavenger receptors (SR) constitute a family of surface receptors that bind to and internalize a variety of polyanionic ligands and deliver them into the endolysosomal pathway.[202] SR can be grouped into three different gene families, SR-A, B and C, with partly different ligand specificities.[203] SR-A was the first SR to be cloned; it exhibits a broad ligand-binding specificity and it is expressed on monocytes and macrophages.[204] SR-A is upregulated during differentiation from monocytes into macrophages[205] but downregulated by INF-γ or lipopolysaccharide (LPS).[206,207] SR-A plays an important role in degradation of endotoxin[208] and recently it has been shown to play a role in cellular immune responses.[209] Due to a phenylalanine-based internalisation signal (VXFD) in the cytoplasmic tail,[210] SR-A may target ligands not only to lysosomes for degradation but also to PLC for peptide loading onto MHC class II.

CD14 is an LPS-binding molecule found predominantly on monocytes and macrophages, and is believed to serve an important role in the innate immunity. LPS first interacts with a serum protein called lipopolysaccharide binding protein, which transfer LPS to CD14. It is well known that CD14-LPS interactions induce the expression of inflammatory and immune-response gene products (for review see ref. 211), however, CD14 lacks an intracellular domain and is thereby unable to induce signal transductions on its own. It has recently been shown that CD14-LPS is recruited to a larger complex at the plasma membrane involving the toll like receptor (TLR)4 and MD-2. Recent studies have shown that antibodies against CD14 and TLR2 are internalized and processed for presentation to CD4+ T cells (K. H. Western and B. Bogen, personal communications). This indicates that receptors involved in the innate immune system may internalize antigens and probably target them to PLC, and thereby serve as a link to the adaptive immune system. However, the signal for internalization and targeting to PLC remains to be elucidated.

Processing Events in the Endocytic Compartment

Antigens must be internalized by endocytosis for subsequent processing within endosomes/lysosomes before peptides can be loaded onto MHC class II. In addition, processing of Ii is necessary to achieve peptide loading on mature class II molecules.[212,213] Both antigen and Ii processing include proteolysis mediated by a spectrum of different proteases,[214] (reviewed in ref. 215) and disulfide reduction.[216,217] The disulfide reduction seems to occur in high-density, lysosome-like compartments,[218] which are rich in class II molecules.[123,219]

Processing of Ii

During intracellular transport of class II-Ii complexes Ii is sequentially degraded from the luminal C-terminal part, but it remains associated with class II molecules[47,121,122,220-222] in a nonameric complex until only CLIP is left.[53] Iip33 is first degraded to generate Iip22 and further to Iip18 and Iip10. The p22 and p10 Ii fragments have been shown to occur approximately 1 hour after the synthesis of MHC class II-Ii complexes in human B-cells.[223] Addition of the cystein protease inhibitor leupeptin leads to accumulation of both Iip22 and Iip10, which indicates that the formation of Iip10 from Iip22 at least partially depends on cystein proteases. Further degradation of Ii, from Iip10 to CLIP, is inhibited by specific cysteine protease inhibitors, which indicates that also this degradation step depends on cystein proteases.[224]

Although cathepsin (Cat) B was believed to play a role in Ii degradation,[223] recent data show that this is not the case.[225-227] Instead, CatB may play a role in antigen degradation and in controlling what kind of T helper cells that are activated by the MHC class II-peptide complexes.[225] Study of knockout mice deficient in the aspartyl protease CatD showed that this cathepsin is also dispensable for normal Ii degradation and antigen presentation on MHC class II.[224] This means that the two major cysteine and aspartyl proteases expressed in APCs, CatB and CatD, respectively, are not critical components of the machinery involved in antigen presentation by MHC class II molecules.

Nakagawa et al[227] showed that CatL is necessary for the late stages of Ii degradation in cortical thymic epithelial cells (cTECs) but not in bone marrow-derived APCs. In CatL-deficient cTECs there has been reported an increased presence of CLIP and/or CLIP-containing Ii fragments in association with MHC class II at the plasma membrane.[227] Another cyteine protease, CatV, is only expressed in human thymic cells, but has 77% amino acid homology to mouse CatL. It has been suggested to play a role analogous to mouse CatL-mediated Ii degradation in the human thymus.[22]

The cysteine protease CatS has been demonstrated to be essential for efficient Ii proteolysis leading to the formation of SDS-stable complexes.[229] A role for CatS in Ii degradation has been supported by in vivo experiments using CatS negative mice.[230-232] These experiments indicated that CatS is involved in the late stages of Ii degradation in ex vivo-isolated DCs, B cells and macrophages, and mice splenocytes without CatS has been shown to accumulate the two Ii degradation intermediates Iip22 and Iip10.[231] Further analyzes of purified B cells and DCs from CatS knockout mice have shown clear differences in the molecular weight of the minor Ii fragments produced in the two different cell types, and in the kinetics of their accumulation. Nakagawa et al[231] suggest that the processing of Ii in DCs and B cells occurs in endocytic compartments containing different protease activities in addition to CatS, indicating an existence of a cell type specific pattern of Ii degradation.

Experiments with an active-site directed radioactive probe that visualizes active CatS have shown that CatS is active along the entire endocytic route, at least in DCs.[230] CatS deficient mice DCs showed an increased fraction of MHC class II in late endocytic structures, and they were in complex with partially degraded forms of Ii.[230] Further studies showed that the conversion of Iip22 to Iip18 most likely takes place in late endosomes and lysosomes. Other in vivo experiments in CatS negative mice have also supported the need for CatS to process Ii beyond the Iip10 fragment in APCs.[232] However, lack of CatS allows the transport of MHC class II-Ii complexes to the plasma membrane, and MHC class II peptide loading is not blocked even though MHC class II remains bound to Iip10.

Recently, it has been shown that macrophages express two cathepsins not found in splenocytes, DCs or B cells. These are CatF and CatZ.[233] CatF is as efficient as CatS in terms of Ii degradation and the production of CLIP fragments in vitro, and it has been suggested that CatF will substitute for and potentially complement the function of CatS in the processing of Iip10. CatF and CatS have different pH profiles, which make it possible for these two enzymes to function in different antigen processing compartments, with CatF activity mainly in the lysosome and CatS activity throughout the endosomal pathway.[233]

Processing of Antigen

In addition to Ii degradation, both cystein and aspartyl proteases are involved in degradation of proteins endocytosed by the APC. Experiments using mice deficient in CatB, D or L have shown that these enzymes may participate in the generation of antigenic peptides. Individually, however, none of them are crucial. CatD deficient APCs present peptides to T cell hybridomas equally efficient as their wild type counterparts.[226] Similar results were obtained with both CatB- and L-deficient splenocytes.[226,227] Recently, processing of exogenous glutamate decarboxylase was shown to be sensitive to CatD inhibitors in addition to cycteine protease inhibitors,[234] indicating that different proteins may have different requirements for proteolysis.

A novel asparagine-specific cysteine endopeptidase was found to be critical for the processing of tetanus toxin.[235] This enzyme was found to be similar to a mammalian homologue of the asparaginyl endopeptidase, legumain, originally found in plants and parasites.[236] It was also found that N-glycosylation of asparagine residues on tetanus toxin blocked the activity of this protease. This indicates that N-glycosylation could eliminate sites for processing of mammalian proteins, allowing preferential processing of microbial antigens.

The class II molecule itself may also have an active contribution in selecting antigenic peptides. Long polypeptides may interact with the class II groove in early endosomal compartments, and become trimmed by endo- and exoproteases which remove areas that are not buried in the groove.[237] This mechanism would not require any particular protease, as long as the cleavage occures.

Antigens internalized by APC need to be degraded to some extent before loaded onto MHC class II molecules. On the other hand, it is crucial that the antigen is protected from being lost by complete proteolytic degradation. One candidate for such protection is the constitutively expressed heat shock cognate protein (hsc) 73. Overexpression of hsc73 in transfected macrophages was shown to bind and protect peptides from extensive degradation and thereby enhance the presentation of exogenous proteins.[238]

Ii May Regulate Antigen Processing

Ii has sequence similarity to the cystatin family of protease inhibitors, and has been demonstrated to inhibit the enzymatic activity of CatL and H, but not of CatB or S.[239,240] The p41 form of Ii is more resistant to proteolysis compared to the p33 form,[241,242] resulting in a 12kD fragment that remains associated with class II for a prolonged period of time.[243] The peptide fragment encoded by the extra exon (6b) in p41 has been co-precipitated with CatL from human kidneys[244] and found to inhibit the activity of this protease[245,246] as well as the CatL-like enzyme cruzipain.[246] This discovery led to the suggestion that Ii could enhance antigen presentation by providing a mechanism to inhibit otherwise destructive CatL activity.[247] Crystal structure of the p41 fragment bound to CatL shows that p41 inhibits the protease by binding to its active site cleft.[240]

Pierre et al[248] have recently suggested that both p33 and p41 Ii can protect murine H2-M molecules from proteolysis in DC lysosomes, probably due to protease inhibition. At high levels of expression, Ii has been found to cause enlarged endosomes in transfected cells.[104,249,250] The transport rate of both endocytosed material[249-251] and MHC class II[251,252] is significantly reduced upon high expression of Ii. This function of Ii may contribute to endosomal retention as altered proteolytic activity decreases the rate of endocytic flow.[253,254] We have recently found that in a cell-free system the cytoplasmic tail of p33Ii is involved in the regulation of endosome fusion.[311] In live cells we have found that expression of p33Ii causes reduced proteolysis of endocytosed antigen, probably due to delayed transport to the lysosomes.[a] The decreased antigen degradation was abolished by an aspartate to arginine substitution in the cytoplasmic tail of Ii (Fig. 1). Both the Ii wild-type and the mutant enhanced the presentation of a hen egg lysosome derived epitope, however the mutant was never as potent as the wild type. These findings suggest that Ii may in principle regulate the movement of both endocytosed antigen and newly synthesized MHC class II-Ii complexes in the endosomal pathway.

The Ii-induced decrease in proteolysis was also shown to apply to Ii itself. These findings suggest that Ii may in principle regulate the movement of both endocytosed antigen and newly synthesized MHC class II-Ii complexes in the endosomal pathway.

Accessory Molecules in MHC Class Peptide Loading

The analysis of mutant B-cell lines expressing class II, but unable to present exogenous antigens,[255] led to the search of other genes mapping to the class II region of the MHC locus.[256-260] Accessory molecules such as HLA-DM, HLA-DO and tetraspanins have recently been found to play major roles in MHC class II peptide loading.

HLA-DM

The two genes HLA-DMA and HLA-DMB encode the two subunits of the heterodimer HLA-DM that has been shown to play a critical regulatory role in class II-restricted antigen presentation.[261-263] Cell lines lacking HLA-DM are defective in the presentation of a number

[a]Gregers TF, Norgeng TW, Birkeland HCG, Kÿølsrud S, Sandlie I, Bakke O, submitted.

of epitopes derived from intact protein antigens, but not of exogenously supplied peptides. Furthermore, their class II molecules lack the characteristics of mature, peptide-loaded molecules such as SDS-stability and recognition by conformation-specific antibodies. The major fraction of class II molecules in these cells lack a wild type repertoire of endogenous peptides and they are instead associated with the CLIP fragment of Ii.[264,265]

In vitro isolated HLA-DM molecules enhance peptide loading directly by accelerating the off-rate of CLIP[263,266,267] suggesting that HLA-DM is necessary for active removal of CLIP in vivo. HLA-DM is expressed at a low level compared to class II molecules (1:23), but change to 1:5 in PLC.[268] However, as HLA-DM is able to mediate CLIP removal in an enzyme-like fashion, one HLA-DM may facilitate loading of 3-12 HLA-DR molecules per minute, depending on the HLA-DR allotypes and peptides used.[269] In addition, it has been shown that HLA-DM preferentially associates with HLA-DR-CLIP complexes rather than HLA-DR molecules loaded with stable peptides, suggesting that the latter is an unfavorable substrate for HLA-DM.[270,271] However, not all class II molecules require HLA-DM for proper peptide loading; therefore it has been suggested that the HLA-DM dependency might be allele and/or species specific.[272] Recently, Chou et al[273] proposed that HLA-DM interacts with a flexible hydrophobic pocket in the peptide binding groove of HLA-DR1 and converts the molecule into a peptide receptive conformation. However, a more rigid conformation generated by filling of the pocket, is less susceptible to the effects of HLA-DM.

The catalytic effect of HLA-DM involves a transient interaction with class II, but prolonged association may be necessary to prevent aggregation of empty class II molecules[271,274] thus taking over the role of Ii during $\alpha\beta$ assembly and transport to a PLC. Data suggest that HLA-DM physically interacts with class II molecules also during their functional maturation[275] and both HLA-DM and its murine equivalent, H-2M, has been shown to associate with Ii during synthesis.[275,276]

HLA-DM is sorted to the same intracellular location as MHC class II, but is largely absent from the cell surface.[277,278] However, surface expression of HLA-DM have been found on DCs derived from CD34-positive bone marrow haematopoietic stem cells, indicating that HLA-DM can be used as a useful DC lineage-specific marker.[279] The cytoplasmic tail of the β-chain contains a typical tyrosine-based signal (YPTL).[280] Interestingly, Lindstedt et al[281] found that Ii could associate with H2-M and sort the complex to endosomes when the H2-M signal was deleted.

What criteria do HLA-DM use to distinguish between CLIP fragments and peptides destined for stable association with class II molecules? Recent experiments show that it is the off-rate rather than the affinity of a peptide per se that determines whether HLA-DM can remove it from the class II groove.[270,282] A dynamic model for HLA-DM function was suggested by Kropshofer et al[274] where HLA-DM executes its function by proofreading the kinetics of the $\alpha\beta$-peptide association: At low pH, HLA-DM interaction with class II-peptide complexes will stabilize the open state conformation of the peptide-binding groove. In this model, low stability peptides with higher off-rate than HLA-DM will be released by HLA-DM whereas high stability peptides will remain bound, leading to HLA-DM dissociation. Thus, a single class II molecule may undergo several rounds of peptide editing by HLA-DM until a high stability peptide is captured.

HLA-DO

The efficiency of HLA-DM-mediated edition of the peptide repertoire presented by class II molecules relies on several factors such as the class II:DM ratio, the availability of high-versus low-stability peptides and the pH of the reaction milieu. Other molecules may also be involved in the modulation of the peptide repertoire. Another class II-like protein, called HLA-DO, is expressed in B cells, DCs and subpopulations of thymus cells.[283] The mouse equivalent, H2-O is only expressed in B cells.[13] The functions of HLA-DO is still obscure (for a review see ref. 284), but it may associate with Ii[283] and HLA-DM[285] during intracellular transport and codistributes with HLA-DR[283] and HLA-DM[285] in endosomes. Apparently, HLA-DO is a co-chaperone of HLA-DM and depending on the species, it has been found to stimulate[286] or to inhibit[287] HLA-DM-catalyzed peptide loading, possibly by affecting the pH-range of HLA-DM activity.[288]

Tetraspanins

Tetraspanins intersect the membrane four times and display two extracytoplasmic loops.[289] They have a broad tissue distribution and a variety of functions, and recently they have been found to play a role in antigen processing and presentation. Several tetraspanins colocalize with HLA-DR molecules in PLC, especially in multi-vesicular compartments.[290] In addition, the tetraspanin TAPA-1 (CD81) have been found in association with HLA-DR at the B cell surface.[291] It has been suggested that tetraspanins may assist peptide loading in several ways. First, by clustering MHC class II molecules together with other components of the antigen presenting machinery through protein-protein interactions. Second, by target MHC class II to lipid rafts at the cell surface (see below), thereby clustering the class II molecules in distinct plasma membrane patches. Third, tetraspanins could segregate MHC class II and the other accessory molecules, thereby protecting them from degradation by lysosomal proteases.

Cells Surface Expression of Class II Molecules

The pathway(s) for transport of peptide-loaded MHC class II to the plasma membrane remain enigmatic. One route to the cell surface has been described in a study of B-LCL by Raposo et al,[292] in which multivesicular MHC class II positive compartments were observed to fuse with the plasma membrane and release antigen-presenting vesicles. These multivesicular compartments also contained endocytosed BSA-gold that was released together with the class II positive vesicles. However, the release of class II-containing vesicles is slow and therefore an unlikely major pathway by which antigen-loaded class II reach the plasma membrane. By using GFP-tagged MHC class II and confocal microscopy of live cells, Wubbolts et al suggest that MHC class II-rich late endosomes/lysosomes themselves could fuse with the cell surface,[293] representing a second possible MHC class II-peptide routing to the plasma membrane. A third candidate for transport vesicles directed to the plasma membrane could be a set of vesicles described by Stang et al.[127] These are relatively small and contain only mature class II and no endocytic markers and could be transport vesicles. Pond and Watts[294] found that the endocytic apparatus was not required for delivery of functional complexes of peptides and newly synthesized MHC class II back to the surface. However, the transport was sensitive to Brefeldin A, also suggesting the involvement of a vesicular intermediate, but there is still no direct proof of such a transport mechanism.

Transport routes are in general signal mediated. The nature of the transport signal for peptide loaded MHC class II to the plasma membrane is not known. This might be associated with the finding that peptide loading promotes a conformational change in class II molecules,[6,253,295,296] a process shown to be sufficient for transport of class II-peptide complexes to the surface.[297] Moreover, a conformational change induced by peptide binding could activate a signal for membrane transport in the cytoplasmic tail, possible by a mechanism resembling the basolateral transport of MHC class II (reviewed in ref. 298). Once at the plasma membrane, the peptide-MHC class II complex is recognized by CD4+ T cells, which in turn leads to transduction of intracellular signals through the T cell receptor on the T cell [299] and through the MHC class II molecule on the APC.[300] This will eventually lead to activation of the T cell and the APC, respectively. The cytoplasmic tails of MHC class II is dispensable for such class II mediated signaling.[300] However, the transmembrane domain of both α and β chain are important, probably due to interactions with other components in the cell membrane. Huby et al[301] have recently demonstrated that class II molecules associate in membrane rafts that in turn contribute to their aggregation on the cell surface and mediate interactions with intracellular protein tyrosine kinases. Lipid rafts have also been shown to be important for antigen presentation. In a typical immune response, the amount of available peptide-MHC class II complexes is limited, and MHC restricted T cell activation is dependent on the density of antigen complexes on the surface of the APC. Anderson et al[302] have proposed that peptide-class II complexes become concentrated in lipid rafts where they stimulate T cells more efficiently. Disruption of lipid rafts dramatically inhibited antigen presentation at limited antigen concentrations. In addition, MHC class II has been found to associate with the raft-associated protein Thy-1 in intracellular compartments.[303] It is tempting to speculate that MHC class II associate with lipid rafts intracellularly and then traffic to the plasma membrane as a complex with other molecules involved in antigen presentation.

Antigen Presentation on Recycling MHC Class II Molecules

Cells expressing class II molecules in the absence of Ii may present certain antigens efficiently.[80,152,158,304,305] Epitopes within single antigens can show differential dependence on Ii for efficient presentation.[124,139] Peptides that are exposed at the protein surface are more susceptible to proteases localised early in the endosomal pathway than epitopes in the core of the antigen, whose exposure often requires the more reducing environment of late endosomes and lysosomes.[218,306] Mature class II-peptide complexes can recycle from the cell surface to early endosomes and bind peptides generated in these compartments in an Ii independent way.[153] Recycling of class II molecules is mediated by the sorting motif in the class II β chain.[124,125] In contrast, newly synthesized MHC class II molecules are targeted to late endosomes where they bind and present peptides in the Ii-dependent pathway.

Although the major association between Ii and MHC class II is believed to occur in the ER, it has been suggested that newly synthesised Ii also associate with mature class II molecules at the plasma membrane and thus provide a recycling pathway for class II molecules into de novo PLC.[105,307,308] We have found that in the absence of Ii, class II is mainly found at the plasma membrane and to a very little extent in intracellular vesicles in transfected MDCK cells. However, upon Ii expression class II is internalized from the cell surface, and after 3-4 hours class II is mainly found to colocalize with Ii in intracellular compartments. We found that class II molecules were internalized more rapidly upon association with Ii compared to Ii-free class II molecules.[b] This Ii-induced class II clearance from the cell surface was shown to be independent on the cytoplasmic tails of the a and b chains, indicating that Ii somehow interact with class II molecules on the cell surface as well (the pathways for MHC class II antigen presentation are summarized in Fig. 2).

Concluding Remarks

As described in this review the last 15 years have been crucial for elucidating the intracellular molecules and mechanisms behind MHC class II transport, antigen degradation, peptide loading and how antigen is recognized by T-cells. It seems to be a great diversity between how different professional APC (i.e., DCs, B-cells and macrophages) treat the antigen. It is still not known how important MHC class II expression is in other cell types such as various epithelial cells and other cells that upon stimuli induce expression of molecules. The variety of antigens is also enormous and recent data show that there is substantial interplay between the innate and adaptive immune system, which is crucial for the defense against foreign invaders. Lately, it has also become clear that MHC class II may also present antigen when released from the cells as small spheres (exosomes; for review see ref. 312). This suggests that peptide/ MHC class II complexes may stimulate a much wider variety of cells than those directly in contact with the APC. The challenge today is thus to understand how this part of the immune system deals with the variety of antigens and in particular how important MHC class II is for the defense towards invading pathogens.

---

[b]Gregers TF, Nordeng TW, Walseng E, Fladeby C, Bakke O, manuscript.

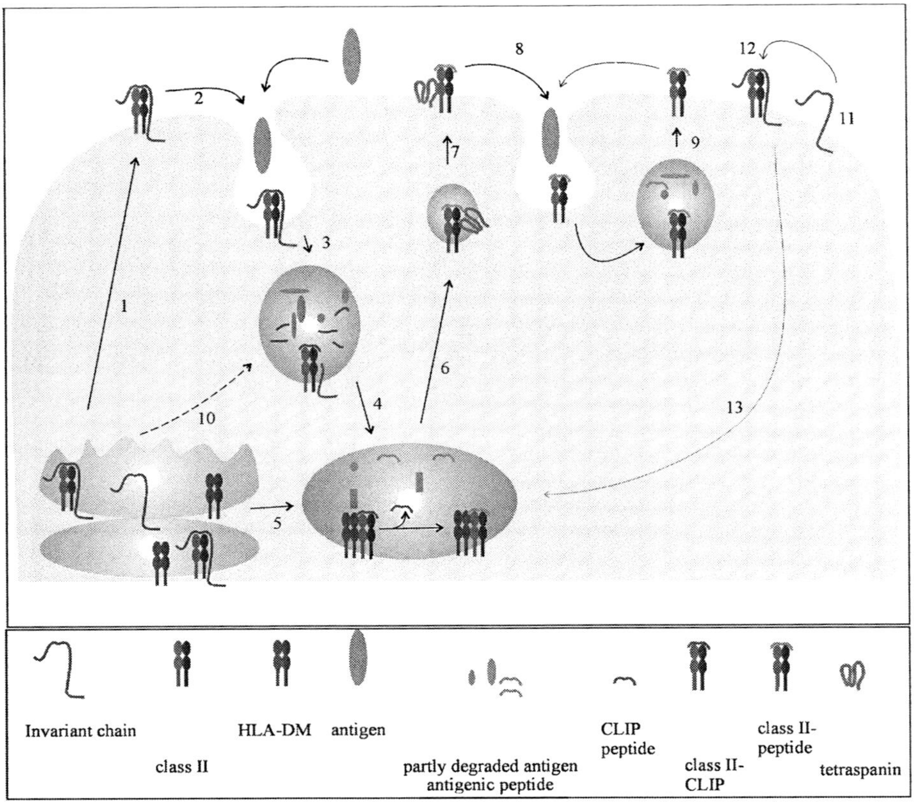

Figure 2. MHC class II/Ii transport route.
Ii makes transient associations with MHC class II in the ER, and the complex is transported to the plasma membrane[1] where it is rapidly internalized due to two leucin based endosomal sorting signals in the cytoplasmic tail of Ii.[2] Antigens are also internalized either non-specifically together with the Ii/MHC class II complex,[2] or specifically via an antigen receptor. Ii/MHC class II is targeted to early endosomes where degradation of both Ii and antigen begins,[3] however both are transported further to later endocytic compartments with higher proteolytic activity.[4] Here, MHC class II-CLIP associates with HLA-DM, which has been transported directly from ER to this compartment due to tyrosine based sorting signals.[5] HLA-DM replaces CLIP with antigenic peptides and a conformational change in the MHC class II-peptide complex induces the transport towards[6] and release to the cell surface.[7] MHC class II may interact with tetraspanins, which can facilitate transport to the cell surface. Mature peptide-MHC class II complexes recycle at the plasma membrane due to a leucin based sorting signal in the β chain.[8] In the recycling compartment, peptides may be exchanged by other antigenic peptides and again released to the cell surface.[9] MHC class II may go several rounds of recycling and peptide exchange before the complex is targeted to lysosomes for degradation. A minor fraction of MHC class II-Ii complexes are transported directly to the early endosomes where the proteolytic activity begins.[10] Ii is produced in excess compared to class II molecules, and several Ii molecules may be released from ER in the absence of class II.[11] These Ii molecules may interact at the plasma membrane with mature MHC class II molecules,[12] and thereby target MHC class II to late endosomes or PLC for new rounds of peptide loading[13] (see text for references).

Acknowledgements

We would like to thank Nicolas Barois for critical reading of the manuscript. The authors were supported by grants from the Norwegian Cancer Society (TFG and TWN) and the Norwegian Research Council (TWN and OB).

References

1. Lindner R, Unanue ER. Distinct antigen MHC class II complexes generated by separate processing pathways. EMBO J 1996; 15:6910-6920.
2. Jaraquemada D, Marti M, Long EO. An endogenous processing pathway in vaccinia virus-infected cells for presentation of cytoplasmic antigens to class II-restricted T cells. J Exp Med 1990; 172:947-954.
3. Malnati MS, Marti M, LaVaute T et al. Processing pathways for presentation of cytosolic antigen to MHC class II-restricted T cells. Nature 1992; 357:702-704.
4. Kaufman JF, Auffray C, Korman AJ et al. The class II molecules of the human and murine major histocompatibility complex. Cell 1984; 36:1-13.
5. Kjær-Nielsen L, Perera JD, Boyd LF et al. The extracellular domains of MHC class II molecules determine their processing requirements for antigen presentation. J Immunol 1990; 144:2915-2924.
6. Wettstein DA, Boniface JJ, Reay PA et al. Expression of a class II major histocompatibility complex (MHC) heterodimer in a lipid-linked form with enhanced peptide/soluble MHC complex formation at low. pH J Exp Med 1991; 174:219-228.
7. Cosson P, Bonifacino JS. Role of transmembrane domain interactions in the assembly of class II MHC molecules. Science 1992; 258:659-662.
8. Sung E, Jones PP. The invariant chain of murine Ia antigens: its glycosylation, abundance and subcellular localization. Mol Immunol 1981; 18:899-913.
9. Kvist S, Wiman K, Claesson L et al. Membrane insertion and oligomeric assembly of HLA-DR histocompatibility antigens. Cell 1982; 29:61-69.
10. Roche PA, Marks MS, Cresswell P. Formation of a nine-subunit complex by HLA class II glycoproteins and the invariant chain. Nature 1991; 354:392-394.
11. Trowsdale J, Ragoussis J, Campbell RD. Map of the human MHC. Immunol Today 1991; 12:443-446.
12. Kelly AP, Monaco JJ, Cho SG et al. A new human HLA class II-related locus, DM. Nature 1991; 353:571-573.
13. Karlsson L, Surh CD, Sprent J et al. A novel class II MHC molecule with unusual tissue distribution. Nature 1991; 351:485-488.
14. Claesson L, Peterson PA. Association of human gamma chain with class II transplantation antigens during intracellular transport. Biochemistry (Mosc) 1983; 22:3206-3213.
15. Koch N, Lauer W, Habicht J et al. Primary structure of the gene for the murine Ia antigen-associated invariant chains (Ii). An alternatively spliced exon encodes a cysteine-rich domain highly homologous to a repetitive sequence of thyroglobulin. EMBO J 1987; 6:1677-1683.
16. Lipp J, Dobberstein B, Haeuptle MT. Signal recognition particle arrests elongation of nascent secretory and membrane proteins at multiple sites in a transient manner. J Biol Chem 1987; 262:1680-1684.
17. O'Sullivan, DM, Noonan, D, Quaranta, V. Four Ia invariant chain forms derive from a single gene by alternate splicing and alternate initiation of transcription/translation. J Exp Med 1987; 166:444-460.
18. Strubin M, Berte C, Mach B. Alternative splicing and alternative initiation of translation explain the four forms of the Ia antigen-associated invariant chain. EMBO J 1986; 5:3483-3488.
19. Koch N. Posttranslational modifications of the Ia-associated invariant protein p41 after gene transfer. Biochemistry (Mosc) 1988; 27:4097-4102.
20. Quaranta V, Majdic O, Stingl G et al. A human Ia cytoplasmic determinant located on multiple forms of invariant chain (gamma, gamma2, gamma3). J Immunol 1984; 132:1900-1905.
21. Volc-Platzer, B, Majdic, O, Knapp, W et al. Evidence of HLA-DR antigen biosynthesis by human keratinocytes in disease. J Exp Med 1984; 159:1784-1789.
22. Long EO. In search of a function for the invariant chain associated with Ia antigens. Surv Immunol Res 1985; 4:27-34.
23. Pober JS, Collins T, Gimbrone MA Jr et al. Lymphocytes recognize human vascular endothelial and dermal fibroblast Ia antigens induced by recombinant immune interferon. Nature 1983; 305:726-729.
24. de Preval C, Hadam MR, Mach B. Regulation of genes for HLA class II antigens in cell lines from patients with severe combined immunodeficiency. N Engl J Med 1988; 318:1295-1300.

25. Hume CR, Lee JS. Congenital immunodeficiencies associated with absence of HLA class II antigens on lymphocytes result from distinct mutations in trans-acting factors. Hum Immunol 1989; 26:288-309.
26. Lisowska-Grospierre B, Charron DJ, de Preval C et al. A defect in the regulation of major histocompatibility complex class II gene expression in human HLA-DR negative lymphocytes from patients with combined immunodeficiency syndrome. J Clin Invest 1985; 76:381-385.
27. Reith W, Steimle V, Durand B et al. Regulation of MHC class II gene expression. Immunobiology 1995; 193:248-253.
28. Boss JM. Regulation of transcription of MHC class II genes. Curr Opin Immunol 1997; 9:107-113.
29. Steimle V, Otten LA, Zufferey M et al. Complementation cloning of an MHC class II transactivator mutated in hereditary MHC class II deficiency (or bare lymphocyte syndrome). Cell 1993; 75:135-146.
30. Zhou H, Glimcher LH. Human MHC class II gene transcription directed by the carboxyl terminus of CIITA, one of the defective genes in type II MHC combined immune deficiency. Immunity 1995; 2:545-553.
31. Riley JL, Westerheide SD, Price JA et al. Activation of class II MHC genes requires both the X box region and the class II transactivator (CIITA). Immunity 1995; 2:533-543.
32. Zhu XS, Linhoff MW, Li G et al. Transcriptional scaffold: CIITA interacts with NF-Y, RFX, and CREB to cause stereospecific regulation of the class II major histocompatibility complex promoter. Mol Cell Biol 2000; 20:6051-6061.
33. Chang C-H, Flavell RA. Class II transactivator regulates the expression of multiple genes involved in antigen presentation. J Exp Med 1995; 181:765-767.
34. Kern I, Steimle V, Siegrist C-A et al. The two novel MHC class II transactivators RFX5 and CIITA both control expression of HLA-DM genes. Int Immunol 1995; 7:1295-1299.
35. Morris AC, Spangler WE, Boss JM. Methylation of class II trans-activator promoter IV: A novel mechanism of MHC class II gene control. J Immunol 2000; 164:4143-4149.
36. Zhu L, Jones PP. Transcriptional control of the invariant chain gene involves promoter and enhancer elements common to and distinct from major histocompatibility complex class II genes. Mol Cell Biol 1990; 10:3906-3916.
37. Brown AM, Barr CL, Ting JP. Sequences homologous to class II MHC W, X, and Y elements mediate constitutive and IFN-gamma-induced expression of human class II-associated invariant chain gene. J Immunol 1991; 146:3183-3189.
38. Cao ZA, Moore BB, Quezada D et al. Identification of an IFN-gamma responsive region in an intron of the invariant chain gene. Eur J Immunol 2000; 30:2604-2611.
39. Moore BB, Cao ZA, McRae TL et al. The invariant chain gene intronic enhancer shows homology to class II promoter elements. J Immunol 1998; 161:1844-1852.
40. Glimcher LH, Kara CJ. Sequences and factors: A guide to MHC class-II transcription. Annu Rev Immunol 1992; 10:13-49.
41. Pessara U, Koch N. Tumor necrosis factor alpha regulates expression of the major histocompatibility complex class II-associated invariant chain by binding of an NF-kappa B-like factor to a promoter element. Mol Cell Biol 1990; 10:4146-4154.
42. Kolk DP, Floyd-Smith G. Induction of the murine class-II antigen-associated invariant chain by TNF-alpha is controlled by an NF-kappa B-like element. Gene 1993; 126:179-185.
43. Noelle RJ, Kuziel WA, Maliszewski CR et al. Regulation of the expression of multiple class II genes in murine B cells by B cell stimulatory factor-1 (BSF-1). J Immunol 1986; 137:1718-1723.
44. Ivashkiv LB, Ayres A, Glimcher LH. Inhibition of IFN-gamma induction of class II MHC genes by cAMP and prostaglandins. Immunopharmacology 1994; 27:67-77.
45. Rohn W, Tang LP, Dong Y et al. IL-1 beta inhibits IFN-gamma-induced class II MHC expression by suppressing transcription of the class II transactivator gene. J Immunol 1999; 162:886-896.
46. Klagge I, Kopp U, Koch N. Granulocyte-macrophage colony-stimulating factor elevates invariant chain expression in immature myelomonocytic cell lines. Immunology 1997; 91:114-120.
47. Nguyen QV, Humphreys RE. Time course of intracellular associations, processing, and cleavages of Ii forms and class II major histocompatibility complex molecules. J Biol Chem 1989; 264:1631-1637.
48. Shih N-Y, Floyd-Smith G. Invariant chain (CD74) gene regulation: Enhanced expression associated with activation of protein kinase Cdelta in a murine B lymphoma cell line. Mol Immunol 1995; 32:643-650.
49. Barois N, Forquet F, Davoust J. Selective modulation of the major histocompatibility complex class II antigen presentation pathway following B cell receptor ligation and protein kinase C activation. J Biol Chem 1997; 272:3641-3647.
50. Bijlmakers ME, Benaroch P, Ploegh HL. Mapping functional regions in the lumenal domain of the class II-associated invariant chain. J Exp Med 1994; 180:623-629.

51. Gedde-Dahl M, Freisewinkel I, Staschewski. M et al. Exon 6 is essential for invariant chain trimerization and induction of large endosomal structures. J Biol Chem 1997; 272:8281-8287.
52. Jasanoff A, Wagner G, Wiley DC. Structure of a trimeric domain of the MHC class II-associated chaperonin and targeting protein Ii. EMBO J 1998; 17:6812-6818.
53. Amigorena S, Webster P, Drake J et al. Invariant chain cleavage and peptide loading in major histocompatibility complex class II vesicles. J Exp Med 1995; 181:1729-1741.
54. Newcomb JR, Carboy-Newcomb C, Cresswell P. Trimeric interactions of the invariant chain and its association with major histocompatibility complex class II alphabeta dimers. J Biol Chem 1996; 271:24249-24256.
55. Ashman JB, Miller J. A role for the transmembrane domain in the trimerization of the MHC class II-associated invariant chain. J mImmunol 1999; 163:2704-2712.
56. Arneson LS, Miller J. Efficient endosomal localization of major histocompatibility complex class II-invariant chain complexes requires multimerization of the invariant chain targeting sequence. J Cell Biol 1995; 129:1217-1228.
57. Hofmann MW, Honing S, Rodionov D et al. The leucine-based sorting motifs in the cytoplasmic domain of the invariant chain are recognized by the clathrin adaptors AP1 and AP2 and their medium chains. J Biol Chem 1999; 274:36153-36158.
58. Bertolino P, Staschewski M, Trescol-Biémont M-C et al. Deletion of a C-terminal sequence of the class II-associated invariant chain abrogates invariant chains oligomer formation and class II antigen presentation. J Immunol 1995; 154:5620-5629.
59. Freisewinkel IM, Schenck K, Koch N. The segment of invariant chain that is critical for association with major histocompatibility complex class II molecules contains the sequence of a peptide eluted from class II polypeptides. Proc Natl Acad Sci USA 1993; 90:9703-9706.
60. Romagnoli P, Germain RN. The CLIP region of invariant chain plays a critical role in regulating major histocompatibility complex class II folding, transport, and peptide occupancy. J Exp Med 1994; 180:1107-1113.
61. Stumptner P, Benaroch P. Interaction of MHC class II molecules with the invariant chain: Role of the invariant chain (81-90) region. EMBO J 1997; 16:5807-5818.
62. Stern LJ, Brown JH, Jardetzky TS et al. Crystal structure of the human class II MHC protein HLA-DR1 complexed with an influenza virus peptide. Nature 1994; 368:215-221.
63. Ghosh P, Amaya M, Mellins E et al. The structure of an intermediate in class II MHC maturation: CLIP bound to HLA-DR3. Nature 1995; 378:457-462.
64. Gautam AM, Pearson C, Quinn V et al. Binding of an invariant-chain peptide, CLIP, to I-A major histocompatibility complex class II molecules. Proc Natl Acad Sci USA 1995; 92:335-339.
65. Malcherek G, Gnau V, Jung G et al. Supermotifs enable natural invariant chain-derived peptides to interact with many major histocompatibility complex-class II molecules. J Exp Med 1995; 181:527-536.
66. Sette A, Southwood S, Miller J et al. Binding of major histocompatibility complex class II to the invariant chain-derived peptide, CLIP, is regulated by allelic polymorphism in class II. J Exp Med 1995; 181:677-683.
67. Lee Cz, McConnell HM. A general model of invariant chain association with class II major histocompatibility complex proteins. Proc Natl Acad Sci USA 1995; 92:8269-8273.
68. Morkowski S, Goldrath AW, Eastman S et al. T cell recognition of major histocompatibility complex class II complexes with invariant chain processing intermediates. J Exp Med 1995; 182:1403-1413.
69. Kropshofer H, Vogt AB, Hämmerling GJ. Structural features of the invariant chain fragment CLIP controlling rapid release from HLA-DR molecules and inhibition of peptide binding. Proc Natl Acad Sci USA 1995; 92:8313-8317.
70. Kropshofer H, Vogt AB, Stern LJ et al. Self-release of CLIP in peptide loading of HLA-DR molecules. Science 1995; 270:1357-1359.
71. Thayer WP, Ignatowicz, L Weber DA et al. Class II-associated invariant chain peptide-independent binding of invariant chain to class II MHC molecules. J Immunol 1999; 162:1502-1509.
72. Wilson NA, Wolf P, Ploegh H et al. Invariant chain can bind MHC class II at a site other than the peptide binding groove. J Immunol 1998; 161:4777-4784.
73. Castellino F, Han R, Germain RN. The transmembrane segment of invariant chain mediates binding to MHC class II molecules in a CLIP-independent manner. Eur J Immunol 2001; 31:841-850.
74. Lang ML, Yadati S, Seeley ES et al. Mutations in specific I-A(k) alpha(2) and beta(2) domain residues affect surface expression. Int Immunol 2000; 12:777-786.
75. Claesson-Welsh L, Peterson PA. Implications of the invariant gamma-chain on the intracellular transport of class II histocompatibility antigens. J Immunol 1985; 135:3551-3557.
76. Layet C, Germain RN. Invariant chain promotes egress of poorly expressed, haplotype-mismatched class II major histocompatibility complex A alpha A beta dimers from the endoplasmic reticulum/cis-Golgi compartment. Proc Natl Acad Sci USA 1991; 88:2346-2350.

77. Schaiff WT, Hruska KA, Jr., Bono C et al. Invariant chain influences post-translational processing of HLA-DR molecules. J Immunol 1991; 147:603-608.
78. Anderson MS, Miller J. Invariant chain can function as a chaperone protein for class II major histocompatibility complex molecules. Proc Natl Acad Sci USA 1992; 89:2282-2286.
79. Viville S, Neefjes J, Lotteau V et al. Mice lacking the MHC class II-associated invariant chain. Cell 1993; 72:635-648.
80. Miller J, Germain RN. Efficient cell surface expression of class II MHC molecules in the absence of associated invariant chain. J Exp Med 1986; 164:1478-1489.
81. Simonsen A, Momburg F, Drexler J et al. Intracellular distribution of the MHC class II molecules and the associated invariant chain (Ii) in different cell lines. Int Immunol 1993; 5:903-917.
82. Humbert M, Raposo G, Cosson P et al. The invariant chain induces compact forms of class II molecules localized in late endosomal compartments. Eur J Immunol 1993; 23:3158-3166.
83. Salamero J, Humbert M, Cosson P et al. Mouse B lymphocyte specific endocytosis and recycling of MHC class II molecules. EMBO J 1990; 9:3489-3496.
84. Simonis S, Miller J, Cullen SE. The role of the Ia-invariant chain complex in the posttranslational processing and transport of Ia and invariant chain glycoproteins. J Immunol 1989; 143:3619-3625.
85. Lamb CA, Yewdell JW, Bennink JR et al. Invariant chain targets HLA class II molecules to acidic endosomes containing internalized influenza virus. Proc Natl Acad Sci USA 1991; 88:5998-6002.
86. Lamb CA, Cresswell P. Assembly and transport properties of invariant chain trimers and HLA-DR-invariant chain complexes. J Immunol 1992; 148:3478-3482.
87. Anderson KS, Cresswell P. A role for calnexin (IP90) in the assembly of class II MHC molecules. EMBO J 1994; 13 (3):675-682.
88. Bikoff EK, Germain RN, Robertson EJ. Allelic differences affecting invariant chain dependency of MHC class II subunit assembly. Immunity 1995; 2:301-310.
89. Degen E, Williams DB. Participation of a novel 88-kD protein in the biogenesis of murine class I histocompatibility molecules. J Cell Biol 1991; 112:1099-1115.
90. Ahluwalia N, Bergeron JJ, Wada I et al. The p88 molecular chaperone is identical to the endoplasmic reticulum membrane protein, calnexin. J Biol Chem 1992; 267:10914-10918.
91. Galvin K, Krishna S, Ponchel F et al. The major histocompatibility complex class I antigen-binding protein p88 is the product of the calnexin gene. Proc Natl Acad Sci USA 1992; 89:8452-8456.
92. Hochstenbach F, David V, Watkins S et al. Endoplasmic reticulum resident protein of 90 kilodaltons associates with the T- and B-cell antigen receptors and major histocompatibility complex antigens during their assembly. Proc Natl Acad Sci USA 1992; 89:4734-4738.
93. Schreiber KL, Bell MP, Huntoon CJ et al. Class II histocompatibility molecules associate with calnexin during assembly in the endoplasmic reticulum. Int Immunol 1994; 6:101-111.
94. Ou WJ, Cameron PH, Thomas DY et al. Association of folding intermediates of glycoproteins with calnexin during protein maturation. Nature 1993; 364:771-776.
95. Arunachalam B, Cresswell P. Molecular requirements for the interaction of class II major histocompatibility complex molecules and invariant chain with calnexin. J Biol Chem 1995; 270:2784-2790.
96. Romagnoli P, Germain RN. Inhibition of invariant chain (Ii)-calnexin interaction results in enhanced degradation of Ii but does not prevent the assembly of alphabetaIi complexes. J Exp Med 1995; 182:2027-2036.
97. Bonnerot C, Marks MS, Cosson P et al. Association with BiP and aggregation of class II MHC molecules synthesized in the absence of invariant chain. EMBO J 1994; 13:934-944.
98. Claesson-Welsh L, Ploegh H, Peterson PA. Determination of attachment sites for N-linked carbohydrate groups of class II histocompatibility alpha-chain and analysis of possible O-linked glycosylation of alpha- and gamma-chains. Mol Immunol 1986; 23:15-25.
99. Spiro RC, Quaranta V. The invariant chain is a phosphorylated subunit of class II molecules. J Immunol 1989; 143:2589-2594.
100. Koch N, Hammerling GJ. Ia-associated invariant chain is fatty acylated before addition of sialic acid. Biochemistry (Mosc) 1985; 24:6185-6190.
101. Sant AJ, Zacheis M, Rumbarger T et al. Human Ia alpha- and beta-chains are sulfated. J Immunol 1988; 140:155-160.
102. Machamer CE, Cresswell P. Monensin prevents terminal glycosylation of the N- and O-linked oligosaccharides of the HLA-DR-associated invariant chain and inhibits its dissociation from the alpha-beta chain complex. Proc Natl Acad Sci USA 1984; 81:1287-1291.
103. Elliott EA, Drake JR, Amigorena S et al. The invariant chain is required for intracellular transport and function of major histocompatibility complex class II molecules. J Exp Med 1994; 179:681-694.
104. Pieters J, Bakke O, Dobberstein B. The MHC class II associated invariant chain contains two endosomal targeting siganls within its cytoplasmic tail. J Cell Sci 1993; 106:831-846.

105. Henne C, Schwenk F, Koch N et al. Surface expression of the invariant chain (CD74) is independent of concomitant expression of major histocompatibility complex class II antigens. Immunology 1995; 84:177-182.
106. Ishikawa S, Kowal C, Cole B et al. Replacement of N-glycosylation sites on the MHC class II E alpha chain. Effect on thymic selection and peripheral T cell activation. J Immunol 1995; 154:5023-5029.
107. Nag B, Wada HG, Arimilli S et al. The role of N-linked oligosaccharides of MHC class II antigens in T cell stimulation. J Immunol Methods 1994; 172:95-104.
108. Nag B, Passmore D, Kendrick T et al. N-linked oligosaccharides of murine major histocompatibility complex class II molecule. Role in antigenic peptide binding, T cell recognition, and clonal nonresponsiveness. J Biol Chem 1992; 267:22624-22629.
109. Lotteau V, Teyton, L Peleraux A et al. Intracellular transport of class II MHC molecules directed by invariant chain. Nature 1990; 348:600-605.
110. Schutze M-P, Peterson PA, Jackson MR. An N-terminal double-arginine motif maintains type II membrane proteins in the endoplasmic reticulum. EMBO J 1994; 13:1696-1705.
111. Aitken A. 14-3-3 and its possible role in co-ordinating multiple signalling pathways. Trends Cell Biol 1996; 6:341-337.
112. Kuwana T, Peterson PA, Karlsson, L. Exit of major histocompatibility complex class II-invariant chain p35 complexes from theendoplasmic reticulum is modulated by phosphorylation. Proc Natl Acad Sci USA 1998; 95:1056-1061.
113. Anderson HA, Roche PA. Phosphorylation regulates the delivery of MHC class II invariant chain complexes to antigen processing compartments. J Immunol 1998; 160:4850-4858.
114. Anderson HA, Bergstralh DT, Kawamura T et al. Phosphorylation of the invariant chain by protein kinase C regulates MHC class II trafficking to antigen-processing compartments. J Immunol 1999; 163:5435-5443.
115. Miller J, Hatch JA, Simonis S et al. Identification of the glycosaminoglycan-attachment site of mouse invariant-chain proteoglycan core protein by site-directed mutagenesis. Proc Natl Acad Sci USA 1988; 85:1359-1363.
116. Sant AJ, Cullen SE, Giacoletto KS et al. Invariant chain is the core protein of the Ia-associated chondroitin sulfate proteoglycan. J Exp Med 1985; 162:1916-1934.
117. Naujokas MF, Morin M, Anderson MS et al. The chondroitin sulfate form of invariant chain can enhance stimulation of T cell responses through interaction with CD44. Cell 1993; 74:257-268.
118. Germain RN, Margulies DH. The biochemistry and cell biology of antigen processing and presentation. Annu Rev Immunol 1993; 11:403-450.
119. Cresswell P. Assembly, transport, and function of MHC class II molecules. Annu Rev Immunol 1994; 12:259-293.
120. Cresswell P. Intracellular class II HLA antigens are accessible to transferrin-neuraminidase conjugates internalized by receptor-mediated endocytosis. Proc Natl Acad Sci USA 1985; 82:8188-8192.
121. Blum JS, Cresswell P. Role for intracellular proteases in the processing and transport of class II HLA antigens. Proc Natl Acad Sci USA 1988; 85:3975-3979.
122. Pieters J, Horstmann H, Bakke O et al. Intracellular transport and localization of major histocompatibility complex class II molecules and associated invariant chain. J Cell Biol 1991; 115:1213-1223.
123. Peters PJ, Neefjes JJ, Oorschot V et al. Segregation of MHC class II molecules from MHC class I molecules in the Golgi complex for transport to lysosomal compartments. Nature 1991; 349:669-676.
124. Zhong GM, Romagnoli P, Germain RN. Related leucine-based cytoplasmic targeting signals in invariant chain and major histocompatibility complex class II molecules control endocytic presentation of distinct determinants in a single protein. J Exp Med 1997; 185:429-438.
125. Simonsen A, Pedersen KW, Nordeng TW et al. Polarized transport of MHC class II molecules in Madin-Darby canine kidney cells is directed by a leucine-based signal in the cytoplasmic tail of the beta-chain. J Immunol 1999; 163:2540-2548.
126. Geuze HJ. The role of endosomes and lysosomes in MHC class II functioning. Immunol Today 1998; 19:282-287.
127. Stang E, Guerra CB, Amaya M et al. DR/CLIP (class II-associated invariant chain peptides) and DR/peptide complexes colocalize in prelysosomes in human B lymphoblastoid cells. J Immunol 1998; 160:4696-4707.
128. Odorizzi CG, Trowbridge IS, Xue L et al. Sorting signals in the MHC class II invariant chain cytoplasmic tail and transmembrane region determine trafficking to an endocytic processing compartment. J Cell Biol 1994; 126:317-330.
129. Bénaroch P, Yilla M, Raposo G et al. How MHC class II molecules reach the endocytic pathway. EMBO J 1995; 14:37-49.

130. Warmerdam PAM, Long EO, Roche PA. Isoforms of the invariant chain regulate transport of MHC class II molecules to antigen processing compartments. J Cell Biol 1996; 133:281-291.
131. Liu SH, Marks MS, Brodsky FM. A dominant-negative clathrin mutant differentially affects trafficking of molecules with distinct sorting motifs in the class II major histocompatibility complex (MHC) pathway. J Cell Biol 1998; 140:1023-1037.
132. Roche PA, Teletski CL, Stang E et al. Cell surface HLA-DR-invariant chain complexes are targeted to endosomes by rapid internalization. Proc Natl Acad Sci USA 1993; 90:8581-8585.
133. Bremnes B, Madsen T, Gedde-Dahl M et al. An LI and ML motif in the cytoplasmic tail of the MHC-associated invariant chain mediate rapid internalization. J Cell Sci 1994; 107:2021-2032.
134. Wang K, Peterson PA, Karlsson L. Decreased endosomal delivery of major histocompatibility complex class II-invariant chain complexes in dynamin-deficient. cells J Biol Chem 1997; 272:17055-17060.
135. Ong GL, Goldenberg DM, Hansen HJ et al. Cell surface expression and metabolism of major histocompatibility complex class II invariant chain (CD74) by diverse cell lines. Immunology 1999; 98:296-302.
136. McCoy KL, Gainey D, Inman JK et al. Antigen presentation by B lymphoma cells: Requirements for processing of exogenous antigen internalized through transferrin receptors. J Immunol 1993; 151:4583-4594.
137. McCoy KL, Noone M, Inman JK et al. Exogenous antigens internalized through transferrin receptors activate CD4+ T cells. J Immunol 1993; 150:1691-1704.
138. Gagliardi M-C, Nisini R, Benvenuto R et al. Soluble transferrin mediates targeting of hepatitis B envelope antigen to transferrin receptor and its presentation by activated T cells. Eur J Immunol 1994; 24:1372-1376.
139. Griffin JP, Chu R, Harding CV. Early endosomes and a late endocytic compartment generate different peptide-class II MHC complexes via distinct processing mechanisms. J Immunol 1997; 158:1523-1532.
140. Carlsson SR, Fukuda M. The lysosomal membrane glycoprotein lamp-1 is transported to lysosomes by two alternative pathways. Arch Biochem Biophys 1992; 296:630-639.
141. Harter C, Mellman I. Transport of the lysosomal membrane glycoprotein lgp120 (lgp-A) to lysosomes does not require appearance on the plasma membrane. J Cell Biol 1992; 117:311-325.
142. Johnson KF, Kornfeld S. The cytoplasmic tail of the mannose 6-phosphate/insulin-like growth factor-II receptor has two signals for lysosomal enzyme sorting in the Golgi. J Cell Biol 1992; 119:249-257.
143. Bakke O, Dobberstein B. MHC class II-associated invariant chain contains a sorting signal for endosomal compartments. Cell 1990; 63:707-716.
144. Kang S, Liang L, Parker CD et al. Structural requirements for major histocompatibility complex class II invariant chain endocytosis and lysosomal targeting. J Biol Chem 1998; 273:20644-20652.
145. Motta A, Bremnes B, Morelli MAC et al. Structure-activity relationship of the Leucine-based sorting motifs in the cytosolic tail of the major histocompatibility complex-associated invariant chain. J Biol Chem 1995; 270:27165-27171.
146. Pond L, Kuhn LA, Teyton L et al. A role for acidic residues in di-leucine motif-based targeting to the endocytic pathway. J Biol Chem 1995; 270:19989-19997.
147. Simonsen A, Bremnes B, Nordeng TW et al. The leucine-based motif (DDQxxLI) is recognized both for internalization and basolateral sorting of invariant chain in MDCK cells. Eur J Cell Biol 1998; 76:25-32.
148. Johnson AO, Lampson MA, McGraw TE. A di-leucine sequence and a cluster of acidic amino acids are required for dynamic retention in the endosomal recycling compartment of fibroblasts. Mol Biol Cell 2001; 12:367-381.
149. Tikkanen R, Obermuller S, Denzer K et al. The dileucine motif within the tail of MPR46 is required for sorting of the receptor in endosomes. Traffic 2000; 1:631-640.
150. Humbert M, Bertolino P, Forquet F et al. Major histocompatibility complex class II-restricted presentation of secreted and endoplasmic reticulum resident antigens requires the invariant chains and is sensitive to lysosomotropic agents. Eur J Immunol 1993; 23:3167-3172.
151. Reid PA, Watts C. Constitutive endocytosis and recycling of major histocompatibility complex class II glycoproteins in human B-lymphoblastoid cells. Immunology 1992; 77:539-542.
152. Anderson MS, Swier K, Arneson L et al. Enhanced antigen presentation in the absence of the invariant chain endosomal localization signal. J Exp Med 1993; 178:1959-1969.
153. Pinet V, Vergelli M, Martin R et al. Antigen presentation mediated by recycling of surface HLA-DR molecules. Nature 1995; 375:603-606.
154. Smiley ST, Rudensky AY, Glimcher LH et al. Truncation of the class II beta-chain cytoplasmic domain influences the level of class II invariant chain-derived peptide complexe.s Proc Natl Acad Sci USA 1996; 93:241-244.

155. Forquet F, Barois N, Machy P et al. Presentation of antigens internalized through the B cell receptor requires newly synthesized MHC class II molecules. J Immunol 1999; 162:3408-3416.
156. Nabavi N, Ghogawala Z, Myer A et al. Antigen presentation abrogated in cells expressing truncated Ia molecules. J Immunol 1989; 142:1444-1447.
157. St-Pierre Y, Nabavi N, Ghogawala Z et al. A functional role for signal transduction via the cytoplasmic domains of MHC class II proteins. J Immunol 1989; 143:808-812.
158. Pinet V, Malnati MS, Long EO. Two processing pathways for the MHC class II-restricted presentation of exogenous influenza virus antigen. J Immunol 1994; 152:4852-4860.
159. Smiley ST, Laufer TM, Lo D et al. Transgenic mice expressing MHC class II molecules with truncated Abeta cytoplasmic domains reveal signaling-independent defects in antigen presentation. Int Immunol 1995; 7:665-677.
160. Sallusto F, Cella M, Danieli C et al. Dendritic cells use macropinocytosis and the mannose receptor to concentrate macromolecules in the major histocompatibility complex class II compartment: downregulation by cytokines and bacterial products [see comments]. J Exp Med 1995; 182:389-400.
161. Racoosin EL, Swanson JA. M-CSF-induced macropinocytosis increases solute endocytosis but not receptor-mediated endocytosis in mouse macrophages. J Cell Sci 1992; 102:867-880.
162. Racoosin EL, Swanson JA. Macrophage colony-stimulating factor (rM-CSF) stimulates pinocytosis in bone marrow-derived macrophages. J Exp Med 1989; 170:1635-1648.
163. West MA, Bretscher MS, Watts C. Distinct endocytotic pathways in epidermal growth factor-stimulated human carcinoma A431 cells [published erratum appears in J Cell Biol 1990 Mar; 110(3):859]. J Cell Biol 1989; 109:2731-2739.
164. Mellman I. Endocytosis and molecular sorting. Annu Rev Cell Biol 1996; 12:575-625.
165. Lanzavecchia A. Receptor-mediated antigen uptake and its effect on antigen presentation to class II-restricted T lymphocytes. Annu Rev Immunol 1990; 8:773-793.
166. Bonnerot C, Lankar D, Hanau D et al. Role of B cell receptor Ig alpha and Ig beta subunits in MHC class II- restricted antigen presentation. Immunity 1995; 3:335-347.
167. Ferrari G, Knight AM, Watts C et al. Distinct intracellular compartments involved in invariant chain degradation and antigenic peptide loading of major histocompatibility complex (MHC) class II molecules. J Cell Biol 1997; 139:1433-1446.
168. Siemasko K, Eisfelder BJ, Stebbins C et al. Igalpha and Igbeta are required for efficient trafficking to late endosomes and to enhance antigen presentation. J Immunol 1999; 162:6518-6525.
169. Snider DP, Segal DM. Targeted antigen presentation using crosslinked antibody heteroaggregates. J Immunol 1987; 139:1609-1616.
170. Casten LA, Pierce SK. Receptor-mediated B cell antigen processing. Increased antigenicity of a globular protein covalently coupled to antibodies specific for B cell surface structures. J Immunol 1988; 140:404-410.
171. Finkelman FD, Scher I, Mond JJ et al. Polyclonal activation of the murine immune system by an antibody to IgD. II. Generation of polyclonal antibody production and cells with surface IgG. J Immunol 1982; 129:638-646.
172. Finkelman FD, Scher I, Mond JJ et al. Polyclonal activation of the murine immune system by an antibody to IgD. I. Increase in cell size and DNA synthesis J Immunol 1982; 129:629-637.
173. Lunde E, Munthe LA, Vabo A et al. Antibodies engineered with IgD specificity efficiently deliver integrated T-cell epitopes for antigen presentation by B cells Nat Biotechnol 1999; 17:670-675.
174. Eynon EE, Parker DC Small B cells as antigen-presenting cells in the induction of tolerance to soluble protein antigens J Exp Med 1992; 175:131-138.
175. Saoudi A, Simmonds S, Huitinga I et al. Prevention of experimental allergic encephalomyelitis in rats by targeting autoantigen to B cells: Evidence that the protective mechanism depends on changes in the cytokine response and migratory properties of the autoantigen-specific T cells [see comments] J Exp Med 1995; 182:335-344.
176. Davidson HW, West MA, Watts C Endocytosis, intracellular trafficking, and processing of membrane IgG and monovalent antigen/membrane IgG complexes in B lymphocytes J Immunol 1990; 144:4101-4109.
177. Knight AM, Lucocq JM, Prescott AR et al. Antigen endocytosis and presentation mediated by human membrane IgG1 in the absence of the Igalpha/Igbeta dimer EMBO J 1997; 16:3842-3850.
178. Lanzavecchia A, Bove S Specific B lymphocytes efficiently pick up, process and present antigen to T cells Behring Inst Mitt 1985:82-87.
179. Rock KL, Benacerraf B, Abbas AK Antigen presentation by hapten-specific B lymphocytes. I. Role of surface immunoglobulin receptors J Exp Med 1984; 160:1102-1113.
180. Lanzavecchia A Mechanisms of antigen uptake for presentation Curr Opin Immunol 1996; 8:348-354.

181. Bonnerot C, Briken V, Amigorena S. Intracellular signaling and endosomal trafficking of immunoreceptors shared effectors underlying MHC class II-restricted antigen presentation. Immunol Lett 1997; 57:1-4.
182. Amigorena S, Bonnerot C. Role of B-cell and Fc receptors in the selection of T-cell epitopes. Curr Opin Immunol 1998; 10:88-92.
183. Zimmermann VS, Rovere P, Trucy J et al. Engagement of B cell receptor regulates the invariant chain-dependent MHC class II presentation pathway. J Immunol 1999; 162:2495-2502.
184. Hunziker W, Fumey C. A di-leucine motif mediates endocytosis and basolateral sorting of macrophage IgG Fc receptors in MDCK cells. EMBO J 1994; 13:2963-2969.
185. Daëron M, Malbec O, Latour S et al. Distinct intracytoplasmic sequences are required for endocytosis and phagocytosis via murine Fc gamma RII in mast cells. Int Immunol 1993; 5:1393-1401.
186. Bonnerot C, Amigorena S, Choquet D et al. Role of associated gamma-chain in tyrosine kinase activation via murine Fc gamma RIII. Embo J 1992; 11:2747-2757.
187. Amigorena S, Salamero J, Davoust J et al. Tyrosine-containing motif that transduces cell activation signals also determines internalization and antigen presentation via type III receptors for IgG. Nature 1992; 358:337-341.
188. Amigorena S, Lankar D, Briken V et al. Type II and III receptors for immunoglobulin G (IgG) control the presentation of different T cell epitopes from single IgG-complexed antigens. J Exp Med 1998; 187:505-515.
189. Pfefferkorn LC, Yeaman GR. Association of IgA-Fc receptors (Fc alpha R) with Fc epsilon RI gamma 2 subunits in U937 cells. Aggregation induces the tyrosine phosphorylation of gamma 2. J Immunol 1994; 153:3228-3236.
190. Shen L, van Egmond M, Siemasko K et al. Presentation of ovalbumin internalized via the immunoglobulin-A Fc receptor is enhanced through Fc receptor gamma-chain signaling [In Process Citation]. Blood 2001; 97:205-213.
191. Berg T, Gjøen T, Bakke O. Physiological functions of endosomal proteolysis. Biochem J 1995; 307:313-326.
192. Johnson KF, Chan W, Kornfeld S. Cation-dependent mannose 6-phosphate receptor contains two internalization signals in its cytoplasmic domain [published erratum appears in Proc Natl Acad Sci USA 1991 Feb 15;88(4):1591]. Proc Natl Acad Sci USA 1990; 87:10010-10014.
193. Johnson KF, Kornfeld S. A His-Leu-Leu sequence near the carboxyl terminus of the cytoplasmic domain of the cation-dependent mannose 6-phosphate receptor is necessary for the lysosomal enzyme sorting function. J Biol Chem 1992; 267:17110-17115.
194. Stahl PD, Ezekowitz RA. The mannose receptor is a pattern recognition receptor involved in host defense. Curr Opin Immunol 1998; 10:50-55.
195. Jiang W, Swiggard WJ, Heufler C et al. The receptor DEC-205 expressed by dendritic cells and thymic epithelial cells is involved in antigen processing. Nature 1995; 375:151-155.
196. Tan MCAA, Mommaas AM, Drijfhout JW et al. Mannose receptor mediated uptake of antigens strongly enhances HLA-class II restricted antigen presentation by cultured dendritic cells. Adv Exp Med Biol 1997; 417:171-174.
197. Engering AJ, Cella M, Fluitsma DM et al. Mannose receptor mediated antigen uptake and presentation in human dendritic cells. Adv Exp Med Biol 1997; 417:183-187.
198. Engering A, Lefkovits I, Pieters J. Analysis of subcellular organelles involved in major histocompatibility complex (MHC) class II-restricted antigen presentation by electrophoresis. Electrophoresis 1997; 18:2523-2530.
199. Schweizer A, Stahl PD, Rohrer J. A Di-aromatic motif in the cytosolic tail of the mannose receptor mediates endosomal sorting [In Process Citation]. J Biol Chem 2000; 275:29694-29700.
200. Stahl P, Schlesinger PH, Sigardson E et al. Receptor-mediated pinocytosis of mannose glycoconjugates by macrophages: characterization and evidence for receptor recycling. Cell 1980; 19:207-215.
201. Mahnke K, Guo M, Lee S et al. The dendritic cell receptor for endocytosis, DEC-205, can recycle and enhance antigen presentation via major histocompatibility complex class II-positive lysosomal compartments [In Process Citation]. J Cell Biol 2000; 151:673-684.
202. Zhang H, Yang Y, Steinbrecher UP. Structural requirements for the binding of modified proteins to the scavenger receptor of macrophages. J Biol Chem 1993; 268:5535-5542.
203. Pearson AM. Scavenger receptors in innate immunity. Curr Opin Immunol 1996; 8:20-28.
204. Matsumoto A, Naito M, Itakura H et al. Human macrophage scavenger receptors: primary structure, expression, and localization in atherosclerotic lesions. Proc Natl Acad Sci USA 1990; 87:9133-9137.
205. Geng Y, Kodama T, Hansson GK. Differential expression of scavenger receptor isoforms during monocyte- macrophage differentiation and foam cell formation. Arterioscler Thromb 1994; 14:798-806.

206. Geng YJ, Hansson GK. Interferon-gamma inhibits scavenger receptor expression and foam cell formation in human monocyte-derived macrophages. J Clin Invest 1992; 89:1322-1330.
207. van Lenten BJ, Fogelman AM. Lipopolysaccharide-induced inhibition of scavenger receptor expression in human monocyte-macrophages is mediated through tumor necrosis factor-alpha. J Immunol 1992; 148:112-116.
208. Hampton RY, Golenbock DT, Penman M et al. Recognition and plasma clearance of endotoxin by scavenger receptors. Nature 1991; 352:342-344.
209. Nicoletti A, Caligiuri G, Tornberg I et al. The macrophage scavenger receptor type A directs modified proteins to antigen presentation. Eur J Immunol 1999; 29:512-521.
210. Morimoto K, Wada Y, Hinagata J et al. VXFD in the cytoplasmic domain of macrophage scavenger receptors mediates their efficient internalization and cell-surface expression. Biol Pharm Bull 1999; 22:1022-1026.
211. Medzhitov R, Janeway C, Jr. Innate immunity. N Engl J Med 2000; 343:338-344.
212. Roche PA, Cresswell P. Proteolysis of the class II-associated invariant chain generates a peptide binding site in intracellular HLA-DR molecules. Proc Natl Acad Sci USA 1991; 88:3150-3154.
213. Daibata M, Xu M, Humphreys RE et al. More efficient peptide binding to MHC class II molecules during cathepsin B digestion of Ii than after Ii release. Mol Immunol 1994; 31:255-260.
214. Vidard L, Rock KL, Benacerraf B. Diversity in MHC class II ovalbumin T cell epitopes generated by distinct proteases. J Immunol 1992; 149:498-504.
215. Villadangos JA, Bryant RA, Deussing J et al. Proteases involved in MHC class II antigen presentation. Immunol Rev 1999; 172:109-120.
216. Jensen PE. Enhanced binding of peptide antigen to purified class II major histocompatibility glycoproteins at acidic pH. J Exp Med 1991; 174:1111-1120.
217. Hampl J, Gradehandt G, Kalbacher H et al. In vitro processing of insulin for recognition by murine T cells results in the generation of A chains with free CysSH. J Immunol 1992; 148:2664-2671.
218. Collins DS, Unanue ER, Harding CV. Reduction of disulfide bonds within lysosomes is a key step in antigen processing. J Immunol 1991; 147:4054-4059.
219. Harding CV, Geuze HJ. Class II MHC molecules are present in macrophage lysosomes and phagolysosomes that function in the phagocytic processing of *Listeria monocytogenes* for presentation to T cells. J Cell Biol 1992; 119:531-542.
220. Marks MS, Blum JS, Cresswell P. Invariant chain trimers are sequestered in the rough endoplasmic reticulum in the absence of association with HLA class II antigens. J Cell Biol 1990; 111:839-855.
221. Newcomb JR, Cresswell P. Structural analysis of proteolytic products of MHC class II-invariant chain complexes generated in vivo. J Immunol 1993; 151:4153-4163.
222. Xu M, Capraro GA, Daibata M et al. Cathepsin B cleavage and release of invariant chain from MHC class II molecules follow a staged pattern. Mol Immunol 1994; 31:723-731.
223. Morton PA, Zacheis ML, Giacoletto KS et al. Delivery of nascent MHC class II-invariant chain complexes to lysosomal compartments and proteolysis of invariant chain by cysteine proteases precedes peptide binding in B-lymphoblastoid cells. J Immunol 1995; 154:137-150.
224. Villadangos JA, Riese RJ, Peters C et al. Degradation of mouse invariant chain: Roles of cathepsins S and D and the influence of major histocompatibility complex polymorphism. J Exp Med 1997; 186:549-560.
225. Zhang T, Maekawa Y, Hanba J et al. Lysosomal cathepsin B plays an important role in antigen processing, while cathepsin D is involved in degradation of the invariant chain inovalbumin-immunized mice. Immunology 2000; 100:13-20.
226. Deussing J, Roth, W Saftig P et al. Cathepsins B and D are dispensable for major histocompatibility complex class II-mediated antigen presentation. Proc Natl Acad Sci USA 1998; 95:4516-4521.
227. Nakagawa T, Roth W, Wong P et al. Cathepsin L: Critical role in Ii degradation and CD4 T cell selection in the thymus. Science 1998; 280:450-453.
228. Riese RJ, Chapman HA. Cathepsins and compartmentalization in antigen presentation [see comments]. Curr Opin Immunol 2000; 12:107-113.
229. Riese RJ, Wolf PR, Bromme D et al. Essential role for cathepsin S in MHC class II-associated invariant chain processing and peptide loading. Immunity 1996; 4:357-366.
230. Driessen C, Bryant RA, Lennon-Dumenil AM et al. Cathepsin S controls the trafficking and maturation of MHC class II molecules in dendritic cells. J Cell Biol 1999; 147:775-790.
231. Nakagawa TY, Brissette WH, Lira PD et al. Impaired invariant chain degradation and antigen presentation and diminished collagen-induced arthritis in cathepsin S null mice. Immunity 1999; 10:207-217.
232. Shi GP, Villadangos JA, Dranoff G et al. Cathepsin S required for normal MHC class II peptide loading and germinal center development. Immunity 1999; 10:197-206.

233. Shi GP, Bryant RA, Riese R et al. Role for cathepsin F in invariant chain processing and major histocompatibility complex class II peptide loading by macrophages. J Exp Med 2000; 191:1177-1186.
234. Lich JD, Elliott JF, Blum JS Cytoplasmic processing is a prerequisite for presentation of an endogenous antigen by major histocompatibility complex class II proteins. J Exp Med 2000; 191:1513-1524.
235. Manoury B, Hewitt EW, Morrice N et al. An asparaginyl endopeptidase processes a microbial antigen for class II MHC presentation [see comments]. Nature 1998; 396:695-699.
236. Chen JM, Dando PM, Rawlings ND et al. Cloning, isolation, and characterization of mammalian legumain, an asparaginyl endopeptidase. J Biol Chem 1997; 272:8090-8098.
237. Villadangos JA, Driessen C, Shi GP et al. Early endosomal maturation of MHC class II molecules independently of cysteine proteases and H-2DM. Embo J 2000; 19:882-891.
238. Panjwani N, Akbari O, Garcia S et al. The HSC73 molecular chaperone: Involvement in MHC class II antigen presentation. J Immunol 1999; 163:1936-1942.
239. Katunuma N, Kakegawa H, Matsunaga Y et al. Immunological significances of invariant chain from the aspect of its structural homology with the cystatin family. FEBS Lett 1994; 349:265-269.
240. Guncar G, Pungercic G, Klemencic I et al. Crystal structure of MHC class II-associated p41 Ii fragment bound to cathepsin L reveals the structural basis for differentiation between cathepsins L and S. Embo J 1999; 18:793-803.
241. Kämpgen E, Koch N, Koch F et al. Class II major histocompatibility complex molecules of murine dendritic cells: synthesis, sialylation of invariant chain, and antigen processing capacity are down-regulated upon culture. Proc Natl Acad Sci USA 1991; 88:3014-3018.
242. Arunachalam B, Lamb CA, Cresswell P. Transport properties of free and MHC class II-associated oligomers containing different isoforms of human invariant chain. Int Immunol 1994; 6:439-451.
243. Fineschi B, Arneson LS, Naujokas MF et al. Proteolysis of major histocompatibility complex class II-associated invariant chain is regulated by the alternatively spliced gene product, p41. Proc Natl Acad Sci USA 1995; 92:10257-10261.
244. Ogrinc T, Dolenc I, Ritonja A et al. Purification of the complex of cathepsin L and the MHC class II-associated invariant chain fragment from human kidney. FEBS Lett 1993; 336:555-559.
245. Fineschi B, Sakaguchi K, Appella E et al. The proteolytic environment involved in MHC class II-restricted antigen presentation can be modulated by the p41 form of invariant chain. J Immunol 1996; 157:3211-3215.
246. Bevec T, Stoka V, Pungercic G et al. A fragment of the major histocompatibility complex class II—Associated p41 invariant chain inhibits cruzipain, the major cysteine proteinase from *Trypanosoma cruzi*. FEBS Lett 1997; 401:259-261.
247. Rodriguez GM, Diment S. Destructive proteolysis by cysteine proteases in antigen presentation of ovalbumin. Eur J Immunol 1995; 25:1823-1827.
248. Pierre P Shachar I, Matza D et al. Invariant chain controls H2-M proteolysis in mouse splenocytes and dendritic cells. J Exp Med 2000; 191:1057-1062.
249. Romagnoli P, Layet C, Yewdell J et al. Relationship between invariant chain expression and major histocompatibility complex class II transport into early and late endocytic compartments. J Exp Med 1993; 177:583-596.
250. Stang E, Bakke O. MHC class II associated invariant chain induced enlarged endosomal structures. A morphological study. Exp Cell Res 1997; 235:79-82.
251. Gorvel J-P, Escola J-M, Stang E et al. Invariant chain induces a delayed transport from early to late endosomes. J Biol Chem 1995; 270:2741-2746.
252. Loss GE, Jr., Sant AJ. Invariant chain retains MHC class II molecules in the endocytic pathway. J Immunol 1993; 150:3187-3197.
253. Neefjes JJ, Ploegh HL. Inhibition of endosomal proteolytic activity by leupeptin blocks surface expression of MHC class II molecules and their conversion to SDS resistance alpha beta heterodimers in endosomes. EMBO J 1992; 11:411-416.
254. Zachgo S, Dobberstein B, Griffiths G. A block in degradation of MHC class II-associated invariant chain correlates with a reduction in transport from endosome carrier vesicles to the prelysosome compartment. J Cell Sci 1992; 103:811-822.
255. Mellins E, Smith L, Arp B et al. Defective processing and presentation of exogenous antigens in mutants with normal HLA class II genes. Nature 1990; 343:71-74.
256. Mellins E, Kempin S, Smith L et al. A gene required for class II-restricted antigen presentation maps to the major histocompatibility complex. J Exp Med 1991; 174:1607-1615.
257. Ceman S, Rudersdorf R, Long EO et al. MHC class II deletion mutant expresses normal levels of transgene encoded class II molecules that have abnormal conformation and impaired antigen presentation ability. J Immunol 1992; 149:754-761.

258. Riberdy JM, Cresswell P The antigen-processing mutant T2 suggests a role for MHC-linked genes in class II antigen presentation [published erratum appears in J Immunol 1992 Aug 1;149(3):1113] J Immunol 1992; 148, 2586-2590.
259. Malnati MS, Ceman S, Weston M et al. Presentation of cytosolic antigen by HLA-DR requires a function encoded in the class II region of the MHC J Immunol 1993; 151, 6751-6756.
260. Ceman S, Petersen JW, Pinet V et al. Gene required for normal MHC class II expression and function is localized to approximately 45 kb of DNA in the class II region of the MHC J Immunol 1994; 152, 2865-2873.
261. Morris P, Shaman J, Attaya M et al. An essential role for HLA-DM in antigen presentation by class II major histocompatibility molecules Nature 1994; 368, 551-554.
262. Fling SP, Arp B, Pious D HLA-DMA and -DMB genes are both required for MHC class II/ peptide complex formation in antigen-presenting cells Nature 1994; 368, 554-558.
263. Sloan VS, Cameron P, Porter G et al. Mediation by HLA-DM of dissociation of peptides from HLA-DR Nature 1995; 375, 802-806.
264. Riberdy JM, Newcomb JR, Surman MJ et al. HLA-DR molecules from an antigen-processing mutant cell line are associated with invariant chain peptides Nature 1992; 360, 474-477.
265. Sette A, Ceman S, Kubo RT et al. Invariant chain peptides in most HLA-DR molecules of an antigen-processing mutant Science 1992; 258, 1801-1804.
266. Denzin LK, Cresswell P HLA-DM induces CLIP dissociation from MHC class II alphabeta dimers and facilitates peptide loading Cell 1995; 82, 155-165.
267. Sherman MA, Weber DA, Jensen, PE DM enhances peptide binding to class II MHC by release of invariant chain-derived peptide Immunity 1995; 3, 197-205.
268. Schafer PH, Green JM, Malapati S et al. HLA-DM is present in one-fifth the amount of HLA-DR in the class II peptide-loading compartment where it associates with leupeptin-induced peptide (LIP)-HLA-DR complexes J Immunol 1996; 157, 5487-5495.
269. Vogt AB, Kropshofer H, Moldenhauer G et al. Kinetic analysis of peptide loading onto HLA-DR molecules mediated by HLA-DM Proc Natl Acad Sci USA 1996; 93:9724-9729.
270. Kropshofer H, Vogt AB, Moldenhauer G et al. Editing of the HLA-DR-peptide repertoire by HLA-DM. EMBO J 1996; 15:6144-6154.
271. Denzin LK, Hammond C, Cresswell P. HLA-DM interactions with intermediates in HLA-DR maturation and a role for HLA-DM in stabilizing empty HLA-DR molecules. J Exp Med 1996; 184:2153-2165.
272. Stebbins CC, Loss GE, Jr., Elias CG et al. The requirement for DM in class II-restricted antigen presentation and SDS-stable dimer formation is allele and species dependent. J Exp Med 1995; 181:223-234.
273. Chou CL, Sadegh-Nasseri S HLA-DM recognizes the flexible conformation of major histocompatibility complex class II. J Exp Med 2000; 192:1697-1706.
274. Kropshofer H, Hämmerling GJ, Vogt AB. How HLA-DM edits the MHC class II peptide repertoire: Survival of the fittest? Immunol Today 1997; 18:77-82.
275. Monji T, McCormack AL, Yates JR et al. Invariant-cognate peptide exchange restores class II dimer stability in HLA-DM mutants. J Immunol 1994; 153:4468-4477.
276. Sanderson F, Thomas C, Neefjes J et al. Association between HLA-DM and HLA-DR in vivo. Immunity 1996; 4:87-96.
277. Karlsson L, Péléraux A, Lindstedt R et al. Reconstitution of an operational MHC class II compartment in nonantigen-presenting cells. Science 1994; 266:1569-1573.
278. Sanderson F, Kleijmeer MJ, Kelly A et al. Accumulation of HLA-DM, a regulator of antigen presentation, in MHC class II compartments. Science 1994; 266:1566-1569.
279. Min YH, Lee ST, Choi KM et al. Surface expression of HLA-DM on dendritic cells derived from CD34- positive bone marrow haematopoietic stem cells. Br J Haematol 2000; 110:385-393.
280. Marks MS, Roche PA, Van Donselaar E et al. A lysosomal targeting signal in the cytoplasmic tail of the beta chain directs HLA-DM to MHC class II compartments. J Cell Biol 1995; 131:351-369.
281. Lindstedt R, Liljedahl M, Peleraux A et al. The MHC class II molecule H2-M is targeted to an endosomal compartment by a tyrosine-based targeting motif. Immunity 1995; 3:561-572.
282. Weber DA, Evavold BD, Jensen PE. Enhanced dissociation of HLA-DR-bound peptides in the presence of HLA-DM. Science 1996; 274:618-620.
283. Douek DC, Altmann DM. HLA-DO is an intracellular class II molecule with distinctive thymic expression. Int Immunol 1997; 9:355-364.
284. Vogt AB, Kropshofer H. HLA-DM—An endosomal and lysosomal chaperone for the immune system. Trends Biochem Sci 1999; 24:150-154.
285. Liljedahl M, Kuwana T, Fung-Leung WP et al. HLA-DO is a lysosomal resident which requires association with HLA-DM for efficient intracellular transport. EMBO J 1996; 15:4817-4824.

286. Kropshofer H, Vogt AB, Thery C et al. A role for HLA-DO as a co-chaperone of HLA-DM in peptide loading of MHC class II molecules. EMBO J 1998; 17:2971-2981.
287. Van Ham SM, Tjin EPM, Lillemeier BF et al. HLA-DO is a negative modulator of HLA-DM-mediated MHC class II peptide loading. Curr Biol 1997; 7:950-957.
288. Liljedahl M, Winqvist O, Surh CD et al. Altered antigen presentation in mice lacking H2-O. Immunity 1998; 8:233-243.
289. Maecker HT, Todd SC, Levy S. The tetraspanin superfamily: molecular facilitators. Faseb J 1997; 11:428-442.
290. Escola JM, Kleijmeer MJ, Stoorvogel W et al. Selective enrichment of tetraspan proteins on the internal vesicles of multivesicular endosomes and on exosomes secreted by human B-lymphocytes. J Biol Chem 1998; 273:20121-20127.
291. Schick MR, Levy S. The TAPA-1 molecule is associated on the surface of B cells with HLA-DR molecules. J Immunol 1993; 151:4090-4097.
292. Raposo G, Nijman HW, Stoorvogel, W et al. B lymphocytes secrete antigen-presenting vesicles. J Exp Med 1996; 183:1161-1172.
293. Wubbolts R, Fernandez-Borja M, Oomen L et al. Direct vesicular transport of MHC class II molecules from lysosomal structures to the cell surface. J Cell Biol 1996; 135:611-622.
294. Pond L, Watts C .Characterization of transport of newly assembled, T cell-stimulatory MHC class II peptide complexes from MHC class II compartments to the cell surface. J Immunol 1997; 159:543-553.
295. Sadegh-Nasseri S, Germain RN. A role for peptide in determining MHC class II structure. Nature 1991; 353:167-170.
296. Germain RN, Hendrix LR. MHC class II structure, occupancy and surface expression determined by post-endoplasmic reticulum antigen binding. Nature 1991; 353:134-139.
297. Thery C, Brachet V, Regnault A et al. MHC class II transport from lysosomal compartments to the cell surface is determined by stable peptide binding, but not by the cytosolic domains of the alpha- and beta-chains. J Immunol 1998; 161:2106-2113.
298. Bakke O, Nordeng TW. Intracellular traffic to compartments for MHC class II peptide loading: endosomal and polarized sorting signals. Immunol Rev 1999; 172:171-187.
299. Klausner RD, Samelson LE. T cell antigen receptor activation pathways: the tyrosine kinase connection. Cell 1991; 64:875-878.
300. Andre P, Cambier JC, Wade TK et al. Distinct structural compartmentalization of the signal transducing functions of major histocompatibility complex class II (Ia) molecules. J Exp Med 1994; 179:763-768.
301. Huby RD, Dearman RJ, Kimber I. Intracellular phosphotyrosine induction by major histocompatibility complex class II requires co-aggregation with membrane rafts. J Biol Chem 1999; 274:22591-22596.
302. Anderson HA, Hiltbold EM, Roche PA. Concentration of MHC class II molecules in lipid rafts facilitates antigen presentation. Nat Immunol 2000; 1:156-162.
303. Turley SJ, Inaba K, Garrett WS et al. Transport of peptide-MHC class II complexes in developing dendritic cells. Science 2000; 288:522-527.
304. Sekaly RP, Tonnelle C, Strubin M et al. Cell surface expression of class II histocompatibility antigens occurs in the absence of the invariant chain. J Exp Med 1986; 164:1490-1504.
305. Nijenhuis M, Calafat J, Kuijpers KC et al. Targeting major histocompatibility complex class II molecules to the cell surface by invariant chain allows antigen presentation upon recycling. Eur J Immunol 1994; 24:873-883.
306. Harding CV, Collins DS, Slot JW et al. Liposome-encapsulated antigens are processed in lysosomes, recycled, and presented to T cells Cell 1991; 64:393-401.
307. Moldenhauer G, Henne C, Karhausen J et al. Surface-expressed invariant chain (CD74) is required for internalization of human leucocyte antigen-DR molecules to early endosomal compartments. Immunology 1999; 96:473-484.
308. Triantafilou K, Triantafilou M, Wilson KM et al. Intracellular and cell surface heterotypic associations of human leukocyte antigen-DR and human invariant chain. Hum Immunol 1999; 60:1101-1112.
309. Bevec T, Stoka V, Pungercic G et al. Major histocompatibility complex class II-associated p41 invariant chain fragment is a strong inhibitor of lysosomal cathepsin L. J Exp Med 1996; 183:1331-1338.
310. Siemasko K, Clark MR. The control and facilitation of MHC class II antigen processing by the BCR. Curr Opin Immunol 2001; 13:32-36.
311. Nordeng TW, Gregers TF, Kongsvik TL, Meresse S, Gorvel JP, Jourdan F et al. The cytoplasmic tail of invariant chainregulates endosome fusion and morphology. Mol Biol Cell 2002; 13:1846-1856.
312. Denzer K, Kleijmeer MJ, Heijnen HF, Stoorvogel W, Geuze HJ. Exosome: from internal vesicle of the multivesicular body to intercellular signaling device. J Cell Science 2000; 113:3365-3374.

CHAPTER 17

Bacteria-Induced Innate Immune Responses at Epithelial Linings

Fredrik Bäckhed and Agneta Richter-Dahlfors

The impact of the innate immune system for survival of flies and other insect has been known for long. During recent years it has become evident that the innate immune system plays a major role for the health of higher organisms as well. Interestingly, the innate defense systems of insects and vertebrates are remarkably conserved. The intention of this Chapter is to give a brief overview of the innate defenses initially identified using insects as model systems, then to focus on the applicability of this knowledge to mammalian systems. In particular, this Chapter deals with the innate responses that occur at epithelial linings of the host upon exposure to bacteria.

Why Don't Insects Die from Infections?

Many insects spend most of their lives in environments densely colonized by bacteria and other microbes that potentially could cause infectious diseases. In the absence of adaptive immune system, insects have developed very potent actions of the innate immune system that provide immediate antimicrobial responses required for survival in this particular niche. One key component is an array of antimicrobial peptides, which are expressed by the insect upon microbial exposure. When the fruit fly *Drosophila melanogaster* is exposed to fungi, an endogenous peptide is produced (Spätzle) which acts as a ligand to the Toll receptor.[1,2] Once activated, the Toll intracellular signaling pathway leads to degradation of Cactus (homologue of I-κB in vertebrates), which allows nuclear translocation of Dorsal, a transcription factor belonging to the NF-κB/Rel family of transcription factors (Fig. 1).[3] Once located in the nucleus, Dorsal immediately initiates transcription of drosomycin, a peptide with high anti-fungal activity.[1] Originally, Toll was identified as a receptor involved in dorso-ventral formation of the *Drosophila* embryo, while later studies revealed its importance in the immune defense of the fly.[1,4,5] Today, we know that *Drosophila* expresses a whole repertoire of receptors similar to Toll, specifically recognizing different groups of microorganisms.[6] Another yet not identified receptor in *Drosophila* is activated by exposure to bacteria, which results in the production of antibacterial peptides via the Imd-pathway.[7] Thus, *Drosophila* is capable to discriminate between different infectious agents in order to mount an effective response specific for the particular microbe.[2]

Similar mechanisms were later discovered to protect wounded frogs living in ponds. The potent antibacterial property of the frog peptide magainin made this peptide suitable for therapeutic applications, and it is now used as an antibacterial drug in humans.[8] A large variety of antimicrobial peptides are also identified in mammals. The role of antimicrobial peptides as effector molecules in mammalian innate defenses is further discussed later in this Chapter.

Intracellular Pathogens in Membrane Interactions and Vacuole Biogenesis,
edited by Jean-Pierre Gorvel.

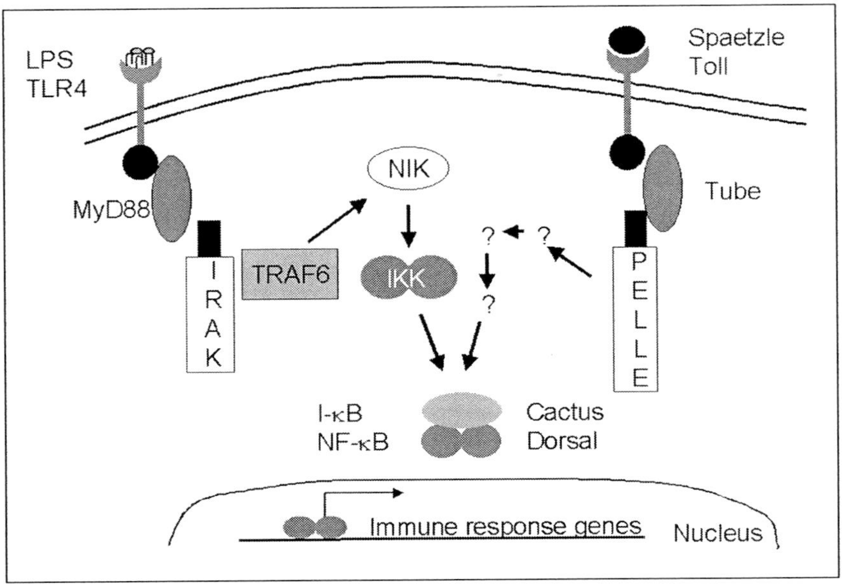

Figure 1. Components of the Toll-like receptor (TLR) signaling pathway in humans (left) and the Toll pathway in *Drosophila* (right) are homologous. Following ligand recogniton, the adapter protein MyD88 binds to the receptor. IRAK, TRAF6, NIK, and IKK are kinases involved in an autophosphorylation cascade, which eventually leads to phosphorylation of I-κB. In a phosphorylated state, I-κB is targeted for proteolytic degradation. This allows nuclear translocation of the transcription factor NF-κB, which regulates transcription of several pro-inflammatory genes.

Adaptive versus Innate Immune Systems

The innate immune system responds rapidly to microbial challenge, often within minutes or even seconds after exposure. This is in sharp contrast to the adaptive immune system, which require several weeks to achieve maximum effect in the host. However, once a response is mounted, the adaptive system remembers the antigen, which enables the host to rapidly respond to the next infection by the same antigen. This memory is crucial for survival of long-lived animals, while short-lived fruit flies do just as well without it. It is clear though that a functional adaptive immune system initiated and directed by the innate immune system is a prerequisite for mammals to achieve their lifetime expectancy.

The adaptive immune system is clonal, meaning that lymphocytes that recognize a specific antigen must proliferate before they differentiate into effector cells. Although this process is time-consuming, it produces tremendous amounts of lymphocytes recognizing the same antigen, and together, these cells can fight the infection.[9] The innate immune system constitutes of many different cells expressing the same oligo-specific receptors, which recognize the same microbial pattern molecules. Typical microbial pattern molecules are conserved molecules shared by groups of microbes, exemplified by lipopolysaccharide (LPS) on Gram-negative bacteria, peptidoglycan and lipoteichoic acid from Gram-positive bacteria, CpG motifs present in bacterial DNA, viral double-stranded RNA, and mannans on fungi. A common denominator for these molecules is that they have a low likelihood of becoming mutated during evolution, often because they are essential for microbial survival. Thus, microbe-specific pattern molecules serve as alarm signals that triggers the innate immune system via pattern recognition receptors.[10] To date, expression and activation of these receptors have mainly been studied on a variety of circulating white blood cells and endothelial cells, however, bacteria are rarely present within the bloodstream unless during pathologic conditions.

Mucosal, Epithelial Surfaces Constitute an Active and Efficient Barrier to the Outside World of Microbes

The human body has mainly three locations that are open for bacterial exposure; the airways, the gastro-intestinal tract, and the urinary tract. In all three locations, the microbes are encountering the mucosal linings, which for a long time were regarded as simple barriers that kept bacteria on the outside by mechanical means. Today, we know that these linings serve a much more delicate function and that they play a central role in the host defense. Mucosal surfaces are constituted of epithelial cells. Their ability to sense an infection and respond to it is crucial for the innate defenses. Furthermore, the epithelial response can be fine-tuned to fit the specific situation in different parts of our body, implicating that epithelial cells are specialized for their particular functions. This can be exemplified by a comparison of the normal, bacterial exposure in the three organs previously mentioned. Throughout our lifetime, the lungs are exposed to approximately 10,000 bacteria per hour through the air that we breath. These must rapidly be eliminated to protect the airways from infections. In contrast, epithelial cells of the gastro-intestinal lining must tolerate the presence of the immense number of bacteria present within the endogenous microflora to avoid ongoing inflammation as is seen in inflammatory bowel diseases. Although the presence of the microflora is permitted, intestinal epithelial cells must still be capable to induce a proper response if exposed to enteropathogens.[11,12] Again, a different situation is seen in the urinary tract. Due to retrograde ascending bacteria, the bladder is regularly exposed to bacteria, while the kidneys are sterile. The differences in the degree of bacterial exposure as well as the functions of the cells suggest that epithelial cells originating from bladder and kidney may differ in their response to bacteria.[13]

Different Mechanisms for Bacterial Recognition at Epithelial Linings

Epithelial cells can recognize the presence of bacteria by several mechanisms (Fig. 2). Components that are spontaneously released or actively secreted from bacteria can freely diffuse to the cell surface where they elicit a response. Among the most studied stimuli is bacterial endotoxin, LPS (lipopolysaccharide). This major component of Gram-negative bacterial outer membranes is continuously shedded during bacterial growth, and is recognized as a microbial pattern molecule by a human homologue to the Toll receptor.[14] Secreted toxins may also interact with target cells thus inducing production of pro-inflammatory cytokines.[15] Although these mechanisms allow the detection of distantly located bacteria, some other bacteria do not elicit a response until they are in close proximity to the target cells, bound via bacterial attachment organelles.[16] Thus, the combination of factors produced by bacteria and receptors expressed on the particular target cell provide numerous possibilities for the host to define a response that fit the specific situation. Several of the signaling pathways involved in pro-inflammatory response leads to activation of the transcription factor NF-κB. Once located in the nucleus, NF-κB regulates the expression of several pro-inflammatory mediators e.g., IL-8, a chemokine that recruits neutrophils to the site of infection.[13]

The Repertoire of Pattern Recognition Receptors Expressed on Epithelial Cells Determines Responsiveness towards Bacteria

Although ten Toll-like receptors (TLR) have been identified in mammalian systems to date, only a subset of them have been ascribed a particular ligand. Despite some early controversies, it is now commonly accepted that LPS is recognized by TLR4, peptidoglycan and lipoproteins by TLR2, while CpG motifs present in bacterial DNA are recognized by TLR9.[14,17-19] Data suggest that dimerisation of receptors occur upon ligand interaction, e.g., homodimerisation of TLR4 upon recognition of LPS, as well as heterodimerisation of TLR2 / TLR6 to mediate recognition of peptidoglycan.[20] While the variable extracellular domains confer specificity to the members of the Toll family of receptors, all of their cytoplasmic domains are closely related to the IL-1 receptor (IL-1R).[21] Thus, the intracellular signal transduction pathway closely mimics

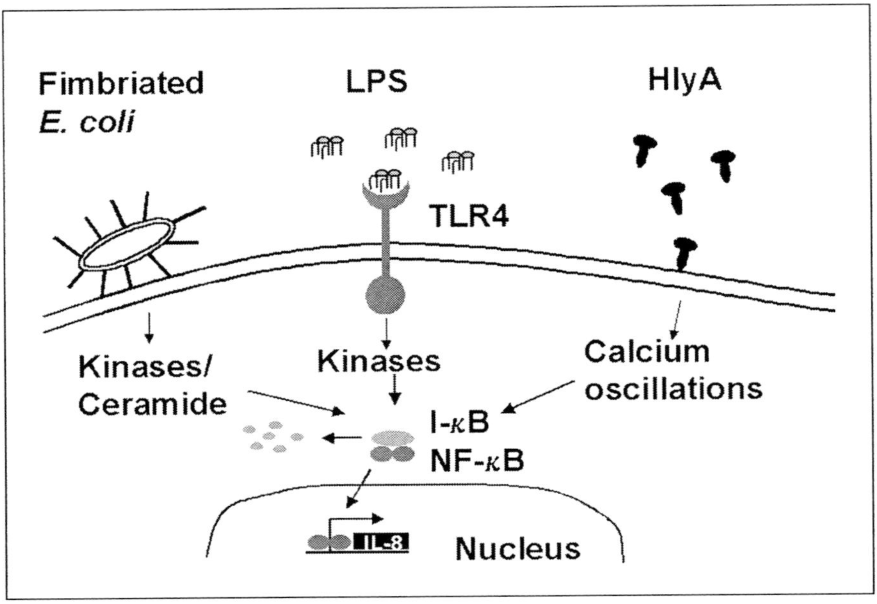

Figure 2. Epithelial cells can recognize the presence of bacteria by several mechanisms. Bacterial binding to host target cells, receptor-mediated recognition of LPS by Toll-like receptor 4 (TLR4), as well as the effect of the bacterially secreted toxin α-hemolysin (HlyA) on target cells induce different intracellular signaling pathways which all leads to degradation of I-κB and the subsequent translocation of NF-κB. Nuclear localization of this transcription factor is required for induction of pro-inflammatory mediators such as IL-8.

the IL-1R signaling pathway, which was recently described in detail elsewhere. A series of phosphorylation events leads to degradation of I-κB, thus allowing nuclear translocation of NF-κB.[22]

Urinary tract infection serves as one model system to study organ- and cell-specific innate responses to bacteria and bacterial components. Many lines of research show that epithelial cells of the urinary bladder promptly respond to bacterial exposure by production of pro-inflammatory mediators. These responses serve dual functions. Clearance of bacteria obviously eliminates the risk of establishing cystitis in the exposed host, but more importantly, the response in the bladder efficiently prevents bacteria from reaching the sensitive upper urinary tract. Innate responses triggered in cultured human bladder epithelial cells after exposure to *E. coli* LPS is attributed to the expression of TLR4, while a non-responsive renal epithelial cell line does not express this receptor (Fig. 3). This was further strengthened by the analysis of primary preparations of human renal epithelial cells, which also lack TLR4 expression. Bladder cells stimulated with different chemotypes of LPS revealed a possible role of the variable O polysaccharide present on the vast majority of clinical isolates of *E. coli*; to shield bacteria from being recognized by TLR4 on target cells, thus to halt the induction of pro-inflammatory response at epithelial linings. These findings demonstrate that epithelial cells in different locals of the body do not behave the same during infection, but rather are adjusted to the specific compartment where they reside. It was hypothesized that some epithelial cells are engaged in functions of such importance that even a low-degree of inflammation would be deleterious, i.e., in renal cortex, and therefore are dependent on defense mechanisms present in other locals such as the bladder. Moreover, lack of TLR4-mediated responses can be partly compensated for by other mechanisms such as bacterial attachment-induced innate responses.[13]

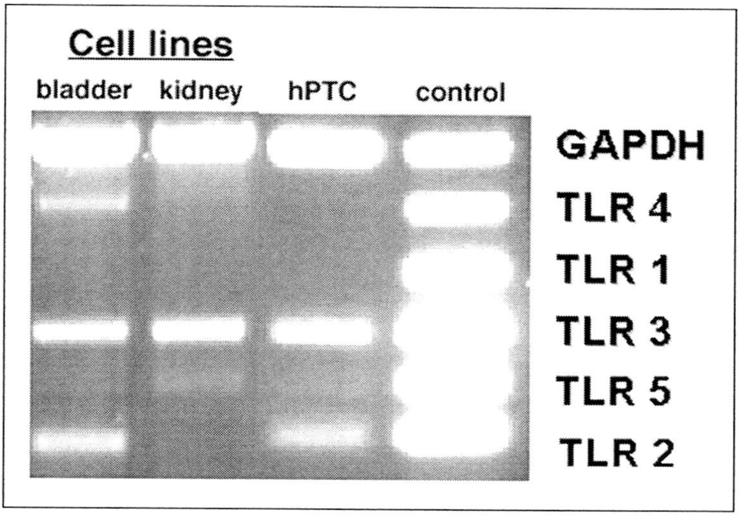

Figure 3. Different expression pattern of TLR1- TLR5 from bladder and renal cells. A human cell line originating from the urinary bladder expresses TLR4 as shown in a Multiplex-PCR. Neither the renal epithelial cell line nor a primary preparation of human proximal tubule cells express TLR4. GAPDH is used as internal control.

Bacterial Attachment to Epithelial Cells Induces Innate Immune Responses

Fimbriae are surface-located proteinaceous, filamentous polymeric organelles that mediate bacterial adherence to host cells. Binding to cells is mediated by an adhesin, a minor component situated at the tip of the pili.[23a] Although P pili and type 1 fimbriae expressed on uropathogenic isolates of *E. coli* have been most extensively studied, attachment organelles play a major role for the pathogenesis of other bacteria as well. Recently, it was shown that binding of uropathogenic *E. coli* via the type 1c fimbriae induced expression of pro-inflammatory cytokines in a renal epithelial cell line.[23b] Moreover it has been noted that P fimbriated bacteria, commonly associated with pyelonephritis, induce significantly more IL-6 as compared to isogenic, non-fimbriated bacterial isolates. The intracellular signaling pathway triggered by P pili attachment utilizes ceramide as a second messenger. Ceramide is released from the lipid portion of globoside, which also functions as the receptor for PapG. NF-κB activation by ceramide is believed to involve a series of kinase activities, however, the exact mechanism is not yet established.[16]

The situation for bacteria expressing the type 1 fimbriae differs in that binding of the adhesin FimH to its receptor, mannose, mediates invasion of bacteria into bladder cells. Invasion is dependent on host actin reorganization, activation of PI 3 kinases, as well as host protein phosphorylation.[24] Although FimH-mediated binding does not lead to cytokine induction per se, it augments the LPS-mediated induction of IL-6.[25] This is probably because FimH mediates intimate attachment of bacteria to the cell, thus exposing LPS more efficiently to the LPS receptors. Alternatively, LPS-molecules may be bound unspecifically to the fimbrial structure, allowing the fimbriae to present LPS by a mechanism similar to CD14-mediated presentation of LPS to target cells.

The Bacterially Secreted Toxin α-Hemolysin Induces Innate Responses via Induction of Intracellular Ca^{2+} Oscillations in Epithelial Cells

Bacteria produce a large variety of exotoxins, which potentially can affect host target cells. One such toxin is α-haemolysin expressed by most uropathogenic isolates of *E. coli*. Recent studies show that this toxin exerts dual effects on target cells: high concentration is cytolytic for target cells, while low concentration induces a second messenger response that stimulates the production of pro-inflammatory cytokines. Interestingly, the toxin activates an intracellular signal-transduction pathway, which induces low-frequent oscillation of the intracellular Ca^{2+} concentration.[15] The increased expression of IL-6 and IL-8 is probably mediated by frequency-modulated activation of NF-κB. Whether this signalling pathway is initiated by other microbes/microbial proteins is currently under investigation.

Indirect Effector Molecules in Innate Immunity—Cytokines

NF-κB plays a central role in transcriptional regulation of a variety of inflammatory genes. Induction of pro-inflammatory cytokines and chemokines are commonly used as a read-out for NF-κB activation. Depending on the definition of effector molecules, one can argue that cytokines are not true effectors, because they are not bactericidal per se. The chemokine IL-8 is expressed by a diverse repertoire of epithelial cells after microbial challenge. Binding of IL-8 to its receptor activates neutrophils, which immediately extravasate from the vasculature into the tissue towards the increasing concentration of IL-8, a mechanism known as chemotaxis. Once located at the site of infection, microbes are phagocytised and killed within various phagosomes by oxidative metabolites.[26]

Direct Effector Molecules in Innate Immunity—Antimicrobial Peptides

Vacuolar compartments of neutrophils also contain antimicrobial peptides, which kill bacteria by targeting their membranes. Also, some peptides act on intracellular targets.[27] The first identified antimicrobial peptide was isolated from the moth cecropia, thus named cecropin.[28] Today, more than 500 peptides are identified from all different kingdoms. According to their structure, peptides are classified in three groups: linear, amphipatic molecules, defensin-like or molecules consisting a repetitive sequence of a few aminoacids. Defensins are salt-sensitive, cationic, broad-range antimicrobial peptides active against Gram-negative and Gram-positive bacteria, viruses as well as fungi. Defensins are sub-divided into α- and β-defensins, which are structurally related (both contain three disulfide bonds) although their amino acid sequences differ.[29] Neutrophils and Paneth cells express mainly α-defensins, while two β-defensins are present in human epithelia. Human β-defensin-1 (hBD1) is highly, constitutively expressed in urogenital tissues, and to a lesser extent in airways and other epithelia.[30,31] Human β-defensin 2 (hBD2) is a host defense molecule whose expression is induced in response to infection and inflammation.[32] hBD2 was originally found in psoriatic skin, but the mRNA is also present in human lung.[33] LPS induction of hBD2 in human tracheobronchial epithelium requires CD14.[34] Moreover, there is evidence for TLR4 being required in bacterial induction of hBD2 expression in Caco-2 cells, an intestinal epithelial cell line (D. O'Neil, personal communication). Upstream from the transcriptional initiation site, the gene encoding hBD2 includes three NF-κB consensus sequences.[35] Taken together, these different lines of data suggests that induction of antimicrobial peptides upon bacterial exposure of the epithelium occur via activation of innate immune pattern recognition receptors, and the following activation of NF-κB/Rel family of trans-activating factors.

Other microbial molecules than LPS can also induce production of β-defensins. L-isoleucin, an essential amino acid that cannot be synthezised by the host, specifically induces the production of epithelial defensins by an as yet unknown mechanism. Because isoleucin is readily transported into cells, it is possible that it is recognized by an intracellularly located pattern recognition receptor.[36]

Both α- and β-defensins are produced as pre-pro-forms that require proteolytic processing to achieve active, mature forms. This step confers an additional level of control for the activation of peptides. Processing can be activated by attachment of bacteria to epithelial cells. This was shown in in vitro experiments where binding of bacteria to host target cells via the type 1 fimbriae activates matrilysin, a metalloprotease required for the processing of enteric α-defensins (cryptdins).[37] In contrast, matrilysin was not activated by isogenic, non-fimbriated strains of *E. coli*. Germ-free and colonized mice exhibited the same processing pattern, suggesting that in vivo, processing is independent of bacterial presence.[38] As expected, knock-out mice lacking the gene encoding matrilysin did not produce active defensins, and were consequently more sensitive to intestinal bacterial infections.[37]

Antimicrobial peptides are not only synthesized locally in peripheral tissues. Recently, the cystein-rich antimicrobial peptide hepcidin was shown to be synthezised within the liver, processed, and eventually secreted into the urine. Although hepcidin is active against fungi as well as bacteria, the role of this peptide in urinary tract infection is not clear.[39]

Functional Coupling of the Adaptive and the Innate Immune System

When antimicrobial peptides are released from host cells, peptides diffuse from the site of infection thereby establishing a chemotactic gradient similar to IL-8 and other chemokines. hBD2 is chemotactic for memory T cells and dendritic cells when bound to CCR6 on these cells.[40] Moreover, human α-defensins HNP1 and HNP2, as well as the human antimicrobial peptide LL-37 are chemotactic for neutrophils and CD4-positive T cells.[41] Thus, antimicrobial peptides play an important role in the integration of the innate and acquired immune responses.

Innate immune pattern recognition receptors also regulate molecular events, which are required for the activation of the adaptive immune system. In addition to transcriptional activation of pro-inflammatory cytokines, activation of TLRs induces expression of co-stimulatory molecules, e.g., CD80 and CD86 on antigen-presenting cells.[42] These molecules mediate binding to CD28, a receptor present on T cells, thereby delivering the co-stimulatory signal required for T cell proliferation and differentiation.[43-45] In fact, the absence of co-stimulatory signals leads to induction of immunological tolerance.

Bacteria or bacterial components (cholera toxin from *Vibrio cholerae* and heat labile toxin from *E. coli*) are among the most effective adjuvants used in vaccines. However, adjuvant-induced inflammation is the cause of the majority of the side-effects of a vaccine (e.g., pain, swelling, fever), a problem which urges scientists world-wide to investigate alternative adjuvants. Recently, it was shown that binding of bacteria to epithelial linings are crucial for induction of antibodies. Fimbriated *E. coli* bound to the tissues of the urinary tract induced antibodies of subclass IgA, while fimbriated *Bordetella* induced antibodies of subclass IgG2a in the respiratory tract. The non-fimbriated derivatives of both bacteria failed to induce any antibodies. Infection with fimbriated bacteria induced protection to both fimbriated and non-fimbriated bacteria, suggesting that other molecules than the fimbriae is recognized as the antigen.[46,47]

Bacteria Can Thwart the Innate Immune System by a Variety of Mechanisms

Several requirements must be fulfilled if a bacterium wants to become a successful pathogen. As examples, one can mention the development of a mechanism for bacteria to circumvent recognition by the innate immune system, to induce resistance to the action of effector molecules, and to enter protected sites thereby avoiding immune presentation. *Phorphymonas gingivalis* and *Helicobacter pylori* are two bacteria, which successfully have applied a mechanism to avoid recognition, and are thus able to establish chronic infections in the mouth (peridontal disease) and stomach (peptic ulcers). Reduced immunogenicity is ascribed to special features of

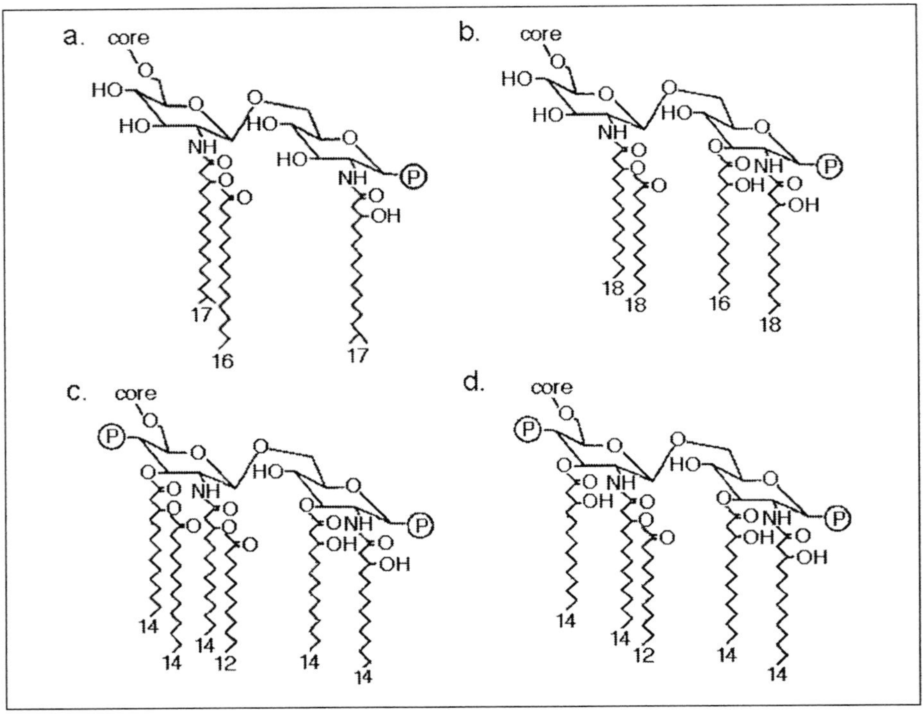

Figure 4. Structure of lipid A isolated from different bacterial strains. The low immunogenicity of LPS isolated from a. *P. gingivalis* and b. *H. pylori* is due to their tri- and tetra-acylated structures respectively. In addition, both LPS's are mono-phosphorylated. c. Structure of native, hexa-acylated lipid A from *E. coli*. d. Penta-acylated lipid A isolated from an *E. coli* strain harboring the *waaN*-mutation also shows reduced immunogenicity.

both strains LPS's, and is shown by a thousand-fold reduction in cytokine production from macrophages as compared to *E. coli* LPS. Moreover, expression of E-selectin, which is required for neutrophil adhesion and migration, is strongly suppressed.[48] The non-immunogenic features of LPS from *P. gingivalis* and *H. pylori* is ascribed to modifications within the lipid A moiety of their LPS structures (Fig. 4). Lipid A from *P. gingivalis* is tri-acylated, while *H. pylori*'s is tetra-acylated.[49,50] The importance of the acylation state of lipid A in inflammation has been demonstrated in numerous studies. *S. typhimurium*, which normally produces hexa-acylated lipid A, can cause chronic infections in a mouse model if a mutation is introduced which render the lipid A penta-acylated. Furthermore, penta-acylated strains of *E. coli* and *S. typhimurium* are unable to stimulate IL-6 and IL-8 from epithelial cells, as well as E-selectin in endothelial cells. Finally, macrophages induce less TNFα and NO in response to penta-acylated lipid A.[51,52] Additional features which differ between non-immunogenic and immunogenic LPS structures is that two of the acyl chains in *P. gingivalis* and *H. pylori* are somewhat longer, and show an unusual phosphorylation pattern.[53]

Salmonella readily survive and multiply within vacuolar compartments of macrophages.[54] Intracellular survival is dependent on the PhoP/PhoQ system, a two-component signal-transduction system, which regulates expression of a variety of bacterial genes including *pagP*. This enzyme is engaged in the lipid A biosynthesis, i.e., it adds palmitate to the structure. The resulting hepta-acylated lipid A confers resistance to cationic antimicrobial peptides.[55] Thus,

Salmonella utilize a combination of mechanisms (intracellular survival and induction of resistance to effector molecules) which in concert aid in this bacterium's survival strategy.

A somewhat different mechanism has been developed by the human pathogen *Staphylococcus aureus*. This Gram-positive bacterium can cause a wide variety of human diseases ranging from wound infections to severe systemic infections, often with fatal outcome. The bacterium expresses fibrinogen-binding proteins which not only facilitate binding to human extracellular matrixes, but also mediate coating of the bacteria in fibrinogen. Covered by host proteins, it turns invisible for the immune system.[56]

Concluding Remarks

Significant progress has been made in the characterisation of molecular events involved in microbial interactions with target cells. Completion of the human genome sequencing project allows intra- and inter-species comparisons for the identification of new molecules invovlved in the innate immune system, such as pattern recognition receptors and antimicrobial peptides. Improvement of analytical instruments enables detection of minute amounts of human effector molecules upon infection. Together, these findings will reveal novel targets for future drug design.

Acknowledgments

We thank Mathias Hornef and Anders Folkesson for critical reading of the manuscript. FB is supported by a felloship from the programme "Glycoconjugates in Biological Systems" sponsored by the Swedish Foundation for Strategic Research. Work in ARD's laboratory is supported by operating grants from the Wenner-Gren Foundation. the Swedish Research Council, AMF Försäkringar and the Cancer Foundation.

References

1. Lemaitre B, Nicolas E, Michaut L et al. The dorsoventral regulatory gene cassette spatzle/Toll/cactus controls the potent antifungal response in Drosophila adults. Cell 1996; 86:973-983.
2. Lemaitre B, Reichhart JM, Hoffmann JA. Drosophila host defense: Differential induction of antimicrobial peptide genes after infection by various classes of microorganisms. Proc Natl Acad Sci USA 1997; 94:14614-14619.
3. Imler JL, Hoffmann JA. Signaling mechanisms in the antimicrobial host defense of Drosophila. Curr Opin Microbiol 2000; 3:16-22.
4. Schneider DS, Jin Y, Morisato D et al. A processed form of the Spatzle protein defines dorsal-ventral polarity in the Drosophila embryo. Development 1994; 120:1243-1250.
5. Morisato D, Anderson KV. The spatzle gene encodes a component of the extracellular signaling pathway establishing the dorsal-ventral pattern of the Drosophila embryo. Cell 1994; 76:677-688.
6. Tauszig S, Jouanguy E, Hoffmann JA et al. From the cover: Toll-related receptors and the control of antimicrobial peptide expression in Drosophila. Proc Natl Acad Sci USA 2000; 97:10520-10525.
7. Klush RS, Lemaitre B. Drosophila immunity: Two paths to NF-kappaB. Trends Immunol 2001; 22:260-264.
8. Jacob L, Zasloff M. Potential therapeutic applications of magainins and other antimicrobial agents of animal origin. Ciba Found Symp 1994; 186:197-216; discussion 216-223.
9. Fearon DT. Seeking wisdom in innate immunity. Nature 1997; 388:323-324.
10. Medzhitov R, Janeway CA, Jr. Innate immunity: The virtues of a nonclonal system of recognition. Cell 1997; 91:295-298.
11. O'Neil DA, Porter EM, Elewaut D et al. Expression and regulation of the human beta-defensins hBD-1 and hBD-2 in intestinal epithelium. J Immunol 1999; 163:6718-6724.
12. Neish AS, Gewirtz AT, Zeng H et al. Prokaryotic regulation of epithelial responses by inhibition of IkappaB-alpha ubiquitination. Science 2000; 289:1560-1563.
13. Bäckhed F, Söderhäll M, Ekman P et al. Organ- and cell-specific expression of TLR4 correlates to induction of innate immune responses within the human urinary tract. Cell Microbiol 2001; 3:153-158.
14. Poltorak A, He X, Smirnova I et al. Defective LPS signaling in C3H/HeJ and C57BL/10ScCr mice: mutations in Tlr4 gene. Science 1998; 282:2085-2088.

15. Uhlen P, Laestadius A, Jahnukainen T et al. Alpha-haemolysin of uropathogenic E. coli induces Ca2+ oscillations in renal epithelial cells. Nature 2000; 405:694-697.
16. Hedlund M, Svensson M, Nilsson A et al. Role of the ceramide-signaling pathway in cytokine responses to P-fimbriated Escherichia coli. J Exp Med 1996; 183:1037-1044.
17. Takeuchi O, Hoshino K, Kawai T et al. Differential roles of TLR2 and TLR4 in recognition of gram-negative and gram-positive bacterial cell wall components. Immunity 1999; 11:443-451.
18. Aliprantis AO, Yang RB, Mark MR et al. Cell activation and apoptosis by bacterial lipoproteins through toll-like receptor-2. Science 1999; 285:736-739.
19. Hemmi H, Takeuchi O, Kawai T et al. A Toll-like receptor recognizes bacterial DNA. Nature 2000; 408:740-745.
20. Ozinsky A, Underhill DM, Fontenot JD et al. The repertoire for pattern recognition of pathogens by the innate immune system is defined by cooperation between toll-like receptors. Proc Natl Acad Sci USA 2000; 97:13766-13771.
21. Rock FL, Hardiman G, Timans JC et al. A family of human receptors structurally related to Drosophila Toll. Proc Natl Acad Sci USA 1998; 95:588-593.
22. Kopp EB, Medzhitov R. The Toll-receptor family and control of innate immunity. Curr Opin Immunol 1999; 11:13-18.
23a. Hultgren SJ, Abraham S, Caparon M et al. Pilus and nonpilus bacterial adhesins: assembly and function in cell recognition. Cell 1993; 73:887-901.
23b. Bäckhed F, Alsen B, Roche N et al. Identification of target tissue glycosphingolipid receptors for uropathogenic, F1C-fimbriated Escherichia coli and its role in mucosal inflammation. J Biol Chem 2002; 277:18198-18205.
24. Martinez JJ, Mulvey MA, Schilling JD et al. Type 1 pilus-mediated bacterial invasion of bladder epithelial cells. Embo J 2000; 19:2803-2812.
25. Schilling JD, Mulvey MA, Vincent CD et al. Bacterial invasion augments epithelial cytokine responses to Escherichia coli through a lipopolysaccharide-dependent mechanism. J Immunol 2001; 166:1148-1155.
26. Eckmann L, Kagnoff MF, Fierer J. Epithelial cells secrete the chemokine interleukin-8 in response to bacterial entry. Infect Immun 1993; 61:4569-4574.
27. Boman HG, Agerberth B, Boman A. Mechanisms of action on Escherichia coli of cecropin P1 and PR-39, two antibacterial peptides from pig intestine. Infect Immun 1993; 61:2978-2984.
28. Steiner H, Hultmark D, Engstrom A et al. Sequence and specificity of two antibacterial proteins involved in insect immunity. Nature 1981; 292:246-248.
29. Bevins CL, Martin-Porter E, Ganz T. Defensins and innate host defence of the gastrointestinal tract. Gut 1999; 45:911-915
30. Valore EV, Park CH, Quayle AJ et al. Human beta-defensin-1: an antimicrobial peptide of urogenital tissues. J Clin Invest 1998; 101:1633-1642.
31. Zhao C, Wang I, Lehrer RI. Widespread expression of beta-defensin hBD-1 in human secretory glands and epithelial cells. FEBS Lett 1996; 396:319-322.
32. Singh PK, Jia HP, Wiles K et al. Production of beta-defensins by human airway epithelia. Proc Natl Acad Sci USA 1998; 95:14961-14966.
33. Bals R, Wang X, Wu Z et al. Human beta-defensin 2 is a salt-sensitive peptide antibiotic expressed in human lung. J Clin Invest 1998; 102:874-880.
34. Becker MN, Diamond G, Verghese MW et al. CD14-dependent lipopolysaccharide-induced beta-defensin-2 expression in human tracheobronchial epithelium. J Biol Chem 2000; 275:29731-29736.
35. Diamond G, Kaiser V, Rhodes J et al. Transcriptional regulation of beta-defensin gene expression in tracheal epithelial cells. Infect Immun 2000; 68:113-119.
36. Fehlbaum P, Rao M, Zasloff M et al. An essential amino acid induces epithelial beta-defensin expression. Proc Natl Acad Sci USA 2000; 97:12723-12728.
37. Wilson CL, Ouellette AJ, Satchell DP et al. Regulation of intestinal alpha-defensin activation by the metalloproteinase matrilysin in innate host defense. Science 1999; 286:113-117.
38. Putsep K, Axelsson LG, Boman A et al. Germ-free and colonized mice generate the same products from enteric prodefensins. J Biol Chem 2000; 275:40478-40482.
39. Park CH, Valore EV, Waring AJ et al. Hepcidin: A urinary antimicrobial peptide synthesized in the liver. J Biol Chem 2000; 11:11.
40. Yang D, Chertov O, Bykovskaia SN et al. beta-defensins: linking innate and adaptive immunity through dendritic and T cell CCR6. Science 1999; 286:525-528.
41. Agerberth B, Charo J, Werr J et al. The human antimicrobial and chemotactic peptides LL-37 and alpha-defensins are expressed by specific lymphocyte and monocyte populations. Blood 2000; 96:3086-3093.

42. Medzhitov R, Preston-Hurlburt P, Janeway CA Jr. A human homologue of the Drosophila Toll protein signals activation of adaptive immunity. Nature 1997; 388:394-397.
43. Koulova L, Clark EA, Shu G et al. The CD28 ligand B7/BB1 provides costimulatory signal for alloactivation of CD4+ T cells. J Exp Med 1991; 173:759-762.
44. Linsley PS, Brady W, Grosmaire L et al. Binding of the B cell activation antigen B7 to CD28 costimulates T cell proliferation and interleukin 2 mRNA accumulation. J Exp Med 1991; 173:721-730.
45. Linsley PS, Clark EA, Ledbetter JA. T-cell antigen CD28 mediates adhesion with B cells by interacting with activation antigen B7/BB-1. Proc Natl Acad Sci USA 1990; 87:5031-5035.
46. Mattoo S, Miller JF, Cotter PA. Role of Bordetella bronchiseptica fimbriae in tracheal colonization and development of a humoral immune response. Infect Immun 2000; 68:2024-2033.
47. Soderhall M, Normark S, Ishikawa K et al. Induction of protective immunity after escherichia coli bladder infection in primates. Dependence of the globoside-specific P-fimbrial tip adhesin and its cognate receptor. J Clin Invest 1997; 100:364-372.
48. Darveau RP, Cunningham MD, Bailey T et al. Ability of bacteria associated with chronic inflammatory disease to stimulate E-selectin expression and promote neutrophil adhesion. Infect Immun 1995; 63:1311-1317.
49. Kumada H, Haishima Y, Umemoto T et al. Structural study on the free lipid A isolated from lipopolysaccharide of Porphyromonas gingivalis. J Bacteriol 1995; 177:2098-2106.
50. Moran AP, Lindner B, Walsh EJ. Structural characterization of the lipid A component of Helicobacter pylori rough- and smooth-form lipopolysaccharides. J Bacteriol 1997; 179:6453-6463.
51. Khan SA, Everest P, Servos S et al. A lethal role for lipid A in Salmonella infections. Mol Microbiol 1998; 29:571-579.
52. Sommerville J. Activities of cold-shock domain proteins in translation control. Bioessays 1999; 21:319-325.
53. Moran AP. Cell surface characteristics of Helicobacter pylori. FEMS Immunol Med Microbiol 1995; 10:271-280.
54. Richter-Dahlfors A, Buchan AMJ, Finlay BB. Murine salmonellosis studied by confocal microscopy: Salmonella typhimurium resides intracellularly inside macrophages and exerts a cytotoxic effect on phagocytes in vivo. J Exp Med 1997; 186:569-580.
55. Guo L, Lim KB, Poduje CM et al. Lipid A acylation and bacterial resistance against vertebrate antimicrobial peptides. Cell 1998; 95:189-198.
56. Wann ER, Gurusiddappa S, Hook M. The fibronectin-binding MSCRAMM FnbpA of Staphylococcus aureus is a bifunctional protein that also binds to fibrinogen. J Biol Chem 2000; 275:13863-13871.

CHAPTER 18

Pathogens and Hosts:
Who Wins?

Jonathan C. Howard

Who wins? is certainly an evolutionary question, and to some extent also the semantic question, what do we mean by winning? The ultimate victory over microbes is Pyrrhic: we have seen the immunodeficient children in their sterile hoods. A sterile life is not a fulfilled one. So that idea of winning has to be put aside and a more biological context for the question analysed. Haldane identified a critical attribute of the host-pathogen relationship in his celebrated paper 'Disease and Evolution',[7] namely the tendency of host-pathogen systems to generate polymorphism in their virulence and resistance genes. Indeed he went so far as to imply, if not to say outright, that persistent biochemical polymorphism in a host species could most plausibly all be attributed to disease resistance genes. This study preceded by only a short while Lewontin and Hubby's exploitation of starch gel electrophoresis to reveal extensive polymorphism in housekeeping enzymes of wild *Drosophila*[15] and the development of the neutral theory of gene evolution (reviewed in ref. 14). The neutral proposition is that most mutations in a gene are nearly neutral so that their rate of penetration into finite populations is more a function of randomness than it is of their selective value. A clear consequence of neutrality is that polymorphism in multiple genes, indeed in essentially all genes, is the expected state, as nearly neutral substitutions wander at random through the population. Indeed because neutral change in allelic frequencies is slow, variations that are under intense positive directional selection enter the population at relatively great speed, carrying with them whatever neutral variants may be linked with the target of selection. Natural selection of this kind thereby purges a genetic region of neutral polymorphism. The importance of polymorphism under selection has been to a degree obscured by the neutrality revolution and it is noticeable that Kimura's book,[14] which established polymorphism as an unavoidable attribute of neutrality, and paid generous tribute to Haldane's contributions to quantitative genetics, failed to refer to his Disease and Evolution paper which, to the contrary, generalised infectious disease as the most important cause of polymorphism under selection.

In 'Disease and Evolution' Haldane was especially concerned with differential evolution rates between host and pathogen. Hosts that were genetically homogeneous with respect to genes coding for resistance functions could expect successful pathogen breakthrough in due course. Once broken through, the host is extinguished unless a resistant variant can be generated. A resistant variant replaces the now-susceptible old one until pathogen transmission rates slow down sufficiently to put pressure on the pathogen to defeat the new variant. Further selection pressure on the host generates a third variant, the third variant stimulates the evolution of a new pathogen virulence gene and so ad infinitem, resulting in a polymorphic population of hosts being infected by a polymorphic population of pathogens. This scenario is the classic model for the evolution of polymorphism in the major histocompatibility complex even though the formal evidence is scant in the extreme. Nevertheless it cannot be doubted that the MHC is a resistance locus, or rather a collection of resistance loci, and that it encodes some of the most polymorphic proteins known, the class I and class II MHC molecules which bind peptides

Intracellular Pathogens in Membrane Interactions and Vacuole Biogenesis, edited by Jean-Pierre Gorvel.

derived from pathogen proteins and present them to the immune system. Furthermore the distribution of variation in these molecules clearly shows that the polymorphism is directed towards residues determining the peptide-binding specificity,[2] while the high frequency of coding substitutions in the alleles shows that selection has been operating at the level of the expressed protein.[10] It is markedly less straightforward to identify the pathogens to which MHC polymorphism is directed. It is well established that certain MHC alleles confer relative resistance to lethal childhood Falciparum malaria,[8] findings which satisfactorily account for the high incidence of the protective alleles in endemic malaria regions of West Africa. It is quite another matter to show that there is a reciprocal polymorphism in the pathogen species, indicating that immune resistance is limiting pathogen transmission. The presumed protective epitope of HLA-B53-restricted resistance to malaria is non-polymorphic in the pathogen.[9] When pathogen polymorphism for an MHC-restricted epitope does occur, as for the HLA.B35-restricted LSA-1 epitope in *Plasmodium falciparum*, the meaning may be much more subtle. In this case pathogen polymorphism is probably favoured by an advantageous consequence of mixed infection by pathogens expressing different alleles at this epitope, where the two polymorphic epitopes interact through peptide ligand antagonism to mitigate the immune response against one or the other.[6] Influenza virus neuraminidase and hemagglutinin variants can be attributed to the development of antibody immunity against individual viral capsid epitopes, but this argument works for the virus rather than for the host since there is no experimental reason to believe that genetic factors play a major role in determining the specificity of antibodies made against these complex glycoproteins. So we have cases where we can apparently explain the polymorphism of the host resistance system, cases where we can apparently explain polymorphism of the pathogen virulence system, but little enough evidence for the reciprocal interactions that one would like to be able to refer to.

Nevertheless the general idea seems still to be sound, and consistent with the general principle. We can frequently find evidence of pathogen avoidance of a host resistance mechanism even if we can't always find the reciprocal polymorphism that ought to follow. Examples are the multiple cases of pathogen inhibitors of the interferon system,[5] or of the MHC itself.[19] The sheer complexity of the confrontation between the well-adapted host and the practised pathogen is breathtaking. The mammalian immune system operates resistance mechanisms at multiple levels, the MHC being only one of these, and there is every reason to seek polymorphism in all of them. The early component of the classical complement pathway, C4, which interacts with pathogen surfaces, is strikingly polymorphic.[3] In the rat the transporter associated with antigen processing (TAP) has a surprising polymorphism in peptide transport specificity[19] (though this may be secondary to polymorphism in the MHC molecules themselves[13]). Recently a potentially selectively significant polymorphism has been documented in the leucine-rich repeat segment of the the Toll-like LPS receptor, TLR4.[21] It is still early days, but the hunt for polymorphism is justified not least on the grounds that once discovered, it tells us about functionally important segments of our resistance molecules.

Structural polymorphism is not the only way of segregating patterns of resistance: much interesting genetic variation is more drastic than that, either having the gene or not. An example from viral immunity is the cell-autonomous resistance factor for several RNA viruses, Mx, a dynamin-related GTPase inducible by type I interferon. In the mouse most inbred strains carry null mutations of one of the two genes encoding Mx proteins and are susceptible to certain RNA viruses including infleuenza,[22] and in the wild, null alleles for this locus segregate at high frequency.[12] The same is true for NRAMP, where the polymorphism is also associated with susceptibility and resistance to a number of intracellular bacteria.[16] Null alleles for the HIV co-receptor CCR5 are associated with resistance to HIV,[4] but the abundant null alleles in certain populations certainly predated the disease, suggesting that selective pressure on these molecules is not new. The principle of gain and loss variation in host resistance genes is echoed in gain and loss type variation in virulence genes from the side of the pathogen. It remains important and unobvious that so many key virulence factors are carried not in bacterial chromosomes but on plasmids. If the common explanation given is that this guarantees rapid spread to other individuals, it also ensures that many, even most, individuals do not carry a given plasmid at a given time. One explanation which would reconcile gain and loss variation in resistance molecules and in virulence molecules is that both resistance and virulence carry a cost to the owner. The challenge in the case of the host would be to identify what that cost is, in the case of the pathogen to understand the different strategies of survival that are favoured by possession or loss of virulence genes.

We can, however, surely agree that both pathogen and host populations generally show extensive polymorphism in genes related to the infectious process. It is likely that the minute subdivision of both sides of this complex war in a patchwork of virulence and resistance types simultaneously guarantees the survival of the pathogen and the host. This principle has been neatly exemplified in plant immunity, where in a single square metre of ground 75 members of a single plant species showed 10 different patterns of resistance to 5 different isolates of the same fungal pathogen.[1]

Who wins? We do not consider it a victory that disease picks off only a fraction of our population that is normally well below our ability to replace, because we care so much about individuals. Our victory at the evolutionary level of our species still brings tragedy to the family of the child dead from infectious disease. There really is only one plausible device by which we can escape from our genetic endowment within the species and develop resistance at the level of the individual, and that is the adaptive immune system itself. We know its successes and its failures, but still the development of ever more imaginative vaccines against significant pathogens provides the best guarantee that these tragedies will continue to get rarer. Where vaccines fail, intelligent use of antibiotics and the development of novel pharmaceuticals still provide vital palliatives.

Haldane recognised that high population densities play into the hands of infectious pathogens. They guarantee high transmission rates and thereby ensure that rare virulent pathogen variants derived by spontaneous mutation from a less virulent strain find an opportunity to infect and prosper. Equally, pathogens that rely on rapid transmision rates to avoid being destroyed by the immune system can only thrive in a dense host population. As stressed above, a host species can become dense if individual resistance types remain rare as a result of multiple segregating polymorphic systems, a principle seemingly used to good effect by many organisms, ourselves included. Nevertheless there are exceptions and one especially striking one was noted with bated breath by Haldane himself. "It is very remarkable that *Drosophila* is as generally immune as it is. I venture to fear that some bacillus or virus may yet find a suitable niche in the highly overcrowded *Drosophila* populations of our laboratories, and that if so this genus will lose its proud position as a laboratory animal".[7] After more than 50 years further intensive rearing of *Drosophila* without any significant outbreak of infectious disease (the only serious problem for Drosophilists is an ectoparasite at a similar taxonomic level to the host, namely mites) some of the determinants of its immunity are now being discovered.[17] So far, they have not differed in type from mechanisms also operating in animals. Indeed the Toll-like receptors, which stimulate key resistance functions in *Drosophila* against a variety of pathogen types, were first identified in this species and only subsequently and as a result discovered in mammals. Disease resistance in *Drosophila* is not a population effect: individual flies are resistant to disease but it is still not really clear why these defenses are so hard to breach. One can predict that further work on non-mammalian models of immunity will bring yet more valuable insights that can be converted into resistance or cure for people. Perhaps the Nod1 and Nod2 genes, recently identified as intracellular LPS receptors[11] whose failure may be responsible for Crohn's disease,[18] and which are structurally remarkably similar to plant resistance (R) genes, will be the next case of insight from beyond the mammals.

References

1. Bevan JR, Clarke DD, Crute IR. Resistance to Erysiphe fischeri in two populations of Senecio vulgaris. Plant Pathology 1993; 42:636-646.
2. Bjorkman PJ, Saper MA, Samraoui B et al. The foreign antigen-binding site and T cell recognition regions of class I histocompatibiliy antigens. Nature 1987; 329:512-518.
3. Campbell RD, Carroll MC, Porter RR. The molecular genetics of components of complement. Adva Immunol 1986; 38:203-244.
4. Carrington M, Dean M, Martin MP et al. Genetics of HIV-1 infection: Chemokine receptor CCR5 polymorphism and its consequences. Hum Mol Genet 1999; 8:1939-1945.
5. Gale Jr. M, Katze MG. Molecular mechanisms of interferon resistance mediated by viral-directed inhibition of PKR, the interferon-induced protein kinase. Pharmacol Ther 1998; 78:29-46.
6. Gilbert SC, Plebanski M, Gupta S et al. Association of malaria parasite population structure, HLA, and immunological antagonism. Science 1998; 279:1173-1177.
7. Haldane JBS. Disease and evolution. La Ricerca Scientifica 1949; 19 Supplemento Anno 19°:68-75.
8. Hill AVS, Allsopp CEM, Kwiatkowski D et al. Common West African HLA antigens are associated with protection from severe malaria. Nature 1991; 352:595-600.
9. Hill AVS, Elvin J, Willis AC et al. Molecular analysis of the association of HLA-B53 and resistance to severe malaria. Nature 1992; 360:434-439.
10. Hughes AL, Nei M. Pattern of nucleotide substitution at major histocompatibility complex class I loci reveals overdominant selection. Nature 1998; 335:167-170.
11. Inohara N, Ogura Y, Chen F et al. Human Nod1 confers responsiveness to bacterial lipopolysaccharides. J Biol Chem 2001; 276:2551-2554.
12. Jin HK, Yamashita T, Ochiai K et al. Characterization and expression of the Mx1 gene in wild mouse species. Biochemical Genetics 1998; 36:311-322.
13. Joly E, Le Rolle A-F, González AL et al. Co-evolution of rat TAP transporters and MHC class I RT1-A molecules. Current Biol 1998; 8:169-172.
14. Kimura M. The Neutral Theory of Molecular Evolution. Cambridge: Cambridge University Press, 1986.
15. Lewontin RC, Hubby JL. A molecular approach to the study of genic heterozygosity in natural populations, II. Amount of variation and degree of heterozygosity in natural populations of Drosophila pseudoobscura. Genetics 1966; 54:595-609.
16. Malo D, Vogan K, Vidal S et al. Haplotype mapping and sequence analysis of the mouse Nramp gene predict susceptibility to infection with intracellular parasites. Genomics 1994; 23:51-61.
17. Meister M, Hetru C, Hoffmann JA. The antimicrobial host defense of Drosophila. Curr Top Microbiol Immunol 2000; 248:17-36.
18. Ogura Y, Bonen DK, Inohara N et al. A frameshift mutation in NOD2 associated with susceptibility to Crohn's disease. Nature 2001; 411:603-606.
19. Ploegh HL. Viral strategies of immune evasion. Science 1998; 280:248-253.
20. Powis SJ, Deverson EV, Coadwell WJ et al. Effect of polymorphism of an MHC-linked transporter on the peptides assembled in a class I molecule. Nature 1992; 357:211-215.
21. Smirnova I, Poltorak A, Chan EK et al. Phylogenetic variation and polymorphism at the toll-like receptor 4 locus (TLR4). Genome Biol 2000; 1:research002.1-002.10.
22. Staeheli P, Grob R, Meier E et al. Influenza virus-susceptible mice carry Mx genes with a large deletion or a nonsense mutation. Mol Cell Biol 1988; 8:4518-4523.

Index

D

E

F

G

H

I

J

K

L

M

U

V

W

Y